D1180559

The Nuclear Barons

The Nuclear Barons

Peter Pringle &
James Spigelman

MICHAEL JOSEPH
London

First published in Great Britain by Michael Joseph Ltd
44 Bedford Square, London WC1
1982

ISBN 0 7181 2061 2

Printed in Great Britain by
Hollen Street Press, Slough, Berkshire
and bound by Dorstel Press, Harlow, Essex

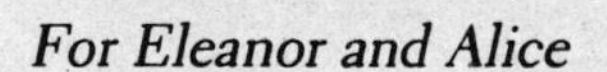
For Eleanor and Alice

Contents

Prologue

Four decades have passed since the power hidden inside the uranium atom emerged from scientific speculation and began to affect the lives of men and women throughout the world. This book is the story of an international elite of scientists, engineers, politicians, administrators, and military officers who brought atomic energy under control.

While the scientists in this elite have made impressive advances in understanding the magic of the atom and the engineers have devised remarkable techniques for harnessing its power, the atomic revolution is marred by: the victims of Hiroshima and Nagasaki; the huge arsenals of the superpowers; the crippled nuclear industry; the victims of excessive radiation in the uranium mines, in atomic laboratories and factories, and in the towns that lay in the fallout paths of the early bomb tests.

With the benefit of hindsight, and drawing on the memoirs by scientists and politicians, official histories, recently declassified documents from government agencies, and the diligent research of many concerned citizens groups, we have examined the momentous choices that faced this international elite. We have attempted to explain, through the actions of the men and women who made the decisions, how and why they have left the world in such a nuclear mess.

This is not an academic history. As a journalist and a lawyer, our aim has been to provide the lay reader with an overview of the atomic revolution, to record the evidence, to focus on the key problems as they arose and to analyze the motives of the people who tried to solve them.

It has not been a simple task. The separate world of nuclear decision making has been a closed one, often hiding behind the cover of national security. From the early days, knowledge of the atom and its powers was something to be hoarded, to be kept from a nation's citizens as well as from its enemies.

The atom offered each country's decision-makers many opportunities to indulge a yearning for power, relish a sense of achievement, proclaim a vision for the future, or to fear the unknown. Their decisions were frequently distorted by personal ambition and institutional self-interest; just as often, they ignored the broader questions of humanity, the cumulative dangers of a runaway arms race and of the unregulated expansion of nuclear power.

One motivation, perhaps more important than any other, was pride—especially that form of pride known as patriotism: atomic bombs were built to demonstrate national manhood; nuclear power stations were built to prove a nation's technological expertise.

Those who wanted to expand nuclear power proclaimed that they had no choice—especially during the early years. The American bomb project was launched by men who sincerely believed they were in a race with Hitler. When it became clear that Hitler was not building a bomb, the Americans found a new target in the Japanese and convinced themselves there was no alternative to bombing Hiroshima and Nagasaki. Four years later, when the Soviet Union exploded its first atomic bomb, the Americans felt they had to build an H-bomb to stay ahead in the arms race. The alternative—to renew the search for ways of limiting the arms race—was subsumed by transitory fears and enthusiasm for bigger and better bombs. It was the same story with the British, French, and Chinese decisions to build a bomb. And it was to be repeated in those and other countries in their decisions about power reactors and in their grandiose projections of nuclear power capacity.

Those who sought to slow the momentum, whether they were physicists, doctors, politicians, lung-diseased miners, victims of fallout, or simply concerned citizens, were told they were not "expert" enough to understand the extent of the problem—and they could not learn any more about it because the details were top secret.

Of all the choices that faced the decision-makers, one was always overriding: whether to build a bomb. More than a dozen nations, at different times and for different reasons, have had to make this decision, and seven have actually produced atomic weapons. They are the United States, the Soviet Union, Britain, France, China, Israel, and India. They have all faced a series of secondary choices, such as whether to build bigger bombs, whether to test them, and where and how often. Some nations say they are not building bombs but have kept open the option to do so by taking their peaceful nuclear power programs beyond the stage of basic research: they include Pakistan, Iraq, South Africa, and Brazil. And a few nations have renounced the weapons option: they include Canada, Sweden, Switzerland, Germany, and Japan.

But even those nations that have renounced the weapons option have faced important and complex nuclear decisions. Each nation developing electricity from nuclear power has had to decide how much to spend on the enormously costly research and development; what type of reactor to build; how much to rely on foreign technology; how to regulate the operation of power plants to ensure public health and safety when the experts differ so widely on what precautions are necessary. Many of these decisions were taken without public debate, made in the name of energy security, national prestige, or the self-interest of the atomic institutions.

As the nuclear revolution expanded, the advocates built these special institutions to keep the atom apart from the checks and balances of the normal political process. As the number of atomic questions grew—questions about the effects of radioactive fallout from bomb tests, or the dangers of power reactors getting out of control, or the problems of disposing of radioactive waste—the nuclear system showed an extraordinary capacity to create or manipulate its own needs. Always seeking a mission that would match the scale of its technical abilities and ambitions—like building a nuclear-powered airplane or a merchant ship or an advanced reactor such as the "breeder," which makes more fuel than it consumes—the system fed off itself.

The momentum continued uninterrupted, fed by a constant stream of self-declared threats and shortages that justified each new commitment; threats of another country doing something first, or better, or bigger, threats which were magnified as easily for the peaceful programs as they had been for the military ones. The "uranium gap" and the "energy gap" of the 1970s were as spurious as the "bomber gap" of the 1950s and the "missile gap" of the 1960s.

The atomic institutions became almost totalitarian in their powers, often requiring scientists and engineers to suppress information that stood in the way of the nuclear revolution. In the mid-1970s, the nature and intensity of the public's opposition to those institutions grew as the information about the past emerged. For the first time, ordinary citizens began to reshape nuclear decisions and limit the choices available to the nuclear barons.

Washington, D.C., and
Sydney, Australia, March 1981

PART ONE

The Forties

1

Original Sin

In some sort of crude sense which no vulgarity, no humor, no overstatement, can quite extinguish, the physicists have known sin; and this is a knowledge which they cannot lose.

—J. ROBERT OPPENHEIMER, 1948

A short, bumptious Hungarian physicist named Leo Szilard is credited with being the first scientist to believe it was possible to make an atomic bomb.

An intellectual gypsy who often found himself without a job, a laboratory, or any great personal wealth, Szilard was thirty-five years old and still near the bottom of his profession in 1933 when he became obsessed with a great vision: on the eve of war in Europe, he was determined to warn the non-fascist powers about the military potential of the atom. He was not the only scientist who tried to influence reluctant governments to start atomic bomb programs, but he was the most tenacious and inventive in breaking through the bureaucratic lethargy and military self-assuredness that characterized this period.

At the time, his mission appeared doomed. The leading physicists of the day haughtily dismissed his ideas, but Szilard was convinced that it was possible; for him it became a scientific emergency, a frantic period of trying to warn, first the British and then the Americans, that if they did not create the "impossible" weapon it would be made by the other side, at the direction of Adolf Hitler.

In 1935 Szilard found himself in London, a refugee from Nazi

Germany. He had sensed earlier than most the problems that were soon to confront Jewish academics under Hitler and, resigning the minor research job he held at the University of Berlin, had moved to London. There he had found a few businessmen willing to give some support to his atomic research, much of which was done, Archimedes-like, in bathtubs in London hotels, where he speculated about atomic reactions.

Some of the current theories were very seductive. Physicists already knew, for example, because Einstein had explained it to them, that nature had cunningly concealed vast stores of energy in the atomic nucleus. What they did not yet know was how to release that energy. Theories abounded, but they were only theories, science fiction, literary whimsy.

The best-known of these wild dreams came not from a scientist, but from a science fiction writer, H. G. Wells. Shortly before World War I, Wells had written a book called *The World Set Free: A Story of Mankind.* In it, he predicted that man would eventually solve the problem of how to build a bomb so powerful it could wipe out an entire city—or even a civilization.

Szilard was very taken by the Wellsian fantasy, and he thought out a way to make such a bomb. Two key theories were involved. They are known today as the "chain reaction" and "critical mass." Put simply, each atom is made of electrically charged particles. These can be positive, negative, or neutral (called neutrons). In most elements, these particles are firmly bonded together and never move, but in some—the radioactive elements—the atoms are in a constant state of change, breaking up and releasing the charged particles. Of all the particles, the neutron is of special interest because it can act like an atomic bullet, smashing up the nuclei of other atoms and releasing more neutrons—and energy.

The theory of the nuclear chain reaction went like this: one neutron splits one atom, which produces two neutrons that split two atoms, which produce four neutrons that split four atoms, and so on. If this chain reaction could be made to happen all at once, Szilard argued, there would be a sudden, huge release of energy. The key was to find an element that released at least two neutrons after its atom had been split. Szilard had no idea at the time that the element would be uranium; he suspected another element, beryllium, might be a potential candidate.

The second theory, of "critical mass," assumed that not all of the neutron bullets released when the atoms were split would find their targets; a minimum, or critical mass, of the key radioactive element would be required before the chain reaction could take place.

Convinced he was on the verge of a great discovery, Szilard decided to patent his ideas. If he found a way of releasing energy from the atom, he would start a scientific revolution. Atomic energy would be used not only for bombs but also to make electricity, to run cars, ships, airplanes, factories—anything that needed energy to move it.

Uppermost in Szilard's mind was the idea of the bomb, however, and it scared him. If he could think of ways to make one, then surely his former colleagues in German research institutes were having similar brainwaves. They might already be at work making a bomb for Hitler. He decided he had to inform the British Army of his theories and alert them to the superweapon. And he would assign his patent to them so that it could be kept secret. They could make the bomb and, when the coming war with Germany was over, the same patent could be used to make nuclear energy.

Szilard was advised by British scientists that he had little chance of persuading the army to accept the patent. For one, he was alone in his conviction that it would work. For another, in Britain he was a small voice bumping against a scientific establishment whose most distinguished physicist, Ernest Rutherford, had only recently dismissed the possibility of harnessing the power of the atom as "the merest moonshine." And finally, the military commanders of the day were not ready for superweapons; as far as they were concerned, wars were still fought with well-trained troops and reliable ships and tanks, not massive bombs.

Nevertheless, Szilard, motivated more by a deep fear of a Nazi bomb than any hope of pecuniary reward, offered his patents first to the army, then to the navy. A British Army ordnance officer, unaware of his future place in history, was not impressed. He wrote to Szilard, "There appears to be no reason to keep the specification secret as far as the War Office is concerned." Somewhat to Szilard's surprise, the navy accepted. Their reason was unclear; they had no more information than the army. Perhaps it was simply institutional pride. In any case, in February 1936, Szilard became the proud possessor of two British patents suggesting ways—which would eventually turn out to be basically correct—of making an atomic bomb. Both patents were classified Top Secret.

Three years later, in 1938, Szilard's basic insights were confirmed.

In December of that year, two Berlin University physicists, Otto Hahn and Fritz Strassmann, conducted experiments proving that atoms of the radioactive element uranium would split when bombarded with neutrons. A few weeks later in Paris, the French Nobel Prize-winner Frédéric Joliot-Curie showed that the uranium atom, once split, released more than one neutron; a chain reaction was indeed possible.

As the scientific evidence mounted, Szilard worked ceaselessly to encourage first the British and then the American governments to make an atomic bomb. The chilling coincidence of the outbreak of the Second World War with the atomic revolution reinforced his anxieties, and despite repeated rebuffs, he continued his entreaties. His persistence finally resulted in the American Manhattan Project, out of which came the two atomic bombs dropped on Japan in August 1945.

As war loomed, other physicists in France, Germany, and Russia also began to fear the outcome of their experiments. In France, Joliot-Curie made urgent requests for materials to expand his research, but he had to wait months for any reply. In Germany, patriotic physicists who tried to interest Hitler's military High Command in the bomb were accused of talking "atomic poppycock." In Russia, atomic scientists were eagerly pursuing the lines of research being pioneered in the West, but their atomic program was cut back to pay for more conventional weapons. And in the Britain of 1939, the atomic scientists themselves were still doubtful that a bomb could be made. The government was even more skeptical. After reviewing the evidence, a member of the War Cabinet wrote of the bomb, "I gather that we may sleep fairly comfortably in our beds."

Only after the outbreak of war would the message of the atomic Cassandras be taken seriously in the corridors of power. Politicians and military strategists belatedly realized that the physicists were developing gadgets that could win wars. Radar was one; the atomic bomb another. No longer scorned as unworldly intellectuals, physicists were drawn abruptly and sometimes unwillingly into a new relationship with government and the military. World War I had been the war of the chemists; World War II would be the war of the physicists. By early 1942, even before the Manhattan Project got off the ground, America had already spent four times as many research dollars on physicists than they had on chemists. Physicists, the undefiled theorists who were supposed to be above anything so vulgar as the marketplace, suddenly became a marketable commodity. As one American professor remarked in 1943, "Almost overnight, physicists have been promoted from semi-obscurity to membership in that select group of rareties which include rubber, sugar, and coffee." No respectable war cabinet could now operate without a scientist, preferably a physicist, as a permanent consultant. By the end of the war, this new relationship between physicist and government had become formalized.

The theories of the atomic bomb had emerged from the interwar period of scientific experiment, the so-called Golden Years of physics. The center of research was Germany.

Students and professors from Europe and America had gathered at the universities of Göttingen, Berlin, and Munich. When not in the classroom, they had spent hours huddled in the bars and cafés, poring over the latest concepts and covering tablecloths with mathematical formulas. They had listened admiringly to Einstein, Max Planck, and Danish Nobel Prize–winner Niels Bohr. Bohr's Atomic Physics Institute in Denmark, Cambridge University's Cavendish Laboratory in England, and the Radium Institute in Paris were also great centers of experiment and debate. The Russians, too, were following Western atomic developments at the Leningrad Physics Institute, and, in America, the universities of Columbia, Chicago, Princeton, and Berkeley had their nuclear research departments.

Most of the physicists were deeply conscious of being in a race to uncover the inner secrets of the atom. They jealously guarded their theories until, like Szilard, they registered them in patent offices. The Germans, with their traditional links to industry, were fanatical about patents; so were the French, who were constantly trying to reduce Germany's lead in science. The exceptions were the British "gentlemen scientists," who believed that their discoveries should be seen for what they were: key steps in the progress of mankind, not some crude commercial deal.

The pace of the race dramatically changed with the Hahn-Strassmann discovery of uranium fission and Joliot's discovery that uranium fission produced the right number of neutrons for a chain reaction. Szilard, who by then had moved on to America because he thought he would be prevented from working for the war effort in Britain, noted that the physics department at Princeton University was "like a stirred up ant heap" at the news of Joliot's triumph. With the French breakthrough, Szilard tried to persuade his colleagues in Britain and France to stop publishing the results of their atomic work so that the Germans would not have access to it. The British agreed, but the French, led by Joliot-Curie, refused. A colleague of Joliot's commented later that it was a matter of national pride for the Frenchman. "France's reputation was for women's fashions and the niceties of life. Joliot wanted to change that." By the middle of 1939, Joliot's team had written a secret patent on a crude uranium bomb and sketched out a workable reactor.

In America, Szilard teamed up with the Italian physicist Enrico Fermi, himself a refugee from Mussolini's fascism, and together they worked at Columbia University to confirm Joliot's neutron experiments. Szilard later described the dramatic night of their own breakthrough.

> Everyone was ready. All we had to do was to turn a switch, lean back, and watch the screen of a television tube. If flashes of light appeared on the screen it would mean that large-scale liberation of atomic energy was just around the corner. We just turned the switch and saw the flashes. We watched for a little while and then we went home. That night there was very little doubt in my mind that the world was headed for grief.

Szilard persuaded Fermi to go to Washington and tell the U.S. Navy the significance of the findings. He thought the navy might offer some funds for research, but the admirals were not interested. After the war, Szilard recalled, "It is an incredible fact in retrospect that between the end of 1939 and the spring of 1940 not a single experiment was under way in the United States which was aimed at exploring the possibilities of the chain reaction."

The German authorities, meanwhile, had taken Joliot's news very seriously. Within a few days of the publication of the French neutron experiments, a professor at Göttingen had written to the Reich Ministry of Education about the possibility of using uranium fission in a reactor to produce electricity. The ministry immediately convened a secret conference of theoretical physicists and set up a small research project that became known as the Uranium Club.

At Hamburg, two physicists—Professor Paul Hartech, a physical chemist who had worked in England under Rutherford, and his assistant, Dr. Wilhelm Groth—sent a warning to the German War Office. "We take the liberty of calling your attention to the newest development in nuclear physics. In our opinion it will probably make it possible to produce an explosive many orders of magnitude more powerful than the conventional ones . . . the country which makes first use of it has an unsurpassable advantage over the others." The Hamburg project eventually won the support of the German military.

Even the Russians were taking the new discoveries more seriously than the Americans. Igor Tamm, a future Nobel Prize–winner and the man who was to become the most senior scientist on Russia's H-bomb project, reportedly said of the French breakthrough: "It means that a bomb can be built that will destroy a city out to a radius of maybe ten kilometers." The U.S.S.R. Academy of Sciences set up a top research committee on what the Russians were to call the "Uranium Problem." And two of the leading Russian physicists, Igor Kurchatov, later scientific head of the Soviet atomic bomb project, and Yulii Khariton, later the foremost weapons designer in the Soviet Union, evolved a plan for achieving a chain reaction, but their work was seriously curtailed by the German invasion in June 1941.

As the war between Germany and the Allies intensified in the spring and summer of 1940, so did the guessing about each other's atomic projects. Fearing the Germans were at work on a bomb, Britain and France were eventually persuaded, despite scarce resources, to start their own projects. Ironically, although the Germans had started the war with a greater experience of atomic physics, they were to make little progress in producing a bomb. They did not have the vast resources available to the Americans, of course; and if they had started building the huge atomic factories needed, these would undoubtedly have been spotted by Allied intelligence and bombed.

One explanation for their failure, much debated after the war, centered on the motivation of the German physicists to build a bomb for Hitler. In his postwar recollections, the leading German theorist, Werner Heisenberg, sought to give the impression that he and his colleagues did not work on the German bomb project as diligently as they might have done because they were reluctant to see an atomic bomb in the Führer's hands. But there is a problem reconciling that view with the facts garnered from captured German documents. These show that at least until 1942 German physicists plunged headlong into work on uranium projects that could easily have ended with a bomb. One of the main reasons for the German failure was professional snobbery. When the time came for the physicists to put their blackboard theories into practice, they refused to let the engineers take over. Engineering had always been regarded by the physicists as a lesser calling; engineers were dull, practical men whereas physicists were bright, exciting inventors. The German physicists thought they alone could build the atomic reactors needed to produce the bomb material. As it turned out, they were wrong; by the end of the war, the Germans did not even have a working reactor.

Once the British and French governments had been persuaded of the importance of the nuclear chain reaction, they acted swiftly to try to prevent Germany from obtaining the essential ingredients needed to make a bomb.

In Paris, Joliot's research team was experimenting with ways of making a nuclear reactor work. They had discovered that a chain reaction could, theoretically, be set up in a lump of uranium if the neutrons released by the splitting of the atoms were slowed down, or "moderated," thereby increasing the likelihood they would find their atomic targets. In 1939, Joliot's team found this could be done if the uranium was surrounded with a substance called heavy water, discovered by the American chemist Harold Urey in 1931. It is indistinguishable in appearance from ordinary water, in which it is present in the tiny fraction of .15 percent. By 1939 the only producer of heavy water on a large scale was the Norsk Hydro Company in Norway. The majority of the company's shares—65 percent—were held in France by the Banque de Paris et des Pays Bas; another 25 percent of the shares had been acquired by the German chemical conglomerate I. G. Farben; the German scientists also suspected heavy water could be used to moderate the speed of neutrons.

In the winter of 1939, both the French and the German governments received urgent requests from their atomic physicists for stocks of heavy water. In Paris, Joliot sent a message to the French Ministry of Armaments asking them to buy up all that Norsk Hydro could produce. At the time, he knew nothing of the German involvement in the company, nor that the German physicists were also interested.

In February 1940, four months after Joliot's request, the French government responded. An agent from the Deuxième Bureau,

Jacques Allier, was sent to Norway to purchase the Norsk Hydro heavy water and bring it back to Paris. Allier was well qualified for the task, having been an employee of the Norwegian section of the Banque de Paris et des Pays Bas before joining the Deuxième Bureau.

The Germans tried desperately to stop him. They followed him to Norway and, when he was about to dispatch his precious cargo by plane to France, they intervened. The plane on which Allier had booked the freight and himself for the first leg of its journey was intercepted by German fighters and forced to land in Hamburg. Allier, however, had outwitted the Nazis. At the last moment and unseen by Gestapo agents, he had switched the cargo and himself to a plane bound for Scotland. Several days later, the heavy water arrived safely in Paris via truck and ship across the English Channel.

Unfortunately for Joliot, the German defeat of France prevented him from making use of it. In the spring of 1940, the Germans moved swiftly across Europe; by the middle of June, they had reached Paris. Joliot and his two assistants, Hans von Halban and Lew Kowarski, fled south with the heavy water to the town of Clermont-Ferrand. There, they hid the heavy water first in a bank vault, and then, as the Germans drew closer, in a condemned cell in the local prison. Before the Germans could get their hands on it, the British completed arrangements to have the stock shipped to England, where it was stored for safekeeping at Windsor Castle. Halban and Kowarski accompanied the heavy water in a coal barge to England, but Joliot refused to leave France; he thought it was his duty to try to keep French science alive during the German occupation.

In London, Joliot's original neutron experiments had stirred the British scientific establishment into a near-panic. Within a few days of the publication of the results from Paris, the British government approved a proposal from its scientific advisers designed to block the German access to the world's known uranium stocks. One major source of uranium, the mines in Czechoslovakia, was already in German hands; the other, high-grade ore in the Belgian Congo, was still available to the highest bidders.

The British plan for acquiring the Belgian uranium called for pulling a few colonial strings. The Belgian company Union Minière de Haut Katanga had a British vice-president, an English aristocrat named Lord Stonehaven. He arranged a meeting between the company's president, Edgar Sengier, and the British government's chief scientific adviser, Sir Henry Tizard. Sengier, however, was an astute businessman with excellent connections. He had long been a friend of Madame Curie and had learned, through her son-in-law, Joliot, of the sudden importance of uranium. As a precautionary measure, he had shipped almost his whole stock of yellow uranium oxide, or yellow-cake, to the United States. There it was stored in a warehouse on Staten Island, a far safer place, Sengier rightly thought, than anywhere on Belgian soil. He told Tizard he could offer the British only a few tons of uranium. But he also told Tizard that, thus far, no one else had sought supplies of uranium in the quantities the British wanted; not even the Germans.

With the Belgian uranium and the French heavy water, the British now had the ingredients for the chain reaction, but they were still reluctant to devote much of their scarce resources to a bomb program because they had identified a special problem concerning the nature of uranium. The element occurs naturally in the earth's crust in three different forms, or isotopes. Two are important. The scientists write them in chemical formula shorthand, U-238 and U-235—the *U* standing for uranium and the number for the atomic weight of each isotope. The isotope containing what scientists call a "fissile" atom is U-235. It readily splits, or "fissions," when bombarded with neutrons, but it is present only in minute quantities—less than 1 percent—in naturally occurring uranium. More than 99 percent of the ore mined is the nonfissile U-238.

No one yet had any idea how much U-235 was needed to form a "critical mass." It was generally assumed that if buckets of uranium ore were heaped on top of each other, eventually there would be enough of the fissile U-235 in the pile to start a chain reaction. At the time, estimates of the size of the pile needed ranged as high as forty tons, suggesting that the overall weight of a uranium atomic bomb would be too great for any aircraft to carry. (Even the blockbuster bombs used at the end of the war weighed only ten tons each.) Thus, the very idea of the atomic bomb seemed to many British scientists totally impractical.

In any case, the British were preoccupied by their own demands for conventional weapons; their enthusiasm for the superweapon waned until two émigré German research physicists, Otto Frisch and Rudolph Peierls, produced a new concept for the bomb. Frisch and

Peierls had been working in Britain at the outbreak of war and had decided not to return home. In a remarkable report, only three pages long, they suggested that instead of trying to increase the amount of U-235 by increasing the amount of uranium, the better approach was to extract a lump of pure U-235 from the U-238. It would be a difficult and expensive operation, but they believed it possible. The two substances were chemically similar, differing only in weight—like the difference between cream and milk. And it would be worthwhile: the two German scientists calculated that the energy liberated from five kilograms of U-235 undergoing fission could equal that of several thousand tons of dynamite.

The Frisch-Peierls Memorandum, as their paper became known, persuaded the previously reluctant British government to finance a bomb project. It was code-named the Maud Committee, and by the end of 1940, significant advances had been made. A team of British scientists had confirmed that the separation of U-235 from U-238 was theoretically possible.

In the early summer of 1941, however, the British came to the conclusion that they could not support research into an atomic bomb and fight Hitler at the same time. Apart from anything else, the huge factories needed to produce the U-235 would be open to air attack. They sent the Maud Committee report, uncompromisingly entitled "The Use of Uranium for a Bomb," to America. The Americans, whose atomic research at this stage was behind Britain's, were very excited by what the report had to say, and eventually it formed the basis of the Manhattan Project.

In America, Szilard had gotten nowhere in his original attempts to convince the U.S. armed forces of the need to build a bomb. Increasingly frustrated, he had decided, as a last resort, to try to persuade Einstein to write a letter explaining the importance of the bomb to President Roosevelt. Szilard, never short of an exotic contact, had found a direct line to the White House through Dr. Alexander Sachs, a member of the Wall Street finance house of Lehman Bros. Sachs, a friend of Roosevelt's, promised to deliver the letter.

Einstein readily agreed to the plan. His letter, dated August 2, 1939, began:

> Some recent work by E. Fermi and L. Szilard, which has been communicated to me in manuscript, leads me to expect that the

> element, Uranium, may be turned into a new and important source of energy in the immediate future. Certain aspects of the situation which have arisen seem to call for watchfulness and, if necessary, quick action on the part of the Administration.

Einstein went on to list the successful experiments of the previous six months that had made the construction of a new bomb possible. "A single bomb of this type," he continued, "carried by boat and exploded in a port, might very well destroy the whole port together with some of the surrounding territory." He ended by suggesting that the United States secure a supply of uranium ore—perhaps from the Belgian Congo—and speed up experimental work on a bomb project. Noting he had heard that Germany had already stopped the sale of uranium from the Czechoslovakian mines, he warned that this was an ominous sign of their own efforts toward a bomb.

In response, Roosevelt created the Uranium Committee to study the prospects of a bomb, but its wheels ground slowly, not least because the committee's military members refused to believe that an atomic bomb could be built in time for use against Hitler. Giving evidence to the committee, Szilard enlisted the help of two other immigrant Hungarian physicists, Eugene Wigner and Edward Teller, to argue his case. It was still no use. At one meeting, a colonel from the Pentagon told them bluntly that it usually takes two wars to know whether a new weapon is any good and, in any case, weapons did not win wars, troops did. Wigner remarked later that he and Szilard felt as though they were "swimming in syrup."

While the Uranium Committee spun its wheels, Szilard continued to monitor the progress of the European scientists, particularly Joliot, through the scientific literature. All this was open to the German scientists, of course. One of Joliot's experiments had mentioned his theory of using heavy water to moderate the neutrons in the chain reaction. Using Joliot's notes, published in *Nature*, Szilard did some quick calculations and came up with a similar system that would use graphite instead of heavy water. He then proposed to Einstein that they threaten open publication of this new method unless the American government took stronger action over uranium research.

In March 1940, Einstein agreed to write a second letter. Addressed to Dr. Sachs, it began: "Since the outbreak of war, interest in uranium has intensified in Germany. I have learned that research there is being carried out in great secrecy." Then, mentioning Szilard's graphite-moderated reactor, he went on: "The question arises whether something ought to be done to withhold publication. The answer to this question will depend on the general policy which is being adopted by the administration with respect to uranium."

The letter was passed to President Roosevelt, and Szilard finally got some results. During the summer of 1940, the United States agreed to an official curtain of secrecy on the publication of research material relating to uranium and the chain reaction, something for which Szilard had fought for so long. In turn, the Allies each agreed to the censorship. The only physicists outside Germany who continued to publish were the Russians, and they soon began to notice the absence of Western research in the standard journals, eventually deducing for themselves that censorship had been imposed because of the military implication of the work.

Roosevelt placed the Uranium Committee under the National Defense Research Committee and the chairmanship of a civilian electrical engineer, Vannevar Bush. Bush, then fifty years old and president of the Carnegie Institution, eagerly accepted the challenge and the moral responsibilities posed by the new generation of atomic weapons. For the rest of the war, Bush would be Roosevelt's chief scientific adviser. He accomplished with extraordinary success the delicate interfacing of the civilian and military sections of the bomb project. The physicists entered a new world of big government and big business that would shatter their academic independence forever.

2

The Act of Obedience

For we cannot command nature except by obeying her.
—FRANCIS BACON, *Novum Organum*

The leadership of the atomic revolution changed hands in the fall of 1942. Until then, the physicists were in charge; afterward, it was the turn of the engineers. What the physicists had worked out on blackboards, slide rules, backs of envelopes, and café tablecloths had proved the theory of the atomic bomb, but it could not be built on a laboratory bench or cooked up in a test tube. As the Manhattan Project gathered steam in America, huge new plants had to be constructed to produce the fissile material, and new metal alloys had to be found to withstand the corrosive properties of new toxic liquids and gases. An entirely new skilled work force of more than 100,000 men and women had to work with the dangerously radioactive uranium and plutonium in the world's first factories operated by remote control. Finely integrated systems of cranes, tongs, cameras, and periscopes had to be linked together in precise harmony. The Manhattan Project was the most expensive single program ever financed by public funds. The physicists' bill for working out the theories had been paid in modest sums of $100,000 here and there, but the engineers' bill to construct the first atomic bomb came to more than $2 billion.

The importance of the engineers, the dull backroom boys with no tradition of talking or writing about themselves or their achievements, has been obscured over the years by an outpouring of physicists' memoirs. The physicists' dazzling mastery of figures and formulas has always given them a special intellectual sex appeal. The engineer's job is lusterless by comparison; it never rests on the one brilliant insight, the single stroke of genius. In retrospect, scientific revolutions have a certain inevitability about them: if one scientist fails to comprehend an undiscovered law of nature, another will. But there is nothing so inevitable about technological development. The engineer knows full well that his achievements are the result of highly organized work, disciplined pursuit of priorities, and tough-minded decisions by practical men.

Throughout 1943 and 1944, American engineers presided over the birth of a whole new industry with the single aim of producing a few kilograms of fissile material for a bomb. The country's recently developed organizational skills and technical talents were tested to the full. Almost overnight, new towns sprang up in isolated places so that work could proceed in secrecy. At Hanford, a small village on the banks of the Columbia River in Washington State, an army of 25,000 construction workers built a new town in less than a year. With 11,000 pieces of major construction equipment they excavated 25 million cubic yards of earth and laid down 158 miles of railway track and 386 miles of road. And on the pine- and dogwood-covered slopes of the Clinch River Valley between the Great Smoky and Cumberland mountains in Tennessee, another new town called Oak Ridge had grown in a year out of nothing.

The new plants were among the largest ever built. The factory where the precious U-235 was separated from the less useful U-238 was four stories high with each side of its U-shaped structure half a mile long. The machinery inside the plant required metal that had been processed on the south coast of England and worked into valves, pumps, and joints in factories in Michigan, New Jersey, Illinois, and Wisconsin. The wartime shortage of copper, needed for the miles of piping and electrical wiring, forced the Manhattan Project organizers to use thousands of tons of silver from the vaults of the U.S. Treasury. Thousands of tons of steel were required for giant magnets, 250 feet long. The list goes on.

Engineers swarmed over the new construction sites and into the laboratories—much to the annoyance of the talented group of international physicists who had been brought together for the Manhattan Project; like the German physicists they preferred to believe that the atomic bomb was 90 percent physics and 10 percent engineering, when, in fact, it was the other way around. They abhorred the takeover by engineers, especially by military engineers. The physicists' freewheeling lifestyles were rudely restricted, taken over by new wartime rules and procedures.

They were also to learn another unpalatable fact of this new life. Although their advice on political and moral issues would often be sought in the councils of Big Government, it would rarely be taken. The bomb they would create was a weapon of war to be used by politicians and military officers. Those scientists who, after the bomb was built, felt it was immoral to drop it on the Japanese were in for a shock; their scruples had come too late. In the beginning, they had begged the politicians to make a bomb to protect the world against Hitler—if necessary, to drop it on him. In the end, they should not have expected the military would ban such a superweapon before they had a chance to use it.

Even as Germany lay defeated in the spring of 1945, the engineer in charge of the Manhattan Project, Leslie Groves of the U.S. Army Corps of Engineers, ordered his men to step up production so that the bomb could be used on the Japanese. What had begun as a race to build a bomb before Hitler now turned into a race to build a bomb before the war ended. It was not a time for doubts or remorse. There were harsh political and military imperatives to be accomplished. Years later, the atomic scientists' ultimate folly was best summed up in the words of the Manhattan Project's chief physicist, Robert Oppenheimer: "When you see something that is technically sweet, you go ahead and do it and you argue about what to do about it only after you have had your technical success. That is the way it was with the atomic bomb."

The U.S. Army Corps of Engineers was the obvious place to look for an engineer overlord for the Manhattan Project. Officers of the Corps, who invariably came from the top ten of their class at West Point, would quickly grasp the engineering needs of the theoretical physicists. Their traditional involvement in building dams and roads had given them experience in working with civilian contractors and also in costing and accounting their own projects. One of the most independent bureaucracies in the federal government, they had a jealously guarded reputation for getting the job done. No Corps engineer, however, had ever had the opportunity for political and military honors that the Manhattan Project would offer to the pompous, portly, forty-six-year-old Colonel Leslie Groves.

One day in September 1942, Groves had just finished giving testimony on Corps expenditures to a congressional committee on Capitol Hill when he was sent for by his superior officer. "The Secretary of War has selected you for a very important assignment and the President has approved the selection," Groves was told. "If you do the job right it will win the war." The job was to make the world's first atomic bomb.

Groves had been desperately hoping for a front-line job. He had been overseas only once, to Nicaragua during an abortive plan to build a second ship canal through Central America. His present job was exceedingly unglamorous for a career army officer: he was in charge of what his colleagues called "cookbook engineering"—building army camps, airfields, and ordnance depots, where ready-made plans existed for each building. Groves wanted to be stretched, to develop his skills and his imagination under pressure. He saw himself in a combat command, building instant roads and runways in unknown terrain with makeshift materials. Like many career officers, Groves was impatient to join the war America had entered nine months before; every soldier knows promotion comes quickly on the front line. Groves, a West Point graduate (fourth in the class of 1918), had climbed ponderously through the peacetime ranks, spending sixteen years as a lieutenant. The prospect of missing combat appalled him, but the laurels he would eventually earn from the Manhattan Project far outstripped anything he could have won in battle. The promotion, too, was instant: Groves became a brigadier general.

From his office in Washington, D.C. (where he had spent the last years overseeing, among other things, the building of the Pentagon), Groves lost no time in realizing the potential of his extended kingdom. Spurred by the consensus that the American bomb program was two years behind the Nazi project and by his own need to establish authority quickly over the scientists—the "long-hairs" as he called them—he set a cracking pace.

Within days of his appointment, he had left Washington bound for laboratories in New York, Chicago, and California. There he would meet the physicists who were working on ways to produce fissile material. The tools of his new trade were a manila envelope containing secret documents and a small Colt automatic pistol, easily concealed in the trouser pocket of his voluminous civilian suit. The manila envelope never left his side: during meals in the train's dining car, he sat on it and, while asleep, he kept it under his mattress. Groves took the meaning of Top Secret very seriously, as the scientists were shortly to find out.

The brigadier general was to burst into their secluded world of theory, experiment, and augury like a blast of Arctic air. He wanted them all in uniform and all taught how to salute and say "Yes, sir." In order to maintain optimum secrecy, there was to be no contact between the laboratories, and idle chatter outside the labs was also proscribed. They were not even to tell their wives what they were up to. Mrs. Groves never knew, the general later admitted, what he was doing until after the bomb was dropped on Hiroshima. The road ahead with General Groves obviously was not going to be easy. The President's chief scientific adviser, Vannevar Bush, recognized that the moment he met him. "Having seen Groves briefly, I doubt whether he has sufficient tact for the job. I fear we are in the soup," he warned Secretary of War Henry L. Stimson.

By the time Groves arrived on the Manhattan Project, the broad range of scientific options for separating the fissile material needed to fuel the bomb had been determined. Two different types of fuel were being investigated: U-235 and plutonium. In the Maud report the British had suggested a method for producing U-235 on an industrial scale, but nothing had been tried out. It was Groves's job to evaluate the methods and fund the ones he thought promising. Possible ways of separating U-235 from the nonfissile U-238 were being researched at three centers: Columbia University, the Westinghouse Research Laboratory in Pittsburgh, and the University of California at Berkeley. Ways of producing plutonium from a nuclear reactor were being studied at the University of Chicago.

At Columbia University, Harold Urey, whose discovery of heavy water had won him a Nobel Prize, was in charge of a technique called gaseous diffusion. This required passing uranium gas, including U-238 and U-235, through a series of porous barriers. Since the lighter U-235 isotope passes through the pores more easily than the heavier U-238, the end result would be a gas "enriched" in U-235.

At the University of California, Ernest Lawrence—the inventor of the cyclotron, or atom smasher, and himself a Nobel laureate—headed the group working on the electromagnetic process for separating U-235. This involved passing uranium gas over a large magnet. The lighter particles of U-235 would be attracted to the magnet at a different rate than the heavier U-238.

In Pittsburgh, Eger Murphree, a Standard Oil of New Jersey engineer, was supervising a project to separate U-235 using a centrifuge. In this method, the lighter U-235 is parted from the heavier U-238 by putting uranium gas into a large rapidly spinning machine and driving off the heavier isotope—a method akin to separating cream from milk.

Finally, at the University of Chicago, Arthur Compton, who had won the 1927 Nobel Prize in Physics for his studies of X rays, was in charge of building what was shortly to be the world's first working nuclear reactor. It would soon produce minute quantities of plutonium.

The scientific odds against one of the four projects proving so much better than any other at producing fissile material seemed impossible to predict before Groves entered the scene, and all four projects had been given the go-ahead by Roosevelt's Uranium Committee. In so doing, the basic production pattern for the Manhattan Project was set: duplication of research effort regardless of cost.

Groves's knowledge of the scientific worth of one project over another was limited to the rapid briefings on atomic theory he had crammed in before the trip; his assessment of the projects would be based largely on his estimate of the caliber of the men in charge, of their ability as leaders to get the job done.

Groves was not at all impressed with his first stop, the Westinghouse Research Laboratory in Pittsburgh. The leader, Eger Murphree, was out ill, and Groves was quick to note that without him the project lacked any real direction. Groves was horrified to find that the laboratory still closed on weekends and public holidays. Had these people no sense of urgency? Did they not know there was a war on? Groves never recovered from the impression gained from this first visit; the centrifuge project was doomed. By the spring of 1943, it was out of the picture entirely.

At Columbia University, he was equally disappointed. He did not see in the fifty-year-old Harold Urey the kind of man who had the single-minded drive to make it work. Urey appeared a vacillating and emotional person. His enthusiasm rose and fell with each new technical problem. There was nothing wrong with Urey the scientist, Groves felt; he had a fine, open, inquiring mind able to see all points of view; but this served only to magnify the difficulties and make him incapable of displaying the overweening self-confidence required to win through. The engineering problems led him to fits of despair.

Working under Urey, however, was a man much more to Groves's liking: the scientist actually getting his hands dirty with the hard work was John Dunning, a thirty-seven-year-old electrical engineer. Groves asked Dunning a series of penetrating questions about the progress of the gaseous diffusion project. Dunning's answers were full of hope and made in a tone of deference that Groves was to experience all too rarely, as far as he was concerned, from some of the other scientists. Groves approved of Dunning. He also discovered that Dunning had a good partner in a hard-headed and successful process engineer from the M. W. Kellogg Company, Percival ("Dobie") Keith. The Kellogg Company was a New Jersey–based multimillion-dollar firm of consultant engineers that had grown up with the oil and chemical industries. Keith was a specialist in the design and construction of huge new industrial plants, and one of his plants had already used the gaseous diffusion process to separate hydrogen gas from methane.

"Dobie is not the type who gets ulcers," a colleague said; "he gives them." Impetuous, dynamic, hard-working, fiery, and authoritarian, he was the general's type exactly. Like Groves, he was prepared to work only if he was in complete control, and his adrenaline pumped harder every time he was told the task was impossible. Keith never belittled the engineering challenges; he knew their worth and liked to compare the work of devising a perfect pump, a perfect seal, or a perfect valve, tedious as it might seem, to the French artisans who devoted a lifetime to a single stained-glass window at Chartres. He passionately believed that the atomic industrial plants would be the cathedrals of the era. For centuries to come, people would marvel at the feats of construction. Groves liked that kind of talk. He thought Dunning and Keith could probably overcome the shaky leadership of Urey.

In San Francisco, the general was met at the train (he rarely flew, for safety reasons) by Ernest Lawrence. "You're going to have a surprise," Lawrence said excitedly; "you've been listening to all those theories, but out here you'll see separation actually going on." Groves warmed to him immediately. Here was a scientist who talked and acted like an engineer.

Lawrence had a hard act to sell. His huge machines—vast magnets using thousands of tons of steel—required equipment on a scale unheard of by Groves in his visits to the other labs, and although Lawrence had produced minute quantities of U-235, the electromagnetic process appeared incapable of making more than thimblefuls. Lawrence's enthusiasm was infectious, however. The scientist long ago had decided that the race to build the bomb could not go on without him. He pursued his personal interests with a relentless drive that would typify many of the scientific administrators in the history of nuclear energy. More of a gadgeteer than a scientist, Lawrence had devoted his life to building bigger and bigger machines.

He was the most American of all the Manhattan Project scientists. The son of a South Dakota teacher, he had the indelible imprint of the small towns of the Plains. Questions of morality were clear-cut, the strict requirements of a punishing God. He believed in honesty, frugality, enterprise, hard work, family life; above all, he was a loyal and proud citizen. He despised idleness, philandering, and bad language, and, like many Americans, he believed the ultimate sin was to lie. There was no subtlety to his style, no awe about the complexity of human motivation. Generous-minded but not open-minded, Ernest Lawrence could only attribute evil motivation to those who persistently disagreed with him. Nothing was more distinctively American.

Lawrence was the first of the "big science" men. His entrepreneurial verve and skills created a distinctive new mode in the way of doing science. At his Radiation Laboratory at Berkeley, he created the first multidisciplinary team of scientists, using his machines to explore not only physics but chemistry, biology, and a range of medical disciplines. He was one of the first to realize that the days of the major breakthrough in isolated laboratories equipped in a haphazard manner were over. Big science required big money, and Ernest Lawrence had the promotional zeal to get it. General Groves would ensure that Lawrence received funds that made his previous grants look paltry.

Finally, there was the plutonium operation at Chicago. Arthur Compton had brought together a loose federation of scientists, among them Enrico Fermi, Leo Szilard, and Eugene Wigner, who were tinkering with no fewer than four different ways of producing plutonium. Compton also had the bomb design team under his command. To Groves, the visible strains of a cooperative creative process at its best seemed like anarchy. He was dumbfounded to hear that the estimates of how much fissile material was actually needed to construct a bomb were correct only to a factor of ten: that is, anything from ten times less to ten times more. Groves commented later: "My position could well be compared to that of a caterer who is told he must be prepared to serve anywhere between ten and a thousand guests." The Chicago scientists, Groves noted, had no conception of the requirements of industrial production, and he doubted the necessary drive and discipline would be imposed by their academic administrator: Compton was too much one of them.

To an engineer like Groves, however, the superiority of the plutonium route was obvious. Even though virtually nothing was known about the new element, it was at least chemically different from uranium; in principle, it could therefore be separated more easily than U-235, which was chemically identical to its cousin U-238. Chicago was a proud and aggressive university, and Compton was eager to accept the challenge. He had tried to concentrate a vast array of talented physicists at that institution; the plutonium project itself, along with the new flow of government funds, attracted a wide range of the most prominent and the most promising scientists in America. Compton was pleased to preside over this seething mass of creativity. (No one could purport to lead it.) His own commitment to the pursuit of truth through science was passionate, and he shared the scientists' fear of a Nazi bomb project. Compton was personally convinced that the plutonium work at Chicago was a serious option worth pursuing, and he tried to impress on Groves his keenness to keep his team alive.

With each of the scientific teams vigorously promoting its own systems, the decision-makers—Groves and the two presidential advisers, Vannevar Bush, now head of Roosevelt's Office of Scientific Research and Development, and James Conant, the president of Harvard who became chairman of the National Defense Research Committee—found it hard to choose among them. Expert assessment of the engineering was tricky because most of the plans were still on the drawing board. In the end, they dropped the one project that was most advanced—the centrifuge method. Groves believed that Eger Murphree, the man in charge of the centrifuge program, was not providing the leadership needed to overcome the problems.*

Before making the decisions on the other three systems, Groves and Bush were treated to a spectacular display of the proof of the theories about uranium. At the University of Chicago on December 2, 1942, Enrico Fermi successfully demonstrated the world's first nuclear chain reaction. Much to the consternation of his superiors, Compton announced the top-secret experiment would take place in a squash court underneath the university's football stadium. Fermi was supremely confident it would work and that there would be no danger.

The reactor, or pile as they used to call it, consisted of 40,000 graphite blocks arranged in a circle; inside it were 50 tons of uranium in metallic and oxide form. Inserted into the uranium were neutron-absorbing control rods that, when withdrawn, allowed the neutrons to start their atom bombardment leading to a chain reaction.

As the experiment started, Fermi, using his slide rule, predicted the exact moment at which the chain reaction would begin. At 3:53 P.M. he announced, "The reaction is self-sustaining." Then he ordered the control rods to be reinserted. The first planned release of nuclear energy in history had been effectively controlled.

Compton rang James Conant in Washington. "Jim," he said, "you'll be interested to know that the Italian navigator has just landed in the New World. He arrived sooner than he expected."

"Were the natives friendly?" Conant inquired.

"Everyone landed safe and happy," Compton replied.

The key principle of the atomic bomb, and the method of producing plutonium, had both been proved in practice, and Compton had ensured a place for his team in the Manhattan Project. The engineers could now proceed with their takeover.

The original hopes that the number of options for making fissile material could be reduced to one or two were soon abandoned, and

*It would be twenty years before the centrifuge project made a comeback, proving, ironically, to be a highly efficient method of producing U-235, one that used much less electricity than the others. The gaseous diffusion plant, which eventually emerged as the most successful of the wartime plants, was a voracious consumer of electricity. By the 1950s, the U.S. diffusion plants were devouring 10 percent of the country's electricity. In the three decades after the war, similar plants were built in Russia, France, Britain, and China, using millions of barrels of oil-equivalent energy.

the President approved full-scale development of the three remaining methods. The enrichment plants, both gaseous diffusion and electromagnetic, were built at Oak Ridge, Tennessee, and the plutonium reactors at Hanford, Washington. Driven, first, by the fears that the project had lost two vital years to Nazi Germany and then, by the prospect that the war would end before they could build a bomb, Groves demanded a furious pace.

Even the most hostile of Groves's scientist critics would grudgingly admit in the end that the general's organizational powers were formidable. He battled, bruised, and bluffed his way through agency bureaucracies and corporate boardrooms to fill the staggering requirements of the program. It was not without its lighter moments. When a Groves aide went to the U.S. Treasury and asked for thousands of tons of silver to use in electric wiring for the huge new factories, he was told, "In the Treasury we do not speak of tons of silver. Our unit is the troy ounce." It was the same for graphite, demanded in a degree of purity never before envisaged, and uranium metal, previously handled in grams. Groves needed tons of everything.

The general enlisted the best of American industrial engineering know-how: du Pont, Union Carbide, Kodak. "We're not looking for scientists," Groves said. "We have so many Ph.D.'s now that we can't keep track of them." Impressed by Groves's drive and determination, the companies—few believing the job was possible—signed up.

Within two years, a new industry had been built. It was a new technology of such scope, variety, and sophistication that, had it not been wartime, its development in two decades rather than two years would have been regarded as extraordinary. The first fully automated factory, the first plant completely operated by remote control, the first totally leakproof industrial system—six million square feet of machinery that had to be kept as spotless as an operating theater. A tiny leak or a speck of dust could have ruined the process. Revolutionary new pumps, valves, and seals had to be designed and manufactured in the thousands. Perfect vacuums, created by machines of radical design, were monitored by control devices of unprecedented precision. Tons of raw materials, hundreds of miles of pipes, and thousands of pieces of equipment had to be plated with aluminum or nickel, with a metallurgical bond of a perfection never before known; all of it designed to produce just a few kilograms of pure U-235 and plutonium.

The urgency of the project required technical development to proceed in tandem with basic research, so that plants were built before anyone knew what was to go in them. The duplication and waste were enormous. The power plant at Oak Ridge, for example, the largest ever built at the time, was geared for an electricity output of five different voltages because no one knew which would be needed. No industrial development since then has had the luxury of so many options.

The ratio of capital and manpower investment to actual results was appalling. Ernest Lawrence's huge magnets could never be made to produce more than minute quantities of U-235. During the first runs of Lawrence's machines, 90 percent of the uranium missed the specially designed collection bins and ended up being sprayed around the walls of the apparatus. Minor metal impurities and lack of precision in the design and construction caused a breakdown of the first units. Vacuum leaks and electric short circuits sprang up everywhere. Undeterred, Lawrence still argued for building more and more plants using his electromagnetic system. He even tried to persuade Groves to drop the diffusion system. "I am no longer even enthusiastic about continuing research on the diffusion method, because I again feel that the electromagnetic system will ultimately far outstrip it," he said. Groves continued to have faith in Lawrence, and the huge electromagnets became the most expensive of the separation projects, gobbling up $600 million of the $2 billion budget. After the war, the project was shut down forever; a dramatic last-minute breakthrough in the gaseous diffusion system—the creation of a thin metal membrane with millions of minute holes through which gas could be pumped—established that approach as the dominant worldwide technology of uranium enrichment.

One major problem was still outstanding: the design of the bomb. Some scientists had tended to minimize the research that would be needed. Lawrence, for example, had told Groves that thirty scientists could accomplish the design of the bomb in three months. But a brilliant thirty-nine-year-old physicist at Berkeley, Robert Oppenheimer, had made preliminary studies of a bomb design and he disagreed. He said it would take much longer and suggested setting up a separate laboratory to do it.

He outlined the difficulties to General Groves. Once the scientists had worked out the size of the "critical mass" of U-235 or plutonium needed to sustain a chain reaction, they then had to devise a method of separating the mass in the bomb casing to prevent a premature explosion. This meant cutting the mass of the U-235 into two halves and working out some way of getting them back together again within millionths of a second at detonation. The whole process required many tricks, Oppenheimer warned. To imagine this could be done in three months was a hopeless underestimate.

Groves did not need to hear any more. He liked the idea of a separate bomb-construction unit: a new, isolated place that would meet the special needs of security, somewhere to test the bomb without anyone watching, someplace the scientists could be kept out of the public eye. Oppenheimer suggested New Mexico, north of Sante Fe, where the Rio Grande had carved deep canyons around flat-topped mesas. The young scientist had had a ranch in the area for some years. On top of one of the mesas was the Los Alamos Ranch School, a collection of log cabins that housed a boys' school, mostly sons of rich easterners. Oppenheimer advised Groves to go and see it. Groves liked it on sight. It was totally isolated. The scientists, the "prima donnas" and the "spoiled brats," as he called them, could be easily corraled and disciplined. One major decision remained: Who should be the director?

Oppenheimer himself was an obvious choice, but the older scientists resisted the idea: he was a brilliant theoretician, they conceded, but he had no administrative experience. More significantly, the FBI had strongly advised against giving Oppenheimer a security clearance because of his espousal of left-wing causes during the Spanish Civil War and because some of his relatives and friends were known members of the Communist party. The more Groves thought about Oppenheimer, however, the more the younger man seemed ideal.

He was already involved in the bomb project and knew everything there was to know about it. He was not as much of a prima donna as the others. Probably, Groves thought, because unlike so many of them, he was not a Nobel laureate. He had a knack of being able to explain things in simple, clear language that Groves could easily understand. Moreover, despite his flirtation with Communists, Oppenheimer was not a political animal. His scientific prowess was matched by a deep knowledge of literature and languages, but he had no real sense of politics or economics.

Most important, Oppenheimer was regarded as an intellectual giant by the other scientists. He would be able to attract the best physicists to the new laboratory. Groves would take the gamble on his ability to organize them. And if he proved to be no good, it did not matter: Groves would whip them into shape himself.

As it turned out, Oppenheimer proved to be an extremely able administrator and much to Groves's liking in the way he tackled his job. Oppenheimer was in touch at all times with everything that was going on up on the mesa. And his only real conflict with the scientists came when he agreed to Groves's outrageous demand that they wear uniforms. The general proposed making Oppenheimer a lieutenant colonel and his senior staff, majors. Oppenheimer seemed to like the idea of dressing up, but his colleagues hated it. Many threatened to resign if they had to do it, and the idea was dropped.

If the engineer-general could not have his "long-hairs" in uniform, he was damn sure he was going to have them "compartmentalized." Groves's biggest headache was security: how to keep the disparate parts of the Manhattan Project—spread across the length and breadth of America—Top Secret. He did it the only way he knew how: he restricted the flow of information to a "need-to-know" basis. There was to be no free contact between scientists from one laboratory to another. Each of them would be given code names. Enrico Fermi was Dr. Farmer; Eugene Wigner was Dr. Wagner; Niels Bohr, who arrived at Los Alamos after escaping from Nazi-occupied Denmark, was Nicholas Baker. Groves liked to explain "compartmentalization" with a sporting metaphor. "Just as outfielders should not think about the manager's job of changing the pitchers, and a blocker should not be worrying about the ball-carrier fumbling, each scientist had to be made to do his own work."

The scientists hated the idea. It was a real intrusion on their normal work methods—the seminars and the coffee-break chats that are all part of the discovery process. Yet despite the objections of a vocal minority, compartmentalization was upheld for the whole project. The one person privy to all information, of course, was General Groves. Knowledge is power, and the general made good use of it. His power was absolute. Even his aides were kept at arm's length at the Los Alamos installation. When Groves sent an aide on a tour of the bomb factory the general telegraphed in advance that the scientists were to tell him nothing.

Toward the end of the project, the only scientist who knew nearly as much about the Manhattan Project as Groves was Oppenheimer, the scientist in charge of putting all the pieces together. The two struck up a most improbable relationship, which was to survive the war. For Groves, it was a friendship based on necessity; for Oppenheimer, on curiosity. Groves used Oppenheimer as a teacher, and because Oppenheimer lacked an independent power base, he was someone Groves could control. Oppenheimer was endlessly fascinated by things he did not understand and happy to have the opportunity to pry deeply into the mind of this highly professional military officer whose outlook and lifestyle were so different from his own. No circumstances other than war could have brought the two together.

With the other scientists, Groves was often stern, gruff, and offhand—a bully. Some of them did not care; others were genuinely afraid of him. Groves, of course, did not see himself in quite that light.

A "soft-spoken major-general with a flair for the impossible and a way of getting things done," was how the official army handout, approved by the general, introduced him to the waiting press after Hiroshima. It only half-apologized for his interference: "While some of his decisions in the scientific field admittedly have extended much further than he would have liked, it was his responsibility to make them."

The scientists must have chuckled at the paragraph beginning: "A pleasant-mannered gracious officer who outwardly never shows the strain and worry of his job, General Groves is a constant source of amazement to his associates and subordinates." As to his well-known and singular lack of tact, the handout read, "firm and blunt when the occasion demands."

When it finally came to assembling the bomb in the spring of 1945, the principle of detonating the fissile material followed Oppenheimer's original outline to Groves. In the case of U-235, the critical mass amounted to about 15 kilograms. The metal was machined into two hemispheres; each half was then attached to the end of a gun barrel and the two assembled, facing each other and a short distance apart, inside the bomb casing. When the gun was fired the two halves would come together and the chain reaction would begin. At least, that was the theory.

The same method could not be used for plutonium, the scientists found, because the critical mass of plutonium was too unstable; the gun-type assembly would not prevent a premature chain reaction. The scientists devised a new method of detonating the plutonium, known as implosion. In this process, the outside of two hemispheres of plutonium, together weighing less than the critical mass, are coated with explosive. On detonation, the plutonium is squashed inward, forming a mass that, although subcritical in actual weight, becomes critical because of the huge pressures on it from the explosive. The timing of the detonation is crucial. All of the explosive must detonate at precisely the same moment so that the pressure will be equal all the way around the plutonium.

By early summer of 1945, both the gun-type assembly and the implosion method had been perfected. The scientists were confident that the gun-type assembly, using U-235, would work. They decided they needed to test only the implosion device, using plutonium.

The morning of July 16, 1945, began with steady rain over the Journado del Muerto, a desert area on part of the Alamogordo bombing range in New Mexico. Atop a one-hundred-foot metal tower sat the plutonium bomb. At three points ten thousand yards (nearly six miles) from the tower, observation dugouts were filled with nervous scientists and military men. Groves had given the order for the test to go ahead at 5:30 A.M. At 5:29:45 precisely, a pinprick of searing white light pierced the desert darkness from the top of the tower. Within a fraction of a second a terrifying fireball had vaporized the tower, turned the desert sand into glass, and was climbing into the dawn sky. The temperature at the fireball's center was four times that at the center of the sun and more than ten thousand times that at the sun's surface.

Oppenheimer's face, tense and drawn over the final hours, at last appeared relaxed. His thoughts, he said later, turned to the sacred Hindu epic, the Bhagavad Gita:

> if the radiance of a thousand suns
> were to burst at once into the sky,
> that would be like the splendor of the mighty one . . .
> I am become death,
> the shatterer of worlds.

Groves, sitting on the ground in a dugout between Bush and Conant, shook hands with them in silence. His more pedestrian mind could only summon up the image of Blondin's successful crossing of

Niagara Falls on his tightrope. "Only this tightrope had lasted three years," he observed.

As the shock wave reached the observation dugouts, Enrico Fermi let fall a few pieces of paper he had kept in his pocket for the occasion. Watching and measuring how far they were blown by the shock wave, he pulled out his slide rule and made a rapid calculation. The paper had sailed two and a half yards away. "That corresponds to a blast produced by ten thousand tons of TNT," Fermi announced.

Groves turned to his aides. "We must keep this thing quiet," he ordered. "Sir," said an aide, "I think they have heard the noise in five states."

Indeed, some alarmed citizens had actually witnessed the flash of light from as far away as 150 miles. It was like the sun coming up and going down again, one woman observed. For quizzical citizens, the army had concocted a story about an explosion at an ammunition dump. It produced such a brilliant flash, they said, because the dump had contained a mixture of high explosives and pyrotechnics. Groves had prepared an excuse that would both explain and forewarn should the radioactive fallout from the bomb become hazardous enough to force evacuation of the locals; weather conditions affecting the content of some gas shells that had been exploded by the blast might blow poison gases over populated areas. As the radioactive cloud drifted north from the test site, however, Groves determined that no evacuation was necessary, and therefore no one had cause to ask why the army had been so idiotic as to store poison gas shells near an ammunition dump—or, indeed, what they were doing with the poison gas.

Several months before the Alamogordo test, the defeat of Nazi Germany had become a reality, and some of the Manhattan Project scientists, particularly the immigrant physicists from Europe like Szilard and Wigner, lost their original motivation to make a bomb. Moreover, at Chicago, where Szilard worked, the scientists had completed their part of the project earlier than most and had had time to reflect on their achievement. Groves, sensing their growing frustration at having little to do and wary of their developing sudden moral qualms about their work, wanted Arthur Compton to cut back the size of his staff. But Compton resisted such a change and managed to keep the team together with research on new types of reactors. What

Groves had feared soon became a reality: Chicago emerged as the center of opposition to America's use of the bomb against Japan.

As before, Szilard sounded the first warning. His objections to the use of the bomb were based on moral grounds, but he also believed—and in this he was not alone—that its use would lead to an awesome postwar arms race. Accordingly, he prepared a document outlining his case to Roosevelt, but the President died on April 12 before the appeal could reach him. Szilard then tried to see Truman, but Truman declined. He got Einstein to write a letter, as before, to the new President, but even that did not open the door to the Oval Office. Szilard presented his case to James Byrnes, soon to become Truman's Secretary of State, but once again had no luck. Byrnes, an astute politician, had already warned the White House, before Roosevelt's death, of the congressional repercussions if the $2 billion Manhattan Project failed. He was not open to arguments about the immorality of using the bomb.

Groves, on learning of Szilard's attempt, was furious. How dare this Hungarian upstart meddle in affairs of state without his permission? Seeing Byrnes was a grave breach of security, apart from anything else.

Meanwhile, a presidential committee was appointed at the beginning of May to advise on a future policy for atomic energy. Secretary of War Henry L. Stimson called it the "Interim Committee" because he thought Congress probably would wish to appoint a permanent commission at a later date to supervise, regulate, and control the future development of atomic energy. This interim body was to advise only on the bomb, but it never had in its terms of reference the key question: Should the bomb be dropped? Only the scientists at Chicago had confronted the issue.

When the committee met in Washington at the end of May, Compton, a member of the committee's scientific panel and fully conscious of what his scientists back in Chicago were thinking, asked the committee's chairman, Stimson, if it would be possible to arrange something less than a surprise atomic attack. Perhaps a bomb could be exploded out at sea or on an island, something that would give the Japanese an example of the power of the new weapon.

No record exists of what precisely followed from Compton's suggestion, because the discussion took place over lunch. But there are minutes of what followed that afternoon. Considering that this was the only time an alternative to dropping the bomb was seriously discussed by the presidential committee, the minutes are perfunctory and, indeed, stunning in their callous disregard. The discussion is recorded under paragraph eight of the minutes, entitled, "Effect of the bombing on the Japanese and their will to fight."

> After much discussion concerning various types of targets and the effects to be produced, Secretary Stimson expressed the conclusion, on which there was general agreement, that we could not give the Japanese any warning; that we could not concentrate on a civilian area; but that we should seek to make a profound psychological impression on as many of the inhabitants as possible. At the suggestion of Dr. Conant the Secretary agreed that the most desirable target would be a vital war plant employing a large number of workers and closely surrounded by workers' houses.

Both at lunch and afterward, the scientific advisers, including Compton, Fermi, Lawrence, and Oppenheimer, had failed to come up with a suitable way of demonstrating the bomb without actually dropping it on people. For his part, Groves, also a member of the committee, was irked that scientists, rather than military men, should be discussing what sort of target to select. He had already chosen Hiroshima, and Hiroshima it would be.

To underline his annoyance, Groves entered a strong objection into the minutes under what was called: "Handling of undesirable scientists." The Manhattan Project, Groves said, had been "plagued since its inception by the presence of certain scientists of doubtful discretion and uncertain loyalty." Groves was referring specifically to Szilard and the Chicago physicists. His attack on the scientists was more than the bloodlust of a general with victory in sight, wanting to give the enemy one last decisive blow. There was a serious prospect that the war might end before the bomb was ready to be dropped. Groves's sense of urgency is well illustrated by the orders he sent out on the eve of Germany's surrender in May 1945. The Manhattan Project, he said, "will continue and increase after VE Day with Japan as the objective. It is suggested that pre-educational programs for project employees be considered at this time. These should stress the *continuing urgency* of the work and *increasing tempo* after VE Day, focusing attention on Japan as our ultimate objective . . . the avoidance of lost time on riotous celebration on VE Day itself should also be stressed if deemed necessary."

In June, the Chicago scientists responded. A group of six scientists, calling themselves the Committee on Political and Social Problems and led by the Nobel laureate James Franck, who had fled Germany in 1933, issued their report. In it, they again argued the case for a demonstration of the bomb in an uninhabited place, a place where the Japanese could appreciate its terrifying force without experiencing it directly. The report stressed that killing thousands of civilians with the new weapon would prejudice America's postwar image and undermine any effort to reach an agreement on international control of the atom. Such a momentous decision should not be left to military tacticians alone, the scientists concluded. The document was rushed to Washington for the final meeting of the Scientific Panel of the Interim Committee. The Chicago scientists had not mentioned the key problem of their recommendation: that it would prolong the war. Other points had to be considered, too. There was a fear, for example, that if a technical demonstration was announced, the Japanese would move American P.O.W.s to any island marked for the demonstration. In the end, persuaded, above all, by their obligation to save American lives, the Scientific Panel saw no acceptable alternative to direct military use. Their advice was passed on to Truman. It included a suggestion that the United States approach its principal allies before using the new weapon and ask them for plans for postwar control. What was to happen afterward was a strictly political question, however, and it was not part of the discussions at presidential level that led up to America's use of the bomb.

Once Truman inherited the scattered strands of partial information about Roosevelt's deferred decisions on the issue, he had never been in any doubt about one thing: if the bomb could be built, it would be used. Until he became President, he had known nothing about the bomb; he had never been briefed on it as a senior congressman, or as Vice-President. When finally told, he reacted with enthusiasm. It seemed a total confirmation of all his basic faith in the American system: it was the ultimate better mousetrap.

It also presented him with a momentous decision, and that was the job of the President of the United States: to make decisions. It was how Harry Truman explained his first day in office to his mother. Later, he would title the first volume of his memoirs *The Year of De-*

cision. That revealing sign THE BUCK STOPS HERE had not yet appeared on the presidential desk, but the concept was already firmly in place. Conscious of nothing more than his sense of inadequacy and sharing the belief in the inevitably invidious comparison with his predecessor, Truman was determined to rise to the challenge that fate had placed before him. He would make decisions—as firmly and as clearly as possible. That was his job. Not so much his constitutional prerogative as his constitutional duty. Harry Truman had a strong sense of duty.

Acutely aware, more so than any President before or since, of the distinction between himself as President and the institution of The Presidency, he was determined that his unsought occupancy of the position would do nothing to diminish the stature of the office. Whatever doubts and fears he might have himself, he would at least give the appearance of decisiveness. He had no doubts about the bomb. He dutifully signed the crucial piece of paper. "This is the greatest thing in history," he declared when he heard the news of Hiroshima on his way back from the Potsdam Conference. It would definitively shorten the war and save American lives. He believed this passionately, as did every person who had an opportunity to advise him about the use of the bomb. This was the dominant overriding consideration for all the key actors in this decision-making process. The bomb was a weapon of war to be used in war. The sense of moral outrage that had existed in 1939 about the bombing of civilian centers had all but disappeared in the moral numbness of holy war. Truman's "decision" was not based on the "mature consideration" Roosevelt and Churchill had regarded as necessary in September 1944, when the two leaders had signed a memorandum—not shown to Truman—that treated the decision to drop the bomb as open. The question had long since become not "Why?" but "Why not?" No one ever gave Truman a good reason why not.

In retrospect, the scientists who had argued against its use in political terms, in particular in terms of the postwar atom race, had failed to build a case on their own particular competence. The one thing they knew, which in their naïveté never struck them as *politically* relevant, could have been a key factor: the effects of radioactivity from the bomb were known to be delayed. Many people died from the blast and immediate radiation effects of the explosion, but the scientists knew that many more would die from delayed effects in the years ahead.

No one had told any of the political decision-makers this basic fact. Among the thousands who would die immediately and the millions who would be threatened by an arms race, the later injuries did not seem significant at the time, yet they were to prove decisive. It is why Hiroshima and Nagasaki are remembered whereas the much deadlier conventional attacks of World War II, such as Tokyo and Dresden, are all but forgotten. Good politicians like Byrnes and Truman could have understood the difference. If people were still to be dying from the effects of the bomb a decade or more later, politically it had to be treated in different terms than any other means of ending the war. We will never know whether this would have been a determining factor.

The advantages of hindsight show how easy it might have been to avoid the tragedy of Hiroshima and Nagasaki. Intercepted cable traffic was already showing the desire of senior administrators in Japan to end the war through Russian mediation. The psychological effect of the secretly planned Russian declaration of war, which came the day after Hiroshima, was expected to be profound. So, too, was the impact the Russians would have on the one surviving significant Japanese military force, the Manchurian Army. Senior American military commanders still reaffirmed their faith in the ability of their own commands to win the war: the air force believed conventional strategic bombardment would do it and the admirals believed the naval blockade would do it. Even the option everyone was trying to avoid, direct military invasion, was not planned to begin for almost two months. All of these factors might have been given more prominence, or at least given cause for delay, if the political leadership of the U.S. government had had the long-term effects of radiation put to them as a reason to avoid dropping the bomb. The question was not presented in those terms.

Throughout the world, except possibly in the Soviet Union, the atomic bomb was seen as the definite final act that won the war: an act that brought to an end years of misery and death and the most stunning manifestation of man's control over nature that anyone had ever experienced. An entire generation would marvel at the ingenuity of the achievement. The rhetoric of the atomic age began on August 6, 1945, at Hiroshima. It gave rise to wonder and enthusiasm. It would be a decade before the aftereffects of that day, the insidious, invisible consequences of radioactivity, began to balance exhilaration with fear. It would be two decades before the powerful image of the mushroom cloud, representative of man's Promethean urge to control nature, would need to adjust to an equally powerful image: the first picture of spaceship earth taken from outer space, which so dramatically stressed the limits of man's environment. And the scientific and technological control of the energy in the atom would never be separated from the impact of the decision to drop the bomb.

3

"A Most Deadly Illusion"

There was a belief, a most deadly illusion, that we could retain a monopoly of the facilities and the knowledge for the production of fissionable material.

—CHESTER BARNARD in correspondence with Walter Lippmann, February 1950

The first shaky steps toward international control of the atom were taken well before the bombing of Hiroshima. Niels Bohr put his own proposal to Winston Churchill at the British Prime Minister's office in London in May 1944. Bohr's plan was the product of the physicist's fiercely logical, yet politically naïve mind. America, he told Churchill, would soon have the bomb; after the war, the Russians would build one, too. A terrifying arms race based on nuclear weapons would be inevitable. One way, perhaps the only way, to prevent this race was to let the Russians into the secret of the Manhattan Project; this would allow discussion of the grave international consequences of unleashing atomic energy on the world and of the need to set up an international controlling agency. To wait until after the bomb was dropped would be too late: the Russians already had a deep suspicion of the Western alliance. To keep them out of this supreme scientific achievement until its awful destructive power had been demonstrated to the rest of the world would simply increase their doubts, reinforce their suspicions, and render postwar cooperation unlikely, perhaps impossible. To Bohr, the scientist, the proposal was simple and obvious, like the steps of a scientific equation. To

Churchill, the politician, who saw the Russians as potentially an even greater enemy than the Germans, it was rubbish and the Prime Minister said as much.

Bohr was shattered. Even Lord Cherwell, Churchill's scientific adviser, who had gone to the meeting hoping to support Bohr with some thoughts of his own about the atom in the postwar world, was taken aback by Churchill's rough handling of the famous Dane. But Churchill's atomic policy was rigid: he refused to discuss the bomb project even within his own cabinet. To tell the Russians about the forbidden subject was unthinkable. Bohr's short speech, prepared and rehearsed many times because of his faltering English, only seemed to annoy the Prime Minister. Churchill simply did not want to hear such talk, especially not from a scientist; the atomic bomb was not going to make any difference to postwar diplomacy—it was simply a bigger bang. All diplomatic problems could be settled directly between himself and his friend President Roosevelt. In rejecting Bohr's proposal, Churchill chose to ignore completely the Dane's papal status among the nuclear physicists. Indeed, the Prime Minister thought the fellow was so soft on the Russians he might even be a spy. When Bohr tried to salvage something from the wreckage of the meeting by offering a memorandum on his idea, Churchill replied, contemptuously, that he would be honored to receive a communication from the professor, but he hoped it would not be about politics. "We did not speak the same language," Bohr commented.

Bohr was so convinced of the importance of his mission, however, that he tried once more—this time with President Roosevelt. Their meeting seemed to go well. Roosevelt was more cordial than Churchill and chatted with Bohr for more than an hour about the bomb and its implications for world peace. Again Bohr urged an approach to Russia and, to his surprise, the President appeared to agree, saying he would raise it at his next meeting with Churchill. But Bohr had been duped: Roosevelt, like Churchill, never had any intention of telling the Russians anything about the top-secret Manhattan Project. He even agreed with Churchill that Bohr was a possible security risk.

After Bohr's failure, some of the Manhattan Project scientists—particularly in the Chicago laboratories—took on the task of trying to persuade the American leadership to bring the Russians into the bomb secret. And the President's own scientific advisers, Vannevar Bush and James Conant, warned Roosevelt that a policy of secrecy could only lead to an arms race. By the spring of 1945, even the

President's aging Secretary of War, Henry Stimson, was suggesting that America might try to trade the secrets of the bomb for liberalization of Stalin's secret police state. But when Truman became President in April 1945, he continued Roosevelt's policy; he would have no truck with the Russians over the bomb secrets. Rather than share the secrets with Stalin, he intended to use them to his own advantage, to bolster his own confidence and his own image. A few weeks before the last of the great tripartite wartime conferences, held in the Berlin suburb of Potsdam in July, Truman ordered General Groves to rush preparations for the bomb test at Alamogordo so that the news could come through while the three leaders were meeting. Groves sent the message to his scientists: "The upper crust want it as soon as possible."

It was just the kind of order Groves was waiting for, and the scientists did not fail him. The world's first plutonium bomb exploded in the desert of New Mexico, on July 16, 1945. Truman was already in Potsdam where he received the following message, "Operated on this morning. Diagnosis not yet complete but results seem satisfactory and already exceed expectations." The President was overjoyed. Within a week, the word from Los Alamos, where scientists had been busy evaluating data from the explosion, was even better. Not only had the bomb been more powerful than the scientists had predicted, but another could be ready for use in a matter of days. At Potsdam, Truman, who was becoming increasingly irked by Stalin's intransigence in the peace negotiations but was reluctant to show his anger, suddenly gained confidence. Churchill noted he started telling the Russians "just where they got on and off." On July 18, Truman had lunch alone with Churchill and discussed the successful bomb test. "Decided to tell Stalin about it," the President noted later. The question of whether to tell Stalin about the bomb had never been more pressing. Should Truman broach the whole subject of international control of the atom now? Or should he wait?

In the end, Truman did nothing. He never even told Stalin explicitly that America had the bomb. In a brief exchange between the two leaders at the end of one Potsdam session, the President simply told Stalin that the United States had developed a new weapon of unusual power. Stalin indicated he was pleased and that he hoped the Americans would use it on the Japanese. That was the end of it. Truman was not even sure that Stalin had understood that he was referring to the bomb.

The first bout of the new atomic diplomacy had left the American secret intact. The way was now open for the nuclear arms race. On August 9, the day the plutonium bomb was dropped on Nagasaki, Truman closed the door on any immediate hopes of international control. "The bomb is too dangerous to be let loose in a lawless world," he declared. Most Americans agreed with him. According to opinion polls, more than 75 percent of citizens and more than 90 percent of congressmen believed that the atom secrets should remain in America. The bomb was a homespun achievement, a piece of pure magic at which the rest of the world had rightly marveled. It was not something to give away. In the coming months, this attitude would take such a grip on American foreign policy that not only Russia but also Britain and Canada were to be excluded by act of Congress from sharing the secret of the bomb.

The enthusiasm, indeed the euphoria, for the new invention became all-embracing. Politicians saw it as a new tool. Truman's hot-headed Secretary of State, James Byrnes, went sailing off to the postwar peace negotiations in London in a cocky mood, with the bomb "in his hip pocket." At home, the U.S. military set about building bigger and better bombs in the knowledge that they could smash any country's army by the dispatch of a single bomber. In the atomic laboratories, scientists dreamed of ways of harnessing the awesome power they had created, of turning it into something peaceful—like electricity.

Truman began to speak of the bomb as a "sacred trust" that had been bestowed on the nation—almost as if by divine right. And it seemed as if the trust might endure for decades. General Groves, eager to maintain his overlord status in the world of the atom, fed the Truman administration with the idea that an American atomic monopoly might endure for twenty years.

Why did Groves think this? It was not what the scientists thought. Their estimate was four to five years—the same time it had taken America. Groves based his guess on two things: the Russian lack of uranium supplies and his own bellicose chauvinism—he thought the Russians incapable of building a Jeep, let alone a bomb.

Almost a year before the end of the war, Groves had asked a team of geologists to draw up a list of the world's known uranium reserves. The best-known deposits were in the Belgian Congo, and in July 1944, on Groves's recommendation, America and Britain created a special agency to buy up the Belgian ore plus any other deposits they could find. Information about Russian reserves was scant, but experts had advised Groves that the Russians might still find uranium within their own borders. Russia, after all, was a vast territory. The general dismissed their advice; why, he asked, had the Russians twice during the war requested that the United States sell them uranium for research purposes if they had enough of their own? Groves chose to believe that the Russians did not have enough uranium.

Even though Groves's prediction had little basis in fact, it permeated the Truman administration to such an extent that the President himself became convinced that the Russians would not have the bomb for some years, if ever. In the spring of 1946 he had a conversation with Robert Oppenheimer, who, like other scientists, subscribed to the five-year theory.

"When will the Russians be able to build the bomb?" asked Truman.

"I don't know," said Oppenheimer.

"I know," said Truman.

"When?"

"Never."

It was self-deception on a grand scale. In August 1949, four years after Hiroshima, Russia surprised the world by exploding her first atomic bomb. Britain followed in 1952. The nuclear weapons club, it was soon discovered, was not founded on such a supersecret after all. Admission of the self-deception did not come until 1950, when the president of the New Jersey Bell Telephone Company, Chester Barnard, a successful businessman appointed by the Truman administration to the panel of experts who devised the first abortive plan for international control of the atom, publicly commented, "There was a belief, a most deadly illusion, that we could retain a monopoly of the facilities and the knowledge for the production of fissionable material."

The supreme arrogance of General Groves and the Truman administration concerning the capabilities of other nations to make an atomic bomb permitted one brief, but important, diversion from the policy of secrecy. Several months before the bomb project had been completed, Groves realized the need for some kind of technical report about the Manhattan Project. It had cost a staggering $2 billion and when the war was over, the American public would quite reasonably demand to know how the money had been spent. They would also be curious to know at least something about how atomic energy could be made into a bomb. Groves and the President's scientific advisers decided that they could tell some of the basic details while still keeping the more important ones secret.

Henry Smyth, a physicist who had worked on the Manhattan Project in Chicago, was chosen to write the report. Even before the President had approved it, Groves had mobilized a corps of stenographers and made use of the Pentagon's reproduction facilities to run off a thousand copies. He was anxious to make the report public; apart from anything else, he wanted to give prompt recognition to those scientists and engineers who had worked so hard and so long in total anonymity—including himself, of course. The thousand copies were rushed into public view on August 12, 1945, only a week after Hiroshima. The report did not explain how to make a bomb, but it did say how not to make it. Included were a number of important hints that proved invaluable to a country like Russia, which was starting the engineering phase of its program from scratch. It was useful, for example, to know that a plutonium bomb actually worked and that the fissile U-235 could be produced in sufficient quantities by the gaseous diffusion process. In the years to come, the report was recognized as a helpful guide to bomb making. In 1948, in congressional hearings, David Lilienthal, the first chairman of the U.S. Atomic Energy Commission, would describe the Smyth Report as the "principal breach of security since the beginning of the atomic energy project." But the breach was only temporary.

The Smyth Report opened the door only to slam it shut again for the next eight years. Its publication signaled the end of the period of U.S. atomic cooperation with the rest of the world and even with America's wartime partners, Britain and Canada. What followed was a policy of total secrecy that, within the year, was mandated by congressional legislation. Not until 1953, when President Eisenhower launched his Atoms for Peace program, would that policy be changed in favor of the free flow of information about what would then be called the peaceful atom. In the intervening years, the arms race was to become a terrible reality and the worst fears of Professor Bohr and the rest of the atomic scientists were realized.

In the immediate postwar period, two distinct tasks confronted American atomic policymakers: how to deal with the atom in foreign affairs, and how to deal with the development of the peaceful atom at home.

In the international arena, there were those in the Truman administration who thought that mere possession of the bomb would be a useful diplomatic tool in the postwar peace negotiations. James Byrnes, Truman's new Secretary of State, was one of them. Byrnes was soon to discover, however, that the bomb had no diplomatic value. The Russians, far from being cowed by America's possession of it, quickly perceived that it was politically impossible to use. As Vannevar Bush pointed out, "There is no powder in the gun for it could not be drawn."

The Byrnes position vis-à-vis the Russians was not shared by all members of Truman's cabinet. Stimson, the Secretary of War, still thought some accommodation with them was possible. And, in the area of atomic energy, Stimson believed that anything less than an equal partnership with the Russians would "irretrievably embitter" the Kremlin. "The chief lesson I have learned in a long life," Stimson wrote to Truman, "is that the only way you can make a man trustworthy is to trust him and the surest way to make him untrustworthy is to distrust him and show your distrust."

The kind of cooperation Stimson had in mind was an extended version of the Russian connection that Niels Bohr had first suggested to Churchill. The essential difference was that whereas Bohr had thought Stalin should be told about the bomb to see if he was ready to cooperate in its postwar control, Stimson thought the United States could go much further and offer an actual exchange of scientific information linked to an agreement on international control.

For the first time, the great atomic "secret" was broken down into stages. As Stimson suggested in the privacy of the Oval Office (and as Truman would later explain publicly), there were three stages. The first was the scientific knowledge that resulted in the theory of the bomb. The second was the technical know-how—the engineering secrets learned during the Manhattan Project—which turned the theory into practice. The third was the combination of industrial capacity and resources that a nation needed to make the explosive device. Stimson's idea was that each stage could be separately traded for Russian concessions.

Since the bombing of Hiroshima, Truman had come under increasing public pressure to produce a comprehensive plan for atomic energy. He decided to give Colonel Stimson the floor to present his proposal at the cabinet meeting of September 21, 1945—the Secretary of War's seventy-eighth birthday and his last official appearance.

Without notes, Stimson gave what Secretary of Commerce Henry Wallace said was an "unusually fine and comprehensive statement." Truman seemed to agree with the colonel's stand but gave each cabinet member a chance to talk. The cabinet split instantly. Hard-line conservatives, like the Secretary of the Navy, James Forrestal, interpreted the Stimson proposal as literally handing Russia the bomb on a plate. Liberals like Wallace argued that the physics of the atom was international: there were scientists in Europe and Russia who had worked on the idea of a bomb before the war and they already knew the basic equations. He warned against developing a "scientific Maginot Line" that allowed Americans to think that other nations could not make their own bombs. Forrestal was unmoved. "It seems doubtful," he said later, of the Stimson proposal, "that we should endeavor to buy their understanding and sympathy. We tried that once with Hitler . . . there are no returns on appeasement." The Undersecretary of State, Dean Acheson, summed up the debate later. "The discussion is unworthy of the subject, no one had a chance to prepare for its complexities," he said.

The lack of a consensus made little difference. Truman had, in fact, already made up his mind. In early October he clarified his position during an impromptu press conference at Tiptonville, Tennessee. He would allow any nation to have the first part of the secret, the scientific knowledge, but no one, not even Britain and Canada, would be allowed access to the second part, the engineering know-how. "If they catch up with us on that, they will have to do it on their own hook, just as we did," he said.

Truman was giving nothing away: the scientific knowledge was already known by the British and Canadians, and the French, the Germans, and the Russians had a very clear idea about it. It was becoming apparent that the President favored a U.S. trusteeship of the new weapon, "a sacred trust." He did not rule out some form of worldwide control, however; the United States would work toward creating an international agency, in partnership with all countries of the United Nations. On a trip to Moscow, in December 1945, Byrnes secured Russian agreement to discuss the matter in the public forum of the new world organization. Then he appointed a committee, led by his deputy, Dean Acheson, to produce an American plan. The committee included the two White House scientific advisers, Bush and Conant, a Stimson aide, John McCloy, and General Groves.

It was an inflammable mix. In the two months since the Stimson farewell cabinet meeting, Acheson's views on international control had become well known: he had come to support Colonel Stimson's plan of seeking a partnership with Russia. He had little time for the input of Byrnes or, for that matter, Truman. As far as he was concerned "they didn't understand anything about the bomb." And he was decidedly uncomfortable about the special position General Groves had carved out for himself. "By [using] the power of veto on the grounds of 'military security,' [Groves has] really been determining and almost running foreign policy," Acheson commented.

The Undersecretary suggested that a panel of consultants be formed to advise the main committee. Groves, predictably, resisted. There was no need, the general protested. He, Bush, and Conant already knew more about the bomb "than any panel that could be assembled." But Groves was losing his wartime power. He was soon to learn that the bomb was no longer his own technological kingdom. Acheson insisted on advisers, and in January 1946 a consultants panel was chosen. The chairman was a forty-six-year-old lawyer named David Lilienthal. For the past six years, he had run America's most ambitious water resources project, the Tennessee Valley Authority; he knew nothing about atomic physics. During the Manhattan Project, Lilienthal had lived only a few miles from the Oak Ridge U-235 separation plant, but all he knew about its operation was that the army had asked the TVA to provide unlimited power and some good-quality water. The other panelists were Robert Oppenheimer, Chester Barnard, the president of the New Jersey Bell Telephone Company, Harry Winne, a vice-president of the General Electric Company, and Charles Thomas, a Monsanto Chemical Company vice-president.

The panel's task was formidable. Its ground rules had been laid in November 1945, when America, Britain, and Canada had resolved to set up a U.N. Commission to deal with the international development of atomic energy. The commission was to fulfill a dual-purpose goal: to prevent the use of atomic energy for destructive purposes, and to promote the use of recent and future advances in scientific knowledge, particularly in the utilization of atomic energy, for peaceful and humanitarian ends.

This dual-purpose goal, with its elusive link between ridding the world of the bomb, on the one hand, and using its power for peaceful purposes, on the other, was the crux of the problem. In meeting the first goal (through an agreement to outlaw bombs) how could you then spread the peaceful atom to other nations without giving them a chance to make a bomb should they so choose? It was a problem that was to bedevil similar panels that followed the Lilienthal group for the next thirty-five years.

The atoms-for-peace and atoms-for-war issue was one without precedent. Never before had nations been faced with the problem of outlawing a weapon of war while creating an international structure to propagate its use for peaceful purposes. It presented a beguiling vision that has retained its appeal ever since: somehow "peaceful uses" could become a substitute for military use. For the first time in history, intelligent men would be able to discuss a military matter without any reference to relations between states, to territorial disputes, or to the balance of armed forces. Only the special image of the atom could have led to so blinkered a vision. At the international level, a separate organization, the U.N. Atomic Energy Commission, was established, an agency unnaturally divorced from all other postwar discussions over territory and arms. At the domestic level, too, the new vision gave birth to such separate institutional arrangements as the U.S. Atomic Energy Commission. Each country that developed the atom would create similar agencies. Every one of them was launched with exaggerated hopes for "peaceful uses," for it was only if the alternatives were grand enough that the bomb could be replaced.

The task before the Lilienthal panel required more than a smattering of optimism. But they recognized that unilaterally outlawing the bomb would never work. No nation would forgo the most destructive weapon known to man simply because the United States declared it would never use its bombs, not even if the price for such refusal was having to start its own nuclear program from scratch. And even were a nation to agree to forgo a bomb project in exchange for the necessary information, how would you police such a state? Inspection of nuclear facilities to make sure they were being used for peaceful rather than warlike purposes was both politically delicate and scientifically difficult. It would require highly trained technical staff, so highly trained that they probably would not be content with the role of policemen.

Lilienthal fully appreciated the enormity of the problem and insisted that his team not "fall into the illusion that there is a solution, one answer that we must seek that will answer everything." Despite the problem, it seemed to this group of extremely earnest and dedicated men that solutions could be found, compromises could be struck. Working long days and sometimes well into the night over the next few weeks, the Lilienthal team flew around the country visiting the remaining parts of the Manhattan Project and listening to lectures on the atomic arts. Lilienthal became enthralled by the magic of the atom. "No fairy tale that I had read in utter rapture and enchantment as a child, no spy mystery, no horror story, can remotely compare," he wrote in his diary.

In the end it was Oppenheimer who came up with a formula. The world's atomic energy programs would be managed through an International Atomic Development Authority, with a clear distinction being made between "dangerous" and "harmless" phases of atomic energy. The "dangerous" activities would be carried out by the international authority, which would also own or control all uranium mines and all plants for the production of fissile material. In one clean sweep not only Groves's secret but also Truman's "sacred trust" would disappear.

"Harmless" activities would not extend much further than the laboratory bench; only small-scale research reactors, which could not produce enough fissile material for a bomb, would be uncontrolled. Everything else—the mining and production of uranium, separating the fissile material, and engineering an explosive device—were classified as "dangerous." As an added safety measure, Oppenheimer came up with an ingenious concept that rendered fissile material useless as an explosive. He called the process "denaturing." It involved introducing into the material a contaminant that prevented a chain reaction from occurring. It was like adding water to the gasoline in a car, which prevents the gas from igniting and the car from working. Of course, just as the water could be removed, so could the contaminant in the fissile material, but that would require large chemical or isotope separation plants, and the scientists felt that the technology would be either too complicated or too costly to contemplate. They had to admit, however, that "denaturing," though valuable in adding to the flexibility of a system of controls, could not of itself eliminate the dangers of atomic warfare: a country with the necessary political will to make bombs could overcome "denaturing" within a year.

The Oppenheimer plan presupposed two things that, in retrospect, were quite untenable. First, that the capitalistic interests that owned the mines would somehow not object to their being taken over—or "internationalized"—by the new controlling agency and, second, that the leaders in the Kremlin would permit roving bands of agency inspectors to tramp all over Russian territory looking for illicit uranium mines or other plants. Members of the Lilienthal panel were well aware of these problems, of course, but they did not think they were insurmountable. Indeed, it is extraordinary to record how sanguine, even confident, the panel became. When the plan was delivered to Acheson, he thought it was "brilliant and profound."

For ten days the Lilienthal panel met with the full Acheson Committee at Dumbarton Oaks, a large mansion in the section of Washington known as Georgetown. The group sat at a long table surrounded by magnificent tapestries, and a Byzantine ebony cat in a glass case watched over the proceedings. Dominating the room was El Greco's *The Visitation*. Outside, workmen would, from time to time, pass the window of the conference room. Lilienthal's diary recalls his deep concern over the burden of his responsibilities to those workers: they were the people who had most at stake, he noted, "and too little to say as to whether someday the order is given and an atomic bomb, perhaps a thousand times greater than Nagasaki, starts on its way against other workmen."

Although the plan was well received, there were some doubters. Groves seriously doubted that an international authority could control the mining of uranium. It might be possible to police the known deposits of high-grade ores, but it would be easy for anyone to mine low-grade deposits and not be caught. Conant agreed with him. Both of them wanted to put more emphasis on the right of the new authority's inspectors to freedom of access—the most politically sensitive part of the whole plan. Vannevar Bush wanted the plan to be presented in specific stages, with a well-defined quid pro quo from Russia each time the United States released scientific and technical information. Bush also wanted provisions for America to pull out of the agreement if the exchanges became one-sided: that was the only way to avoid being double-crossed. After all, commented Bush, there was Russia with its huge army; the United States could not afford to throw the bomb away without cast-iron guarantees that there would be no risks to national security. Groves agreed.

But the Lilienthal panel was not happy with Bush's approach. They argued that their plan already encompassed stages, beginning with the control of the bomb's essential ingredient, uranium; to spell out additional stages at this time would imply mistrust of the Russians' intentions and only fuel their suspicions. But Bush insisted on his stages and Lilienthal gave in. The official history of the U.S. Atomic Energy Commission records the panel's dismay: "Distinctly unhappy, fearing they were blighting the spirit of their work, they decided to undertake the revision."

In the final plan, Bush's separate steps were to extend over fifteen years. As soon as the U.N. Atomic Energy Commission was established, the U.S. would release enough of the atomic secret for others to understand how the next stages would work. First would come free access of scientists to research laboratories. This would be followed by free interchange of scientific information, providing a country agreed to use fissionable material for peaceful purposes only and to permit international inspection. Significantly, the plan did not require the United States to stop making bombs during the establishment of the international authority.

Finally, after many drafts and redrafts, the proposals were ready for Secretary of State Byrnes—"not as a final plan, but as a place to begin, a foundation on which to build." The Acheson-Lilienthal group, as the full committee with its panel of consultants became known, were confident enough to think that some version of their plan might eventually succeed, but they had not bargained on the next move: Byrnes's insensitive choice of Bernard Baruch, a seventy-five-year-old Wall Street speculator, as the American representative who would present the plan to the United Nations. Lilienthal was horrified. "When I read this . . . I was quite sick," he wrote. "We need a man who is young, vigorous, not vain, and whom the Russians wouldn't feel is simply out to put them in a hole, not really caring about international cooperation." Baruch was a disaster: he turned the Acheson-Lilienthal plan, already of dubious acceptance value to the Russians, into what Henry Wallace called a "rigged poker game."

Acheson, also appalled at the choice, tried to persuade Byrnes against the appointment, but without success. After five days of bitter wrangling between Baruch and the Acheson-Lilienthal committee, Byrnes admitted to the committee, "This is the worst mistake I have ever made, but we can't fire him now."

Throughout the spring of 1946, as the two sides argued over the plan during marathon sessions at Dumbarton Oaks, East-West relations steadily deteriorated. Truman himself was growing tired, as he put it, of "babying the Soviets" over the peace negotiations. In February, the existence of a Canadian-based wartime Soviet atomic spy network burst into the open and twenty-two people were arrested in Canada. Then, on March 5, at Fulton, Missouri, Winston Churchill made his celebrated Iron Curtain speech. Churchill wanted to revive the wartime Anglo-American pact, this time against a new enemy, the Soviet Union. With the combined resources of the British Commonwealth and the American nation, there would be, he declared, no "quivering, precarious balance of power to offer its temptation to ambition or adventure." With Truman by his side, the Prime Minister cautioned the West: "It would be criminal madness to cast the bomb adrift in this still agitated and ununited world."

The Cold War was going public, with important effects on atomic energy policy. The bomb was never a determining factor in the Cold War; but it became a recurring symptom of the growing bitterness between East and West. Suspicion between the two sides had been deep-seated enough even before the bomb became a reality. There is no doubt, for instance, that the Russians not only knew of the Manhattan Project but also knew a great deal about it from their American-based spy rings. The refusal of Roosevelt and Churchill to tell Stalin at least of its existence can only have fueled the Russian dictator's doubts and suspicions of the West, but bomb or no bomb, the wrangle over the future of postwar Europe would have worsened the East-West relationship. Indeed, even before Truman had been told of the existence of the Manhattan Project, he had already demonstrated his determination to take a tough line with Stalin over Poland. The bomb monopoly bolstered Truman's confidence, but it did not determine his basic policies.

The scientists' warning that America's atomic monopoly was an illusion never became the starting point for discussion, however. If it had, President Truman and Secretary of State Byrnes could not have continued to believe that mere possession of the weapon would assist their bargaining posture. The proposals for international control in "stages," with American weapons being surrendered only after all the Soviet concessions had been made, would have appeared more biased than it then seemed to them. The almost self-righteous belief in their generosity in proposing to share information might have been deflated.

As it was, the U.S. plan, based on the assumption of a monopoly, was a take-it-or-leave-it proposal, without a basis for negotiation. It thus slid imperceptibly into a Cold War propaganda gambit.

In late March 1946, Bernard Baruch set up his own advisory panel to review the Acheson-Lilienthal plan. He had no intention of simply accepting what he had been given, no matter how good or bad it was. He refused to play messenger boy. He intended to make a big splash at the United Nations and could already see himself, lithe, six feet four inches tall, standing imperially before the assembly and unfolding the most important proposal for peace on earth the world had ever heard. He took scientific advice from Bush, Conant, and Compton, of the Chicago nuclear laboratory. His engineering consultant was General Groves. In order to boost his own position, he even attempted to play down the Acheson-Lilienthal work. Asked by a reporter what he thought of their plan, he turned off his hearing aid and said: "I can't hear you." Baruch wanted no counsel. His attitude to scientific advice about the bomb was: "I knew all I wanted to know. It went boom and it killed millions of people."

Predictably, as an old capitalist, he objected vehemently to the idea of international ownership of the uranium mines, and Groves supported him. From the beginning, Baruch and his men adopted a "banker's, not a believer's," approach to international control. Their attitude to the Acheson-Lilienthal report is, perhaps, best summed up by some comments penciled in the margin of a copy of the report that survived in Baruch's papers. One note declared: "This is a capitalistic country, how can this plan be handled within our free enterprise structure without nationalization—which would endanger our whole way of life?" Another note said the plan was "asking for an adventure in human nature that has never succeeded . . . international control implies an acceptance from the outset of the fact that our monopoly cannot last . . . this is the toughest fundamental to face." Yet another posed the basic question, "Is [the] premise right—should there be international control?"

In the final version of the plan, as amended by Baruch, the original function of the Atomic Development Authority—originally conceived as the sole owner and distributor of uranium ore—was reduced to ownership only of processed uranium. The ore itself, the mining of it, and the plants used to refine it would remain in independent hands.

As a first stage, Baruch proposed a worldwide survey of uranium sources, especially behind the Iron Curtain—thus requiring the Russians to play the only card they possessed before the United States was prepared to do anything. Both Oppenheimer (who had refused an invitation to serve on the Baruch panel) and Lilienthal were distraught. They noted with horror one of the more outrageous proposals: to keep a U.N. stockpile of bombs ready to drop on anyone who joined the new agency and did not obey its rules. Lilienthal feared that Baruch would put forward proposals that would ensure their refusal and his fear was soon to become a reality.

At eleven o'clock on the morning of June 14, 1946, Bernard Baruch, dressed somberly in a dark double-breasted jacket and pinstripe trousers, took the rostrum in the gymnasium of Hunter College, New York, and addressed the eleven delegations to the U.N. Atomic Energy Commission. He began with a rhetorical flourish about the apocalypse. "We are here to make a choice between the quick and the dead. That is our business. We must elect world peace or world destruction." It was heady stuff. He talked of "winning weapons," which could not be relinquished without positive reassurances that they were not needed and of "condign punishment" for those who broke the rules of the new game. Such grand eloquence struck a receptive chord in an America still charmed by the atom. One survey showed 98.5 percent of press opinion was positive. Dissenters were either on the extreme right, viewing the Baruch plan as tantamount to surrender to the Russian bear, or they were on the radical left, considering the "rigged game" to be atomic blackmail.

The Russians waited five days to reply and then, on June 19, the Soviet Foreign Minister, Andrei Gromyko, rejected the plan. His negative response came in the form of a counterproposal that was, in fact, the Baruch plan in reverse. First outlaw atomic weapons, Gromyko suggested, then nations could agree to limited inspection of plants and factories. The Russians did not want any part in Baruch's game.

In the weeks that followed, the gap between the two sides grew. On the Bikini Atoll in the Pacific, the United States exploded its second atomic test device. *Pravda* said it was clear evidence that the U.S. aim was not to restrict atomic weapons but to perfect them. "If the atomic bomb did not explode anything wonderful," *Pravda* commented on the Bikini test, "it did explode something more important than a couple of out-of-date warships; it fundamentally undermined the belief in the seriousness of American talk about disarmament."

On July 5, less than a month after the Baruch offer, the United States formally rejected the Russian proposal. America would continue to make bombs until some form of international control system had been agreed on. The plan had failed. It did not fade away, however. After much tinkering with its proposals, it formed the basis of the U.N. plan for the control of atomic energy, which was approved by the General Assembly in November 1948 by an overwhelming majority. But that, too, came to nothing.

What had eluded the American policymakers on the international front was soon to be established at home, however. During the winter of 1945–46, a long and bitter congressional fight had taken place over plans for domestic control of atomic energy. Even before the bomb had been dropped on Hiroshima, the U.S. Army had drafted its own "military" bill. It proposed a new agency, a commission of nine—five civilians and four serving military officers—to be the guardians of America's domestic nuclear future. As the details of the bill became known in the fall of 1945, they provoked vigorous opposition from a wide cross section of the citizenry. Politicians objected to the idea of serving officers on the commission and, in particular, to the proposal to incorporate a military veto. The whole thing reeked of "military fascism," they argued. The atomic scientists complained bitterly about what they saw as intolerable restrictions being placed on laboratory research in the name of security. To them, the military presence on the commission underlined an atomic energy future linked inextricably to bombs and still more bombs.

At the end of October 1945, the army's congressional sponsors made huge efforts to push—some said railroad—the bill through the House of Representatives Military Affairs Committee after only five hours of hearings. But they failed, and the bill finally died after President Truman withdrew his support in favor of other legislation introduced by a freshman senator from Connecticut, Brien McMahon.

Seizing on the emotional issue of military control, McMahon proposed instead a commission composed of cabinet officers and other federal officials to "conserve and restrict the use of atomic energy for the national defense, to prohibit its private exploitation and to preserve the secret and confidential character of information concerning the use and application of atomic energy." For eleven months, the new bill was shunted from one committee to another on Capitol Hill, drafted and redrafted until it was finally signed into law, as the first Atomic Energy Act, on August 1, 1946.

Just as the army bill had become the symbol of military control, so the McMahon bill became the symbol of civilian control. It introduced a new classification in domestic security known as "restricted data," which was to cover all information about the manufacture and use of atomic weapons. To protect this data, it provided some harsh penalties for transgressors: huge fines, up to twenty years in prison, and even death. It included an all-embracing clause to cover exchange of information with foreign countries. "Until effective and enforceable international safeguards against the use of atomic energy for destructive purposes have been established, there shall be no exchange with other nations with respect to the use of atomic energy for industrial purposes." That included America's wartime allies Britain and Canada and, in particular, Russia.

4

The Atomic Archipelago

Russia's atomic bomb was built in a series of scattered towns, factories, and laboratories spread across the vast territory of the Soviet Union, yet so closely integrated as to appear a single unit. It was a world of its own, totally isolated, with huge buildings constructed in the middle of nowhere by slave labor. The men who entered this world—there is no mention of any women—ceased to exist. They were just numbers in a closed society, the most confined part of a nation already riddled with restriction.

Over the years, only a few landmarks of this Soviet network have been revealed: there is no complete map. It ranged from uranium mines in Saxony, East Germany, to seaside laboratories in converted resorts on the Black Sea, to research institutes at Obninsk, southwest of Moscow, at Dubna on the Volga, and at Kharkov in the Ukraine. Plutonium for the first Russian bomb came from a reactor near Sverdlovsk in the southern Urals. The bomb was exploded near Samipalatinsk in Soviet central Asia in August 1949.

There is no official history of the project. Only one of the scientists connected with the administration of the program, Vassily Emelyanov, has published a series of articles about it and only the project's chief physicist, Igor Kurchatov, has a biography. Three of the German atomic scientists who volunteered to help the Russians build their bomb after the war have published memoirs, but there are less than a half-dozen books in the West that attempt to take more than a cursory glance at the Soviet work.

For many years, one of the great mysteries surrounding the Russian project was the identity of the scientists and engineers who worked on it, and although their names are now known, none of them has even achieved popular international recognition. Immediately after the first successful test, stunned Western observers believed that the leading scientist must have been Peter Kapitsa, the physicist son of a czarist general. Kapitsa was well known in the West between the two world wars; he had worked with Rutherford at Cambridge in the 1920s. But the observers were wrong; Kapitsa had refused to work on the project, apparently on moral grounds.

Western analysts, reluctant to admit Soviet atomic expertise, were also quick to suggest that the atom spies must have made success possible, and this theory gained considerable credence from the timely confession of the German refugee scientist Klaus Fuchs. He had worked on important research on the Manhattan Project and, after the war, in England. In February 1950, seven months after the Russians had exploded their first bomb, Fuchs was arrested after admitting he had been passing atomic information to the Kremlin since 1942. Another atom spy ring, operating in Canada during the war and exposed in 1946, was taken as evidence that the Russians had relied on original Western research.

Much of this was just sour grapes. It was known in August 1939, for example, that the Russians had been keenly interested in atomic physics and that some of their scientists had been following the crucial prewar German and French experiments that proved a chain reaction using uranium was possible. The Russians had also completed atomic experiments of their own. What was not known until the mid-fifties, however, was that late in 1942, when the Russians had started to drive Hitler's invading armies back to Germany, the Soviet State Defense Committee established its own atomic program. It began officially only a few months after the Manhattan Project was pushed into top gear in the fall of 1942—although war-shattered Russia was unable to spare anything like the resources available to General Groves.

Igor Kurchatov was taken off war work and told to set up a base in Moscow. Kurchatov, a tall, broad-shouldered man with a long wispy beard, was known as much for his success as an organizer as he was for his scientific achievements. He had been living in Kazan, five hundred miles to the east of Moscow and a safe distance from the German advance. Leaving his wife behind, he went to Moscow and began to search for a suitable building to turn into a laboratory. Many large buildings had been evacuated in 1941, and Kurchatov had little difficulty in finding one that suited him. A three-story brick building, it stood on the outskirts of the city, a mile from the Moskva River, and was bounded by a pine grove and huge potato fields. There he established the top-secret "Laboratory No. 2 of the Academy of Sciences, U.S.S.R." By the end of the 1940s, a small town had grown up around the laboratory and it was given a new, equally uninformative title: "The Laboratory for Measuring Instruments." Not until the mid-fifties, when its most secret work had been completed, was it given its real name: "The Institute for Atomic Energy."

The State Defense Committee appointed a special task force to look after Kurchatov's staff and his personal needs. Members of Laboratory No. 2 quickly realized that they were part of a new elite. Food and clothing were strictly rationed throughout Russia at the time, but Kurchatov was given special ration cards and housing allocations; one phone call from him was enough to obtain a pass for a staff member at the cafeteria in the House of Scientists on Kropotkinskaya Street. He could even summon a car any time he needed it from the garage of the Council of People's Commissars.

Compared with the huge industrial plants of the Manhattan Project, Laboratory No. 2 was a small-scale effort of basic scientific research, pursued with no great sense of urgency. The Russians knew they could not make the bomb in time to use against Hitler; but accounts of this period do indicate that the scientists set about their task with determination and patriotic enthusiasm equaling that of their Manhattan Project counterparts. There was no question of refusing to make a bomb for Stalin. Although many of the leading Russian scientists had been victims of the purges in the thirties, the shock of the German invasion had overcome any hostility toward or fear of the Stalin regime. It was still important, however, in these early years, that the project was directed by a man like Kurchatov. He was not a political figure; he did not join the Communist party until 1948.

A hard worker, Kurchatov pursued his task with energy, persistence, and demanding leadership. Under pressure, he was prone to outbursts of violent, frequently crude language. But in lighter moments he was a practical joker and a pleasant companion. He had the great facility of being able to see his way through a mass of complex mathematical formulas and to extract the fundamental principle. "Hold it," he would cry, "you are getting too scientific. Let's analyze this in a simpler way." He had studied physics at the Crimea State University and, in 1925, at the age of twenty-two, was invited to join the prestigious Leningrad Physical and Technical Institute. In the early thirties, he was attracted by the excitement of the new atomic physics and kept pace with the rapid development of knowledge outside the Soviet Union, repeating and verifying the experiments of others and, occasionally, making an original contribution. In 1937, Kurchatov built the first cyclotron in Europe, seven years after Ernest Lawrence had built his. Although not well known among his peers in the West, by the late 1930s Kurchatov and his team were near the forefront of nuclear physics. During the war years Kurchatov concentrated on building up three lines of research: a chain reaction in an experimental pile using natural uranium as a fuel and heavy water as a moderator, methods of isotope separation, and the bomb design.

As the Red Army chased the Germans back into their own territory, the officers made considerable efforts—as the Allies were doing in the West—to contact German atomic scientists and recruit them for their own laboratories. One who agreed to go was Baron Manfred von Ardenne, a maverick physicist with a private laboratory who had worked under a German Post Office grant at the beginning of the war and developed, unsuccessfully as it turned out, the beginnings of the electronic magnetic separation of U-235.

In May 1945 a colonel general in the Red Army arrived at the baron's laboratory in Berlin with a handful of physicists from Kurchatov's team. They were impressed by the baron's work and recommended that efforts be made to recruit him. Von Ardenne, his staff, and his laboratory were moved to Sukhumi, a resort on the Black Sea, where he would spend the next ten years working mostly on isotope separation. The German scientists had a good life, but they were virtual prisoners, part of the first circle of Lavrenti Beria's secret police, the NKVD. They were not allowed to travel elsewhere in Russia without permission, and only rarely did they visit other parts of the project. Their major contribution was to help the Russians solve the problems of dealing with the extraordinarily corrosive uranium gas. The Russians, like the Americans before them, used the gaseous diffusion technique for separating the U-235.

By mid-1945, Kurchatov's team had been at work on the physical theories of the bomb for eighteen months. With the demands of war still taking their toll on Soviet resources, the pace of the project was slow. Not until America demonstrated its stunning achievement at Hiroshima and Nagasaki did the pace change.

It was in the middle of August 1945 that Stalin summoned his top physicists and engineers to the Kremlin and told them, "A single demand of you, comrades: provide us with atomic weapons in the shortest possible time. You know that Hiroshima has shaken the world. The equilibrium has been destroyed. Provide the bomb—it will remove a great danger from us."

Beria was in political control of the project. Kurchatov remained its chief scientist. Two engineers were to assume great importance. The first was Boris Vannikov, the Soviet equivalent of General Groves. The second was Avraami Zavenyagin, one of Beria's deputies and the man who recruited von Ardenne. These two engineers were members of an elite group of administrators, called by Western scholars, the Red Specialists. The group consisted of young Bolsheviks who were handpicked for technical training in the 1920s and the 1930s to build Russia's industrial base—much of it with slave labor.

There was no immediate danger: the American capacity to produce a few atomic bombs a year was not as forbidding as the military might of Nazi Germany had been. Yet every year that passed increased the threat. The challenge now was not scientific, for both in quality and in size the Soviet scientific capacity was adequate. The real challenge lay in the production of the fissile material on an industrial scale. It was here that Boris Vannikov and Avraami Zavenyagin came to prominence.

A huge engineering effort was launched to produce the hundreds of new pieces of equipment that had to be designed and manufactured with a precision that allowed no room for error. Dozens of new materials had to be mined and produced in quantities and of a quality that no one had hitherto believed necessary. A recurring problem, as in the West, was the supply of uranium. The available evidence indicates that, for several years, even with the postwar acquisition of mines in Saxony and Czechoslovakia, the shortage of adequate amounts of natural uranium was a major concern, and the development of existing deposits and the search for new fields was a high priority.

Vannikov and Zavenyagin, as Red Specialists, were part of a distinct caste of Soviet society, one that permeated the central decision making of Soviet politics as completely as lawyers invade the process in America. The Red Specialists are the facilitators, the intermediaries, the implementers, and, from time to time, the ultimate decision-makers. This elite, created intentionally by government policy, was to become one of the more enduring legacies of the Stalin era. Social origin and party loyalty were the qualifications for entry into the group. The members, who ran the atomic bureaucracies for the next three decades, shared similar backgrounds. All were born of humble parents at the turn of the century; all were in their late teens at the time of the Bolshevik Revolution. All but one, Vyacheslav Malyshev, who was to take charge of the bomb project in the early fifties, joined the Bolshevik party in the early years and fought in the civil war. In the 1920s and early 1930s, they were chosen, as loyal party men, to attend courses in a series of newly established technical colleges. They emerged from these as plant directors in heavy industry. After the purges of the late thirties, they were promoted to central posts as ministers or deputy ministers responsible for the planning and supervision of segments of heavy industry, each segment related in some way to defense. There was nothing accidental about these parallel careers; they had been selected explicitly for training as a new elite. Unquestioning acceptance of the official ideology and the actions of the party dominated the lives of this new cadre from their early membership through their participation in the civil war, their technical training, and their work assignments. They were exceptionally conscious of their unique role. One of the nuclear administrators, Vassily Emelyanov, recalled, "We were the country's only hope, everything depended on us."

The ideology of the Red Specialists was explicitly material: production was the ultimate value. They glorified large-scale construction projects and believed passionately in the primacy of heavy industry. Their strongest belief opposed a centuries-old Russian cultural tradition of caution and humility in man's relationship with the power of nature. They were most definitely "progressive." Their job was to conquer nature, especially the more forbidding challenges of the

Russian environment: the greater the challenge, the more glorious the triumph. Throughout the thirties, this distinct group shared a spirit of optimism as they built dams, roads, railway lines, and factories in the most inhospitable outposts of their country. In the forties, the ultimate challenge was mastering the energy of the atom.

They brought to their tasks a managerial style that was the product of their training and experience. It was brusque, impersonal, authoritarian, and arbitrary. Perhaps inspired by the harsh lessons of Stalin's personal power, they created the same mixture of awe and fear in their subordinates as they felt toward their superiors. "One-man management" was the rule. At the 1934 Conference of Industrial Managers, the new elite—just about to go into the field—were told precisely what was expected of them:

> The director is the sole sovereign in the plant. Everyone in the plant must be subordinate to him. If the director doesn't feel this, if he wants to play the "liberal" and be a "little brother," to busy himself with persuasion, then he is not a director and should not be directing a plant. Everything should be subordinate to the director. The earth should tremble when the director walks around the plant.

The director was also responsible for everything: the penalty for failure to meet the plan or deadline could be brutal. Pyramids of decision were kept short; the director had to be concerned with the most trivial details; nothing could be taken for granted. It was a world of total insecurity.

In the twenties and thirties, the new Russian technocrats had experienced two periods of great terror: from 1928 to 1931, their teachers, the old "bourgeois specialists" whom they would replace, were liquidated. From 1937 to 1939, their patrons, the old Bolsheviks who had taken over the economy, the so-called Red Directors, were also purged and a new elite moved into their positions in the Moscow bureaucracies that directed the entire economy. During the war years, the sense of insecurity dissipated a little; everyone was needed. And, as the war drew to a close, the atomic program above all others offered an extension of this sense of security. Always, Beria himself was in control. Later, Emelyanov would recall without moral indignation, "What would have happened if we hadn't made it then. They would have shot [us]. Just shot."

Boris Vannikov, the leader of the project, was born in 1897 in the industrial city of Baku. Described in official Russian works as "of working-class origin," the teen-age Vannikov worked as a shipyard fitter and turner and, in his off hours, as a clarinetist in a cinema orchestra, accompanying the silent movies of the day. In 1919, already a shipyard foreman, he joined the Bolshevik party and was made secretary of the local metalworkers' union. During the civil war he served with the Red Army. After his demobilization in 1920, Vannikov became one of the earliest of the party men selected for technical training. He graduated from the Bauman Technical Institute in Moscow in 1926 and was sent out to direct a series of large industrial factories.

A short, squat, strongly built man with a square face and prematurely bald head, Vannikov was extremely bright, inquisitive, strong-willed, with definite opinions and a reputation for personal integrity and exceptional technical and organizational skills. A good family man with a close family life, he even spent his vacations with his wife, an unusual phenomenon among the Soviet elite. Although he was a devoted Bolshevik, he never pushed himself into the political arena; perhaps his Jewish background made him cautious. He was apparently pleasant in person but, like his peers, was a tough industrial manager, hectoring and yelling at his subordinates to get results.

The second key administrator, Avraami Zavenyagin, was fully Beria's man. He had been deputy minister of Beria's own department, the Ministry of Internal Affairs, and for some years had supervised the entire economic empire of this sprawling conglomerate. The son of a locomotive driver, Zavenyagin was born near Moscow in 1901. At the age of sixteen, he joined the Bolshevik party and for the next decade concentrated on party work in his home province and in the Ukraine. He became a candidate member of the Central Committee of the party as early as 1934. In the interim, he had been sent for technical training to the Mining Institute in Moscow. His established party position was such that, even as a first-year student, he was given senior administrative tasks in directing the institute.

Zavenyagin's commitment was passionate. All students enrolling in the Mining Institute were required to fill out a questionnaire. One of the thirty-three questions asked provocatively: "What is your attitude to Soviet power?" "I believe in it" or "I support it" was the common answer. But this was not enough for Zavenyagin. "I am ready to die for it," he declared.

In 1933 he designed one of the largest Soviet projects of that decade: the huge steel plant at Magnitogorsk in the southern Urals. Shortly after he assumed a senior ministerial post in the heavy-industry department. In 1938, when this department was annexed by Beria, Zavenyagin was sent, as part of a loyalty test, to Norilsk, in the Arctic Circle, to build a huge mining industrial complex with slave labor. In 1941 he was appointed deputy minister under Beria, probably becoming the most senior official of the Gulag.

Solzhenitsyn, in *The Gulag Archipelago, 1918–1956*, notes with frustration the lack of personal anecdotes about Zavenyagin. "The newspaper hacks wrote of him: 'the legendary builder of Norilsk'! Did he lay bricks with his own hands? Realizing, however, that from up above Beria loved him dearly and that from down below him the MVD man Zinovyev spoke highly of him, we suppose he was an out-and-out beast. Otherwise he would never have built Norilsk."

Whatever his moral qualities, Zavenyagin was a loyal servant of the regime; the long hard years of economic development had shifted the focus of his loyalty from the ideals of a young Bolshevik to the institutions he served. He was a good Stalinist. A rare glimpse of his managerial style was described by scientists who worked on the Soviet enrichment program. At a seminar, a German project director had remained silent while one of his scientists was severely criticized. Afterward, Zavenyagin had chided the German, "One has always to be a patriot of his own enterprise, which means even to defend mistakes to the utmost." He had a reputation as a protector of scientists and writers and, despite his unsavory institutional connections, his private dacha was a frequent gathering place for members of the intelligentsia whose opinions he sought. One of the few anecdotes about Zavenyagin in the Soviet literature has him arranging the release of a jailed writer.

A fascinating, yet uncorroborated insight into Zavenyagin is provided by an American. John Scott, a young idealistic adventurer who joined the Soviet work force in the early thirties, was drafted into Zavenyagin's Magnitogorsk construction complex as a laborer. In his later memoir, he says he was eventually given unusual access to the technical and financial records of the project. They showed that Zavenyagin had been readily tempted by his absolute power. He had ordered a series of palatial homes for himself and his senior staff, using plans copied from the most recent American architectural catalogs.

Zavenyagin's own house was a three-story, fourteen-room brick home, which included a study, a music room, a billiard room, and a play room for his two sons. Surrounded by a high wall, his accommodation had a small deer park in the back and in front a luxurious garden, with topsoil that had been transported many miles by truck. The rent he was obliged to pay for this establishment came, like his chauffeur-driven car and other perks, from "administrative expenses" of the plant. He had the house furnished at a cost of 170,000 rubles. At the time, his own income was only 2,000 rubles a month and the average worker's wage at the plant was 300 rubles a month.

Tough, determined, disciplined, and reputed to have a crushing wit, Zavenyagin spoke to subordinates in a military command fashion. He worked hard, rose early, and had a precisely organized day. He was proud of his knowledge and his technical competence. The years of sacrifice, of hardship, of work in the most bleak Soviet environments, may have been the cause of his somewhat diminished proletarian consciousness. Besides a regulation desk covered with the compulsory green billiard-table cloth, his office sported a thick Persian carpet. The German scientists who met him during the atomic project were impressed by his sartorial elegance: always clean-shaven and dressed in an immaculate business suit.

In Laboratory No. 2, Kurchatov seemed to be everywhere at once, popping in and out of the workshops and often taking the lead in seminars. Because he was so obviously in command, the scientists nicknamed him "The General." But Kurchatov, though still in charge of the scientific work, was under the overall control of Red Specialist Vannikov in the same way Oppenheimer had been under Groves.

Reflecting the urgency of the task ahead, the project was now called "The First Main Administration of the U.S.S.R. Council of Ministers." Vannikov had no illusions as to the enormity of the effort required. He told a colleague, "We as engineers are accustomed to seeing things with our eyes and touching them with our hands. . . . But in this particular case, all these customs are useless. One cannot see the atoms, let alone what they consist of. And on the basis of what we cannot see or touch, we are called upon to set up factories and to organize industrial-scale production."

The publication in August 1945 of the Smyth Report—the "official" history of the Manhattan Project—gave the Russians useful answers to basic questions. How much this actually helped the Soviet program is hard to judge, but the Russians lost no time in devouring its contents. Copies were quickly distributed to all scientific teams and, in January 1946, it was published in Moscow with a first print run of thirty thousand—a run far exceeding those in the United States or Britain. One German scientist who worked on the Soviet project said that until the report appeared everything was "hazy. . . . It was very valuable for the Soviet Union . . . the most important information at that time was that the Americans had used the gaseous diffusion method to produce U-235 in large quantities. . . . Until then we were in the dark about it."

As in the Manhattan Project, a number of different routes had to be pursued simultaneously, and even within each option, the scientific teams were duplicated to ensure success. As in the Manhattan Project also, the basic scientific work had to be pursued at the same time as engineering development and the massive construction projects. But the military requirement to make a bomb overrode even the priority of postwar reconstruction in a nation devastated by total war.

By 1946, the Russians had successfully operated their first experimental reactor, but they were still having a problem finding uranium. The precise location of the Soviet uranium mines remains secret, and little is known about when they were discovered and developed, let alone the quantities produced in the early years. It is known that Vannikov gave the task of ensuring adequate uranium supplies to the team bully, Yefim Slavsky. When an official Soviet work calls Slavsky "temperamental," it is easy to believe his detractors who called him a hysteric. He apparently displayed the full range of the worst characteristics of the Red Specialists. He was totally ruthless, would never listen to reasons for delay or failure, and, unlike Vannikov and Zavenyagin, was not respected by either scientists or engineers for his technical knowledge.

The son of a peasant family from the Ukraine, Slavsky had joined the Communist party in 1918 and served in the Red Army for a decade before being sent by the party to the Moscow Institute for Nonferrous Metallurgy and Gold. During the war, he became a director of the Urals aluminum plants. In 1945, Slavsky was given the title Deputy People's Commissar of Nonferrous Metallurgy.

The Russians had been searching for uranium deposits, albeit not very earnestly, since 1940. Geological teams had been sent to explore the known deposits in the Fergana basin to the south and east of Tashkent. There is no doubt, however, that some of the most important reserves were found in the areas of Eastern Europe occupied by the Red Army—particularly in Czechoslovakia and East Germany. The Joachimsthal mine in Czechoslovakia was the oldest source of uranium and radium in the world—Marie Curie had gotten her radium from it. In East Germany, there was a series of undeveloped mines in the region known as Saxony.

The record of the Russians' exploitation of Saxony is not a pretty one. The man in charge was a capable construction engineer, Major General Mikhail Maltsev. He had rejected an offer to become a general in Beria's NKVD in favor of remaining in the corps of engineers. "I am an engineer," he asserted. Solzhenitsyn notes that he "was, supposedly, a good man," but he ran a mean operation. Headquartered in Aue, Saxony, Maltsev proceeded to establish new mines under the cover name of the Vismut Corporation. Gradually, a series of some seventy mines were opened and production rose from 135 tons of ore in 1946 to 900 tons in 1948. Everything was done to meet the new urgency of the program. Because of his technical knowledge of the existing Saxony mines, a committed Nazi was released from prison and appointed chief adviser to Maltsev's team. An extensive recruitment program for miners was implemented. Mining construction was rapid and paid little heed to safety. One German mining engineer described conditions in the mines as "criminal." They were susceptible to flooding, and the provisions for radiation protection were primitive. In 1947, a store of dynamite exploded, leaving 70 dead and 170 injured. But work continued. Perhaps nothing indicates the urgency of the operation better than the fact that the first uranium ores were flown at huge cost rather than trucked to the Soviet Union.

Under the supervision of Vannikov and Zavenyagin, two entire new towns were built to receive the ore and to produce U-235 and plutonium. The Russians, adopting the Smyth Report's suggestion that the best way to make U-235 was via the gaseous-diffusion method, built a massive plant somewhere in Siberia, again using slave labor.

The plutonium plants were built near Sverdlovsk in the Urals. In the fall of 1947, Kurchatov left Laboratory No. 2 in Moscow and joined Vannikov at the new site. He lived with the engineer in a railroad car close to the reactors because Vannikov had only just recovered from a heart attack and wanted to avoid the daily journey to and from the new town ten kilometers away. Kurchatov's official biography says he endured the "discomforts of the frosty winter without complaining," which is praise indeed, since it must have been singularly uncomfortable.

By 1947, the project had fallen seriously behind schedule. Supplies of uranium were not coming in fast enough and a host of problems was encountered in Kurchatov's plutonium-producing reactors, mainly in scaling up the design from the research stages. The graphite, which had to be of a special purity, was found to be contaminated from the linoleum it was stored on. The gaseous-diffusion plant was faring no better, and in 1948 Beria personally dismissed its director. The plant had encountered serious corrosion problems; it was not until 1950 that these were sufficiently resolved to produce enough U-235 for a bomb. Everything, therefore, depended on the plutonium reactors.

At the same time, work was also continuing on the design of the bomb. The early designs had been conducted under Kurchatov's direction, but Kurchatov himself did not have the theoretical knowledge to complete it. Apparently, this was done by Yulii Khariton and Yakor Zeldovitch. Even before the war, these two physicists had published papers on the basic theory of chain reactions and of atomic explosions.

Kurchatov and his team finished the graphite reactor in 1948. A year later, in April 1949, another team completed a second plutonium-producing reactor, this time moderated by heavy water. By the time the first bomb was tested in August that year, the Russians had enough plutonium for two bombs.

The first device was detonated on August 29, 1949, in the deserts of Kazakhstan in Soviet Central Asia near the city of Semipalatinsk. The desert was strewn with the wreckage of buildings, old tanks, and artillery pieces used to test the effects of the blast, plus the bodies of several experimental animals used to test the effects of radioactivity. In concrete bunkers, scores of scientists and military officers watched the test. In the command post were leaders of the Soviet government; Kurchatov and Zavenyagin were the last men on the metal tower from which the device was detonated. The blinding flash of light and the familiar mushroom cloud brought the same mixture of pride and scientific relief that the Americans had experienced at Alamogordo four years before.

The bomb, as Khrushchev would later admit, was not a weapon that could be dropped from an aircraft; that refinement was still to come. It was, however, the culmination of an extraordinary scientific and engineering effort—and the beginning of more feverish work to produce a significant nuclear arsenal.

5 "The Poor Relation"

A Poor Relation—is a preposterous shadow, lengthening in the noontide of your prosperity . . .
—CHARLES LAMB, "Poor Relations"

The bombing of Hiroshima and Nagasaki, in which 100,000 people died and another 100,000 were injured, was the climax of a military philosophy that accepted mass killing as a good way of winning wars. The theory is brutally simple: Kill as many civilians as possible and the ones who survive will be so demoralized they won't want to fight. It had been put into practice on a small scale several times before World War II began. There had been Mussolini's air offensive on Ethiopia in 1935, Japan's sorties on Chinese cities in 1937, and the German attack on the Spanish town of Guernica. The world was outraged. In 1939, President Roosevelt made a futile appeal to both sides to refrain from such attacks.

Hitler started the war in Europe with tactical air strikes supported by mechanized ground forces. Then came the London blitz. The German planes were supposedly bombing the docks, but their desperate random strikes hit residential areas. The Luftwaffe had not yet developed a full-blown strategy of indiscriminate killing of civilians—that would come later with the V-rockets—but neither did they have any rules against such practice.

After victory in the Battle of Britain, RAF bombers went on to retaliate with raids over Germany, trying to destroy Hitler's capacity to make war by knocking out his industrial base. The effectiveness of this doctrine depended on the accuracy of the bombing, and during 1941 it became clear to RAF strategists that the British warplanes were incapable of accurate enough strikes: they could not be sure of hitting an oil refinery or a railway yard or a factory. In fact, the RAF bombardiers found it difficult to drop their bombs within a radius of five miles of the target.

This was devastating news to the leaders of the RAF's Bomber Command. They had created a distinctive role—and a handsome budget—for themselves on the basis that they could win wars on their own. Their evident failure worried Churchill to the point that he decided, at the end of 1941, to halt their sorties into Germany.

The RAF High Command reacted like a child deprived of its biggest toy. As the junior service, with infant status anxieties, the RAF had won its independent role for Bomber Command in bitter wrangles against the ambitions of the British Army and the Royal Navy, each of which wanted its own separate air arm. A youthful, self-conscious, technological elite, the RAF was not going to give up easily.

In the center of the battle was the archetype of the new breed of air force commanders, the founder of the RAF himself, Hugh Trenchard. In less than a decade after World War I, he had developed a strong sense of camaraderie and elitism in his men, an esprit de corps that challenged even the centuries-old traditions of the Royal Navy. "Air mastery can be gained only by offensive" had become the fighting doctrine of the new service as surely as "engage the enemy more closely" had been the basic battle cry of the navy. Before and during World War II Trenchard had sought to maximize the budget allocations for Bomber Command, so much so that, had his wishes been carried out, neither of the other two RAF commands, Fighter and Coastal, would have had the resources they needed to fight the Battle of Britain or the Battle of the Atlantic. "Boom" was Trenchard's universal nickname, and he and his pugnacious senior officers were determined to keep the bombs falling despite their ineffectiveness. If they couldn't hit factories, they would hit civilians. Accordingly, the RAF set out to justify the indiscriminate bombing of civilians. The problem was that British surveys had shown the Luftwaffe blitz on London had produced no marked effect on the resolve of the British civilians to continue the war; indeed, their morale had remained high. Spurred on by the increasingly doctrinaire Trenchard, however, the RAF—in particular, a Trenchard disciple called Charles "Peter" Portal, the chief of the Air Staff—tried to persuade Churchill to allow the RAF to "area bomb." It meant hitting civilians while aiming roughly in the direction of Germany's war-production plants. In effect, do what Hitler had done when he was supposed to be aiming at London docks, but make it a definite policy. The "consensus of informed opinion," Portal told the Prime Minister, was that Germany's civilian morale was much more vulnerable than Britain's. Portal enlisted the support of Churchill's court scientist, Frederick Lindemann, later Lord Cherwell, himself already a supporter of "dehousing" the German population. Lindemann had done a series of mathematical calculations, including willful distortions of scientific evidence, to "prove" his point. Other scientists who produced evidence to the contrary were ignored.

With Lindemann's "proof," Portal managed to turn Churchill around. The bombing embargo was lifted, and on St. Valentine's Day, 1942, Portal issued a secret order to his bomber pilots identifying a new primary objective: the morale of enemy civilians. In 1945, the U.S. Air Force would follow suit. The memorials of this strategy became the devastated cities of Dresden, Tokyo, and, finally, Hiroshima and Nagasaki. The bombing of Hiroshima in August 1945 carried the RAF's St. Valentine's Day order into the atomic age. It would be only a matter of time before the RAF wanted its own nuclear bombs plus bombers capable of carrying them deep into enemy territory. In fact, throughout the war, Britain's eventual membership in the nuclear weapons club had always been assumed by the Americans; the only question was whether Britain would build a bomb on her own or with American help. As it turned out, America's obsession with its "sacred trust" left Britain on her own. And when she set out on the road to the "independent deterrent," much of the momentum for the project—and, indeed, some of its leaders—would come from the same men responsible for the St. Valentine's Day order.

Britain entered the postwar atomic race with a lot of scientific know-how about the bomb, but no money. The country was virtually bankrupt. The most influential economist of the day, John Maynard Keynes, warned the newly elected Labour party that "the economic basis for the hopes of the people were non-existent." Yet in 1945 the

Labour leadership took a vast gamble with its meager amounts of money and resources and began building an independent nuclear deterrent. It was an act of vainglory, a desperate attempt to preserve Great Power status when it was clear Britain was no longer in a position to aspire to such. Despite the import of the decision—probably the most momentous in the country's postwar history—the British people did not know about it for a decade. The degree of secrecy was extraordinary even for a country whose government proceedings have been traditionally cocooned by the comprehensive Official Secrets Act.

Discussion of the project, even at cabinet level, was minimal. The formal decision to build the new weapon was taken without an economics minister's having a chance to caution fiscal restraint, let alone say "No." The extent of the parliamentary debate was restricted to the following exchange, of less than one hundred words, between a Labour member of Parliament, George Jeger, who had been instructed to plant a question, and the Minister of Defence, A. V. Alexander in May 1948.

> MP: [Is] the Minister of Defence satisfied that adequate progress is being made in the development of the most modern types of weapons?
>
> MINISTER: Yes, Sir, research and development continue to receive the highest priority in the defence field, and all types of weapons, including atomic weapons, are being developed.
>
> MP: Can the Minister give any further information on the development of atomic weapons?
>
> MINISTER: No, I do not think it would be in the public interest to do that.

Thus quietly, and without fuss, the British bomb project was put on the record—actually sixteen months after the explicit decision to build a bomb had been taken. The British press, bound by the Official Secrets Act and a specific gag order on the bomb program, took the minister at his word: no more questions were asked and no more details printed. The government ensured silence by issuing what is known as a D-Notice, a cozy self-denying ordinance with no legal force by which editors agree not to mention or discuss a topic said by the government to be of importance to national security. In this case, the government included such things as uranium shipments, location of atomic weapons work, and identities of the people involved.

There was no public debate; parliamentary questions went unanswered. One MP complained: "When we ask questions about [the atomic bomb, it is] as if one had asked about something indecent." Another MP complained that the press had become as restrained on the subject of the bomb and atomic energy as they had once been about sex. Even three years later, so well-informed a journal as the *Economist* could confidently dismiss news-agency rumors that Britain might be running a bomb program—something the *Economist* felt simply could not be justified in terms of possible benefit measured against the expenditure of desperately scarce resources.

Over the next fifteen years, before the United Kingdom became an appendage of the U.S. nuclear program, Britain was to spend £1,000 million on making atomic bombs. The first bomb, tested in 1952, worked flawlessly, but the program never produced the strategic and diplomatic dividends the British leaders felt the new weapon could secure. They had sought the weapon to enhance national prestige, but they found the Empire nonetheless continued to disintegrate along with Britain's position as a world power.

The subtle political process that finally led to the decision to make the bomb gained its momentum from the military chiefs of staff, notably from Portal, the author of the St. Valentine's Day order, from the mandarins of the civil service who assumed that Britain's future lay in becoming a nuclear power, and from the scientists and engineers who gave the project their encouragement because they had pride in their own scientific capabilities. "Are we so helpless," asked the British physicist James Chadwick, "that we can do nothing without the United States?"

Prime Minister Clement Attlee was particularly sour about the way the Americans had turned their backs on Britain; he was especially bitter about the McMahon legislation that had closed the door on U.S.-U.K. atomic cooperation. "That stupid McMahon Act," Attlee would recall later. "They [the Americans] were rather apt to think they were the big boys and we were the small boys; we'd just got to show them they didn't know everything."

The irony of the McMahon Act is that, in part, the British brought it upon themselves. When Senator McMahon drew up the legislation, he did not know the extent of the wartime cooperation between Britain, Canada, and the United States; if he had known, he said later, he would not have pushed for such tough restrictions on the flow of information. Yet, when James Chadwick, the leader of the British scientist team in America, tried to hold a private meeting with McMahon to discuss that cooperation, the British diplomatic staff in Washington advised against it.

Both Britain and Canada had made significant contributions to the Manhattan Project. Britain had supplied the key Frisch-Peierls memorandum on the feasibility of the bomb. That document stands, as Margaret Gowing, the official historian of the British project, wrote, "as the first memorandum in any country which foretold, with scientific conviction, the practical possibility of making a bomb and the horrors it would bring."

The primary blame for congressional ignorance in the matter lay with the Truman administration. In the debate on the McMahon bill, three congressmen asked if there were any secret agreements on atomic cooperation between the United States, Britain, and Canada. They were never told that such agreements existed. Yet solemn presidential commitments had been made to the British, both in writing and in oral accords, to include them in the postwar development of atomic energy.

In August 1943, a secret meeting between Roosevelt, Churchill, and Canadian Prime Minister William Mackenzie King had resulted in the Quebec Agreement. It outlined the way in which the three parties would cooperate on the building of the bomb. Referring to the bomb by its British code name of "Tube Alloys," the agreement declared that no party would use "this agency" (the bomb) against each other; would not use it against third parties without each other's consent; would not give away any information to any third party (that is, to the French or the Russians). And, because most of the work was to be done by the United States, the agreement stipulated that the postwar commercial advantages would be dealt with on terms "to be specified by the President of the United States to the Prime Minister of Great Britain." This last point was a startling disclaimer, in effect leaving the U.S. President in control of the postwar development of atomic energy in Britain. Some of the top scientific advisers in London viewed it as signing away Britain's atomic birthright.

In addition, however, the Quebec Agreement had given British scientists a chance to work on the bomb assembly at Los Alamos, on isotope separation at Berkeley, and on the heavy-water reactor project in Canada. They picked up invaluable experience. Furthermore, just before he died, Roosevelt had agreed in a private aide-mémoire with Churchill that "full collaboration" between the United States and Britain should continue after the defeat of Japan, "unless and until terminated by joint agreement." This document, known as the Hyde Park Agreement, seemed a guarantee of postwar cooperation but, with Roosevelt's death in April 1945, the British realized the "full collaboration" clause could die, too. The McMahon Act eventually killed it.

The internal workings of the British Civil Service—the discreet phone calls, the lunches in the Athenaeum Club—are almost as impenetrable as the Soviet Politburo. Nonetheless, the evidence indicates that the nuclear decision-making process was directed by those who regarded it as being above politics. Even if members of Attlee's cabinet, burdened with more pressing domestic priorities, had wished to make their voices heard in the bomb decision, the rapid footwork of the Whitehall mandarins and the acquiescence of Attlee himself ensured that the drive toward the bomb was not diverted by political interference.

The atomic baptism of Clement Attlee, a balding, small man with a military mustache, was crude and abrupt. He had been elected a few days before Hiroshima, and its terrible specter filled his mind with foreboding about the postwar use of the atomic weapon. Like the American leaders, he had immediately turned to a consideration of ways in which atomic energy might be controlled. He believed joint action by the United States, Russia, and Britain was the "only hope of staving off imminent disaster for the world." He wanted the three nations to come forward with a declaration that "this invention has made it essential to end wars." Such sentiments did not last long. Less than a month after Hiroshima, Attlee would be leading a special British cabinet committee down an inexorable road to a British bomb. His guide in this delicate and highly secret task was the Secretary of the Cabinet, Sir Edward (later Lord) Bridges. The Secretary of the Cabinet is a civil servant, not a political appointee, and therefore does not change with administrations. Traditionally, he is the nursemaid of newly elected Prime Ministers.

Clement Attlee was the kind of man who looked as though he could do with a helping hand. "My opponent is a modest man and with reason," Churchill once said of him during the election campaign. A Fabian Socialist, Attlee's one fervent hope when he joined Churchill's coalition war cabinet in the summer of 1940 was to see socialism in postwar Britain; to see a country with cake for none until all had bread. Since 1935, he had led Britain's Labour opposition in the House of Commons. Born in 1883, he had watched the Labour movement grow from a membership of thirty to over eight million and its representation in Parliament from five to three hundred. To him, socialism was the dominant issue of the twentieth century, although his personal brand owed more to the Bible than to *Das Kapital*. He was once described as having a "benevolent air reminiscent of a safely-birthed schoolmaster." Anyone less like his predecessor would be hard to imagine.

Sir Edward Bridges would lead Attlee through the troubled waters of the postwar atom with the smooth professionalism of the civil service. Bridges was the shy, but determined, son of the poet laureate Robert Bridges. He was a trained and highly successful civil servant who had risen through the Treasury to become Secretary to the Cabinet. He had taken the job in the late summer of 1938, a time when most people felt war was inevitable. Tall, slim, and scholarly, with a boyish face that masked an extraordinarily competent and cool bureaucrat, he had a flair not only for oiling the wheels of the decision-making process but also for such bureaucratic refinements as reading documents upside down. Under Churchill's centralized secretariat during the war, Bridges had smoothed feathers ruffled by clashes between the military and civilian staffs, a job he had fulfilled, according to Churchill, with force, ability, and personal charm.

Churchill made great demands on him. No memorandum for the Prime Minister's eyes could be more than a single page of typescript, and Bridges spent many hours pruning the wordy thoughts of others. Churchill's erratic working hours—with conferences often being called in the middle of the night—meant Bridges frequently slept in his office. Under him, the Cabinet Office became a machine of unrivaled efficiency, but few got to know the man who ran it. Bridges, who could often be found lunching alone at the Athenaeum, was reserved and, to strangers, austere. Lord Moran, Churchill's personal physician, recalled: "His steady, appraising gaze behind thick spectacles seems to register disapproval. When I used to run into him at No. 10, I thought he took unnecessary steps to frown on a stranger. But his uncompromising integrity, his dislike of intrigue of any kind and his rugged honesty are known to everyone."

With the election of the first Labour majority government in British history, Bridges faced the greatest challenge of his career. It was a profound test of the hallowed tradition of his service: political impartiality. As a civil servant, his job was not to make policy but to advise, and no one believed in the strict division of the two functions more strongly than Bridges. There were, however, matters of state that were considered to be above politics: those that went to the core of Britain's heritage and future. These special topics had to be insulated from the vagaries of party politics, and the British bomb was one of them.

Before the end of the war, no one had given much thought to the special nature of the weapon, to the extent of the resources it would require to build it, how to build it, or, indeed, once built, how it would fit into military strategy. Although it had been taken for granted throughout the war that, if the Manhattan Project worked, Britain would build a bomb, the surprise Labour victory introduced a note of uncertainty into what had seemed a very simple equation. Would the Labour government have the same priorities? The best thing to do was to act as if nothing had happened, to proceed with a cool determination and be careful not to raise any fundamental questions of policy. If Bridges was merely an efficient engineer to Churchill, his role was somewhat more positive with Attlee, at least in atomic matters. Almost immediately, he set the Attlee cabinet a definite course on atomic energy with three crucial suggestions. Attlee agreed to them all, effectively leaving atomic energy policy in the hands of those who had been running it under Winston Churchill.

The first came three days before Hiroshima. Bridges suggested that the new Prime Minister "might care to consider," as the official historian, Gowing, delicately put it, taking on Churchill's atomic expert, the former Chancellor of the Exchequer, Sir John Anderson. He was a Tory, a civil servant turned politician. But he was also the man who knew most about the bomb; he could act in an advisory capacity, reporting to the Prime Minister on matters where ministerial decisions were required. It was Anderson alone, noted the cabinet secretary, who was "fully seized of the long complicated history" of the bomb project. When the Bridges suggestion was raised on August 10, 1945, at the first meeting of the ad hoc ministerial committee on atomic matters—known simply as "Gen (for General) 75"—the proposal was quickly approved. Anderson's specialist knowledge overshadowed the unpalatable fact that he was a former Tory minister and an avowed opponent of the Labour movement. Thus, Anderson, the brilliant overlord of domestic policy in Churchill's war cabinet, became a quasi-minister in the Labour cabinet. He saw all the cables and the memoranda on atomic energy. No member of the Labour cabinet had such a complete knowledge of the British bomb project's day-to-day progress. He even played his historically unique role from a room in the cabinet office itself.

The second of Bridges's suggestions was that Attlee should not, as was his inclination, appoint a full-time minister for atomic energy. The bomb was too important to be dealt with by a junior minister, Bridges advised, and, in any case, there would not be enough work for such a minister. Policy could be coordinated by a cabinet committee chaired by Anderson. The committee could make policy recommendations for both domestic and international control, but decisions would be taken by the Prime Minister. Again, Attlee agreed, apparently believing he would always have the inclination and, more important, the time to attend to these matters. Bridges must have been pleased at the smoothness of the operation thus far; had a special minister been appointed, he might have done too much rather than too little and upset the pro-bomb commitment.

The third suggestion of the tireless Bridges was that the atomic project should be taken away from its wartime handlers, the Department of Scientific and Industrial Research, and given to the Ministry of Supply. Having organized Britain's ordnance factories during the war, this ministry had the necessary experience to build a bomb, not just organize research. It was also a junior ministry with, at the time, a weak minister, a fact the cabinet secretary would have known. In October, the Prime Minister and his Gen 75 "inner cabinet" agreed to this last proposal. To date, all the essential parts had clicked neatly into place.

Bridges had apparently been quite clear on what he wanted from the beginning, and in November, armed with bureaucratic power—the Prime Minister still controlled major policy decisions, of course—Bridges recommended that a start be made on building a bomb. Britain's chiefs of staff needed no encouragement; they had never been in any doubt what they wanted. Britain, they said, should have a stockpile of bombs as soon as possible, "to be measured in hundreds rather than scores." Attlee's Gen 75 was not quite ready for that quantum leap; Britain had come a long way down the road, but things had to be taken carefully, step by step.

The first step was the creation of a research organization to look into "all aspects of the use of atomic energy." This was set up at Harwell, an old RAF airfield fifty miles from London. Its director was John Cockroft, the Cambridge physicist who had been in charge of Canada's wartime heavy-water reactor.

The second step, taken in November 1945, was a trip by Attlee to America to see Truman and the Canadians about international control and, especially, about future U.K.-U.S. cooperation. Accompanying Attlee was his new chief atomic adviser, John Anderson.

They made an unlikely duo: Attlee, short and with a small, thin face; Anderson, big with the long jowly face of a bloodhound. The son of a London solicitor, Attlee had spent his life trying to escape from his conservative past. Anderson, the taciturn son of a Scottish publisher, had zealously embraced the privileges of his upper-class birth. He loved cigars, billiards, golf, and roses. He had joined the colonial office in 1905, at the age of twenty-three; fifteen years and several unusually rapid promotions later, he was sent as a "strong man" to Ireland to quell the rebellion. In 1932, when Britain was trying to suppress rebellion in Bengal, Anderson was sent out as governor to restore law and order. After five years of his iron hand, Bengal was quiet and Anderson returned home, releasing some political prisoners as a departing act of clemency. "You see, he has a heart after all," said one Indian. "Not at all," said his companion, "there was no more room in the prisons."

In Churchill's wartime cabinet, Anderson took over the Home Office and built thousands of bomb shelters of concrete and corrugated iron in gardens all over England. They became known as Anderson shelters. He was made Chancellor of the Exchequer, but always retained full responsibility for atomic energy. Rarely to be seen in anything but a dark top coat and pinstripe trousers, he was always formal, even with his intimates. At the Home Office, they called him "Pompous John," and in the cabinet he was known as "Jehovah." "Here, we are all Jehovah's witnesses," Attlee once said while waiting impatiently for Anderson. General Groves loathed him. In America, Groves said, Anderson behaved "as high-handedly as he had done in Ireland during the Black and Tan era." Despite Anderson's reputation for toughness, Groves was to get the better of him.

The purpose of Attlee's Washington visit was to seek continued cooperation from the United States and to lift the presidential restrictions placed by the Quebec Agreement on the commercial exploitation of atomic energy in Britain. At the end of their meeting, Truman, Attlee, and Canadian Prime Minister William Mackenzie King affirmed "full and effective" cooperation between the three countries in the development of atomic energy. But, at Groves's insistence, the final wording of the resolution restricted the cooperation to scientific data only. The British already had a good grasp of the scientific data; what they needed was the technical know-how, the engineering details of producing fissile material. The Americans never gave them this information, despite Attlee's repeated requests, which invoked the "special relationship" between the two countries.

Finally, in December 1945, Gen 75, or "my atom bomb committee," as Attlee called it to insiders, met to decide on a recommendation from Anderson's committee to build a large-scale reactor to produce plutonium. The committee's case, which Gen 75 accepted, was that by making plutonium the government was doing no more than keeping the bomb option open. That is not how others saw it, however. The engineer who was put in charge of building and operating the plant, Christopher (now Lord) Hinton, an employee of Imperial Chemical Industries, Ltd., was left in no doubt that the job was to make fissile material for bombs. But such was the degree of compartmentalization and secrecy that Hinton only discovered, by reading the official history three decades later, that the *explicit* decision to build a bomb was not made until January 1947.

The man brought in to oversee the production of the plutonium was the then retired Chief of the Air Staff "Peter" Portal. He was reluctant to take the job. He knew little about atomic energy, and when he had resigned from the RAF at the end of 1945, he had hoped to fade into retirement and live off a handful of directorships in the City of London. But Attlee prevailed on him to take the job; he would be called Controller of Atomic Energy. Although, in theory, he worked for the Minister of Supply, Portal was directly responsible to Attlee. At first he had no written terms of reference and insisted on restricting his duties to the fissile-material production plants. But in his own quiet, unruffled way, which many had admired during the war, Portal, an extremely able commander, gradually assumed a much more important role in the bomb project.

It is widely accepted by those who served under him that Portal had lost much of his wartime drive and enthusiasm by the time he took over atomic energy. But he still commanded tremendous respect. He had completed a superb RAF career: from the moment, as a World War I ace, he had taken on five enemy aircraft and destroyed three of them, to his cool, firm, detached leadership during Bomber Command's time of crisis when they found they could not bomb accurately.

His prime interest was in machines rather than people, but, like Lindemann, he had an "acute understanding of the techniques of manipulating men." In his new job, Portal had almost unrivaled access not only to the Chiefs of Staff but also to key figures in the government. The official British history concludes: "People in the project probably never knew how tenaciously Portal fought with Ministers and Chiefs of Staff for a continued top priority rating for atomic energy."

One of Portal's last acts as chief of the Air Staff was to deliver the military commanders' report recommending the bomb project to Gen 75. By the end of 1946—a year into his new job and shortly before the January 1947 decision to build a bomb was taken—Portal, anticipating that decision, had added to his responsibility for plutonium full control of atomic weapons research.

Attlee's "atom bomb committee," Gen 75, did not, in the end, make the final, formal decision. Throughout early discussions in Gen 75, ministers with economic responsibilities, notably Hugh Dalton, the Chancellor of the Exchequer, and Stafford Cripps, the President of the Board of Trade, had expressed some reservations about the economic implications of the large-scale nuclear program. Their views reflected the refusal of the key British contractor, ICI, and another huge company, English Electric, to have anything to do with the project. The investment needs of postwar reconstruction were so heavy, the companies believed, as to preclude any involvement in nuclear matters. It had been the economic ministers, for example, who had ensured, in December 1945, that the construction of only one plutonium reactor was approved, instead of the two recommended by the Anderson Committee and approved by the Chiefs of Staff.

In December 1946, Gen 75 was wound up and, in the early days of January 1947, another committee surfaced for one vital meeting. It was called Gen 163, and at its first and only gathering, it decided to build the bomb. The composition of Gen 163 was the same as Gen 75 with the pointed exclusion of Dalton and Cripps, the two economic ministers. There is no direct evidence of the hand of Edward Bridges in the choice of the membership of Gen 163, but it is precisely on matters of this kind—the composition of the cabinet committees—that he, as Cabinet Secretary, would have had great influence with the Prime Minister.

Throughout the eighteen months leading up to the bomb decision, only one man inside the secret world of Britain's nuclear program had the courage to put his dissent on record. That extraordinary figure was Professor Patrick Blackett, a member of the government's Advisory Committee on Atomic Energy. Successively naval officer at Jutland, Nobel Prize–winning physicist, virtual inventor of operations research in warfare, Blackett in 1945 fought hard and unsuccessfully against the development of a British bomb. Everything Blackett foresaw of the internal consequences of a British bomb came true: its development deprived other needy areas of industry of both physicists and resources, and it distorted the development of peaceful nuclear energy.

Blackett always maintained that Britain did not and never would have the resources to develop a nuclear force that would be militarily worthwhile. His opposition to the bomb first appeared on the record in a secret memorandum in November 1945. A British bomb project, he advised the government, would not only strain limited resources but would also encourage the spread of nuclear weapons. He further suggested that, once having decided against the bomb, it was essential that Britain open her atomic establishments to inspection by, say, the United Nations; this would prove her sincerity.

Such talk was far from the thinking of the government's insiders, and Blackett got much the same rebuff from Attlee as Niels Bohr had from Churchill in 1944. Dismissing Blackett's report, Attlee commented: "The author, a distinguished scientist, speaks on political and military problems on which he is a layman." The Chiefs of Staff simply ignored the report.

Adding insult to injury, Blackett and another physicist, Henry Tizard, were both excluded from the vital bomb decision-making committees. Tizard's exclusion had its roots in prewar clashes with Churchill's personal scientific adviser, Frederick Lindemann, later Lord Cherwell. In several heated and public squabbles, Tizard had opposed the development of Cherwell's more absurd air defense gadgets—such as developing bombs that could be attached to parachutes and dropped into enemy bomber formations—in favor of giving priority to radar. During the war, the two had clashed again over the effectiveness of strategic bombing, and Tizard, a nonbeliever in the doctrine, was excluded from wartime atomic affairs. When, in 1946, Attlee made him the scientific adviser to the newly created Ministry of Defence, the wary Bridges instantly recognized a warning light. His subsequent actions help to illustrate the tight grip Bridges and Portal had on atomic matters.

Bridges thought Tizard might resent his exclusion from wartime affairs and cause unwelcome waves in the smooth passage toward the bomb. He advised Attlee not to let Tizard's influence grow beyond the Defence Ministry and to make sure that Portal, now firmly ensconced as Controller of Atomic Energy, was the undisputed leader of the program. Attlee agreed, at the same time assuming that Tizard, as scientific adviser, would continue to receive atomic energy information from Portal. He was wrong.

In June 1947, Tizard presented a paper on atomic matters to the Chiefs of Staff. To the chiefs' amazement, he told them that no decision had yet been made to build atomic weapons—when they knew the decision had been taken, in secret, five months earlier. Similarly, Blackett was kept in the dark about the bomb decision. Issuing a paper in February 1947, he argued against the bomb. He talked to Attlee twice about his plan, but the Prime Minister by this time was not interested; he wrote no comments about Blackett's proposals and there was no discussion of his paper at any ministerial meeting. The final irony of the rejection of the sound advice of these two scientists came in 1949 when, after much badgering by Tizard, the Chiefs of Staff at last accepted the consequences of bomb building in Britain. Realizing, too late, that the project meant Britain had no resources to spare for the development of other weapons—notably, missiles and aircraft to deliver the bomb—they actually asked that the project shed some of its priority.

Perhaps no more pointed prediction of Britain's postwar decline was ever written than Tizard's comment when he finally came out in direct opposition to the British bomb project. In a secret memorandum in 1949, Tizard wrote: "We persist in regarding ourselves as a Great Power, capable of everything and only temporarily handicapped by economic difficulties. We are not a Great Power and never will be again. We are a great nation, but if we continue to behave like a Great Power we shall soon cease to be a great nation. Let us take warning from the fate of the Great Powers of the past and not burst ourselves with pride (see Aesop's Fable of the Frog)."

6

The One-Eyed Kings

On September 3, 1949, the special air-filter system of a U.S. weather reconnaissance plane of the air force's Long-Range Detection System, flying at 18,000 feet over the North Pacific east of the Kamchatka Peninsula, registered an increase in atmospheric radiation. It was large enough to trigger an official alert: it seemed as though an atomic bomb had been exploded somewhere in the Soviet Union. Further samples produced substantially higher counts.

A special committee of scientists, chaired by Vannevar Bush, was convened to evaluate the data. More samples were obtained from the radioactive cloud as it drifted eastward across the American land mass. As the cloud moved out across the Atlantic, the British also took samples. Finally, an analysis of the composition and age of the radioactive fission products persuaded the scientists that the Russians had indeed exploded their first atomic device. The precise date was hard to fix—probably August 29, 1949.

As chairman of the Atomic Energy Commission David Lilienthal had the task of taking the bad news to the President. Truman was deeply shocked. As much as any of America's leaders, he had chosen to disregard the advice of atomic scientists when they had told him, after Hiroshima, that the Russians would probably make the bomb within five years. He preferred to believe it would be more like a decade, perhaps even two decades. Truman jumped to the conclusion that German scientists were responsible for helping the Russians do it so quickly—"probably something like that," he told Lilienthal. Despite the overwhelming evidence, Truman delayed making the news public until Lilienthal and the other experts had signed a confirming statement. On September 23, the White House told the world what they had discovered. The bomb was nicknamed "Joe One," after Stalin.

The news generated a renewed passion in America for anything and everything nuclear—from bombs to nuclear-generated airplanes, submarines, and electricity. American policymakers responded to Joe One as though their undisputed status as the world's most powerful nation had been seriously undermined: they panicked into frenetic action, forming new alliances in support of atomic power. The nuclear zealots in the Pentagon sought out the nuclear zealots in the Congress and in the atomic laboratories. During the next six months, these new alliances lobbied energetically for any project that was nuclear. The wide range of programs that resulted was to dominate the U.S. posture in military strategy, foreign policy, and energy options for decades to come.

For some, the Russian test seemed to create an opportunity to revive the Baruch Plan—or any other proposal to control the atom; but such hope was short-lived. Within six months, Truman had agreed to major increases in the production of fissile material and approved a wide-ranging policy of general rearmament for America and its allies. He would follow this with the most awesome development of all: the decision to go ahead with the H-bomb, a weapon in a class of its own.*

For many American policymakers in these immediate postwar years, a showdown with Russia seemed inevitable. They believed the

*Unlike the atom bomb, in which the explosive energy is based on the splitting of atoms of heavy elements like uranium and plutonium, the H-bomb is based on fusion: the joining together of isotopes of the lightest element, hydrogen. These hydrogen isotopes need a massive burst of heat and tremendous pressure before they will explode, but if these temperatures and pressures are achieved, there is, theoretically, no limit to the size of the bomb. The limitation, in these early days, was the size and weight of a *deliverable* bomb—and, hence, its blast capacity. The nuclear arms race, so feared by some of the atomic scientists, had become a reality.

Soviet Union would stop at nothing to achieve world hegemony. With the defeat of Japan and Germany and the disintegration of the British and French empires, only the United States had the capability of defending civilization itself. And if the development of more terrifying weapons of mass destruction brought the prospect of total annihilation closer to millions of people, what was the alternative? In 1950, a U.S. National Security Council policy document declared, "On the one hand the people of the world yearn for the relief from the anxiety arising from the risk of atomic war. On the other hand, any substantial further extension of the area under domination of the Kremlin would raise the possibility that no coalition adequate to confront the Kremlin with greater strength could be assembled."

Each contemporary event—Joe One, the fall of China to the Communists, the Alger Hiss affair, the Soviet blockade of Berlin, the confession of the atomic spy Klaus Fuchs—reinforced the conviction of these policy planners that keeping the Russians at bay meant staying ahead in the nuclear arms race. Truman agreed. At the end of each meeting with his five AEC commissioners, he would warn: "We must keep ahead."

There were no more eager devotees of this policy than the leaders of the U.S. Strategic Air Command, in particular the belligerent SAC commander, General Curtis LeMay. Recalling nostalgically the days of the American atomic monopoly, LeMay wrote: "That was the era when we might have destroyed Russia completely and not even skinned our elbows in doing it." Having lost that monopoly, the air force generals stressed the need for continued superiority and each year demanded a larger slice of the total military budget. Atomic bombs became the centerpiece of U.S. military strategy.

The cries of the military hawks were echoed in Congress. Even before the Russian bomb exploded, Senator Brien McMahon would write, in July 1949, that strategic bombing with nuclear weapons was "the keystone of our military policy and a foundation pillar of our foreign policy as well." McMahon agreed with the air force generals: there should be no limit to the number of atomic bombs in readiness in America.

Hawk policy secured a series of expansions in atomic bomb production that engorged the nuclear arsenal. It culminated in Truman's decision in March 1950 to go ahead with the H-bomb. Throughout the history of the atomic age, man has been presented with decisions that have been surrounded with an aura of inevitability, obscuring the freedom of choice. Not one, however, was as important as those taken during the second term of the Truman administration when the President of the United States abdicated his responsibility to make decisions on these matters and allowed the course of world history to be determined by transitory fears and enthusiasms.

Truman's decision to build the H-bomb went against the advice both of the Atomic Energy Commission's General Advisory Committee and of three of the five AEC commissioners, including David Lilienthal. It marked the beginning of the end to the spirit of open-mindedness and independence that had until then characterized the Atomic Energy Commission—and also the end of the considerable influence of the GAC. Before the decision was finalized, Lilienthal, aware it was imminent and deeply depressed at the H-bomb outcome, would resign from his powerful position as AEC chairman.

As the first chairman of the AEC, Lilienthal had established an effective government institution, an independent organization free, as far as possible, from political interference by Congress or the military. He had taken his mandate—to keep the development of atomic bombs and nuclear power under civilian control—very seriously. Moreover, he had given Americans good reason to hope that the power of the atom would one day be put exclusively to peaceful uses. A dream, perhaps, but Lilienthal held his convictions high.

As the production lines were set up for more atomic bombs and for the superweapon, Lilienthal commented sadly: "Where this will lead us is difficult to see. We keep on saying, 'We have no other course.' What we should be saying is, 'We are not bright enough to see any other course.' "

David Lilienthal had come to Washington in the fall of 1946 as an outspoken champion of democracy in action. In the previous decade he had been, first, a director and, then, chairman of the world's largest publicly owned water resources project, the Tennessee Valley Authority. It had been a massive and highly successful undertaking, and Lilienthal was the embodiment of what it had achieved. His TVA experience had convinced Lilienthal that men could learn to work in harmony with the forces of nature. There was a choice, he once said, to use science for evil or for good.

Of the five AEC commissioners appointed by Truman toward the end of 1946, Lilienthal was the only one who considered himself a political independent. The other four were Republicans: Sumner Pike, a former member of the Securities and Exchange Commission; Lewis Strauss, a Wall Street financier; William Waymack, editor of the Des Moines *Register and Tribune*; and Robert Bacher, one of the key physicists from Los Alamos. Truman was proud of the fact he never inquired into the political affiliations of the men he chose for the first AEC.

The commissioners bore heavy responsibilities in the postwar period. They were among the roughly two dozen men in America who knew the number of atomic bombs that existed in the country's nuclear arsenal, where they were stored, and how many were being added to the stockpile each year. The other members of this elite were the President, the four secretaries of the armed forces, eleven high-ranking military officers, and the three directors of the atomic factories. The commissioners were among an even smaller number who had access to all the production secrets of the atomic bomb.

They had a special relationship with Congress because of the McMahon Act of 1946. The act had established the commission and had also provided for a congressional watchdog committee called the Joint Congressional Committee on Atomic Energy. It was the only committee of Congress ever established by statute, and it had statutory powers. Eventually, the committee's influence would become so extensive that it overturned the traditional division between the executive and the legislative branch. Traditionally, the executive decides policy with the advice and final consent of Congress; in atomic energy the Congress would decide policy with the advice and consent of the executive. But in the beginning, Lilienthal headed an independent and, to some extent, aloof organization that was not subordinate to the JCAE.

Because of this aloofness, relations between the AEC and the military were, from the start, highly charged. The air force, in particular, pressed for custody of the bombs, a move Lilienthal successfully resisted: he was also determined the AEC would decide the size of the arsenal. The air force pressed for an increase in the capacity to make bombs; with the war over, production of U-235 had leveled off and the amount of plutonium being made had actually decreased. The three plutonium production reactors at Hanford were in need of an overhaul: one had been closed down and the other two were being run at reduced power. Bombs were still being produced one by one; there was no such thing as a bomb assembly line: the metal parts of each weapon had to be individually made and put together by hand. Also, scientists were still experimenting with the trigger mechanism, the amount and degree of enrichment of the fissile explosive, and the size and weight of the bomb casing.

In February 1947, a month after the AEC took control of the $1.4 billion properties of the army's Manhattan Project, the Joint Chiefs reported that the supply of atomic weapons was "inadequate to meet the requirements of the United States." That was something of an understatement. As Lilienthal was later to reveal, there was no stockpile. There was not a single operable atomic bomb in the Los Alamos atomic arsenal. It was the biggest atomic secret of the day, so secret that Lilienthal never made a note about it, even for the AEC's secret archive. On the afternoon of April 3, 1947, Lilienthal told Truman the news. Not only were no bombs assembled, but production of critical materials—plutonium and U-235—was "badly out of balance," and a shortage of uranium seemed inevitable in the near future. Lilienthal concluded his report by stating, "Whatever corrective measures are adopted, they are certain to call at the earliest possible moment for extraordinary efforts in terms of money, materials, equipment and human energy. In several critical fields there is need for full steam ahead."

Truman was shocked. Lilienthal recalled:

> We knew just how important it was to get these facts to him, we were not sure how he would take it. He turned to me, a grim, gray look on his face, the lines from his nose to his mouth visibly deepened. . . . The President was rather subdued and thoughtful-looking—his customary joke on parting was missing. Then in that rather abrupt, jerky way of talking, he said, "Come in to see me any time, just any time. I'll always be glad to see you. You have the most important thing there is. You must make a blessing of it or" (and with a half-grin he pointed to a large globe in the corner of his office) "we'll blow that all to smithereens."

Atomic tests were being planned and executed continually. By the summer of 1947, the air force had completed its evaluation of the first postwar bomb tests, which had been exploded at a safe distance from the American mainland on Bikini Atoll in the Pacific. Air force commanders were pleased with the results: the power and efficiency of the new weapon was beyond doubt. Believing the United States should possess "the most effective atomic striking force possible," they suggested that four hundred bombs would be enough to ensure military superiority. To meet this target, Truman and the AEC agreed to increase the production of plutonium at Hanford and approved the construction of another gaseous-diffusion plant at Oak Ridge to make more U-235.

Behind these initial increases—more were on the way—was a total acceptance of the doctrine of strategic bombing. Like the Royal Air Force before it, the U.S. Air Force had a long tradition of asserting an independent strategic role for itself. *Strategic* was the crucial word. The job of air power was to win wars, not merely to assist the army and the navy. The early leaders of the U.S. Army Air Corps, the forerunner of the Air Force, believed air power could win wars by taking the offensive. The objective of total war was to destroy the enemy's domestic military capacity by strategic bombing. The Americans became so committed to this doctrine that the word *strategic* even crept into the title of the group that embraced the U.S. bomber wings. It was called Strategic Air Command.

Up to 1944 the Americans had concentrated, as the British had done, on "precision" targets, linking their effort to the armies on the ground. But toward the end of the war, with the Allied armies temporarily stalled, the U.S. bomber commander, General Carl Spaatz, the first to use the word *strategic* for his group, swept aside this secondary, backup role and launched a campaign to bomb area targets in Germany, first Berlin, then Dresden. This new pattern was repeated in Japan, beginning with the fire bombing of Tokyo on March 9, 1945, under the command of Curtis LeMay.

The American air commanders carried this policy of holy war into the postwar era, declaring that the purpose of the postwar world was "to make the existence of civilization subject to the good will and the good sense of men who control the employment of air power."

In 1948, LeMay was appointed head of the Strategic Air Command, the nuclear-armed air force. He had no thought of creating a force with second-strike (defensive) capability; it was the first strike that mattered. Offense was everything. The last war had ended with the United States having the capacity to inflict total destruction and that, as far as LeMay was concerned, was how the next one was going to begin.

Strategic bombing had a special, clinical appeal to American policyholders, too. The isolationists, opposed to "foreign entanglements," eagerly adopted the concept that wars could be won by bombing from a distance. The Strategic Air Command was the new "Great White Fleet"—America's insulation from direct contact with European intrigue and conflict. Postwar America accepted the "air power" dogma with the same yearning for a simple, but total, doctrine to erase the complexities of military strategy and diplomacy that had characterized the adoption of the geopolitical "sea power" dogma at the turn of the century.

Confident and aggressive air commanders, for whom the question how much is enough had no answer, pushed for the largest share of the military budget. The army and navy were in large measure reduced to arguing for an ill-defined "balance in forces."

Toward the end of 1947, the strategic role of the air force was confirmed by a presidential commission. Testifying before the commission, Carl Spaatz argued that the "barest minimum for our security" was an air force of seventy combat groups reinforced by twenty-two separate and specialized squadrons, but the budget would support only fifty-five combat groups. The ensuing report was uncompromisingly entitled: "Survival in the Air Age," and it fully endorsed Spaatz's outlook and requirements. From this point on, with the concept of the nuclear "deterrent" barely formulated in military circles, U.S. policy never deviated from total acceptance of mass bombardment of cities. The only center of significant resistance to this trend was to be found among those who had custody of the weapons—the members and advisers of the Atomic Energy Commission.

On Capitol Hill, the Joint Committee was taking an unprecedented interest in the activities of the AEC. Senator Brien McMahon, chairman of the JCAE, was consolidating his position as a nuclear zealot, promoting himself as "Mr. Atom." He thought the bombing of Hiroshima had been, as he said on the Senate floor, "the greatest event in world history since the birth of Jesus Christ." America had to stay ahead in the nuclear arms race. If the Russians should, by some mishap, overtake the United States, "total power in the hands of total evil will equal total destruction."

Helping McMahon as the Joint Committee's director was a bright young ex-wartime bomber pilot named William Borden. Lilienthal knew that they would make a formidable pair. Borden's advocacy of nuclear weapons was well known; he had stated his case in his book,

There Will Be No Time, published in 1946. Borden foresaw the day when the American nuclear monopoly would disappear—an unfashionable thought in those days—and argued that the correct course was to build a huge stockpile of weapons that could be delivered by either planes or rockets. The McMahon-Borden axis believed that America could never have enough bombs.

The hawks' basic complaint was that military requirements for bombs were reflecting an estimate of what the AEC was capable of producing with its existing or planned facilities rather than an independent judgment of what was really needed. By January 1949, working with a strategic strike plan that called for SAC to hit 70 Soviet urban areas with 133 bombs within 30 days, the Joint Chiefs told the AEC that "it is now evident that the currently established military requirements for scheduled bomb production should be increased substantially and extended."

Among the services, the only dissenting voice came from the navy. Reacting to the estimated casualties that would follow implementation of the strike plan (2.7 million Russians killed, 4 million injured, and life for the remaining 28 million residents of the seventy target cities rendered "vastly complicated"), Rear Admiral Daniel Gallery, the assistant chief of naval operations for guided missiles, asserted it was wrong for a civilized society like the United States to have as its broad purpose in war "simply [the] destruction and annihilation of the enemy." That kind of war was not as simple as the prophets of the "ten-day" atomic blitz seemed to think, he said, adding the memorable line that "leveling large cities has a tendency to alienate the affections of the inhabitants and does not create an atmosphere of goodwill after the war." It was only partly facetious. And among the hawks, it had no effect. They increased their efforts to undermine civilian control of atom bomb making. As Lilienthal had feared, the military was beginning to order atomic weapons "like mess kits and rifles," and he decided that it was time for the President to intervene.

Lilienthal had every reason to believe that his growing doubts about the drift of American nuclear policy would receive a sympathetic hearing from Truman. In July 1948, at Lilienthal's urging, Truman had rejected a united military effort to acquire direct custody of the nuclear stockpile from the AEC. Moreover, Truman had expressed his own anxiety about the bombs. "It is a terrible thing to order the use of something . . . that is so terribly destructive, destructive beyond anything we have ever had. You have got to understand that this isn't a military weapon. . . . It is used to wipe out women and children and unarmed people and not for military uses. So we have got to treat this differently from rifles and cannon and ordinary things like that." This was the man who, on hearing the news of Hiroshima, called it "the greatest thing in history." This was the man who, on hearing Robert Oppenheimer say he felt he had "blood on his hands," told Dean Acheson, "Don't you bring that fellow around here again. After all, all he did was make the bomb. I'm the guy who fired it off." Harry Truman had come a long way.

Until the day he died, Truman never expressed any self-doubt about Hiroshima directly. Yet, his doubts surfaced in one striking way: he never made a nuclear decision on his own again. He institutionalized his abdication by establishing a three-man committee composed of the Secretaries of State and Defense and the chairman of the AEC. They were to advise him on all aspects of nuclear policy as these arose. All matters connected with the nuclear program were "decided" by Truman only after a clear consensus had emerged from this committee.

Before the committee could make their report on the latest military proposal to expand production of fissile material, however, the Russians set off Joe One, and U.S. policymakers panicked. In the tense atmosphere that followed, Lilienthal did not push for a showdown with Truman over the production expansion. The three-man committee reported on October 10, 1949, that "the proposed acceleration of the atomic energy program is necessary in the vital interests of national security." No rational choice about "how much is enough" seemed possible at such a time. In any case, an even more momentous choice had appeared on the agenda: whether the United States should build a new superweapon, the so-called hydrogen bomb.

In late 1949, the terrifying specter of thermonuclear devastation pervaded the top-secret, high-level Washington debate about the H-bomb. And David Lilienthal's anguish about the broader implications of his work reached its greatest intensity. He found that the choice to use science for good or evil was being blurred by political ideology. He himself never succumbed to such influences—he had always been a true "independent"; even at fifty, when he retired from government service, he still maintained an invincible idealism. He never made a decision without afterward reflecting deeply on its effect on those with whom he worked and on mankind in general. Almost nightly, he would review his day's work in a carefully considered personal diary, an extraordinary, self-effacing document in which he noted his doubts and his anxieties. The H-bomb debate would leave Lilienthal in despair; he not only had to face the military hawks and the hardliners in Congress, but also the opposition within his own commission. And by 1949, there was opposition coming from a new quarter—a handful of atomic scientists.

Within the AEC, Lilienthal had a key opponent the Wall Street financier Lewis Strauss. Personally stunned by the Soviet tests, Strauss was quite sure what the response of the commission should be: build the H-bomb. He knew he could not rely on the support of his fellow commissioners, so he looked outside to find friends. He found them soon enough in Senator McMahon, William Borden, and General Kenneth Nichols, who had been Groves's number two in the Manhattan Project and was now commander of the Armed Forces Special Weapons Project.

The atomic physicists had their hawks, too. Leading them was Edward Teller, a Hungarian refugee from Nazism. Teller had been fascinated by the possibility of the thermonuclear device since the early days of the Manhattan Project, and after the war he had continued to conduct feasibility studies of the superweapon. With him was Ernest Lawrence, the patriotic, conservative director of the Radiation Laboratory at Berkeley, and Lawrence's protégé, Luis Alvarez, a young physicist.

Opposing the H-bomb were members of the AEC's General Advisory Committee (GAC), led by Robert Oppenheimer. "We all hope," they reported, "that by one means or another the development of these weapons can be avoided. We are all reluctant to see the United States take the initiative in precipitating this development. We are all agreed that it would be wrong at the present moment to commit ourselves to an all-out effort toward its development."

The kernel of their dissent centered on the effect of the weapon on civilians, the very nub of the pro-bomb argument. They noted there were no limits to the explosive power of the bomb except those imposed by the requirements of delivering it to its target. "It is clear," they added, "the use of this weapon would bring about the destruction of innumerable human lives; it is not a weapon which can be used exclusively for the destruction of materiel installations of military or semi-military purposes. Its use therefore carries much further than the atomic bomb itself the policy of exterminating civilian populations."

Although against the H-bomb, the GAC did not oppose an expansion of atomic weapons. Indeed, they recommended that research should be extended to include smaller tactical atomic weapons—ones that could be used on the battlefield. Oppenheimer, in particular, pushed for this diversified nuclear arsenal.

Despite unsolved technical problems, the hydrogen bomb probably could be built, the GAC advised, perhaps within five years. They favored research into the basic concepts of the thermonuclear process—the ignition of a mass of hydrogen isotopes under exceedingly high temperatures and pressures—but not the final development of a bomb.

The GAC's argument went beyond a moral limitation: if the Russians went ahead with an H-bomb and America did not the GAC felt it would still be possible to defend America with a stockpile of fission weapons. "Reprisals by our large stock of atomic bombs would be comparatively effective," they said. This was a fundamental point: they believed that it was not necessary to develop the superbomb to ensure the security of the United States, and this central fact allowed them to make moral judgments about the horrifying destructive power of the weapon. Finally, their report emphasized that the H-bomb decision offered a real opportunity to open talks with the Russians; talks that could, perhaps, lead to a breakthrough in controlling the nuclear arms race.

Lilienthal fervently hoped that would come true. Even as he was encouraged by the GAC report, however, he was equally depressed by the pro-bomb forces gathering in the other camp. "Reports from Los Angeles and Berkeley [from the scientists Lawrence and Alvarez] are rather awful," he wrote: "[there] is a group of scientists who can only be described as drooling with the prospect [of a positive H-bomb decision] and bloodthirsty."

Lilienthal also was appalled by the simplistic cold warriors on Capitol Hill, led by McMahon. "What he is talking about is the inevitability of war with the Russians," wrote Lilienthal, "and what he says adds up to one thing: blow them off the face of the earth, quick, before they do the same to us."

On November 6, 1949, Lilienthal wrote in his journal:

> I'm really fed up to the ears . . . the H-bomb discussion in the emotional atmosphere following on the President's announcement of the Russian bomb on September 23 confirms my feeling that we

> are all giving far too high a value to atomic weapons, little, big, or biggest; that just as the A-bomb obscured our view and gave a false sense of security, we are in danger of falling into the same error again in discussion of [the H-bomb]—some cheap, easy way out.

A day later, November 7, 1949, Lilienthal offered his resignation to Truman. Truman was reluctant to let him go. Could he not stay on until the end of his term the following June? Lilienthal refused; he would announce the decision at the end of December and stay on a month or two to see in his successor. Before he left the White House, Lilienthal promised a memo about the H-bomb to help the President make a decision. "Yes," said Truman, "I want the facts; of course, it's a policy question." Lilienthal warned that McMahon and his JCAE would "try to put on a blitz to get a quick decision." "I don't blitz easily," Truman assured him with a grin.

Two days later, Lilienthal handed his AEC memorandum to Truman. The commissioners had approved of the GAC recommendation not to build the bomb by three to two; the dissenters were Strauss and a new member of the commission, a partner in McMahon's law firm, Gordon Dean. The memo was not presented as a recommendation; each commissioner had given his own views.

In December, Truman turned the matter over to his three-man advisory committee made up of Secretary of State Dean Acheson, Secretary of Defense Louis Johnson, and Lilienthal. At their first meeting, on Christmas Day, 1949, they reached no agreement, and it was clear that none would be reached. Lilienthal wanted to delay the decision and make time for a diplomatic push with the Russians—a new attempt at international control—but Johnson would countenance no delay. In the end, Acheson, who acted as mediator between the two, did not support Lilienthal, and Johnson's impatience won the day.

Dean Acheson was the most important single architect of all the major American policy initiatives in the Cold War: the Truman Doctrine, the Marshall Plan, NATO, and the National Security Council review of April 1950, which recommended across-the-board rearmament. Acheson firmly advised Truman to intervene in Korea and promoted a Europe-first policy.

Since the war, Acheson had been learning about the "minds in the Kremlin." In 1950, looking back over the postwar years, he said: "It has been hard for us to convince ourselves that human nature is not pretty much the same the world over." By 1950, he saw the differences between East and West as a question of total evil and total virtue; it required what he liked to call "total diplomacy." He had sought to reconstruct the nineteenth-century Pax Britannica under a new Atlantic alliance, seeing the threat to Europe from Russia as similar to the one Islam had posed centuries before. Islam with its combination of ideological zeal and fighting power was like the Russian bear. He refused to understand or even perceive that the Russians themselves might fear the new Pax Americana; to him it was self-evidently benign.

Lilienthal was distraught not to have Acheson's support in the committee. The two men had always had a good relationship: they respected each other. In 1946, they had shared a common objective in drafting the Acheson-Lilienthal plan; in 1949, they still had an instinctive hostility to the primitive nuclear zealot. Both believed that the United States could not rely entirely for its security on the nuclear arsenal. The precise role of the nuclear arsenal had yet to be defined, and, as rational men, both agreed a better analysis was required. But the Secretary of State no longer believed in moral limits on the use of nuclear bombs: the international system in which such idealistic luxuries could be indulged had broken down, probably forever. In the country of the morally blind, Acheson had joined the ranks of the one-eyed kings, and the Russian bomb had simply reinforced that new position.

After the turn of the year, events moved rapidly toward a decision on the H-bomb. On January 13, 1950, the Joint Chiefs sent their critique of the GAC report to Secretary Johnson. The moral objections raised by the scientists were irrelevant and dangerous, and it was folly to argue whether one weapon is more immoral than another. Secretary Johnson made sure a copy of their report got to the White House, and, on January 19, the President told his Central Intelligence chief, Admiral Souers, that the Joint Chiefs' memo "made a lot of sense." The same week, he indicated at a White House staff meeting that he had reached a decision on the H-bomb. What happened from then on was a formality.

On January 27, at his afternoon White House press conference, the President confirmed officially what had been leaked several times in past weeks: yes, there was a debate over a "superbomb"; yes, a decision had to be made whether to produce it or not. The public clamor for the "super" became a roar. The aging Bernard Baruch jumped in with his support. Then Harold Urey, well respected in the American scientific community, added his. Lilienthal became more and more depressed.

The following Tuesday, January 31, 1950, the special three-man committee of Acheson, Johnson, and Lilienthal met with their aides to prepare the final advice for the President. There was little to do. Acheson read the key recommendation: continue the development of all forms of atomic weapons including the hydrogen bomb. When the three men called on Truman shortly after twelve-thirty that day, Acheson opened the discussion by giving a brief summary of the recommendation. Despite the formal recommendation, Lilienthal repeated his concern about U.S. reliance on atomic weapons and the need for that policy to be reexamined. Before he could finish, Truman interrupted: he had decided to go ahead with the H-bomb feasibility studies. The production decision would wait until they had been concluded. He had no real alternatives, Truman said, implying that pressures from the public, the military, and the Congress were now too great for any delay. The meeting took seven minutes.

Within forty-eight hours came the news of the arrest in Britain of the atom spy Klaus Fuchs. Fuchs had worked on the H-bomb at Los Alamos. His treachery produced a new burst of hysterical speculation. Might the Russians have been making bombs since 1943, the year Fuchs went to work on the Manhattan Project? Might the Russians have a weapons production capacity equal to the United States? Might they already have an H-bomb? The decision to build, rather than just research, the H-bomb became inevitable, and Truman gave the go-ahead on March 10.

Even that did not quiet the hawks. Pressure mounted for yet another expansion in production of fissile materials. The McMahon-Borden axis began to campaign with renewed vigor, with Borden supplying the "facts" and McMahon expressing the horrors about the Russians catching up so soon.

Lilienthal had now left the AEC and, without his restraining influence, the AEC joined the Defense Department in suggesting that two more reactors, of the advanced heavy-water type, should be built. Borden was still not satisfied. He wanted nothing less than a second Hanford with three to five of the wartime-type graphite reactors: double the existing strength. Borden put his case uncompromisingly in terms of life and death. "If we act to increase our supply of atomic weapons and they turn out to be unnecessary, we may lose a few hundred million dollars. If we fail to produce these weapons and they do turn out to be necessary, we may lose our country."

When the Korean War broke out in June 1950, Lilienthal's successor as chairman of the AEC, Gordon Dean, had to face an increasing barrage of demanding voices from the JCAE, demanding a further production expansion. One formidable politician who joined the quest for more and more nuclear weapons was Henry ("Scoop") Jackson, then a representative from Washington State. He thought anything short of a doubling of the authorized output of fissionable material would undermine the security of the country. In October, the hawks won as Truman approved not two, but five, new heavy-water reactors.

In the spring of 1951, as the Korean War raged on, McMahon asked for yet another expansion. He requested the AEC prepare cost estimates for increasing production by 50, 100, and 150 percent. The costs of the proposed expansion were in a new league of defense expenditure on a single weapons system. For the 50 percent expansion, it would be about $2.8 billion, with annual operating costs running to about $220 million. For the 150 percent expansion, the costs would be $7 billion and $774 million. But McMahon was hypnotized by what he thought a gargantuan nuclear arsenal could do for America. Never mind the cost. Huge nuclear arsenals would buy "peace power." He thought that if the United States could amass a big enough stockpile to deter the belligerent Kremlin for long enough, it would buy time for the Russian people to rise up against the Communist dictatorship. Only after it was overthrown could there be peace on earth. McMahon wanted to "go all-out in atomic development and production"; the whole of the armed forces should be converted to nuclear energy for everything. That October, news of the second Soviet atomic test gave McMahon's fantastic notion a nuclear boost.

Like Lilienthal before him, Gordon Dean fought to fulfill the mandate he had been given: to keep atomic energy development under civilian control. But McMahon and the army were going to win.

In January 1952, Acheson, Robert Lovett, the new Secretary of Defense, and Dean went to the White House for a discussion with

Truman on the proposed expansion of production facilities. Lovett opened the discussion: The development of new nuclear weaponry, such as artillery shells, had increased the requirement for fissile material. It made good sense to increase production to meet this new challenge. From the economic point of view, fissile material was a cheaper explosive than conventional alternatives. Moreover, if the extra material was never used in bombs or shells it could be used for peaceful purposes. Dean protested: This was hardly a justification for the proposed expansion. But Truman's interest had been aroused. "Is it not true," he asked Dean, "that the nuclear components could be converted to civilian uses?" Dean had to admit they could. It must have been a small comfort to Truman, amid the awesome implications of the proposals before him, but it would not be the last time that a politician found such reassurance in the peaceful atom. Acheson acquiesced in favor of Lovett: If the Russians seemed to be building more bombs, the United States should do so, too. Dean said the AEC was not against the expansion; it merely wanted to make sure it had fulfilled its duty in extracting some justification for it from the military. Little more was said, apparently, before Truman ended the discussion with this remark: "In view of these considerations, does anyone feel we should not undertake this?" There was no response. The President nodded his assent.

In the two and a half years following Joe One, the United States had approved three plans to increase its capacity to produce fissile material, and, by the mid-fifties, it had produced more bomb material than even the Pentagon thought it needed. Indeed, there was so much excess that in 1953 President Eisenhower was able to offer material equivalent to more than 5,000 bombs' worth in support of his plan for "Atoms for Peace."

The age of overkill had been born. The United States—and eventually the Soviet Union—would possess a destructive capability for which it was impossible to find a need. Thirty years later, it seems as if the decisions that led to this excess were made in response to trivial and predictable events that are banal in contrast to the results. Joe One—an event everyone knew would occur sooner or later—should never have caused the panic it did. Klaus Fuchs's espionage should never have been regarded as significant: how the Russians made their bomb was irrelevant. They had it. But, above all, at no time was there any serious attempt to justify the size of the arsenal the production expansion would create. The vision of the one-eyed kings was limited: produce as much as is technically possible; to what end was as understood as it was disregarded. The handful of those who questioned this philosophy, men like David Lilienthal, were simply overwhelmed by consecutive spasms of hysteria. The world has grown so accustomed to the legacy of those years that the fact that real choices existed has been forgotten. The motives of the Kremlin decision-makers is still unknown, but some of the most respected men in the higher echelons of American government understood and articulated the implications of the decisions and the alternatives—in the end to no avail.

PART TWO

The Fifties

7

The Fates and the Furies

PROMETHEUS: *Craft is far weaker than necessity.*
CHORUS: *Who then is the steersman of necessity.*
PROMETHEUS: *The triple formed Fates and the remembering Furies.*
—AESCHYLUS, *Prometheus Bound*

When the Wall Street banker Lewis Strauss took over as chairman of the U.S. Atomic Energy Commission in the summer of 1953, his first task called for precisely the kind of secretive manipulation at which he excelled. It was to carry out the promise of the Eisenhower election campaign to "weed out the Reds" in the federal bureaucracy.

AEC security men soon noticed that the new chairman was demanding to review a wide range of long-closed security files. Without exception, they were files of individuals whose security clearances had been reviewed when Strauss had served on the first AEC under Lilienthal from 1946 to 1949. The reason soon became known: Strauss had apparently made a promise to the FBI director, J. Edgar Hoover, that he would rid the commission of a number of people who had been "bones in Hoover's throat" for a long time.

Once targeted, these "security risks" were given the full benefit of Strauss's manipulative ways. He used his connections in industry, in the foundations, and in the universities to arrange for new job offers for them. Most of them went quietly. Others were confronted with the "facts" of their past in the hope that they would resign. Most did.

The Strauss technique worked in every case but one—by far the most important and, perhaps, the one the FBI director felt most strongly about: J. Robert Oppenheimer. After twelve years of loyal and dedicated service, during which he had built the two atomic bombs that had ended World War II, Oppenheimer was deemed unreliable because of his prewar links with Communists and his security clearance was taken away. Effectively, he was sacked.

The Oppenheimer file was not new to Strauss. In March 1947, as the first AEC was settling into office, Hoover had sent the commission a memorandum on Oppenheimer's past. It contained information about his close friends and relatives: his wife had been previously married to a Communist; his brother, Frank, had been a card-carrying member of the Communist party; he, himself, had contributed to party funds. The commission had been forced to review Oppenheimer's clearance. It was an agonizing decision to make about a man still regarded as the "Father of the A-bomb." The commission, with Strauss concurring, reconfirmed his security clearance. Since then, nothing had changed; there was no new evidence about Oppenheimer's past. Indeed, his record of service to the United States had been exemplary. Only those who doubted the validity of his advice on the H-bomb had complained. There was, however, a considerable change in the political climate. With a large number of Americans convinced that the Russian possession of the A-bomb was due to spies like Klaus Fuchs, any atomic physicist was particularly vulnerable.

In the United States itself, scientists were constantly harried by the security agencies and plagued with appearances before the House Committee on Un-American Activities (HCUA). Robert Oppenheimer's younger brother, Frank, also a physicist, was one of them. He admitted to being a former member of the U.S. Communist party and immediately lost his job as assistant professor of physics at the University of Minnesota. Another physicist, Edward Condon, had worked with Oppenheimer before becoming director of the National Bureau of Standards and a president of the American Association for the Advancement of Science. He had advocated greater cooperation with the Russians and had come under attack from HCUA. In October 1954, Condon was hunted down by Richard Nixon, then Vice-President, and deprived of his security clearance.

In this litany of harassment, the case of Robert Oppenheimer stands out for its viciousness and its tragedy. Not only did the United

States throw away one of its most experienced and sensitive nuclear advisers, but the case spread fear throughout the scientific community. If a man with Oppenheimer's record was not safe, who else would have the courage to express his views freely? The destruction of the intellectual integrity of the AEC—which began with the failure of David Lilienthal to stem the Cold War tide of the later Truman years—was completed with the "show trial" and purge of Oppenheimer. Never again would the AEC display the open-mindedness of its early years. It became an organization preoccupied with bureaucratic infighting and the pursuit of narrow institutional self-interest. It was not until the late 1970s that this poisonous heritage began to dissipate. In the interim, the course of nuclear history—both in the United States and abroad—was dominated by the collective stupidity of the Cold War and McCarthyite legacy.

The fate of Robert Oppenheimer marks the transition. It was sealed on December 3, 1953, when Eisenhower summarily ordered a "blank wall" to be placed between the scientist and any access to sensitive information. In his acceptance of the twisted logic of the day, Eisenhower confided to his diary that the security information on Oppenheimer had been reviewed many times and that "the overall conclusion has always been that there is no evidence that implies disloyalty." However, the President concluded, "That does not mean that he might not be a security risk."

Before the United States government decided that its most famous atomic scientist was an undesirable, they employed him on every committee, military study, or governmental review that he could find time to serve. Few could have faulted the nature of his advice, which in every case supported and expanded the military use of nuclear energy. In 1945 he had chaired the panel that finally disposed of the suggestion that the bomb should be demonstrated in an uninhabited area before being dropped on Hiroshima. He supported the further development of weapons technology, the expansion in fissile material production, and, eventually, he accepted the development of the H-bomb. At one point, Oppenheimer had even given evidence, based on hearsay and innuendo, against scientists then being pursued by the House Committee on Un-American Activities.

Since he first became associated with the U.S. atomic program, Oppenheimer had on many occasions been faced with a choice between his personal doubts and the demands of the military machine.

With the sole exception of the 1949 H-bomb report of the General Advisory Committee, where he supported a unanimous recommendation not to develop the superweapon, Oppenheimer never took a stand against the wider interests of the Pentagon. He had always relished his access to the corridors of power. For all his private anguish, his public record was that of a hawk. Oppenheimer's admirers always expected him to take a stand on principle that would have shown a different history, but he never did: his insistence on examining all sides of the argument—his open mind—made him a pragmatist.

It would, indeed, have taken a closed mind to find a pattern of treason in any of his advice, but there was no shortage of closed minds in the higher reaches of the American government. Few were more dogmatic than the proponents of strategic bombing as the way to win wars. The key to Oppenheimer's coming trials was that he had undermined the future institutional security of the Strategic Air Command. As early as 1949, senior air commanders indicated their doubts about Oppenheimer's enthusiasm for their cause. In the fall of 1952, the Secretary of Defense, Robert Lovett, a long-time strategic bombing fan, became concerned about the strained relations between the air force and the AEC. He asked one of his aides to investigate the matter. The short answer the aide discovered was Robert Oppenheimer. As a widely used consultant to the AEC and to a range of defense committees, Oppenheimer, it appeared, was opposed to everything the air force wanted.

Lovett dug out Oppenheimer's security file. "It's a nightmare," he commented dramatically. "The quicker we get Oppenheimer out of the country, the better off we'll be." He was not the only one who was worried. A number of men had expressed doubts to Lovett about Oppenheimer—including Lewis Strauss.

The air force idealogues, unschooled in the more subtle techniques Strauss employed to liquidate a problem, believed that nothing less than being burned at the stake was good enough for anyone who sought to undermine the strength of the strategic bombing wings. And, in their eyes, Oppenheimer had done exactly that. First, he had advocated the development of "tactical" weapons, ones that could be used on a small scale in limited wars. Inevitably, this meant diverting some of the nuclear material away from the bomber force. Second, he had compounded this "error" by suggesting that part of the air force's role be changed to the defense of American airspace rather than committing all to an offensive position. Finally, he had questioned the desirability of the ultimate strategic weapon, the hydrogen bomb. Two studies in which Oppenheimer was involved put the air force on the alert. The first was Project Vista, started in 1951. The aim of Vista was to study ways of adapting the atomic bomb to tactical battlefield weapons. Bomb tests in the spring of 1951 had shown that it was technically possible to make small bombs.

The report recommended that the U.S. supply of fissionable material, hitherto kept solely for strategic bombs, should be split three ways: one part would remain with the Strategic Air Command, one part would be set aside for smaller tactical weapons for use by the army, and one part would be kept in reserve for the development of new technologies. Many officers made it clear through leaks to journalists that they were unhappy with the report. They referred to it as the work of "long-haired scientists," regarding it as a major blow to the "big bomber" school of thinking that dominated the air force.

The second report, the Lincoln Summer Study, came in 1952. Since the Russian Joe One explosion, increased attention had been given to the air defense of the United States—a combination of a radar early-warning system and fighter-plane interception. To members of the Strategic Air Command, such a concept was unthinkable: it was like hiding behind a "Maginot Line" when the purpose of the exercise was to deal the enemy a knock-out blow. SAC was horrified when a group of scientists, including Oppenheimer, produced an armed services–sponsored report from the Lincoln Laboratory at the Massachusetts Institute of Technology that recommended investing several billion dollars in an elaborate air defense system. The air force began to develop a certain paranoia about Oppenheimer. "I think it is a fair general observation," said David Griggs, the air force's chief scientist, "that when you get involved in a hot enough controversy, it is awfully hard not to question the motives of people who oppose you." Looking back on Oppenheimer's record, especially his opposition to the development of the H-bomb, there were some military men who became convinced that the scientist was attempting to destroy SAC altogether. The air force case against Oppenheimer was intensifying.

In the spring of 1952, Dean Acheson appointed Oppenheimer to head a special Panel of Consultants on Disarmament. The team included Vannevar Bush and the CIA's Allen Dulles. The secretary to the group was a young Harvard academic, McGeorge Bundy. The panel's conclusions would make Oppenheimer no friends among the nuclear hawks.

The panel quickly resolved that it had nothing hopeful to say about disarmament. As the U.S. stockpile of bombs climbed into the thousands, the Russians would not be far behind. "The time when the Russians will have material to make one thousand atomic bombs may well be only a few years away and the time when they have enough to make five thousand only a few years further on." Dismally, the panel noted that, with the huge increases in the production of fissile material, there would never again be a possibility of verifying a nuclear disarmament plan. No inspection of a country's nuclear facilities could ever make sure that all the fissile material produced would be accurately counted in a manner ensuring none had been held back. In any case, what mattered now was not the number of weapons in the stockpile but the accuracy with which they were delivered. Missiles, the panel concluded, would soon become as important as bombers, hardly something SAC wanted to hear.

It was a crucial turning point in the history of nuclear weapons. Never again would there be a serious proposal for nuclear disarmament. From this time forward, all talk of nuclear arms regulation would focus on delivery systems—as in the Strategic Arms Limitation Talks (SALT). Efforts would be made to limit the testing of atomic weapons and the spread of them to other countries, but the prospect of achieving the abolition of nuclear weapons—along the lines of the abortive Acheson-Lilienthal plan of 1946—disappeared, probably forever. The panel was forced to report that "it would not be useful to attempt a new and modernized version" of the old plan. The most that could be hoped for was some sort of "strange stability" in which both sides believed it was suicidal to "throw the switch."

The major conclusion offered by the panel was not a specific measure of arms control at all, but a simple call for the government of the United States to become more candid with its own people about the dangerous implications of the new military situation. In the past, official secrecy about Russian capabilities had served its purpose in avoiding unnecessary fears. Now, said the report, "the present danger was not of hysteria but of complacency." The U.S. government should provide precise facts and figures about the escalation in weapons stockpiles on both sides and emphasize the fact that "no matter how many bombs we may be making the Soviet Union may fairly soon have enough to threaten the destruction of our whole society." The report specifically attacked the continued secrecy about the size of America's own stockpile.

This section bore Oppenheimer's personal stamp. At a time when he could see no hope, he remained convinced that a solution could emerge only from open, free inquiry and debate. Without candor about the existing dilemma, arms control could not work. The panel's study soon became known as the "Candor Report." It was one of the first national security memoranda to land on President Eisenhower's desk after his inauguration in January 1953, and it quickly became apparent that most of its suggestions were unacceptable to the new administration.

To the man who was about to take a central role in Oppenheimer's downfall, Lewis Strauss, the panel's suggestion of telling the Russians the size of the U.S. stockpile amounted to only one thing: saving the Russians trouble in their espionage activities. It was utterly preposterous: typical of Oppenheimer, a man he had grown to loathe. A few days after Strauss had taken up his post as AEC chairman, a friend of Oppenheimer's received a call from a former colleague at the AEC: "You'd better tell your friend Oppie to batten down the hatches and prepare for some stormy weather."

Many people, especially scientists, were to blame Strauss for Oppenheimer's downfall. Certainly, throughout the affair Strauss was the man in control. He knew that FBI director Hoover and the air force were out to get Oppenheimer. He knew from past experience on the Lilienthal AEC that he did not agree with Oppenheimer on many things. The solution should have been a simple one: at the time Strauss took over the AEC, Oppenheimer was retained only as a consultant. He had already lost much of his former influence. At the first signs that the hawks were not going to give up their witch-hunt until Oppenheimer was publicly expelled, Strauss could have simply canceled his consultancy contract. Instead, he subjected Oppenheimer to a long and terrible ordeal. After it was all over, Strauss remarked, "A tragic thing—I shall have to live with it as long as I live." Even at this distance in time, it is a remark that rings hollow.

When the two first met in the summer of 1945, Strauss was enormously impressed with Oppenheimer, seeing in him a man with an extraordinary mind and a compelling personality. Many felt like that about Oppenheimer. The clarity and precision of his thought process, the elegance of his speech—when you could hear him, for he mumbled a lot—had a magnetic fascination. Some people would say they were spellbound by him.

He was a man of awesome breadth and diversity of interest. From his accomplishments as a cook to his mastery of many languages, including Sanskrit, Oppenheimer appeared to be the quintessential intellectual. Whether dealing with humanity, science, or art, he was never content with only a glimmer of understanding. He searched for total knowledge as if that were still possible in the twentieth century. Oppenheimer always believed in relying on example and persuasion. It was the basis of scientific progress, and he transported it into a general theory of politics. In that, he and Strauss did not speak the same language; in fact, it is hard to see how they could have been more different.

Strauss, a self-made man, had grown up in Richmond, Virginia, where his father worked with his uncle in a small shoe-jobbers' firm. The family had no wealth. They were deeply religious Jews. In 1917, Strauss left home to work for Herbert Hoover in Washington. He went on to make his first real money in the Wall Street banking firm of Kuhn, Loeb. Through it all, he was the epitome of conservatism.

Oppenheimer, on the other hand, was born into wealth. He grew up in New York, where his father was a successful importer of men's suit linings. His parents were also Jews, but he was not brought up in the faith. By the time Oppenheimer was ready to go to Harvard, the family's New York apartment contained three Van Goghs, a Picasso, and a Renoir. He graduated in physics, summa cum laude, and went on to study at Cambridge, England, and Göttingen, Germany. In 1936, he was made professor of physics at the California Institute of Technology. It was a comfortable, privileged childhood and a brilliant, successful adolescence. At an early age, he had achieved all that he wanted as a physicist except recognition for original work.

Of his early years, Oppenheimer said later: "I was an unctuous, repulsively good little boy. My childhood did not prepare me for the fact that the world is full of cruel, bitter things. It gave me no normal healthy way to be a bastard." It was different for Strauss, at first. He could not have climbed his way swiftly through the ranks of Kuhn, Loeb to be made a partner with a million-dollar salary at the age of thirty-three without crushing a few competitors underfoot. He was said to be extremely rough in battle.

While Strauss was amassing a fortune in New York and winning the hand of the boss's daughter, Oppenheimer was embracing left-wing causes on California campuses and courting the widow of a member of the International Brigade who had been killed in action in Spain in 1937. She was Katherine Puening, and he married her in 1940.

Oppenheimer was a lanky six-footer who never weighed more than 130 pounds in his life. He walked with an odd gait and uncontrolled limbs, looking sometimes like an overgrown choirboy. Strauss was portly, small, dynamic, and dapper, with a penchant for well-tailored suits. He was once described as a "well-dressed owl."

When the war came, Oppenheimer joined the Manhattan Project and Strauss joined the navy. They were both fortunate in their military bosses. Oppenheimer's past links with the Communist party would have excluded him from bomb work if General Groves had not stepped in and protected him. The stern, matter-of-fact, practical army officer was much taken by this cultured genius who was able to explain in words of one syllable the basics of nuclear physics without ever talking down to him. Groves simply overruled the objections of the security agencies.

Strauss was given a desk job in the Bureau of Ordnance in Washington, D.C. In this position, he ran into the Navy Secretary, James Forrestal, whom he had known on Wall Street, and Forrestal quickly made him his personal assistant. It meant rapid promotion.

Both men ended their wartime service with distinction. Oppenheimer was showered with laurels, much greater recognition than he would have had being just another Nobel Prize–winner. For a time at least, he was a household name. Rear Admiral Strauss, as he had become, achieved nothing so grand as making the world's first atomic bomb, but, in his own way, the brash financier was as proud of his admiral's stripes as Oppenheimer was of his bomb. When he returned to civilian life, Strauss always liked to be called "the Admiral"—it added that certain respectability money could not buy. He also liked his name to be pronounced "Straws," some said to disguise his German-Jewish descent.

When he entered atomic energy as a member of the first AEC, Strauss was never troubled by the feelings of doubt that many of the atomic scientists, including Oppenheimer, felt about the bomb. "The atom is amoral," Strauss said; "the only thing that makes it immoral is man." Strauss never thought of nuclear physics as anything but progress. "No, I would not wipe out any part of it, not the bomb nor any part of it if I could," he once reflected. "I believe everything man discovers, however he discovers it, is welcome and good for his future. In me this is the sort of belief that people go to the stake for." The atomic bomb was a perfect weapon to cope with the Russians. Strauss believed that the nuclear stockpile was a fundamental necessity of life to "discourage the use of bombs against us by a government that doesn't make any pretense of morals." He despised the peace movement. "I have noticed that heaven helps those who help themselves," he said. "Somebody else may think I am all wrong. Gandhi would—you lie down in front of a juggernaut. There may have been people in antiquity who adopted Gandhi's position against the Huns or the Tartars—but history doesn't retain any record of them."

Oppenheimer, on the other hand, frequently hinted at his personal doubts. "The physicists have known sin," he said after Hiroshima. "I feel we have blood on our hands," he told President Truman. "The atomic bomb is shit," he once told Leo Szilard. "This is a weapon which has no military significance. It will make a big bang—a very big bang—but it is not a weapon which is useful in war." Oppenheimer always recognized that the bigger the bombs became, the greater their capacity would be to kill civilization. In the end, the moral issue would be paramount.

Yet it was not Oppenheimer's moral objections to big bombs that gave his enemies the necessary evidence needed to convict him. It was his connection with Communism and, in particular, an apparently trivial incident that had occurred at Berkeley, just before he took up the directorship of the Los Alamos bomb factory.

He had been approached in 1943 by his friend Haakon Chevalier, a professor of French at Berkeley and an ardent left-winger. Chevalier had told Oppenheimer he had an unnamed contact who was in touch with the Soviet vice-consul in San Francisco. The vice-consul was seeking information about the work of the Berkeley Radiation Laboratory. Chevalier, in his memoirs, says he did not ask Oppenheimer for technical secrets; he was simply passing on a plea from the vice-consul for better scientific cooperation between Russia and

America. Testimony before the House Committee on Un-American Activities suggested that he had asked for specific information. Whichever it was, Oppenheimer turned him down without a second thought, but he failed to report the incident to Groves or the Manhattan Project security men for several months. Then he only mentioned the name of the contact, thereby seeking, it seemed, to protect his friend Chevalier. Groves did not set much store by the event but told Oppenheimer that he should tell everything he knew. Pushed by the security officers to name names, he did something that was to wreck his career. In an act of what can best be described as carefree arrogance, he made up a fantasy tale of spies and contact men. Chevalier's contact, whom he had already named earlier, had, in fact, approached three people, he told the security officers, but he declined to identify any of them. He said the contact had asked for information on microfilm and he gave the security agents the general impression of the existence of a full-scale spy ring. Oppenheimer had, apparently, made up the whole story. When Groves insisted that he name the persons involved, Oppenheimer could only come up with the identity of his friend Chevalier.

Oppenheimer's behavior had shown, said General Groves, "the typical American schoolboy attitude that there is something wicked about telling on a friend." In the scientist's own defense, Groves emphasized that Oppenheimer had at least done his duty in relating the incident, and the affair was forgotten—until after the war.

When, in 1953, he reexamined Oppenheimer's file, Strauss bore a special grudge against him. Over the eight years of their acquaintance, the fundamental differences between the two men had come out. When Strauss was on the Lilienthal AEC, he had clashed on more than one occasion with the scientist, but there had been one most bitter duel. It was over a proposal to export radioactive isotopes for medical purposes. The all-embracing isolating effect of the McMahon legislation excluded giving away *any* of the atom "secrets," including information that could be used in a nonweapons application. Strauss fully supported the letter of the act, even suggesting the isotopes might be diverted to military use.

When the proposal to export small quantities of isotopes came before the five AEC commissioners, four of them voted for it. Only Strauss was against. The exports went ahead. They did so mainly because of Oppenheimer's advice that the isotopes were not worth a jot militarily. Strauss was furious that he had lost. Two years later, in 1949, when the hardliners on the Congressional Joint Committee forced hearings on what they called the "incredible mismanagement" of the Lilienthal regime, Oppenheimer was called to give testimony to support the AEC's decision. The dispute with Strauss came out into the open, with Oppenheimer disdainfully dismissing what he saw as the silly fears of Lewis Strauss that the isotopes might be used in bombs.

"No one," he told the Joint Committee, "can force me to say that you cannot use these isotopes for atomic energy. You can use a shovel for atomic energy; in fact, you do. You can use a bottle of beer for atomic energy. In fact, you do. But to get some perspective, the fact is that during the war and after the war these materials have played no significant part, and in my knowledge, no part at all." All the commissioners of the AEC were present, of course. One observer recalled: "Strauss' eyes narrowed; the muscles of his jaws began to work visibly; color rose in his face."

Oppenheimer continued: "My own rating of the importance of isotopes in this broad sense is that they are far less important than electronic devices, but far more important than, let us say, vitamins, somewhere in between." There was laughter. When the hearing was over, Oppenheimer turned to the AEC counsel and asked how he had done. "Too well, Robert, much too well," came the reply. Gordon Dean later recalled "the terrible look" on Lewis Strauss's face. Lilienthal would note in his diary that Oppenheimer had been too harsh with the powerful Strauss for his own good; it was as though he did not know the kind of man he was dealing with. "If he did [he was] just too courageous to give a damn."

The clash of personalities had been inevitable. Both were intellectually arrogant men, but in atomic matters Oppenheimer had a lot more to be arrogant about. He was not prepared to suffer what he saw as the stupid prejudices of Lewis Strauss with anything less than total disdain. At the same time, Strauss was not prepared to tolerate disagreement from his colleagues. As one of the commissioners said of him: "If you disagree with Lewis about anything, he assumes you're just a fool at first. But if you go on disagreeing with him, he concludes you must be a traitor."

In the end, it was not Lewis Strauss who actually charged Oppenheimer with betrayal; that was left to William Borden, the staff director of the JCAE. On November 7, 1953, Borden, dredging up Oppenheimer's past yet again, sent a letter to Hoover at the FBI. It contained the damning accusation that "more probably than not he [Oppenheimer] has been functioning as an espionage agent" and "more probably than not he has acted under a Soviet directive in influencing United States military, atomic energy, intelligence and diplomatic policy."

Hoover immediately set some agents to work compiling a new dossier on Oppenheimer. He sent one copy each to the President, Lewis Strauss, and the Secretary of Defense, Charles Wilson. Wilson was a former president of General Motors. "When that simple valve-in-the-head character got the letter," a presidential speechwriter, C. D. Jackson, wrote later, "he practically exploded with terror, clapped on his hat, and rushed to the White House to see Eisenhower, clamoring for action." Wilson wanted all defense installations notified immediately that Oppenheimer's clearances were to be withdrawn, and up went Eisenhower's "blank wall" between the scientist and the nation's secrets.

Oppenheimer was sent a list of charges by the AEC and given the opportunity of resigning or facing a board of inquiry. He refused to resign; as far as he was concerned, there was nothing, apart from the Chevalier incident, that could be used against him. But it was the absurd Chevalier affair that became the nub of the case. The AEC played up the fact that Oppenheimer had met Chevalier again during the summer of 1953, while on vacation in France. Here, they suggested, was evidence of the dangerous continuing relationship between the two men. They also made Oppenheimer admit to producing a "tissue of lies" over the incident when he had first informed the authorities about it. Strauss noted triumphantly that Oppenheimer had "cracked."

Oppenheimer's defense was his unworldliness. As to his past associations with Communists, he said: "I am not defending the wisdom of these views. I think they are idiotic." On the Chevalier incident itself, he said again: "I was an idiot."

His enemies in the Pentagon would probably have accepted his "apology," but it was not the Chevalier incident, of course, that troubled them. It was the advice Oppenheimer had been giving on the superbomb, on the diversion of fissile material away from the Strategic Air Command; it was his emphasis on the development of tactical weapons and an air defense system. Borden's letter had unashamedly used his past links with Communism to suggest that Oppenheimer's motives in giving this kind of advice on security matters was to further the interests of the Soviet Union. It was precisely the kind of calumny that typified the witch-hunts: there was no evidence for it.

Many of his scientist friends appeared at the hearing to give supportive testimony, only a few gave unfavorable evidence. The most significant of those was Edward Teller, who had more of his prestige invested in the development of the H-bomb than any of the others. The testimony he gave impugned Oppenheimer not for being unpatriotic, or disloyal, but for being too "confused and complicated" to be trustworthy and lacking the "wisdom and judgement" needed in a person so central to the future security of the United States. He deliberately appealed to the maturity and good sense of the three-man board of inquiry, while emphasizing, without actually having to say so, the obvious immaturity of Oppenheimer that the previous testimony had so clearly shown. It was a masterful courtroom stroke, more reminiscent of criminal-trial tactics than a loyalty hearing. The words were expertly chosen. "In a great number of cases," said Teller, "I have seen Dr. Oppenheimer act—I understand that Dr. Oppenheimer acted—in a way which for me was exceedingly hard to understand. I thoroughly disagreed with him in numerous issues and his actions frankly appeared to me confused and complicated. To this extent I feel that I would like to see the vital interests of this country in hands which I understand better, and therefore trust more. In this very limited sense I would like to express a feeling that I would personally feel more secure if public matters would rest in other hands."

Oppenheimer's inquisitors voted two to one against him. He was a victim of McCarthyism, but he was also a dupe of his own liberalism, a man whose deepest wounds were self-inflicted. In the moment of postwar victory, he had reveled in the glitter, the admiration, and the uncomplicated and enduring force of popularity based on achievement. He found the lure of the power center and its courtiers irresistible. Yet he was hopelessly ill-equipped, as he himself readily admitted, to deal with the deviousness of the ambitious Lewis Strauss or to handle the rough confrontations of the self-interested leaders of the Strategic Air Command. Oppenheimer's friends noted: "Washington's influence upon him was greater than his influence on Washington." His exclusion from public life was a personal blow he did not take lightly. Hans Bethe, a German émigré physicist who worked on the Manhattan Project, recalls: "He was a changed person and much of his previous spirit and liveliness had left him."

While Strauss was watching Oppenheimer being destroyed with one eye, his other was concentrating on Eisenhower's persistent desire to create a major policy statement out of the scientists' Candor Report. Since the report had come out, early in 1953, Eisenhower's speechwriter and propagandist C. D. Jackson had been tinkering with its cardinal suggestion: that a candid assessment of the dangers inherent in the nuclear arms race should be given to the American people. Jackson had produced a few hamfisted scaremongering drafts of a speech, which Eisenhower had rejected with the comment: "We don't want to scare the country to death."

In July, Jackson produced a draft he thought would cover the presidential objection. The speech linked the appalling devastation that would result from a Soviet atomic attack on the United States and an even bigger U.S. counterattack on the Soviet Union. Jackson dubbed this latest effort the "Bang! Bang!" papers. Eisenhower was still not satisfied. "This leaves everybody dead on both sides, with no hope anywhere," he commented. "Can't we find some hope?"

The next month, August 1953, the Russians exploded their first H-bomb. Eisenhower's search for that dwindling hope took on a greater urgency. He had an idea—a single step toward disarmament. Would it not be possible, the President asked hesitantly of Strauss and Jackson, for the weapons powers to hand over some fissile material to an "atomic pool" from which it could then be distributed for peaceful uses? "The amount," the President suggested, "could be fixed at a figure which we could handle from our stockpile but which would be difficult for the Soviets to match."

Strauss and Jackson, both seasoned politicians, seized on the suggestion—it sounded like excellent propaganda. But they doubted it could become a serious implement of disarmament, or even arms limitation, because it had some built-in faults. For one thing, U.S. intelligence of the size of the Russian stockpile and of the production rates of fissile material and bombs was abysmally poor. Russia might be producing more of both, for all the Americans knew. And if that was the case, equal donations of fissile material to the "atomic pool" would eventually put the United States at a disadvantage. For another, the relative importance of a stockpile of fissionable material had been greatly diminished by the successful explosion, by both sides, of the H-bomb, in which fissionable material is used only to trigger the fusion explosion.

On October 3, 1953, Eisenhower, Strauss, Jackson, and top officials from the State Department and the military met over breakfast at the White House to discuss the latest ideas for the speech. The plan for an "atomic pool" gave the old Candor Report project a new direction—the element of hope that Eisenhower had been looking for. The breakfast timing of the meeting gave the project a new name: Operation Wheaties.

What Eisenhower sought was a modest proposal; what he was about to achieve was a Cold War propaganda triumph. "Atoms for Peace" was swiftly turned into a "buzz phrase," in the best traditions of Madison Avenue, firing the imaginations of leaders of nations seeking the magic of the atom. America, eager to capitalize on this public relations success, unlocked the secrets of nuclear technology and, in an expansive gesture, offered them free to anyone interested in pursuing the atom's power to make peaceful things like electricity. The aftermath of the speech saw a worldwide spread of nuclear materials and know-how. The original aim, the one Eisenhower had sought, a first step to disarmament, was quickly, and conveniently, forgotten. More important, the long-term effects of the spread of atomic knowledge and of giving other countries the materials to build a bomb were simply ignored. It was as if America had never considered the basic problem of what is now called "nuclear proliferation"; yet the link between the peaceful and the warlike atom had been the central point of the Acheson-Lilienthal plan for international control, written (mostly by Oppenheimer) seven years before.

In the speech, finely crafted by Jackson, these fearful implications were lost in a cocoon of stirring rhetoric. From the day of the Wheaties breakfast, C. D. Jackson eagerly pursued the public relations angle of Eisenhower's plan. His agenda for the breakfast meeting showed how overwhelmed he was by it. "This could not only be the most important pronouncement ever made by any President of the United States, it could also save mankind."

On November 3, the first of eleven drafts of the speech, due for delivery at the United Nations on December 8, came out of Jackson's office. The new international authority, which would hold the "atomic pool," would supervise the withdrawal of the fissionable material and allocate it to power nuclear reactors that would be built throughout the world, "to help people in the underdeveloped areas of the world where the lack of power keeps them in primitive bondage." Jackson was warming to his theme. By the fourth draft he had the "atomic pool" making the "deserts flourish," and helping "to warm the cold, to feed the hungry, to alleviate the misery of the world."

Eisenhower delivered the speech to a silent audience of three thousand men and women in the U.N. General Assembly on December 8, 1953. The first half was devoted to a candid assessment of the stockpile, including a number of horror facts. Any single one of the air wings of the U.S. Strategic Air Command could deliver in one operation atomic bombs with an explosive equivalent greater than all the bombs that fell on Germany through all the years of World War II. Any one of the aircraft carriers of the U.S. Navy could deliver in one day atomic bombs exceeding the explosive equivalent of all the bombs and rockets dropped by Germany on the United Kingdom through all the years of World War II. The second half of the speech was devoted to the "hope" for the creation of an "atomic pool." "It is not enough to take this weapon out of the hands of the soldiers. It must be put into the hands of those who will know how to strip its military casing and adapt it to the arts of peace," declared the President.

In a frenzy of public relations activity following the speech, Jackson produced a pamphlet containing the speech in ten languages. Two weeks after delivery, over 200,000 copies had been sent out by U.S. firms in their foreign mail; 350 U.S. foreign-language newspapers and related ethnic organizations started a campaign to ensure that excerpts of the speech were sent by immigrants to their relatives and friends abroad. Newspapers and radio and TV stations were deluged with briefings, magazine articles, and special advertising. The United States Information Agency distributed the speech and an endless stream of follow-up feature stories. Voice of America radio recordings were given to newspapers and radio stations throughout the world. The sphere of the peaceful atom expanded rapidly as the American broadcasts produced talks entitled "Nuclear Device in Fight Against Cancer," "Forestry Expert Predicts Atomic Rays Will Cut Lumber Instead of Saws," and "Atomic Locomotive Designed." A range of new promotional films was conceived: *Atomic Green-*

house, Atomic Zoo, Atom for the Doctor. By February 4, 1954, Jackson's operation reported considerable success. "The President's speech . . . by focusing attention on the prospects for peaceful development of atomic energy, initially placed the U.S.S.R. in a defensive position. The speech captured the hopes of the common man and the interest of the scientific and intellectual classes."

The Russians, predictably, accused the Americans of ignoring the key question of banning the bomb—a stance they had taken since the end of the war. There was no official Kremlin reply to the speech: Moscow radio said the President had failed, once again, to make clear the U.S. position on banning the bomb and had simply attempted to scare the world with threats of atomic war. But the world believed Eisenhower. It wanted to believe him. The eloquent speech was hailed as a "moving argument for peace."

In fact, of course, the "atomic pool" was even less of a viable step to disarmament than the old Baruch plan. It did not even address the problem of one superpower verifying the other's atomic stockpile with inspections; everyone agreed that those inspections were impossible. In the end, the United States became a prisoner of its own propaganda triumph: the "atomic pool" never materialized, and America was left having to promote the peaceful atom on its own by hastily writing a series of bilateral agreements to transfer nuclear technology and fissile material and to set up a new International Atomic Energy Agency to promote nuclear power. But the public relations diversion of "Atoms for Peace" could not have come at a better time for the Eisenhower administration: it had just made the important policy decision to rely exclusively on nuclear bombs for U.S. defense. It was called the New Look.

8

The Power of "Les X"

Frédéric Joliot-Curie spent a lonely and frustrating war in occupied France. Deprived of his co-workers, Hans von Halban and Lew Kowarski, who had helped him to the forefront of atomic research in 1939, he waited eagerly for the war to end and an opportunity to return to his work. The prospects for the revival of French science looked bleak, however. France possessed none of the special ingredients needed for a nuclear project. The Anglo-American monopoly of uranium had cut off France's access to the valuable Belgian Congo reserves. Joliot's precious supplies of heavy water had been taken first to England and then to Canada. The French scientists who had worked on the Manhattan Project were having problems getting home; the security-conscious General Groves was reluctant to let them go, fearing that whatever secrets they had learned would be all over Paris in a day and, given Joliot's Communist ties, in Moscow by nightfall.

Even when the French scientists were eventually reunited with Joliot, their usefulness proved limited. They had worked in Canada on the heavy-water reactor project and they had no more than theoretical knowledge of the huge uranium-separation plants. Finally, there was nowhere for them to work; the war had left France without laboratories and without such basic laboratory equipment as glassware and machine tools.

Charles de Gaulle's postwar provisional government was as generous as it could be. Within two months of Hiroshima, it had set up a state monopoly French atomic energy commission, the Commissariat à l'Energie Atomique (CEA), but a shortage of funds and resources set the scale of its operations at about one-tenth the size of Britain's nuclear project and one-hundredth the size of the U.S. program.

Joliot was the effective head of the CEA and, despite the incredible obstacles, he was determined to succeed. With some ingenious scavenging of the American war surplus camps in Europe, a few of his energetic henchmen managed to liberate supplies of radio components; they also looted machine tools from the French zone of occupied Germany.

On the outskirts of Paris, in a fortified compound called the Fort de Chatillon, a dark, dank structure where the Germans had stored explosives during the occupation and where Nazi collaborators were imprisoned and executed after the liberation, Joliot started to build a research reactor. He hired young geology students and old Resistance fighters armed with Geiger counters to roam the French empire, especially Africa, in search of uranium.

At the start, there was no mention of making a bomb; Joliot professed interest only in peaceful nuclear power and tried in vain to bring in scientists from countries other than the United States to help develop it. In June 1946, the French declared their official policy at the United Nations: "The goals which the French government has assigned for the research of its scientists and technicians are purely peaceful. It is our hope that all the nations of the world may do the same as swiftly as possible." Fourteen years later, in 1960, in the Sahara, France would proudly explode her first atomic bomb. It had been an entirely covert operation, just like the British bomb project.

In those early days, it was easy for Joliot to make bold public statements about not building bombs, but the reality of the French project was far more complicated. From the start, the CEA's program included a number of reactors similar to the ones built by the Americans at Hanford for the Manhattan Project. All had the capacity to produce bomb-grade plutonium, and within five years—certainly by the mid-fifties—France, in theory, would have accumulated enough plutonium to make a bomb. If Joliot appeared reluctant to admit this to himself, his colleagues certainly realized it. His co-worker Kowarski noted later: "There was a certain ambiguity in his [Joliot's] statement about what the aim of this [plutonium-producing reactor] would be."

As the project grew, the Americans were to become increasingly concerned that the French project was in the hands of a Communist, but in 1945, they were still bathing in self-congratulatory glory over their own achievement and did not pay much attention to the meager activities in the Fort de Chatillon. On a trip to the United States in 1946, Joliot met the flamboyant Bernard Baruch, who told the Frenchman that it was "madness to try atomic energy in France. A pile—two piles—you'll never get them [given] the state your country is in."

As the Cold War intensified, however, attitudes changed. In 1948, when it became clear that the French would soon be producing significant amounts of plutonium, the Americans began raising objections to Joliot's leadership of the project. On a lecture tour to America, his wife, Irène, was arrested by U.S. immigration authorities and detained for a day. *Time* magazine commented: "All communists in a democracy are potential spies and traitors." The U.S. State Department even asked the British to try to remove Joliot with a direct approach to the French premier. But the British refused to have anything to do with such a plan: it was out of the question to dislodge Joliot—he was a famous Frenchman, a Nobel Prize–winner, the son-in-law of Marie Curie, and a Resistance hero. In the end, Joliot himself sealed his own fate.

Sooner or later, he would have to make the choice between his political commitment to the Communist party and his deeply felt desire to lead France's postwar scientific revival. The moment the first gram of plutonium was produced by his "peaceful" reactors, conservative forces within France would want a bomb; indeed, the majority of the French people, according to a poll taken in 1946, wanted France to build one. In a very personal way, Joliot was caught by the basic dilemma about atomic energy: the inseparable link binding the peaceful and the warlike atom permitted no simple distinction between the two.

The turning point came in late November 1949, when Joliot's team extracted the first tiny amounts of plutonium. The timing could not have been worse for Joliot. A few months earlier, in August 1949, the Russians had exploded their first bomb. America, in a state of panic about its own nuclear readiness, was combining with other countries in Western Europe, including France, to form the North Atlantic Treaty Organization. The French Communists were opposing their government at every turn: Communist-led dockers refused to load ships with arms for the war in Indochina, shipments were sabotaged by railroad workers, and the constant theme of party speeches was peace and nuclear disarmament. Joliot moved closer and closer to the Communist party line and, as Kowarski put it, "was carried away before our eyes." In March 1950, in Stockholm, Joliot was the first to sign the Stockholm appeal, a call for the banning of nuclear weapons. Eventually signed by millions around the world, it was used by the French Communists as the central item of their ban-the-bomb rhetoric.

A month later, Joliot took the final step to self-martyrdom at a party rally. "The French people," he declared, "do not and never shall make war against the Soviet Union. . . . Progressive scientists and communist scientists shall not give a jot of their science [for this purpose]." Joliot told his colleagues: "If the government doesn't fire me after what I've said, I don't know what more they need." Three weeks later, French Prime Minister Georges Bidault, responding to pressure from both right-wing and moderate deputies in the National Assembly to rid the nation of its most famous scientist, dismissed Joliot from government service.

Joliot, the man who had led the team that had first confirmed the basic principle of the chain reaction, was cut off from his life's work. In professional exile, he continued to support French science and lectured all over the country, especially about disarmament. He died in 1958, the year of de Gaulle's triumphant return to power. Much to the disgust of the Gaullists, the general gave Joliot a state funeral. In death, the glory Joliot had brought to French science transcended politics.

The doubts about Joliot's original intentions—what he really thought would happen to the plutonium produced in his reactors—lingered. Twenty years later, they were brought into sharp focus when a French cabinet minister claimed that de Gaulle had once told him Joliot had agreed to build a bomb for France even before the setting up of the CEA in 1946. If that is true, Joliot worked for four years in the vain hope that the pro-bomb forces in France would somehow disappear. Nothing could have been further from reality.

As the pro-bomb lobby grew, the French physicists would go through the same agonies about the bomb as their American colleagues. Not only would they have their victim of McCarthyism in Joliot, they would also have their work taken out of their hands as completely as General Groves had snatched control of the Manhattan Project. After Joliot's departure, anti-Communist politicians in the French National Assembly demanded more blood from the ranks of the CEA. The commissariat's budget was cut substantially and a number of Joliot's leftist colleagues were fired—officially, on the grounds of economy. The influence of the scientists began to wane rapidly, and the engineer-administrators took over: a repeat of what had happened in the Manhattan Project, but with a special French flavor.

The takeover was made not by the army, but by the most prestigious of a group of semifeudal clans, or corps as the French call them, which make up the country's industrial and political elite. It is the Corps des Mines, a freemasonry of power originally organized for mining specialists. It still operates under a charter approved by Napoleon and consists of the best graduates of the Ecole Polytechnique, a conservative military-style academy for engineers. They are not army officers like the commanders of the U.S. Corps of Engineers, and they are politically more important than the Red Specialists of Russia. Their takeover reinforced the rightward political drift of the CEA after the purges and paved the way for the key decision to build a French bomb.

The Ecole Polytechnique had always turned out formidable graduates. Established after the Revolution, it was militarized by Napoleon to produce engineers for the French Army. It has never lost its military orientation and to this day reports to the Ministry of Defense rather than the Ministry of Education. Students live, barrack-style, in the heart of Paris's Latin Quarter, wear their uniforms with pride, participate in military ceremonies, and have their day governed by bugle calls: when to get up, when to eat, when to go to bed. Graduates are universally known as "Les X" after the crossed-cannon symbol of the school.

The molding of this self-consciously superior sect begins with a fiercely competitive nationwide entrance examination. New students are greeted with initiation ceremonies and rites invoking the full panoply of physical and personal humiliation—the well-known hallmarks of authoritarian education systems that claim to build strong moral character. Freshmen are called conscripts—the ultimate military term of abuse. The curriculum emphasizes science, logic, and mathematics. The graduates are treated as specialists, or experts, capable of discovering the correct formula for the most vexing problem. An enterprising Polytechnician once devised a mathematical formula for the existence of God. "They emerge," Marshal Pétain once said of the Polytechnicians, "knowing everything, but knowing nothing else." The passionate loyalty among members and a strong alumni association result in easy access to some of the most powerful men in the land. The shared sense of superiority, instilled by the acute competitiveness of entry and training, stretches across generations.

Many of them develop a distinctively French technocratic ideology: a belief in efficiency and progress, with strong emphasis on the role of technology. Combined with their pragmatism, this makes most Polytechnicians impatient with the "inefficiencies" of politics. They do not serve sectional interests, but the general interest, intangibly represented by the French concept of "L'Etat." The sphere of atomic energy and the rights and privileges of the CEA would serve as a fertile ground for putting such an ideology into practice.

The Corps des Mines is the superelite of this group, open only to the top ten graduates of the Ecole Polytechnique. Over the years, they have monopolized access to a key range of positions in major sectors of the government and, increasingly, top positions in private industry. In 1951 the Corps des Mines annexed the French atomic energy commission. Their leader in this operation was Pierre Guillaumat, the young, aggressive son of a famous World War I general. It was a perfect opportunity for Guillaumat to fulfill his ambition of reversing the decline and fall of France that he had witnessed in his formative years. As a staunch patriot, he longed for the return of the days of "La Gloire," as French national rhetoric had dubbed them, of World War I, in which his father had played such a leading role as a military commander. Such illusions would be revived only when French patriotism entered its second childhood with the return of de Gaulle in 1958, but two components, atomic energy and the bomb, would bloom on soil well prepared by Pierre Guillaumat.

Guillaumat could not inherit his father's reputation and, because there was no money in the family, the best access to wealth and power was through the Ecole Polytechnique. He grew up in uniform as a pupil at a bleak military high school and went on to graduate tenth in his Polytechnique class, thus allowing him to choose the Corps des Mines. He served briefly in the government's mining department in the French colonies in Indochina and Tunisia. When de Gaulle returned to lead the Provisional Government after liberation in 1944, Guillaumat, then thirty-five, convinced de Gaulle that he should be given the key position of Directeur des Carburants (Director of Fuels) in the reconstruction of French industry. His strongly held view of the vital link between national independence and energy supply had impressed de Gaulle. And, from the day of his new appointment, he was to become the personification of the Gaullist political economy.

He was relentless in his demands for exactness, efficiency, speed, toughness, and discretion. "A machine that never misfires," one colleague said. He was adept at institution building and he exercised power without doubts or fear. He never made compromises, sought advice, courted popularity, or felt it necessary to account for his actions—either in private or in public. Above all, he never interfered in anyone else's business. That was one thing on which he never wavered; he expected the same in return. Parliamentarians and especially journalists were meddlers—to be avoided. One of France's leading business magazines once apologized to its readers for publishing an interview with Guillaumat, the draft of which he had changed beyond recognition. He did not give his first open interview until 1974, after thirty years at the center of French public power.

Guillaumat's extraordinary administrative skills gave the CEA a style and inner confidence it maintains to this day. It was not a refusal of the military applications of atomic energy that caused him to rebuff the aspirations of the French military, who wanted a greater say in the machinery of the CEA. (Guillaumat, in fact, was one of the most vigorous promoters of the French bomb.) It was Guillaumat's insistence on absolute authority, his obsession with secrecy and control of information.

In his pursuit of the bomb, he never had much problem with Joliot's scientific replacement, Francis Perrin. The son of a Nobel Prize–winning physicist, an intimate and neighbor of the Curies, and a formidable physicist in his own right, Perrin had grace, charm, and reasonableness. Such virtues, noticeably absent in the aggressive young Guillaumat, put the scientist at an immediate disadvantage in the bureaucratic infighting that characterized the takeover. Like Joliot, Perrin was opposed to the bomb option, but he failed to prevent weapons studies or the construction of facilities that could be used for either peaceful or warlike purposes.

Shortly before Perrin's arrival at the CEA, the first Five-Year Plan, including the building of two high-powered plutonium-producing reactors, had been approved. To Perrin and the scientists under him, the size of the proposed reactors and the priority given to them seemed to be directed toward only one thing: the bomb. But they consoled themselves with the thought that the plutonium might eventually be used for peaceful purposes, probably in a breeder reactor—the type that produces more plutonium than it uses. The weapons option was always there, however, and later, the semiofficial historian of the French project, Bertrand Goldschmidt—a CEA scientist who would work on the extraction of plutonium for the bomb—admitted the option was "undoubtedly predominant" in the minds of those who had inspired and were responsible for the plan.

Guillaumat was not only concerned about the bomb, however; he shared the conviction of the moment that "economic" nuclear power was only a question of time. The same quest for energy independence that had dominated his drive to build the French oil empire was repeated in his involvement with the CEA. To ensure a nuclear future for France, he continued to explore uranium reserves within France and in its former colonies, he encouraged private industry to develop its own nuclear technology, and he created the new industrial department of the CEA, a state-run engineering group empowered to construct and operate nuclear establishments.

But it did not take long for Guillaumat to raise the question of the bomb with the government. At first, he somewhat obliquely referred to the "importance and urgency of taking the preliminary decisions in the military area." When he failed to extract a positive decision from the faction-ridden governments of the Fourth Republic, he began to exploit the political paralysis by taking initiatives on his own.

Without any official sanction other than approval of research projects, the CEA under Guillaumat would continue for years on a full-scale program of bomb development. Goldschmidt wrote later: "Guillaumat played a predominant role in the evolution. Indifferent to governmental fluctuations, he always pursued the same goal: the creation of a powerful national nuclear industry as an indispensable precondition to permit the country to tackle not only civilian development but also the military phase. For the latter, he became a ferocious defender of the CEA." Throughout the turmoil of the Fourth Republic's governments, no political leader had the strength to veto Guillaumat's pet bomb project, which had the full support of his dedicated associates, the fledgling Gaullists.

The first significant political move came in December 1954 when the Socialist Prime Minister, Pierre Mendès-France, pressured by the CEA and the pro-bomb sections of the military to clarify the country's nuclear policy, called in his advisers to find out what aspects of atomic research had an exclusively military orientation. Later, he would recall how the representatives of the CEA—Guillaumat, Perrin, and Goldschmidt—had requested time to consider the question and had retreated into a corner of the room for a whispered discussion. When they returned, they gave Mendès-France some wholly misleading information.

They told him that, since the first Five-Year Plan of 1951—which had called for the building of plutonium-producing reactors—there had been no research of an exclusively military nature. Mendès-France went away persuaded that a firm distinction could be drawn between peaceful plutonium-producing reactors and warlike uses. His advisers had simply neglected to tell him that the plan was also perfectly in tune with a bomb program—should anyone decide that the weapon ought to be built. The only extra work required was the design of the bomb.

It was one of several examples of Guillaumat's exercising his independent authority. Another came later that same month, on Christmas Eve, 1954, when Mendès-France called another meeting, this time to discuss more specifically whether France should build a bomb. Guillaumat, of course, was present. It was a long meeting ending with no consensus. Mendès-France summed up with an overall directive instructing the CEA to keep the bomb option open and continue to pursue basic research that might be useful. In the years to come, Guillaumat would put his own gloss on the Prime Minister's remarks. In CEA documents, the meeting was reported as having approved not only research but also the "preparation of a prototype nuclear weapon," yet there was never any official government order or minutes reflecting such a decision.

Moreover, Mendès-France would later deny that he had made any such decision, and the external evidence supports him. The CEA, for example, continued to ask the government for more specific terms of reference—which they would not have needed if the outcome of the Christmas Eve meeting had been as clear as they had recorded it.

Without hesitation, Guillaumat took the Christmas Eve meeting as a definite order. He swiftly established a bomb-design unit; an army general, Albert Buchalet, was called in to head the project. It was set up in the utmost secrecy under the code name "The Bureau of General Studies."

Guillaumat knew that he could not rely on the full support of the French atomic scientists. Suspecting that some secret move would be made to build a bomb, a third of them had already petitioned the chief scientist, Perrin, protesting the existence of the bomb option on the grounds that much-needed funds would be diverted from the peaceful program. Guillaumat neatly sidestepped their resistance by appointing the only senior pro-bomb physicist, Yves Rocard, to the new secret bureau.

In February 1955, the Mendès-France government fell and the shady days of the transition to the new coalition under Premier Edgar Fauré gave Pierre Guillaumat and the Gaullist politicians the chance to push the bomb project along. Within days, Guillaumat signed a secret agreement with the chiefs of staff of the new Ministry for Defense and Atomic Energy, who were pro-bomb. Four months later, this agreement received the necessary blessing of the minister of finance and, on May 20, 1955, approval was given for the secret transfer of funds from the Ministry of Defense to build a third plutonium-producing reactor. On the same day, the Fauré government announced a doubling of CEA funding for, it was said, long-term electricity production; Fauré told the National Assembly that the government had decided not even to study how to make a bomb, let alone build one.

At the beginning of 1956, Fauré's successor, Guy Mollet, discovered the secret bomb project. He wanted to stop it immediately, but was persuaded not to by a threat of the withdrawal of Gaullist support for him, which would make the fall of his government inevitable. After the 1956 Suez crisis, which highlighted French military weaknesses, even Mollet's doubts disappeared, and Guillaumat acquired a new secret protocol with the Ministry of Defense. The protocol established the military timetable that led to the first bomb test in 1960.

Despite the feverish preparations, however, there was still no formal top-level order to build the weapon.

That order was eventually signed by the last premier of the Fourth Republic, Félix Gaillard. In 1951, as minister responsible for atomic energy, Gaillard had launched the so-called peaceful plan of plutonium production, which was then covertly directed at the ultimate military objective. Gaillard's government fell in May 1958, leaving the way open for de Gaulle to come in a month later. Gaillard actually signed the order in the interim period and was persuaded by his Gaullist Minister of Defense, Jacques Chaban-Delmas, to backdate it, thus making it legal. It was a fitting end to years of covert preparation of the French bomb.

The bomb was exploded in the Sahara on February 13, 1960. As a final protest over how completely the scientists had been outmaneuvered by the power of the engineers of "Les X," Francis Perrin refused to attend the test.

9

"A Separate World"

The atom had us bewitched. It was so gigantic, so terrible, so beyond the power of imagination. . . . So greatly did it transcend the ordinary affairs of men that we shut it out of those affairs altogether; or rather tried to create a separate world, the world of the atom.

—DAVID LILIENTHAL, 1963

By the mid-fifties, the world of the atom had become officially enshrined in new institutions in five countries: America, Britain, France, Canada, and Russia. Other nations, catching up later, would do the same. In most cases, the institutions took the form of a commission of a half-dozen worthy public servants, most of them without scientific qualifications. In all except Canada, their task was to develop both the peaceful and the warlike atom, and, because of the military connection, they became closed societies. Pleading national security, the powerful commission members were able to deflect legitimate and concerned public inquiry about their work; in the years that followed, they would escape the normal checks and balances of democratic control. Even in Canada, which was not making bombs and never had any intention of doing so, information about the development of the peaceful nuclear power program was kept behind closed doors.

In Britain, Lord Cherwell, Churchill's scientific adviser, had always believed that the atom was too important to the country's future as a world power to be left in the cumbersome bureaucracy of the Ministry of Supply, where it had been since the end of the war. Single-handedly he promoted the idea of a new organization that could slice through red tape and, in particular, pay salary scales higher than the civil service. If this were not done, Cherwell believed Britain's atomic project would never attract the country's top scientists and engineers.

In Canada, Clarence Howe, an industrialist-turned-politician, had similar thoughts about a Canadian nuclear power industry. In twenty-two years in public affairs, he had built up the most wide-ranging government-controlled industrial empire in the democratic world. He wanted nuclear power to be a part of it, and the best way to develop it rapidly and with the minimum of outside interference was to put it in a separate organization.

In Russia, the death of Stalin in March 1953 and the subsequent fall of his secret police chief, Lavrenti Beria, meant that the job of exercising political control over the Soviet atomic program was up for grabs. Such evidence as exists suggests that a major construction program for nuclear power was launched in 1954 under a new cover, the Ministry of Medium Machine Building, and a relatively unknown Red Specialist engineer-administrator named Vyacheslav Alexandrovich Malyshev emerged as the minister in charge. These three, Cherwell in Britain, Howe in Canada, and Malyshev in the Soviet Union, all powerful men, moved to center stage in international nuclear development with the creation of these national atomic shrines.

When Winston Churchill returned to No. 10 Downing Street as Prime Minister in 1951, he was amazed to discover the size of the defeated Labour government's atomic program. He was particularly impressed, as an old parliamentarian, that his predecessors had managed to spend nearly £1 million on the British bomb without telling Parliament. If he had done such a thing, he would surely have been branded a warmonger.

Despite America's stockpile of hundreds of bombs, Britain had found it necessary to spend much of its scarce resources on its own secret program. The alliance was in disarray; even with Churchill's return to power, no immediate assurances came from Washington to say that Britain would be able to crawl under America's nuclear umbrella. Though the Prime Minister expected such an invitation, Lord Cherwell was much less sanguine about it. "If we have to rely entirely on the United States army for this vital weapon," Cherwell said, "we shall sink to the rank of a second-class nation, only permitted to supply auxiliary troops, like the native levies who were allowed small arms but not artillery."

Cherwell, born in Germany but brought up in England, had readily assumed the patriotism of an adopted son. Like many leading British and American officials, he had been shocked by the fact the Russians had made the bomb before Britain; like them, he was unable to credit the Russians with ingenuity and hard work and preferred to place credit for the bomb with the atom spies.

The Russian lead in the bomb race had a profound effect on Cherwell; a strident anti-Communist, he decided that Britain's own atomic project was too tardy and must be speeded up. The remedy, he thought, lay in a government-sponsored corporation that could bring together the disparate parts of the program—production of the fissile material, weapons design, and research—under one roof.

Cherwell wanted bombs made quickly and he also wanted atomic power fed into the national grid. "Our prosperity in the Victorian era," he wrote, "was due largely to the men who had the imagination to put and keep England ahead for sixty to eighty years in the use of steam power for industrial purposes. It is quite likely that our prosperity in the coming century may depend on learning how to exploit the latent energy in uranium (one pound equals one thousand tons of coal)."

The British public was already well acquainted with the pallid, stooped figure of Lord Cherwell. In a dark suit and bowler hat, he was often seen shadowing Churchill in wartime photos. In the early 1950s, he was also a formidable figure in Britain's corridors of power: his long friendship with the Prime Minister, dating back to Churchill's years of political exile in the twenties, and his devoted war service had ensured that. In recognition, Churchill made him first a baron and then a cabinet minister.

After a childhood spent in England, Cherwell had returned to Germany between the wars to study physics. These were the "golden years" when the exciting mysteries of the atom were being discovered, but he was to spend more time in the drawing rooms of the rich than with a slide rule and a blackboard. Back in England before war broke out, he became a professional guest of the English gentry and, in the words of a fellow baron, "the biggest snob I have ever known." Privately wealthy, he lived alone in rooms in Christ Church College in Oxford and, on weekends, would climb into a Rolls-Royce with a chauffeur and a valet, drive out to some country house, and expound the wonders of scientific discovery to adoring hosts and hostesses. When Cherwell died, Robert Watson-Watt, the "father" of radar, wrote an appreciation of him, calling him "a gifted and highly influential vulgariser of science in high political circles." His friends knew him as "the Prof," but, in retrospect, an earlier nickname, "the witch doctor," seems more apt.

In wartime, without an executive position, Cherwell stalked No. 10 Downing Street in a hawkish mode, which endeared him to Churchill. With his undoubted gift for rapid and succinct analysis of scientific problems, he gave the Prime Minister invaluable service in reducing long and complicated scientific memoranda to one-line précis. But because he had no executive power, no one took much notice of him.

In 1951, Cherwell returned from his own temporary exile brought on by six years of Labour rule to be Churchill's personal adviser. He left his rooms in Christ Church with one specific aim in mind: to create a new agency to develop atomic energy. He met considerable resistance from the formal head of the civil service, Lord Bridges, the man who had so smoothly steered the atomic bomb project through its early stages with the Labour government. As far as Bridges was concerned, the bureaucratic arrangements he had then devised were working well; the bomb was on schedule. All his instincts argued against any kind of change, especially in lines of control, unless the need was overwhelming. Furthermore, he believed that each new independent statutory corporation, such as the one Cherwell was proposing, threatened the coherence of the entire bureaucratic edifice of the civil service, because they were exempted from its (Bridges's) control.

Cherwell, ever confident of his influence over the Prime Minister, felt that pushing his atomic corporation through Parliament was only a matter of time, but he was wrong. He failed to take account of the fact that the new Tory cabinet, governing with a small majority, was reluctant to give priority to such a costly and controversial bill. The atomic scientists and engineers working on the bomb program were quite happy, moreover, to continue under the Ministry of Supply. Even Churchill himself was not prepared to force the changeover just for his old friend's sake. In 1952 he told Cherwell bluntly to "wait and see whether your bomb goes off or not" before pushing ahead with the plan. Happily for Cherwell, the British bomb did go off. A plutonium device, code-named Hurricane, was exploded at Monte Bello, an island off Australia's western coast, on October 3, 1952. Cherwell returned to the attack with renewed vigor.

For the first time, the imperious Cherwell could be found lobbying cabinet ministers in search of his aim; eventually, they gave in to his demands. In January 1953 the cabinet set up a committee of experts under John Anderson, now Lord Waverley, to prepare the machinery for transferring the entire program to a new, separate organization. The U.K. Atomic Energy Authority emerged in 1954. It had nearly 20,000 employees and was one of the larger government agencies. One of its first military tasks was to carry out Britain's decision—taken, unlike the U.S. decision, with no internal debate—to build a hydrogen bomb.

There is no reason to believe that Britain's bomb program would have foundered or that her nuclear power projects would not have been built if the Atomic Energy Authority had not been created. But the political autonomy of the AEA meant that the flow of information between the authority and the cabinet minister nominally in charge of it was seriously curtailed. Twenty-five years later, Britain's Labour Secretary of State for Energy, Anthony Wedgwood-Benn, would say, "It's not a question of control, it's a question of the integration of this very high technology with the democratic process. . . . In the end, however complicated a technology is, the public have the right in a democratic society to say no." Such was the aura of secrecy surrounding the AEA because of its military role that the British public never had that right.

In theory, Clarence Howe faced no such delicate problems of democratic rights in setting up Canada's atomic institution. As Minister of Supply in 1965, he told the Canadian House of Commons, "We have not manufactured atomic bombs, we have no intention of manufacturing atomic bombs, and therefore I should not like to have it suggested that any great sums of money have been spent in that occupation."

Although this was strictly true, for the next decade and more, Canada would sell all her plutonium—made in heavy-water research reactors—to the U.S. bomb program. Indeed, assurances that the

United States would buy more plutonium was to give the Canadian peaceful nuclear power project a much needed boost in the fifties. But this military umbilical cord was never mentioned at the time. As far as Clarence Howe's public was concerned, this was a peaceful undertaking, a Canadian project. Above all, it was *his* project: he liked to live up to his name as Clarence "Dictator" Howe and the "Minister for Everything." In 1952 he decided it was good for the Canadian nuclear industry—and good for the Howe empire—to put the Canadian nuclear program into a separate organization.

Howe's wartime Ministry of Munitions and Supply had run Canadian industry. He had created a new manufacturing base: not only guns and ammunition but aircraft, aluminum, synthetic rubber—and the country's abundant supplies of the new precious metal uranium, which the United States sought so vigorously. From 1942 onward, the United States was to have access to the entire Canadian uranium production except for the small amount needed for Canada's own research reactors.

Operating virtually without oversight or review, Howe established a series of quasi-independent nuclear corporations; all the shares were owned by the government. From 1944, all the Canadian uranium mines were included. He staffed these corporations with his own personal contacts from industry. "Howe's Boys," as they were known, were given policy guidance from the "Czar of Czars," as the press dubbed him. For two decades, beginning with the Manhattan Project, Howe and his "Boys" would be intimately connected with the development of nuclear technology.

The stocky, silver-haired, stern-faced Howe sat on a gigantic pyramid of government-owned or -controlled enterprises. He ran them all from a sparsely furnished office in a shabby, wartime building on the outskirts of Ottawa. It was known as Temporary Building No. 1. Howe was still there when his extraordinary reign of twenty-two years came to a close in 1957. Such a run was unprecedented for a minister in a parliamentary system of the British model.

Before he entered Parliament, Howe had been the "best-known builder of grain elevators in the world." Between the wars, his firm of consultant engineers had changed the international skyline, building those huge structures as close to home as the Canadian shore of the Great Lakes and as far away as the waterfront of Buenos Aires. The engineering monuments were no less impressive than the political empire he raised when, in 1935, he was lured into politics with the promise of an immediate portfolio by the triumphant leader of the Liberal party, William Mackenzie King. Howe's abuse of power and contempt of Parliament was eventually the most important single issue that caused the overthrow of the Liberals in 1957.

He loved to build institutions. As Minister of Transport in the late 1930s, he reorganized the Canadian National Railways, established the Canadian Broadcasting Corporation, and founded the predecessor company of Air Canada. Some wrongly confused the empire-building with nationalization. When a Conservative member of Parliament called the government-owned airline "a step towards socialism," Howe snapped back: "That's not public enterprise, that's my enterprise."

Howe's introduction to the atomic age came in February 1942, when British officials from "Tube Alloys"—the U.K. bomb program—proposed that the Canadians, who had known supplies of uranium, should enter into a joint atomic-power project. Besides uranium, the Canadians also had a plant that could be easily adapted to produce heavy water—the only such plant outside Norway. Unable to start their own atomic laboratory because of heavy bombing, the British plan was to set up a research facility in Canada, where it would not be swamped by the massive American program.

Canada's chief scientific adviser, Chalmers Mackenzie, advised Howe that it was an unprecedented chance for Canada to "get in on the ground floor" of a major new technological advance. "It is an opportunity that Canada as a nation cannot afford to turn down," he said. Howe needed no persuading: he saw the potential of a vast new industry and he trusted the judgment of Mackenzie, who was also his close friend.

In April 1944, a joint nuclear project was set up in Montreal. An Anglo-Canadian team, with a group of Free French scientists—including Joliot's colleagues Halban and Kowarski—began building the first heavy-water-moderated, plutonium-producing, natural-uranium reactor at a small village called Chalk River, on the south bank of the Ottawa River, 130 miles west of the capital. Professor John Cockcroft, the British radar specialist, was released from his work at the Cambridge Cavendish Laboratory, put in a bomber, and flown across the Atlantic to take charge of the project. The scientists in Canada knew they could not possibly produce anything for use in the war; indeed, their experimental heavy-water reactor, code-named ZEEP, only went critical a month after Hiroshima, in September 1945. It was, however, the first controlled chain reaction outside the United States and a positive test of a promising technology that was to form the basis of Canada's nuclear effort.

In July 1947, the ZEEP reactor was followed by a second heavy-water reactor, thus firmly establishing a Canadian reactor technology and starting the flow of plutonium that heavy-water reactors produce. The project, costing millions of dollars a year, was already more expensive than all the other Canadian research activities put together, and because of the country's huge hydroelectric potential, it was of dubious value. But with Howe's continued support it flourished.

The scientific crews operating the reactors became increasingly confident in their product and wanted to press on toward a power-production reactor. When the U.S. Atomic Energy Commission agreed to buy all the plutonium a new prototype reactor could produce, earlier doubts about the future faded away. Chalmers Mackenzie, who immediately after the war had thought it would take two decades to develop reactor power in Canada, now accepted that large-scale industrial production was "closer at hand than had been expected." Howe, demonstrating his faith in the long-term viability of the new project, created a separate government-owned corporation, Atomic Energy of Canada, Ltd., to run it. The corporation's head was his friend Mackenzie. A Canadian nuclear power industry was born.

There is scant information on the Russian equivalent of Lord Cherwell or Clarence Howe. Although it is now possible to identify the original Soviet scientific and engineering leaders, much less information is available about the administration of the atomic program. What is known, because it has been widely hailed by the Russians themselves, is that their early peaceful atomic effort proved highly successful. In 1954 it produced the world's first nuclear power station, feeding electricity into the homes of the citizens of Obninsk, a town 100 kilometers southwest of Moscow. Soviet histories chronicle the event with pride, even though the output of the reactor was tiny—only 5,000 kilowatts; today's reactors are two hundred times bigger. The man in charge of this considerable achievement was Vyacheslav Malyshev, the new Minister for Medium Machine Building.

His activities are so obscured by Soviet secrecy that one of the key

Russian atomic administrators, Vassily Emelyanov, wrote a biographical sketch of him without ever mentioning his nuclear role. Soviet citizens were given their first glimpse of his newfound importance when he made a surprise appearance with members of the Soviet Politburo at the Moscow Opera in June 1953. The one missing member of the Politburo that night was Malyshev's predecessor, the secret-police chief, Lavrenti Beria. During the performance Beria was arrested. He was later executed, and his huge empire, which included political direction of the Soviet atomic project, was dismembered. The atomic program was given to Malyshev: a Red Specialist had arrived at the pinnacle of state power.

Malyshev took part in Russia's original atom bomb project, but even there his exact role is still obscure. Apart from occasional glimpses that show him at recruiting meetings with the engineer-administrator Boris Vannikov and one general account that acknowledges him as one of the "heads of the project," the Russian literature is silent. The memoirs of a German scientist who worked on Russia's program, however, indicate that although Malyshev was apparently junior to Vannikov, he was superior to the other engineer-administrator, Avraami Zavenyagin, and frequently chaired technical meetings.

In the hard-nosed cadre of the Red Specialist, Malyshev was a calm, even modest character. Compared with the others, he led an ascetic life, devoted to his work and his ideals. He was a good listener, seldom if ever cross or moody, yet still capable of holding his own in the rough and tumble of Soviet decision making in which bitter attacks and bravado are dominant modes of behavior. Malyshev never felt the need to raise his voice to get results or, like his Red Specialist colleague Zavenyagin, to display his position through ownership of a large private dacha.

He was an idealist and a serious student of Marxism. He never lost a youthful, touchingly naïve faith in the virtues of technology and the human advantages that could be derived from it. As a young engineer in 1931, he had given an impassioned speech to his superiors on the prospects of hydroelectric development in Siberia. "We shall induce 4 or 5 million people from the European sector to pack up and move to the Angara, where we shall build world centers of a new Communist industry," he declared. "The myriads of lights from the east will attract hosts of young men and women like magnets. We shall bring wonderful cities into being there." Such dreams had been central to the Bolshevik progressive image from the beginning: electricity had a special cachet in Soviet mythology. Lenin himself had defined Communism for H. G. Wells as "Soviet power plus the electrification of the whole country." Malyshev was a good student of that ideology.

He had joined the Communist party in 1926 when he was working as a locomotive driver at the Moscow railway depot. Born in the village of Syktyvkar, halfway to the Arctic Circle from Moscow, he was a teacher's son, but after the revolution he had been forced to develop a series of highly proletarian credentials, of which the work on the railroads was a part. He was selected by the party for special training at the Bauman Technical Institute in Moscow, under the supervision of Georgi Malenkov, who would later become premier after Stalin's death. Many future top administrators were trained at the Institute in the early thirties. After graduation, he went back to the railroads, but as a design engineer. He was elected to the party's Central Committee and appointed minister in 1939; during the war years, he was a leading administrator of Soviet defense industry with special responsibility for tank production. During the Battle of Stalingrad, the decisive turning point in the war with Germany, Malyshev was in charge of the economic administration of the besieged city, a post that undoubtedly helped him to rise to political prominence. He had to keep the rail links open and the boats moving on the Volga River, but most important of all, he was responsible for maintaining a shaky tank repair line in a converted tractor plant only a few hundred yards from the front. It was one of those testing moments of history that becomes the stuff of patriotic folklore and, in the years to come, those who had participated would, with good reason, be honored; Marshal Grigori Zhukov and Nikita Khrushchev were among them.

With the removal of Beria and the oversight function of the secret police, Malyshev's ministry was ready for a major expansion of its nuclear activities. The tasks on the drawing board required new industrial plants, even new cities. A nuclear submarine project was under way; a new research institute was being built in Dubna. But the most ambitious plans had been reserved for the generation of nuclear power. Putting Lenin's words into action, any new form of producing electricity had a special attraction for Soviet political leaders.

In the wake of the successful Russian bomb tests, the nuclear scientists' senior spokesman, Igor Kurchatov, had made good use of his new status and pushed hard for research into all types of reactors. The first to be completed—the one at Obninsk—was a graphite-moderated, water-cooled reactor, a descendant of the first plutonium-producing reactor at Sverdlovsk.

In 1958, a heavy-water reactor was completed in Siberia but kept secret, presumably because it had the dual function of producing electricity and plutonium. Of more long-term significance was the program for developing reactors to produce "peaceful" nuclear power. The new scientists' enthusiasm for peaceful uses was shared at the top political level. Premiers Nikolai Bulganin and Georgi Malenkov spoke of the "century of atomic energy"; the Russian peaceful program now had its own momentum. Soon, Russia would reorganize and strengthen its projects, carving out of Malyshev's Ministry of Medium Machine Building a new department to be known as the Central Directorate for the Utilization of Atomic Energy. Russia too had decided that the atom was so special that it had to be enshrined in its own special organization.

10

The Admiral

"It would be nice," Enrico Fermi once said of atomic energy, "if it could cure the common cold." Fermi was joking, but such a claim would not have appeared out of place in the mid-fifties as an infectious wave of nuclear enthusiasm spread from the United States throughout the world. The public's imagination was captivated by the dream of electricity "too cheap to meter" and a treasure trove of amazing gadgets—from nuclear-powered airplanes and ships to ways of preserving food, sterilizing sewage, and even curing cancer.

Against the appalling vision of thermonuclear annihilation came a persistent cry for some assurances that people could live in peace. What better place to find the elixir than in the very source of the evil itself? "Atoms for Peace," the propaganda buzz phrase of Eisenhower's 1953 speech, became a term of magic, of witchcraft; it seemed capable of driving off the evil spirit of atomic warfare.

At first there were few doubters; not even the growing sense of alarm over the harmful effects of radioactivity from the tests of bigger and bigger bombs dampened the enthusiasm. The peaceful atom was, by definition, benign; it seemed only a matter of time, and a short time at that, before this philosopher's stone would save the world.

In the midst of these public fantasies, the man who was actually presiding over the birth of the American nuclear power industry had no such illusions. He was Captain Hyman George Rickover of the U.S. Navy. A hardheaded, practical engineer, he believed two things about nuclear power: it could save the navy and it could save America. But there were no easy solutions to the problems ahead.

With the dominance of air power at the end of World War II, the navy was looking for a new role. One possibility was the nuclear-powered submarine, which could remain submerged for long periods, making it less dependent on refueling bases and less vulnerable to attack. If the navy needed nuclear power to ensure its continued existence as a viable independent fighting force, America needed nuclear power to maintain its prestige in the atomic revolution and provide a viable energy alternative when fossil fuels ran out.

As an engineer, Rickover had realized, as early as 1946, that it would take at least five and possibly eight years to build a shipboard reactor; even a decade later, he saw the dream of nuclear power "too cheap to meter" as an empty atomic-age jingle; he knew there were too many unanswered questions.

Although the basic principle of nuclear power had been proved in practice by Enrico Fermi in Chicago in December 1942, there was a huge engineering leap between that relatively simple experiment and the production of electricity. The heat developed during the chain reaction had to be turned into steam and that steam used to drive turbines; such a heat-exchange system required a whole new set of blueprints, coolants, pumps, valves, and control systems. It would take years to develop and years to perfect. Despite the problems, Captain Rickover was supremely confident he could do both; he needed to do both.

The universal boundless enthusiasm for the magic of the atom gave Rickover the chance to elevate himself from a competent but humdrum engineer into one of the most outstanding figures in the postwar development of atomic energy. Rickover's mastery of the potential of nuclear power, his determination to better himself, and his ambition to hold power drove him to accept the challenge to build one of the most extraordinary military and civilian empires ever run by a serving officer in peacetime.

Rickover became to American nuclear power what J. Edgar Hoover was to U.S. law enforcement, yet unlike the gruff-looking FBI director, Rickover, a slip of a man with a thin, wispy frame and a sparrowlike face, seemed wholly unsuited to the role of caudillo. He had always gotten bad marks for military bearing at the U.S. Naval Academy; in the beginning, there seemed no way the Polish-Jewish tailor's son from Chicago was going to make it to the top in the heart of WASP country, the U.S. Navy. His only hope was to gain expertise—preferably, expertise that only a few could obtain. Nuclear power was the answer.

One of the few navy officers chosen to do atomic work after World War II, Rickover quickly mastered the rudiments of the new science, skillfully manipulating the fledgling atomic institutions to his own advantage. He soon maneuvered himself into an extraordinary bureaucratic position as chief architect of the nuclear navy and head of the navy reactors branch of the Atomic Energy Commission—the office that approved or disapproved development of the nuclear navy. In effect, he ordered himself to build nuclear submarines. Wearing these two hats, he created an elite work force of mostly dedicated followers whose loyalty he ensured by bullying. He made them feel guilty if they did not work as hard as he did—most of them did not—and he made them feel inferior because of their lack of expertise about every aspect of the atom that he had mastered.

If he had behaved like that as the captain of a ship, the crew would have mutinied; but the magic of nuclear power allowed him to run a Manhattan Project in peacetime. Like General Groves, Rickover spent only part of his day as a military officer; he was also a politician and a businessman, the godfather of the atomic establishment.

Rickover became the driving force behind the light-water reactor, the type that would become the most popular in the world. It uses ordinary water to cool and to moderate the chain reaction. Two types were developed: the pressurized-water reactor (PWR) and the boiling-water reactor (BWR). Rickover built the first light-water reactor for the first atomic submarine, *Nautilus*, named after Jules Verne's fantasy craft and launched in 1954. He built another for the first U.S. nuclear power station, which began operating in 1957 at Shippingport, Pennsylvania.

If Rickover displayed a masterful understanding of nuclear engineering, he was equally adept at exploiting the manipulative power that is inherent in any office disbursing federal funds. In letting civilian contracts, he always played one company against another, constantly threatening to cancel all naval work unless he had his way. They were no idle threats. In the companies he dealt with, his displeasure led to the firing or transfer of employees and the cancellation of projects.

Rickover became one of the most powerful wheeler-dealers in what Eisenhower called the "military-industrial complex." He awarded military contracts worth millions of dollars to billion-dollar companies like Westinghouse and General Electric. By his selection of Westinghouse over its rival GE to build the first submarine reactor, and, later, the first civilian reactor, Rickover was actually dictating the course of corporate nuclear development.

In 1946, when Rickover got his first glimpse of atomic energy, nothing could have foretold he would become so powerful. At forty-six, after twenty-seven years of service, Captain Rickover seemed almost at the end of his career. It was not that he had failed; far from it. He was a well-qualified submariner, an Annapolis graduate with a master's degree in electrical engineering from Columbia University. But he had chosen to become an "Engineering Duty Only" officer, which barred him from command afloat and probably, it seemed, from further promotion. It would not be long, he thought, before he would be forcibly retired.

Atomic submarines offered Rickover an opportunity to stay. Ever since news of the Germans' successful fission of uranium had reached America in 1939, physicists at the Naval Research Laboratory had been dreaming of the world's first nuclear-powered submarine. The army's control of the Manhattan Project had effectively cut the navy out of atomic matters until, a year after the war's end, they were invited to send a delegation to Oak Ridge to learn about reactor development.

Shut out though it had been, the navy fully recognized the potential of the nuclear-powered submarine. Existing submarines of World War II vintage had to operate on the surface most of the time because of the basic need for oxygen to keep their diesel motors working. When they submerged, they used batteries that could drive the vessel only a short distance before needing a recharge. Few naval officers believed, however, that a submarine atomic reactor would ever warrant top priority; it would clearly be an enormously costly project and, in the immediate postwar years, cash for the military budget was in short supply. Rickover alone remained committed. "There were so many people who were certain the project would fail and they didn't want to be associated with it," he said. "I was kept on [instead of being retired] and I was assigned this job all to myself—and given an office in a former ladies powder room."

Rickover would come to run the Western world's largest nuclear corporation, but in the early days, his headquarters, the Naval Reactors Branch of the Bureau of Ships, was hardly the powerhouse of a multibillion-dollar enterprise. It was lodged on the top floor of a seedy, gray "temporary" wartime structure on Washington's Constitution Avenue. Paint was peeling off the walls, and the floors were covered with bilious brown linoleum. It suited Rickover's carefully cultivated image of an underdog, continually harassed, even persecuted; his shabby quarters were just another example of his oppression.

He liked a stark, spare operation, a no-frills show; there was no provision in his budget for staff amenities. His own office was no different; a desk and a few tables overflowing with papers, reports, and reactor models on a bare floor. The walls were lined with bookcases crammed with volumes on physics, engineering, history, and philosophy. The captain's only hobby was reading; his work was his life.

Behind his desk was a framed biblical quotation: "Where there is no vision, the people perish"—a message at once messianic and humanitarian, inspiration for a zealot. In another frame were lines from Shakespeare's *Measure for Measure*:

> Our doubts are traitors,
> And make us lose the good we oft might win,
> By fearing to attempt.

Hyman Rickover's life seemed to be a continual struggle against his inner self-doubt. He worked a sixteen-hour day. A secretary brought him his lunch: broth, a hard-boiled egg, fruit, and skimmed milk. He seldom drank alcohol. At the end of each day, his secretary deposited a pile of pink sheets on his desk, copies of every note, letter, and memo that had been sent out of his offices that day, from a letter to the President to an order for paper clips. Each typist was under threat of instant dismissal to produce her "pinks," as they were called. The next day they were often returned with terse comments in the margin correcting grammar, careless expressions, vague terminology, any sign of sloppiness. Rickover demanded the best from everyone:

the project must not fail because of human weakness. He demanded to know what was going on the instant it happened. His field officers had to phone in daily inspection reports.

Rickover carefully cultivated his image as a man who cut red tape, a man who got the job done by ignoring rules and regulations, by going over the heads of his superiors and behind the backs of his equals. "A straight line is the shortest distance between two points even if it bisects six admirals" was one of his favorite epigrams. For all that, he actually had a passionate belief in red tape, which he used to strangle those who got in his way. A master at bureaucratic politics, he used every subtle nuance of institutional lines of authority, every bifurcation of an organization chart, every ambiguity in terms of reference, every overlap in jurisdiction. He insulated his own bureaucratic empire with all the skill and eye for detail of one designing a complex diagram of electrical circuitry. He used red tape to prevent any inquiry, attempt at supervision, or even serious discussion of his own activity.

As a chief executive, he spent a lot of time on what many in his position would regard as trivial and unimportant matters; but Rickover had to have complete control. He insisted, for example, on interviewing personally each applicant for a job directly under him and, later, also those officers who wanted to join the growing fleet of nuclear submarines. The interviews became a legend, an unforgettable experience for hundreds of men. The official historians of the U.S. nuclear navy noted that for some the interview "was a shattering event leaving scars that would last a lifetime." His opening thrust was always aimed to intimidate. "Everyone who interviewed you tells me you are extremely conservative and have no initiative or imagination." Each applicant experienced the lash of his irascible tongue. "You're either dumb or lazy, which is it?" They felt his guile. Admiral Zumwalt recalls in his own interview:

R. Are you resourceful?

A. Yes, sir.

R. Suppose you're on a sinking ship with five other men. The conditions are that one and only one of you can be saved. Are you resourceful enough to talk the other five into letting you be the one?

A. Um, yes, sir.

R. All right, son. Start talking.

A tortured, hectoring examination of a man's deepest motivations and sources of pride, these interviews were frequently punctuated with summary dismissals for no particular reason to an adjoining room, a bare cell known as "the tank." It had a table and chair facing a blank wall: solitary confinement for thirty minutes, seldom longer. Rickover is also said to have put his interviewees in a special chair with an inch or two cut off the front legs so that the applicant tended to slide forward. The chair was so positioned that Rickover could flick the venetian blinds and direct sunlight into the applicant's eyes while continuing his relentless penetrating questioning.

"Did you do your best?" was a basic theme. For those who admitted they did not always do their best, "Why not?" was the next question. For those who said they had, the process of establishing guilt had just begun.

R. Did you study as hard as you could?

A. Yes, sir.

R. Do you say that without any mental reservations?

A. Yes, sir.

R. Did you do anything besides study?

A. Yes, sir.

R. In other words you didn't study as hard as you could.

A. I gave my answer in the context of what I thought was a reasonable balance between academic and extracurricular . . .

R. Stop trying to conduct the interview. You're acting like a damn aide. I told you for the last time I am conducting the interview. Now shall we go ahead on that basis or do you want to get out of here?

A. I am ready to go ahead on that basis.

R. Now, what are these special extracurricular activities you are so proud of?

A. I was a debater, an orator and . . .

R. A debater (sneering). In other words you learned to speak equally forcefully on either side of a question. Doesn't make a damn bit of difference what you believe is right—just argue the way someone tells you—good training for an aide. . . . I am sick of talking to an aide that tries to pretend he knows everything.

The candidate was sent to "the tank," this time for an hour.

Rickover searched for personal strengths in his employees. He wanted—needed—people he could depend on, people who would fight for themselves, fight for him, even fight against him, in case he was ever wrong.

By the end of each interview he knew whether he could rely on the applicant or not.

> R. What are you going to do when you get back to the Pentagon, run up and down the E ring and tell everybody about this interview?
> A. Admiral, I'm going to say it was the most fascinating experience of my life.
> R. Now you're being greasy. Get out of here.

A sign in the waiting room read, "After having been interviewed by Admiral Rickover no one is to return to this room except to pick up his personal possessions and leave immediately without talking of interviews."

Rickover was born in Russian Poland in 1900. He arrived in New York at the age of six. His family soon moved from the city's Lower East Side Jewish ghetto to Chicago, having saved enough to buy a house in the respectable Woodlawn district. The grinding poverty of Poland and the early immigrant years behind them, the family worked hard and by 1919, Abraham Rickover, a tailor by trade, was able to start his own small clothing company. He was still working there when he died at the age of ninety-three.

In the years ahead, Hyman Rickover would not encourage writers or interviewers when they sought to fill out the picture of his background. He spoke a little about how hard his father worked. He never spoke about his mother at all. "Forget the personal stuff," he would say; "the real issue is what a man accomplishes." If others had to judge him, they would do so, as with everything else, on his own terms.

Within the authorized snippets of his background, however, there are signs of a harsh, disciplined, unsupportive family. There are no signs of parents who sacrificed everything for the welfare and, especially, for the education of their children. The familiar warm tales from thousands of other immigrant Jewish families are simply absent. Rickover's elder sister was sent to work at the age of thirteen, as soon as she had finished elementary school. Hyman was permitted to attend high school but had to work, first as a delivery boy, then as a Western Union messenger, giving all his earnings to his father. A small, frail, and sickly child, he nonetheless took his cables by bicycle around the streets of Chicago from three in the afternoon to eleven each night. The pressure of work made him a tired, indifferent, friendless student. He had no time for parties, games, sports, or shows. He was a loner, not only at school but even in his own family.

There was no drive to send him to college. When Rickover was graduated from high school, his father was almost ready to start his small factory. As for so many other immigrant families, it was the local congressman who provided the escape route, arranging for the nineteen-year-old Hyman to sit for the entrance examinations to the United States Naval Academy. He was accepted.

Again, he proved an indifferent student, his scholarly results further depressed by poor marks for "military bearing." On graduation in 1922, Rickover saw what was in store for him: that year's number-two graduate, who had been in a neck-and-neck race for the top position, was a recluse like himself, a swot, a grind—in midshipman's parlance, a "cutthroat." The class of '22 purposefully obliterated all mention of this man in the yearbook—the kind of public degradation for which Rickover would always be on the alert. The fact that the "cutthroat" was also Jewish was not lost on the less illustrious Rickover, who may have owed his luck at missing the same treatment to his not making even the top one hundred of his class. That luck wouldn't hold, and in the future he would also be subject to the navy's brand of pseudo-aristocratic anti-Semitism.

For almost twenty years, he pursued a regular navy career. He served on various types of ships, qualified for submarine duty—apparently because promotions were quicker in that service—and, in 1937, finally achieved his first command. After that, he applied for designation as an "Engineering Duty Only" officer. Perhaps the seclusion of a mechanical world appealed to the loner. Or perhaps it was a need for some measurable objective criterion of achievement, something that was absent in the unrewarding activity of a peacetime navy.

In June 1939 he was appointed to the Washington-based Bureau of Ships, the unit responsible for design, construction, and repair of the entire naval fleet. As head of the Bureau's Electrical Section throughout the war, he first displayed the special skills he could bring to specific projects. At last he had a definite, clear-cut set of responsibilities, and he pursued them relentlessly. This soon drew comments on his apparent incapacity to see a broader picture, to understand priorities other than his own. Nevertheless, his energy, resourcefulness, and success were appreciated as he fought for staff and authority, oblivious to any other aspects of the navy's war effort.

Rickover's staff completely rewrote the navy's electrical manual, pruning huge numbers of wasteful spare parts and redesigning others. He shocked company presidents by ringing them up and demanding new designs, threatening to cancel all their existing navy contracts. These men had never been talked to like that. He refused to accept the inexorable implications of bureaucratic growth that turned other sections of the Bureau of Ships into general contract administration units. He insisted on close and continuous supervision of design and production. The management style he later brought to building the nuclear navy had already emerged.

When Rickover started on the atomic energy course at Oak Ridge in 1946, his most decisive insight was intuitive: he knew that practical reactor development was 90 percent engineering. The physics community, ignoring the lessons of the Manhattan Project, still saw it in terms of scientific development. Rickover had to fight his bureaucratic battles on two fronts: first, against a navy administration disturbed by his patent single-mindedness; and, second, against the scientists in the Atomic Energy Commission who wanted to pursue research and investigate several reactor options before deciding on definite lines of development. But he skillfully transmitted a sense of urgency to his superiors, and the submarine reactor project was launched.

In 1948, Rickover, formally in charge of naval reactors, set himself a deadline for having the first nuclear-powered submarine in the water and working: January 1, 1955. After seven years of driven, tortured, frenzied work, the *Nautilus*, with Rickover on board, cast off from the pier of the Electric Boat Company's yard at Groton, Connecticut. It was January 17, only sixteen days over the deadline. On April 22, *Nautilus* sailed into the navy's bureaucratic domain and, after a few days' delay because of a minor leak, it began a series of record-breaking runs. In eighty-four hours, it steamed 1,300 miles under water—ten times farther than any submarine had previously traveled submerged. Other records followed: *Nautilus* maintained a submerged speed of sixteen knots for longer than an hour, a record for any combat submarine.

Rickover's success lay in his compulsive pursuit of technical excellence: nuclear engineering did not allow for the margins of error often acceptable in conventional projects. "Millionths of an inch exist only in the mind," one of Rickover's harassed private contractors complained. Likewise, the self-imposed deadline, unrelated to any technical or strategic need, was a means of imposing order, ensuring urgency and determination, a measurable touchstone of achievement. It also reinforced his resentment of any interference, any questioning or attempt to audit the program, which would slow it down. And it reinforced Rickover's sense of total personal responsibility as well as his sense of possible failure, and that, in turn, increased the frenzy of the task. In its tenor, the Rickover program was much like the Manhattan Project, but this was not wartime. Rickover did not have the extraordinary powers of General Groves to expropriate what he wanted in the name of war. He did not have the privilege of forgoing accountability—at least in theory.

At the time he set his deadline for the launching of *Nautilus*, Rickover had entered his remarkable institutional arrangement that, in effect, gave him precisely these extraordinary powers. As head of the navy's nuclear unit, he was responsible for building *Nautilus*; as head of the AEC's Naval Reactor Branch, he was responsible for building it as well and as cheaply as possible. He could cut red tape simply by writing letters to himself. Sometimes he would act as a naval officer and sometimes as an AEC official; his superiors never knew which it would be. He kept his lines of authority so fluid that no precedents were ever set; there were no expectations of review or approval. Rickover set his own priorities and his own timetable.

It was not always that easy, however. After the war, the AEC had attempted to centralize reactor research at the Argonne National Laboratory, outside Chicago. The laboratory was the successor to Arthur Compton's University of Chicago group. The scientists of Argonne favored looking at a broad range of reactor types and resisted Rickover's gamble on the pressurized-water reactor, the PWR.

Choosing any specific reactor design at this stage was bound to pose risks, but Rickover felt that there were enough bright people working on the problems of the PWR to overcome them. Light-water reactors, such as the PWR, use ordinary water both as a moderator, to slow down the neutron bullets, and as a coolant, to cool down the reactor core. In the PWR, the water is prevented from boiling as it surrounds the reactor core by keeping it at very high pressures of about 1,500 atmospheres.

Initially, there was some caution at the AEC about giving Rickover free rein over the PWR; then, in the summer of 1949, the Russians exploded their first atomic bomb. Suddenly, all reservations were swept aside. Rickover's lines of authority were swiftly untangled. From that period onward, Captain Rickover got everything he wanted from the navy—except promotion. In mid-1952, he had to face the ignominy of being passed over a second time for promotion from captain to rear admiral. Under navy rules, Rickover had to retire the following year.

Why had he been turned down by a military branch that had put him in charge of one of their most vital departments? Put simply, Rickover was not a good military man; he could lead, but he could not follow. To his superiors, he was always a problem. He was a self-conscious black sheep, reveling in his lack of orthodoxy; worse, he was a specialist in a world that had always placed emphasis on ability to perform general tasks. His respect for tangible achievement had always been matched by total scorn for the illusions of status and sociability. He ignored, indeed defied, protocol, was repelled by conformity, and showed no respect for tradition, rank, procedure, or "official channels." His career had been studded with personality clashes. He had left in his turbulent wake an array of personal animosities, spite, and jealousy. Clinton Anderson, a powerful congressman on the Joint Committee on Atomic Energy, recalled meeting two top navy men at a dinner party. One said to him, "That Jew bastard will never get to be an admiral," and the other nodded in agreement. As Rickover once told a congressional inquiry, "This business of being on the team is not all it's cracked up to be."

Rickover refused to retire and turned to Congress for support. It came with a speed that only the magical world of atomic energy could have summoned. All members of the Joint Committee on Atomic Energy rallied behind the fifty-two-year-old captain. He was their kind of man: he shared their atomic zeal and he got things done.

The central role in obtaining Rickover's promotion was played by an original member of the Joint Committee, Henry ("Scoop") Jackson, recently elevated from the House to the Senate. It was virtually his first act as a senator. Rickover poured his heart out to Jackson, and Jackson arranged for the Senate Armed Services Committee to hold up ratification of all other naval promotions, usually rubber-stamped, until something was done about Rickover. The navy swiftly gave in; Rickover became rear admiral. From this time on, all future extensions of his tour of duty as architect of the nuclear navy would pass without problem; only Rickover himself, constantly possessed of a persecution mania, would continue to believe there were plots afoot to unseat him.

He became a regular witness at congressional inquiries, giving crisp, concise, frank, and pointed testimony, exuding self-confidence, often arrogance, but always giving the impression of total honesty. He pandered to the congressional self-image of watchdog; congressional listeners mostly loved him for it and gave him what he wanted. They adored his little epigrams, like, "The more you sweat in peace the less you bleed in war." But most of all, they knew Admiral Rickover was one person out in that mush of officers and officials who got the job done. That was enough to make it possible to ignore all the stories about his running roughshod over others to get what he wanted. And the Rickover who appeared in public was always courteous.

In 1953, with his new lines of responsibility to Congress established, Rickover was ready to do their bidding; to turn his attention from nuclear-powered naval submarines to civilian nuclear-powered plants to generate electricity. Rickover turned to the peacetime project with the enthusiasm of a warrior.

It began as a proposed reactor system for an aircraft carrier—the first step away from submarines in Rickover's emergent technical dream of an all-nuclear navy. In 1950, Rickover already had a team working on a technical design for his propulsion reactor. However, many questions loomed about the vulnerability of carriers under conditions of modern warfare. These, combined with the Eisenhower administration's commitment to cut spending, led to the cancellation of the project in April 1953.

As with many such budgetary events in the U.S. government, cancellation of the project was by no means the end of it. It had been known from the beginning that, if the navy project was cancelled, Rickover could quickly turn the reactor into a civilian program. And Westinghouse, which had developed the submarine reactor, would be only too happy to build it.

Pressure on the United States to build a civilian reactor had recently increased. The Russians were claiming they had an almost completed power reactor; the British were about to start operating one that would feed electricity into their national grid by 1956. Above all, President Eisenhower was compiling his Atoms for Peace speech, and the chairman of the AEC, Lewis Strauss, wanted a power reactor built rapidly to keep the United States ahead in atomic matters. Strauss thought it would be best—and get the quickest results—if the whole project was put under Rickover's control.

Rickover rose triumphantly to the occasion. Just before the AEC commissioners made their final choice of contractor, Rickover dramatically brought his Westinghouse prototype of the submarine PWR up to full power at the testing station at Idaho, demonstrating the success of the project by charting the "progress" of the reactor as it propelled an imaginary submarine across the Atlantic. By the end of the test run, he noted, the submarine had "reached" Ireland.

The AEC was impressed. Rickover's ally on the commission, a New York industrialist, inventor, engineer, and nuclear zealot named Thomas Murray, was convinced Rickover was the man for the civilian project. He used his considerable influence to get AEC approval of Rickover at a closed meeting of the commissioners; the objections of the AEC staff—that Rickover was becoming impossible to deal with—were swept aside.

The busy rear admiral applied his successful formula of tight personal control and technical excellence to the new site of the civilian reactor at Shippingport, on the Ohio River. In the wake of the Atoms for Peace speech, he had no problem persuading President Eisenhower to attend the ground-breaking ceremony. To heighten the sense of magic, the President was given a radioactive wand to pass over an electronic gadget that would start the first bulldozer. It was September 1954. Only three years and three months later, the reactor was operational and the Duquesne Light Company of Pittsburgh added 60 megawatts (Mw) of nuclear power, the first in the United States, to their electricity capacity. Westinghouse Electric had a new product line, the pressurized-water reactor. It could have been said, with equal justification, that it was Admiral Rickover's new line, for he had sought the same measure of control in Westinghouse as he had done everywhere else. And he succeeded.

Westinghouse was known as "the engineers' company," an image of stolid reliability. It never seemed to have the flair or originality of its arch rival, General Electric. Westinghouse executives were conscious of nothing more strongly than the fact they were "Number

Two"—second not only in size but in profitability and growth. The company lacked glamour, a sense of excitement, a feeling of fast-moving progress. This was reflected in its management system. Unlike the General Electric policy of rapid rotation of executives, the "engineers' company" tended to allow its top men to stay in the same field for years. It would be like that in nuclear power, too. For the next thirty years, three men successively held the top position in the company's atomic division: Charles Weaver, John Simpson, and Joseph Rengle. All three came up through the Rickover school; from naval propulsion to power reactors, their progress up the corporate hierarchy was tied to Rickover's technical program and his budget—about $4 billion during the decade of the fifties.

In 1947, at Rickover's urging, Westinghouse, seeing a new market with a touch of glamour, established a separate atomic power division with authority to recruit top-flight people throughout the company. Its first head was Charles Weaver, an engineer-turned-salesman. Rickover knew he could control him; as head of Westinghouse's marine department during the war, Weaver had cooperated with the demands of Rickover's Electrical Section of the Bureau of Ships. Rickover's lines of communication to Weaver would be simple and direct.

Weaver became a malleable implementer of Rickover's priorities. His nuclear division, and its specially created laboratory on the site of Pittsburgh's old airport at Bettis, became more closely tied to Rickover's team than to the Westinghouse corporation. Weaver accepted without demur Rickover's demanding behavior, reorganizing his division and assigning personnel as Rickover requested. He also acquiesced in the total probing interference of Rickover's inspection teams. Rickover once complained that a Westinghouse machinist looked sloppy and his overalls were dirty, and the worker was immediately taken off naval contracts.

Under Weaver at the reactor division was an Annapolis graduate, younger than Rickover, named John Simpson. Rickover had first met him at Oak Ridge, when they were both on the reactor induction course and had shared quarters. By 1953, Simpson had already been Westinghouse's chief engineer on the Idaho submarine project and then project manager on the abortive aircraft carrier reactor. In 1955, when Weaver became a vice-president of Westinghouse—the first to ride the atom to this position within the company—Simpson took his place as director of the Bettis laboratory. Joseph Rengle, another Annapolis graduate, took Simpson's place. They were both Rickover's boys.

The admiral tolerated no disloyalty, or, for that matter, disobedience in his acolytes. When, in the course of time, Simpson succeeded Weaver as vice-president of the Westinghouse power-reactor division, he decided to set up the astronuclear laboratory for the National Aeronautics and Space Administration to pursue research into nuclear-propelled rockets. Rickover was furious and reacted with venom, vetoing any transfer of Bettis-trained personnel from naval contract work to the NASA project. When Simpson went ahead, Rickover imperiously removed the naval contract for the most advanced nuclear submarine reactor and reassigned it to Westinghouse's rival, General Electric. He also placed a ban on any of the Bettis workers who transferred to NASA from ever returning to work on a naval contract.

By this time, however, the naval reactor project had been responsible for the training of hundreds of nuclear engineers and had created for Westinghouse an atomic empire large enough to insulate itself from the admiral's anger. That the Shippingport PWR produced electricity at ten times the cost of the average conventional coal- or oil-fired plants did not matter. Through the AEC, Westinghouse already had approval and the promise of government assistance to build an even larger PWR reactor. Thanks to Rickover, the company that had always been "Number Two" had a real chance of becoming a leader in the latest development on the atom.

Their rivals had not been idle, however. Immediately after the war, General Electric had agreed to take over responsibility for managing the plutonium-producing reactors at Hanford, on one condition: the new Atomic Energy Commission would have to fund a complete nuclear research laboratory for the company. The AEC accepted the terms. The new laboratory was built near the company's headquarters at Schenectady in upstate New York.

Scientists dominated the GE program in those early days, and it is not surprising that they focused on the one technology that seemed to ensure that the company could stay ahead in its field: the breeder reactor. Fueled by plutonium, the breeder produces more plutonium than it consumes.

Rickover was not impressed; breeder reactors were never likely to be used in his submarines or his aircraft carriers. He also objected to

GE's basic approach of placing scientists rather than engineers in control. And he demanded absolute priority for his submarine projects; GE balked at that. In 1950, however, the company ran into technical difficulties on the breeder; it was then willing to negotiate with the admiral on his terms and accept a navy contract. GE built one submarine reactor, sodium- rather than water-cooled; but it never allowed its laboratory to be used, like the Bettis laboratory of Westinghouse, as a single-purpose organization devoted exclusively to Rickover's pet schemes.

GE's position as Number Two in nuclear power development was only temporary. At the government-run Argonne National Laboratory, nuclear scientists were working on the theory of another light-water reactor, the boiling-water reactor, or BWR. This is similar to the PWR in that it uses ordinary, or light, water as a coolant and a moderator, but there are two differences. First, the water is not under pressure and is allowed to boil in the vessel surrounding the core; hence, the vessel does not have to withstand the tremendous pressures of the PWR and there are no external steam generators. Second, the steam produced in the core is used to drive the electricity generators; it does not, as in the PWR, heat a separate water system that produces steam to drive the generators. In fact, the BWR is the simplest concept of all nuclear-power reactors; it is also cheaper than the PWR.

GE eagerly seized the opportunity to develop the new technology—on its own. It spurned the option of government subsidies, created a new laboratory near San Jose, California, specifically to develop the BWR prototype, and found an electric utility consortium led by Chicago's Commonwealth Edison Company eager to buy it. GE offered the consortium a fixed-price contract, accepting total responsibility for nuclear plant construction and start-up. The arrangement made with the consortium became known as a "turn-key" contract: the buyer is sold the finished article; all he has to do is turn the key to start it. A decade later, this sales incentive became the central factor in persuading reluctant power utilities in the United States to go nuclear.

GE's first boiling water reactor, known as Dresden 1, began feeding 200 Mw into the Commonwealth Edison Company's grid in 1959. It was the world's first large-scale, privately financed nuclear power plant. Shippingport, of course, had been built entirely with government funds. Shouldering the development cost was a huge risk for GE, but at least the historical omens were favorable; Commonwealth Edison had placed the order for General Electric's first steam turbine for power generation in 1903. The two famous 5-Mw turbines, powered by coal-fired boilers, were installed at Chicago's Fisk Street Station. From that small start, GE had developed the world's most successful electricity-generating technology. Now it was striding out again, this time into the nuclear business.

Their turn-key contracts would become really important only in the mid-sixties as General Electric and Westinghouse sought to sell bigger and more advanced nuclear reactors. At this early stage, there were only a handful of atomic-power zealots in the private utilities, men who, accepting the puffery of "electricity too cheap to meter," wanted to be in on the ground floor of this amazing technological revolution. The other private companies that joined Commonwealth Edison in accepting the nuclear option at this time were New York's Consolidated Edison, California's Pacific Gas and Electric, and a new consortium of New England utilities. A decade later, the same executives in these companies would place orders for the first round of the larger reactors.

By the end of the Eisenhower period, the key pieces in the pattern of private nuclear reactor development were in place. All that remained was to translate enthusiasm into plants. Yet, with hindsight, there is a final irony in the five-volume study of America's natural resources that was on the President's desk when he took office in January 1953. The document was the report of a special commission headed by CBS chairman William Paley. Forecasting to the year 1975, the study, predicted oil shortages and concluded: "Nuclear fuels, for various technical reasons, are unlikely ever to bear more than about one-fifth the load. . . . It is time for aggressive research in the whole field of solar energy—an effort in which the United States could make an immense contribution to the welfare of the world."

In the intervening years, some $200 billion have been spent throughout the world in attempts to develop nuclear power. Solar has received perhaps one-thousandth that amount. At the end of the 1970s, solar technology was still regarded as "futuristic" and nuclear technology as "mature." Nuclear advocates continue to talk as if there had never been a choice about what to do first, as if there was a natural order for the sequence of human discovery.

11

"The Perennial Fountain"

The revelation of the secrets of nature, long mercifully withheld from man, should arouse the most solemn reflections in the mind and consciences of every human being capable of comprehension. We must indeed pray that these awful agencies will be made to conduce peace among nations, and that instead of wreaking measureless havoc upon the entire globe, they may become a perennial fountain of world prosperity.

—WINSTON CHURCHILL, August 6, 1945

The summer of 1955 was a landmark year for the "peaceful atom." For the first time since the Cold War began, scientists from East and West—seventy-three countries in all—gathered at Geneva for the start of what was to become a series of international conferences on the peaceful uses of atomic energy.

At the outset, the Americans had been determined to turn the conference into a propaganda triumph. They had persuaded the somewhat bewildered Swiss to let them build a small working reactor in the magnificent grounds of the Palais des Nations, and the Russians, who had brought only a model of a power reactor, were easily upstaged.

But the icy competitiveness quickly thawed. Much to the surprise of the Western scientists, their Soviet counterparts were open and cooperative. They gossiped their secrets away on the ornate couches of the Palais des Nations and, after the close of each formal daily session, happily continued their conversations over dinner in Geneva's expensive restaurants. By the end of the conference, the head of the U.S. delegation, the physicist Isidor Rabi, commented, "Right now I would say that the old Iron Curtain is composed of almost equal parts of iron and red tape. Of the two, the red tape is the harder to figure."

Paper after scientific paper covering the whole range of the peaceful atom—from producing electricity to sterilizing sewage—was read with simultaneous translations in several languages. For two weeks, the world's press gave the gathering front-page coverage. Those were euphoric days; the future of nuclear fission seemed assured. New sources of uranium, previously thought to be in short supply, were being discovered throughout the world, and nuclear engineers were suggesting that one ton of uranium would eventually do the work of a million tons of coal.

The hazards of radiation, even at low levels, and the problems of disposing of the highly toxic wastes produced by the power reactors—problems that were to plague nuclear power in the years to come—were put aside; without a stitch of practical evidence, the scientists and engineers declared their boundless faith in their own technological virtuosity. At the final session, the conference president, Dr. Homi Bhabha of India, said, "The feasibility of generating electricity by atomic energy had been demonstrated beyond doubt."

The overwhelming success of the Geneva Conference gave rise to a new international community of physicists, chemists, biologists, doctors, and engineers, all with a common interest in the development of the peaceful atom, and to a never-ending stream of seminars, exchange visits, and more Geneva conferences. At the United Nations, the development of the atom, unlike any other field of science, was given a special organization to itself: the International Atomic Energy Agency. Such treatment reflected the enthusiasm and optimism of the moment, an optimism the new international community nurtured vigorously.

Germany, Japan, Belgium, Italy, Spain, Brazil, and Argentina created their own atomic energy commissions. A huge international nuclear bonanza emerged, with nations vying for bigger and better pieces of the atomic magic. Everyone wanted to know the size and thrust of the other's atomic program: What type of reactor had they chosen? Just as important, which ones had they left out? Who had found new sources of uranium? Everywhere, there was the same refrain: If we don't jump on the nuclear bandwagon now, we'll be left behind forever. To be cautious was to be unpatriotic.

At the time of the first Geneva Conference, only five nations—America, Russia, Britain, France, and Canada—were developing reactors with an export potential. With the exception of Canada, all were developing civilian power reactors as a direct outgrowth of existing military programs, and even Canada's heavy-water reactor program had received a significant indirect subsidy from the U.S. purchase of its plutonium. The British and the French were developing civilian reactors based on the carbon-dioxide gas–cooled, graphite-moderated reactors they had built to produce plutonium for their bomb programs. In so doing, they could continue to increase their plutonium stockpile while producing nuclear power. America and Russia had chosen the pressurized-water reactor for early civilian development after first using it in naval propulsion programs.

Of the five, only Russia had actually produced a tiny amount of nuclear power for civilian use; in 1956, Britain would become the first nuclear nation to feed electricity from a reactor into a national grid. At the time of the Geneva Conference, Britain was the only country to have set itself a definite goal for nuclear energy: twelve gas-cooled reactors in ten years and the production of half its electricity needs from nuclear power by 1975. For Britain, the economic equation was straightforward. Coal was still in short supply, as it had been since the end of the war, and imports of the readily available alternative, oil, were dogged by political upheavals such as the one that led to the nationalization of the Iranian oilfields. Above all, the current optimistic cost estimates showed nuclear power to be competitive with, if not cheaper than, coal.

At Geneva, the question of reactor types was openly discussed. Britain's chief nuclear engineer, Sir Christopher Hinton, championed the gas-cooled reactor, calling it safe and more reliable than the American light-water types. It was certainly more expensive, requiring more sophisticated engineering than the U.S. light-water types. But the British argued that their reactor's by-product, plutonium, would offset the extra costs to some extent; the plutonium could be used not only for bombs but also, eventually, to supply fuel for the more advanced, though as yet unbuilt, breeder reactor, which burns plutonium as a fuel and in the process makes more than it uses. The British were also ahead of anyone in designing a breeder prototype.

Although the Russians were developing a range of reactor types, including the breeder, they were not ready to enter the post-Geneva scramble for the nuclear export market. Until the early 1960s, that market would remain the preserve of Britain and America.

Within six months of the Geneva Conference, the United States had signed twenty-nine "Agreements of Co-operation" with so-called friendly countries to help promote sales. These agreements initially supplied the countries concerned with small research reactors; they were soon extended to include assistance loans from the Export-Import Bank to finance demonstration power reactors in cooperation with the U.S. manufacturers, Westinghouse and General Electric. At the same time, the British were making direct commercial sales through their own General Electric Company and Nuclear Power Plant Company. By the end of the 1950s, Westinghouse had sold one light-water reactor each to Belgium and Italy; General Electric had sold one each to Germany, Italy, and Japan; the British General Electric Company had sold a gas-cooled reactor to Japan; and the British Nuclear Power Plant Company had sold one to Italy.

The enthusiasm of the established atomic powers swiftly infected those countries who, at Geneva, had yet to join the nuclear revolution. Two of the most eager were the defeated powers of World War II, Germany and Japan. A third was India, the first underdeveloped country to attempt a serious nuclear program. In both Germany and Japan, the postwar bans on nuclear research had only recently been lifted, and each had come to Geneva with a sense of technological backwardness and a determination to make up for lost time. For India, going nuclear was part of a fierce determination to exchange the spinning wheel for modern technology. In Germany and Japan, starting a nuclear power program in the 1950s was at best premature and at worst a substantial misallocation of scarce resources. For India, it became a national tragedy.

The prestige of the new industry, boosted by the frolic of confidence and pride at Geneva, attracted industrial gamblers—moguls who were willing to invest huge sums to launch nuclear programs. In Germany, it was Karl Winnacker, a research chemist who had been a wartime "crown prince" of the vast German chemical conglomerate I. G. Farben. After Geneva, he became the driving force of the German Atomic Energy Commission. Within hours of savoring the technical excitement of the first conference sessions, Winnacker was on the phone to his leading technicians in Hoechst, the chemical company of which he was then chief executive, demanding they come to the conference to learn the new technology. He was profoundly moved by the exhibits—especially the Americans' working research reactor. His unflagging support for the birth of an independent German nuclear industry, which eventually resulted in the most significant nuclear technological development of any nation not assisted by a weapons program, was to earn him the title of the "German Nuclear Pope."

Geneva made the Germans acutely conscious of the loss of the atomic prowess they had enjoyed in the heady days of the 1930s. After the conference, German newspaper headlines complained: "Now We Have Fallen Behind" and "The Government Neglects the Future." This put considerable pressure on the government of Chancellor Konrad Adenauer to create its own separate organization to develop the nuclear industry. Two government bodies were created: a Ministry for Atomic Questions, which was to disburse the taxpayers' money, and the German Atomic Commission, which was to spend it. There was also a separate regulatory body known as the Reactor Safety Commission.

Of the three vice-chairmen of the German commission, Karl Winnacker quickly asserted himself as the most powerful. The other two, Otto Hahn, the co-discoverer of nuclear fission, and Leo Brandt, a State government representative, were never a threat to Winnacker's prominence. Winnacker began to shape the commission to his own ends, creating a new nuclear center at Karlsruhe and placing his friends and colleagues in key positions. Karlsruhe's first technical director was Gerhard Ritter, an engineer-administrator from Winnacker's own company, Hoechst. The Reactor Safety Commission was run by Hoechst's chief engineer, Joseph Wengler, a friend of Winnacker's from student days.

Born in the north German town of Bremen—the heartland of the most provincial, puritanical, and bigoted element of German Protestantism—Winnacker was the archetype of the successful German industrialist: sober, aloof, disciplined. Winnacker was one of the first specialists in the new field known as process engineering—the technique of producing new chemicals on an industrial scale. This was the borderland where chemistry, physics and engineering met—the same frontier which would pose the greatest challenges in the development of atomic energy. Because of his wartime association with I. G. Farben, he had been subject to de-Nazification orders and, in the immediate postwar years, the only work he could get was as a gardener. During these years, he wrote a multivolume treatise on chemical technology that remains to this day a classic in its field. In 1952, as the reorganization plan for I. G. Farben was implemented, he was allowed to return to the management committee of the Hoechst plant, one of the three successor companies. By the end of 1952, he had become chairman of the committee and had succeeded in ensuring that Farbwerke Hoechst A.G. would receive a major slice of the I. G. Farben inheritance. Each of the three successor companies was to grow into a major multinational operation, and all would find a place on the list of the top ten companies in Germany. Hoechst eventually became the largest chemical company in the world.

On the eve of the Geneva Conference, Winnacker, an early convert to the future industrial importance of nuclear energy, had already played a central role in grouping the major German corporations into what was called the "Physical Study Society"—the first step in a pattern of industrial support for nuclear research. At Geneva, he was the senior businessman in the German delegation.

Over the next twenty years the German government and private investors would pour nearly 20 billion Deutschmarks into nuclear research. Thirteen different reactor types would be investigated; the aim was to develop independent systems that did not rely on the established technologies, the light-water reactors in America and the gas-cooled types in Britain. Winnacker wanted to perpetuate the I. G. Farben tradition of autarchy.

At first, the best prospect for short-term national independence seemed to be in the development of the heavy-water type of reactor. This used natural uranium as fuel rather than the slightly enriched uranium used by the light-water reactors, which could be obtained only from the American enrichment plants. It also used heavy water as a moderator, and Hoechst had developed heavy-water technology during the war, but the plant was inefficient and not revived after the war. For long-term independence, Winnacker had even grander ideas: he believed that the best hope lay with the breeder reactor, and he plunged Hoechst into reprocessing plutonium for use as breeder reactor fuel, assuring his board and his shareholders that it would be money well spent.

It had seemed to Karl Winnacker that, through nuclear energy, physics would transform the industrial system in the second half of the twentieth century as surely as chemistry had transformed it in the first half. His grasp of the technology told him so. His Japanese nuclear counterpart, Shoriki Matsutaro, had no such training. But he was a gambler. Nuclear energy looked like a good bet in the mid-fifties, and he was to become the most colorful promoter of nuclear power the world had ever seen.

Shoriki was a swashbuckling wheeler-dealer whose colorful list of exploits in the name of Japan include being the father of Japanese baseball, the father of Japanese television, and, finally, the father of Japanese atomic energy. He plunged into atomic energy in 1955, declaring the prospect was no less than "to liberate mankind from poverty and disease." And he vowed that Japan could recover its lost lead in nuclear technology as quickly as it had done in electronics; all that was needed was determined leadership willing to take the huge risks involved—and he was the man to provide it.

His enthusiasm for nuclear energy was the climax to an extraordinary career as press baron, promoter, TV mogul, and cabinet minister. In the early 1920s, he had taken an ailing daily newspaper, the *Yomiuri Shimbun*, and used it to create the foremost press empire in a nation with a huge appetite for the printed word. He had mastered every detail of publishing, absorbing himself in everything from printing costs to ink and paper supplies. Above all, he had kept a close watch on the cutthroat battle of Japan's contorted retail distribution system. He knew every agent's newspaper sales. He was especially good at organizing media spectaculars, from exhibitions of national art treasures to explorations of the inside of a volcano. His greatest successes, however, were sporting events.

Shoriki organized tours of French boxing champions and American tennis aces and baseball teams. When he brought Babe Ruth to Tokyo in 1934, a million people turned out to watch the open cavalcade drive down the Ginza. Shoriki organized the first of the Japanese professional teams, the Tokyo Giants, to play the Americans in the Meiji stadium; for this, he was attacked in the street by right-wing militants who accused him of having defiled the Emperor Meiji's memorial stadium by staging the baseball game. Left for dead in a Tokyo alley, Shoriki spent two months in hospital.

Shoriki had believed from childhood that he was destined for greatness. His grandfather had been a bridge builder, a commoner who was elevated to the minor samurai class and granted the family name for his public service by a grateful lord. His father had risen no further, being a local engineer in a small rural community. But when Shoriki was six, a fortuneteller had prophesied: "Matsutaro . . . will become the greatest man in this part of the country."

Shoriki was plagued by illness as a child, and the first sign of his energy and determination to succeed was shown in his battle for good health. With extraordinary self-discipline, he eventually gained the third highest of the ten black belt judo ranks. He graduated from the law faculty of Tokyo Imperial University—the central training ground for the Japanese administrative and political elite, and even more influential in Japan than Oxford and Cambridge in Britain. His classmates included an impressive array of future ministers and industrialists.

The civil service offered one of the few avenues for upward social mobility for a young man without title or wealth, and in 1913 Shoriki joined Tokyo's police force. He featured in a number of sensational criminal cases—in one, exposing a bogus Christian minister as a mass murderer and, in another, tracing the assassins of a member of the Korean royal family. His efficient control of mass demonstrations attracted the attention of the political leadership. With considerable physical courage—and occasionally serious injury to himself—he quelled riots over the price of rice, demonstrations over the right to vote, and student uprisings, breaking the traditional sanctuary of the campuses to make arrests. In 1921, he was appointed director of the Secretariat of the Metropolitan Police Board, the central post in domestic political intelligence.

He kept a firm hand on the militant movements of Japan and became well known at the highest political levels. But in 1924, his police career came to an abrupt end over an assassination attempt on the then Crown Prince, Hirohito. In the Japanese manner of apportioning blame to the highest rank, Shoriki, then police chief, was one of the men who had to accept responsibility for this breakdown in security, and he resigned. By this time, however, he had accumulated enough influence with the rich and powerful to get financing to buy the ailing *Yomiuri Shimbun*.

Shoriki proved especially adept at exploiting the Japanese public's voracious thirst for military news during the conflicts of the thirties. By World War II, Shoriki, then in his middle sixties, was an energetic supporter of Japanese imperialism. Two days after Pearl Harbor, he was a keynote speaker at a huge Tokyo demonstration billed as a "National Rally to Crush the United States and Britain." Shoriki had finally arrived. He became a leader of the Imperial Rule Assistance Association, which replaced Japan's political parties during the war and, in 1944, was elevated by the Emperor to the House of Councillors. It was for these activities that he was classified by General Douglas MacArthur's occupation administration as a Class A war criminal. Although no charges were ever brought against him, he was subject to purge orders, and this prevented him from holding any public or corporate office. For almost two years, he was an inmate of the Sugamo prison, where he passed the time in Zen meditation—something he had taken up in his youth to improve his judo. His cynical fellow prisoners noted that the American warden exempted Shoriki from hard labor because he thought he was religious.

In turbulent postwar Japan, Shoriki's entrepreneurial ardor never dampened. With the lifting of the purge order, he began launching new schemes. After the 1955 Geneva spectacular, he turned his attention to nuclear energy, inviting two American nuclear power experts, John Jay Hopkins of General Dynamics and Laurence Hafstad of the Chase Manhattan Bank, to lecture on the new atomic age. He organized an Atoms for Peace exhibition on loan from the United States; it drew hundreds of thousands of visitors. He also grouped top Japanese industrialists into a new organization to promote nuclear energy.

Shoriki became a parliamentarian, joining the Liberal party of his old friend Hatoyama Ichiro, who became Prime Minister when the U.S. occupation ended. Hatoyama gave him a choice of portfolios in the new government, and Shoriki chose atomic energy. He modeled the Japanese atomic decision-making process on the German example; the Japanese Atomic Energy Commission distributed government funds, promoted research, and trained personnel. It was the foundation of a nuclear future for Japan.

Shoriki shared the traditional philosophy of the Japanese industrial and political leadership: import foreign technology, then copy it. He was impatient to bring nuclear power to Japan, and that was the quickest way. A minority within the Japanese Atomic Energy Commission and the overwhelming majority of Japan's scientific community wanted to delay any precipitous commitment, favoring instead support of autonomous research, but Shoriki simply overrode them and signed early agreements of cooperation with the United States and Britain.

In 1957, Britain's chief nuclear engineer, Sir Christopher Hinton (whom Shoriki brought to Tokyo at his own expense), advised that it was technically possible, but as yet untested even by the British, to increase the size of the first successful gas-cooled reactor. Shoriki promptly ordered the biggest one he could get. He refused to listen to the other members of the Japanese AEC who, counseling caution, wanted some independent domestic research done before buying the British product. Japanese scientists were quick to point out that the pile of graphite bricks surrounding the core might be an acceptable structure in Britain but it was totally unacceptable in earthquake-prone Japan. Those with long memories recalled the last British effort to introduce "progress" into Japanese society. During the Meiji era, British architects and construction engineers, then regarded as the best in the world, had built massive structures like the Army Staff Headquarters and the Foreign Ministry in Tokyo; both had been completely destroyed by earthquakes. A brief but intense debate ensued in Japan over reactor safety, but nothing deterred the ebullient Shoriki. He helped organize Japanese corporations to take part in a "Long-Term Basic Plan for Atomic Energy Development and Utilization." With extraordinary confidence in the untried technology of breeder reactors, Japan's plan included three large breeders for the 1960s. [Within a year it became clear that the plan was hopelessly overoptimistic, and none of the breeders was built.]

Shoriki's stubbornness led to the early resignation from the commission of Professor Yukawa Hideki, Japan's first Nobel Prize–winning physicist. In the end, Shoriki's enthusiasm would be shown to have been misplaced. The Japanese would have a bitter experience with the British reactor: it took longer to build and cost far more than originally estimated.

If Winnacker and Shoriki led Germany and Japan too hastily into nuclear research, they at least did so in countries that already had an industrial base. India had no such foundation. Indeed, in the midst of the optimistic speeches at Geneva, one note of caution had been introduced: a British delegate, Sir John Cockroft, had warned, "We must recognize that nuclear energy itself is not a magic key to prosperity for underdeveloped countries." But the founder of India's nuclear program, Homi Bhabha, ignored the warnings.

An unusual mixture of the abstract dreamer and the practical engineer, Bhabha was the personification of the age of optimism. Educated in Europe, where he learned the rudiments of nuclear physics, Bhabha was caught in India when World War II broke out. Ahead of his time and his country's needs, he made it clear to his well-connected friends and family that he would leave India when the war was over if a nuclear research institute was not established for him to help steer his backward homeland into the atomic age. His family was connected by marriage with the Tata family, owners of the largest industrial empire in India, and the Tata trust fund provided the money for the institute. In 1948, India established an Atomic Energy Commission; at the time, the only others in existence in the West were in America and France. Bhabha became the commission's chairman, a post he held until his death in 1966. Under the political patronage of Nehru, the Indian AEC established a degree of autonomy that insulated it from the checks and balances of regular democratic government. No questions were asked by other government departments, and the commission worked with a degree of secrecy that would not be breached until the late 1970s. A year after the 1955 Geneva Conference, India's first homemade research reactor started up, the first indigenously built reactor in Asia.

In the next two decades, India spent hundreds of millions of dollars following Bhabha's grand strategy of leading India toward an independent nuclear industry. (It was not an exclusively peaceful program: the option for an independent deterrent was always kept open.) Such faith in the power of the atom would have been staggering even in a nation with a greater margin for error in allocating scarce resources of capital and skilled manpower. Homi Bhabha never lived to witness the extent of the tragedy that resulted, and if he had, it is unlikely he would have been remorseful. He saw the huge capital outlay as an act of high patriotism, an act that would simultaneously bring India out of the spinning-wheel era and ensure its national security.

India's terrible poverty had never touched Bhabha. He was born the son of a wealthy Parsi family and, though a native of Bombay, he was spared the social torments of his city. A sign on Little Gibbs Road, where he grew up, forbade even the cries of street hawkers. His school lunches were taken at the ancestral home of the fabulously wealthy Tata dynasty. He lived in a patrician world of books, music, and painting.

The Parsi community of Bombay, abiding by the faith of the

monotheistic Persian prophet Zarathustra, maintained a close-knit family life, but there was nothing especially Indian about Bhabha's experience or his education. His youth was dominated by European classical music and opera, and he was educated at Cambridge, England, and in Switzerland. During those formative years abroad, Bhabha poured all his adolescent enthusiasm and passion into the excitement and beauty of the new theory of atomic physics.

After the war, he cultivated the image of a superpatriot. Throughout his public life, however, it was never clear how much he really was motivated by his loyalty to the idea of national independence and sovereignty, and how much by his pride and self-interest as a scientist and as a scientific administrator. Although his determination to build a totally independent nuclear industry was rationalized in terms of nationalism, its real roots appeared to lie in his personal egoism. Bhabha was determined to establish his reputation as a creative scientist and, later, as a driving, successful administrator with a large indigenous and self-sustaining program (nothing was ever to be imported twice).

His determination was nurtured in a sympathetic environment. The Indian elite had a deep sense of grievance; they shared with the Chinese a consciousness of a civilization of great antiquity and continuity that had been subjected to foreign domination and was suffering from economic backwardness. Measured against the European cultural values the Indian elite embraced, their society was a failure. They held an unquestioned faith that the key to spanning the centuries of backwardness lay in science and technology. Their shared sense of anxiety for a miraculous cure made them susceptible to the exaggeration of the nuclear zealots; like political elites in other emergent countries, they relished the thought of a shortcut to progress, a magic formula for a great leap forward. Just when they thought they were being most independent, they were in fact at their most subservient to the intellectual fads of the West. One example of this insecurity was the matter of solar power. When Bhabha embarked on his nuclear drive, there was great enthusiasm in India for solar power. Bhabha himself thought it was an "urgent" alternative source of energy; but because the technology was not being developed by the industrialized West, the urgency dissipated and solar power became subsumed by the nuclear age.

No one believed in scientific advance more strongly than the Prime Minister, Jawaharlal Nehru. Under his patronage, Bhabha established atomic research institutions of acknowledged international stature. His relationship with Nehru was a natural one: sharing an aristocratic family background, study at Cambridge, and an interest in Western culture, Bhabha and Nehru dined frequently, and Bhabha always had immediate access to him.

Nothing indicates more completely Nehru's disengagement from the philosophy of Mahatma Gandhi than his patronage of nuclear energy. While acknowledging the power of Gandhi's idealistic world of "appropriate technology"—the concept of small is beautiful—the new examples of Western science and technology made this philosophy seem impractical. The change in emphasis was clear from the first days of independence, when Nehru's government decided that Gandhi's symbol of the spinning wheel was inappropriate for the centerpiece of the new national flag, and it was replaced by a circular flower.

Confidence in India's nuclear future was so overwhelming that no one inside the country seriously questioned the fact that its atomic research employed as many scientists and received as much money as government research for either industry or agriculture. Doubts about this internal resource drain came later, and only in retrospect does the distortion of economic priorities seem so huge as to be bizarre.

Bhabha's nuclear drive was based, in large part, on the fact that India had the world's largest deposits of the element thorium. Under certain conditions, thorium can also be used as a reactor fuel. He saw these thorium deposits as the foundation stone of energy independence for a millennium. While the rest of the world concentrated on uranium, Bhabha adopted a theoretically viable long-term strategy leading to thorium reactors. The first stage was to include natural uranium reactors producing plutonium. The plutonium would be burned in breeder reactors with a thorium blanket, producing the less common fissionable isotope of uranium, U-233. This, in turn, would fuel thorium breeders. The whole plan was a manipulation of untested technology and uncosted investments. The ambition was staggering. Yet, for many years Western observers were almost unanimous in their praise for the Indian program under Bhabha's leadership. Indian scientists and technologists maintained positions at the forefront of international technological development. Indeed, their achievements were admired particularly because of the nation's limited resources; few questioned the economic wisdom of the project.

There was, however, one dissident voice at the end of 1955, and it came from I. M. D. Little, one of the leading energy economists of the day. Professor Little pointed out that Bhabha had overestimated the costs of conventional power stations, assumed optimistic availability factors for untested nuclear plants, and underestimated the thermal efficiency of modern coal-fired plants. Little said: "To put any of her own capital resources into buying the early products of this Western research would seem to be a great waste of the very limited savings of the Indian people. As Dr. Bhabha says: electricity is in short supply in India. It is likely to go on being in short supply if one uses twice as much capital as is needed to get more."

Bhabha never commented on Little's analysis. After the second Geneva Conference in 1958, the French physicist Francis Perrin added his voice to Little's criticism, arguing that underdeveloped countries could not expect the full advantages of nuclear technology until they had passed through a phase of industrialization in the traditional way. Bhabha simply asserted: "I do not believe it."

12

"What Are You Doing?"

Tests snowfall Rochester Monday by Eastman Kodak Company give ten thousand counts per minute, whereas equal volume snow falling previous Friday gave only four hundred. Situation serious. Will report any further tests obtained. What are you doing?

—Telegram sent by the National Association of Photographic Manufacturers to the U.S. Atomic Energy Commission, January 1951

The winter is bitter on the industrial shore of Lake Ontario. An icy wind blows relentlessly, and temperatures stay well below freezing, rising only when it snows. On January 29, 1951, heavy snow fell on the lakeside city of Rochester, New York. As the snow floated down to earth, the Geiger counters at the film production plant of the Eastman Kodak Company started to click madly. The snow was radioactive.

Kodak executives immediately suspected fallout from the atom bomb tests at the new test site in Nevada, more than two thousand miles away. It had happened before: after the first atom bomb had exploded in New Mexico in 1945, radioactive particles had "rained out" into the Wabash River in Indiana, over a thousand miles from the test site. The river water had contaminated some paper-stiffening board used to package film, and the film had been ruined. The tests that followed had been run on the remote Marshall Islands in the Pacific and there had been no recurrence. Now, after an interval of almost six years, it was happening again.

Kodak complained to their industry watchdog, the National Association of Photographic Manufacturers, who sent a telegram to the bomb test organizers, the U.S. Atomic Energy Commission, asking, "What are you doing?"

The answer came back loud and clear. In the midst of the Korean War, the Truman administration had decided, as a matter of national security, that it had become necessary to have a nuclear test site on the continental United States. A site, a remote, underpopulated desert region in the state of Nevada, had been chosen as a good place to let off bombs. No assurances could be given to the photographic industry that radioactive fallout would not be carried by wind, rain, or snow to distant parts of the country—such as Rochester, New York. In fact, since the prevailing winds at the Nevada test site were southwesterly, the fallout cloud from any test would move in the general direction of Lake Ontario. Unfortunately for Kodak, Rochester just happened to be in what the AEC bomb-test experts called the fallout path.

The Kodak Company's first inclination was to sue the government for damages. They settled, instead, for privileged status: in an extraordinary departure from a policy of secrecy, the AEC offered to send Kodak, in advance of all new tests, a set of secret fallout maps predicting the areas of possible heavy radioactivity. Several top executives of Kodak and other photographic companies were given security clearances so that they could handle the information.

Those who lived in Nevada and southern Utah, and who were about to be more directly affected by the fallout than the Kodak company at Rochester, were not so privileged. What the AEC decided it had to do for the large photographic companies was something they decided they did not have to do for those communities. They were not given maps telling them where the AEC expected the fallout to come down to earth; nor were they given advance warning of the testing of the "dirty bombs"—the ones with a high explosive force that produced more fallout.

In fact, during the decade of the 1950s, the AEC never developed a fallout protection system to ensure that those people would not receive more radiation than the experts of the day thought to be dangerous. Nor were the most harmful effects of fallout—from ingested radioactive particles that pass into the food chain through cow's milk or crops—ever measured. The sheep and cattle farmers whose animals grazed close to the test site were never told when to lot-feed their animals to prevent their eating contaminated pasture grasses.

When thousands of sheep died mysteriously after one of the tests and hundreds had radioactive burns on their faces, the AEC deliberately played down any causal link to the tests, suppressing reports from their own experts that seemed to confirm this relationship.

In the 1950s, the AEC knew that the public did not understand the weird and wonderful ways of the atom; it was assumed that as long as the public remained uninformed, they would not worry; nor would they ask embarrassing questions. The AEC could continue unchallenged with their higher purpose: the national security of the United States. The civilian members of the AEC and the experts they consulted had an acute sense of acting for the public good: in their case, the public good meant defending the world against the march of Communism. Unlike some of the atomic scientists in the weapons laboratories, they had no moral qualms about the spread of nuclear bombs; any doubts and criticisms about the risks of radiation were put aside as they decided that national security took priority over public health.

Indeed, the AEC encouraged Americans to celebrate the big bangs in the desert, which were exploding at a rate of one, and sometimes four, a month. Such demonstrations of the power of the atom became a comforting confirmation of the country's greatness at a parlous moment in world history. Little concern was given to their being a source of potential harm; "fallout" was not yet a dirty word in the U.S. test program. For those who lived in Nevada and Utah, the bombs were a marvelous spectacle. Parents would take their children at dawn to the nearest hilltop to watch the billowing mushroom cloud drifting toward the heavens. "Good bangs and so pretty coming at sunrise as they did," one Las Vegas resident recalled. Any fears that the pinkish red cloud of dangerous dust particles would damage the city's booming gambling and divorce trade soon disappeared, and in fact the bomb tests boosted the town's image: Las Vegas was a place where risk was at a premium, and waiting for the blast waves that often broke windows in the city of chance became a popular sport.

Bomb voyeurs learned that cancellation of airline flights from Las Vegas to Reno, a route that overflew the test site, was a sure sign that another test was imminent. The casinos would be packed until dawn on those days, and people even drove up from Los Angeles for the dawn bomb parties. When there was only a small bomb—and no windows were broken—Las Vegas residents would complain bitterly. "Bigger bombs, that's what we're waiting for," declared one nightclub owner in 1952. "Americans have to have their kicks."

Even the most ignorant citizen must have sensed all was not well, however. Within hours of a test, the AEC radiation monitors would regularly arrive in the towns and villages nearby, their Geiger counters clicking. Sometimes the fallout was so heavy that the dust particles from the bomb's cloud could be seen collecting on the ground and on the tops of cars. Residents would be advised to stay indoors—at the same time they were being assured there was no danger. The AEC knew best. This policy of obfuscation and mystification came right from the top, continuing as the bombs got bigger and "dirtier." According to the 1953 diary of Gordon Dean, then chairman of the AEC, Eisenhower himself "made the suggestion that we [AEC] leave 'thermonuclear' out of the press releases and speeches. Also 'fusion' and 'hydrogen' . . . the president says 'keep them [the public] confused about fission and fusion.' "

There was cause to worry about fallout, of course, as the AEC knew full well. From 1949 onward—before the choice of Nevada as the continental test site had been made—the AEC had held seminars and studies on the problem of fallout from fission explosions. One of these studies, completed at Oak Ridge in the spring of 1949, had estimated that plutonium, Strontium-90, and another new isotope called Yttrium-90 would be the most dangerous of the fallout substances. They would be distributed downwind from the test site over a 350,000-square-mile area, and, the report warned, "We do not fully understand strontium metabolism in man and factors of absorption and excretion may have to be altered. . . . There is little question but what [*sic*] there is real danger to inhaling particulate radioactive substances in such finely divided particles as to be retained within the lung." Such reports were never made public.

The AEC got away with their bland public reassurances because of the widespread ignorance of the effects of low-level radiation, ignorance that continually worked in their favor. Experts selected by the AEC could soberly assert that low levels of radiation were not harmful and they could not be proven wrong: the burden of proof that radiation was harmful rested not with the AEC but with those who opposed them. And at a time of international crisis, Americans were prepared to believe the U.S. government had their best interests at heart. Complaints would come only if there was proof to the contrary. A radiation protection system was already in place in the nation's nuclear production facilities; no one supposed that the Nevada protection system would be any less vigorous than the system so painstakingly worked out for the atomic factories. But some of the more dangerous radioisotopes measured routinely in the factories were not monitored at all in the wide open spaces dusted by the fallout. And there was no system of regular medical checks for those exposed, no mandatory isolation from further exposures. The practices adopted by the AEC to protect the public health from the tests were not worthy of the name "radiation protection"; at best, there was only a disaster warning system.

This state of affairs continued until the Partial Test Ban Treaty of 1963 forced all tests underground. By that time, the United States had exploded 183 atmospheric nuclear tests in the Pacific and Nevada; the Russians had exploded 118; the British, 18; and the French, 4. (The French, who exploded their first bomb in 1960, never signed the test ban treaty and continued to explode bombs in the atmosphere, as did the Chinese after their first successful test in 1964.) Even after the ban, no less than 31 of the underground Nevada tests released radioactivity into the atmosphere.

Only toward the end of the 1970s, as much of the early U.S. Nevada fallout reports became declassified, did a picture of the radiation effects of nuclear bomb testing begin to emerge. The information came only from the United States. Russia, Britain, and France contributed no significant data on their tests, and virtually nothing is known today about the fallout patterns and wind dispersal of the radioactive clouds from their tests. What little is known suggests the problem was even worse: the Russians have consistently been cavalier about radiation protection for their industrial work force; it is possible, therefore, that their attitude to public protection from test fallout may have been even worse than the U.S. AEC's. Information on the early British tests is protected more by embarrassment than by the needs of national security. In its first program in the 1950s, Britain was unique in gaining the full cooperation of another country, Australia. Between 1952 and 1956, Prime Minister Robert Menzies, an ardent Anglophile and a firm believer in the British empire, permitted the British to conduct nine tests, first on the island of Monte Bello off the west coast of Australia and then in the central Australian desert. Though these areas were sparsely populated, nothing has ever been made public about the path of the fallout clouds from the British tests.

In view of the extent of the knowledge about radiation at the time these countries began their tests, the callous disregard for public health is staggering. But it is also true that expert knowledge of radiation effects evolved slowly as scientists came to understand and measure the different types of radiation and arrived at standards for industrial and public exposure. Experts did not easily agree on what the standards should be. The increasing knowledge of the effects of radiation, particularly of the new isotopes produced during nuclear fission, required a continuous reexamination of the standards. Even when the experts agreed, many were only too conscious that they had, in fact, stepped beyond the limits of their expertise. Because radiation poisoning took so many different forms and also took so long to produce effects—up to twenty years in some cases—there was no hard scientific evidence of its effects at low level. Indeed, the consensus that emerged was that no level was safe. The experts found they could talk only in terms of "risk": it was more risky for the population to be exposed to doses above a certain level than below it. But the population was at risk at all levels, apparently. The act of setting up a standard, therefore, like the AEC's decision not to inform the public about the effects of fallout, was a political judgment, not a scientific one.

In their zealous pursuit of bigger and better bombs, the experts secretly exceeded the limits of the blind trust ordinary people had invested in them; public knowledge of the harmful effects of radiation always lagged behind the accumulated wisdom of the experts.

Before the first atomic bomb was exploded at Alamogordo in 1945, experience of the harmful effects of radiation was half a century old. In 1895, a fifty-year-old German professor named Wilhelm Conrad Roentgen discovered a source of invisible rays that, unlike light rays, could penetrate a shield of black paper. These rays also caused tiny crystals of a barium compound to glow in the dark. Roentgen did not know what the magical rays were, so he called them X rays. They had remarkable properties: they clouded unexposed photographic plates and they could penetrate flesh but not bones. Shortly after this discovery, a French scientist, Henri Becquerel, found that similar penetrating rays were emitted by certain uranium compounds.

Marie Curie named this oddity of science "radioactivity," and shortly thereafter, the Curies discovered the radioactive element radium. They isolated some of the substance and, when Becquerel carried a glass vial of radium in his waistcoat pocket for several days, he found it gave him skin burns. When Pierre Curie exposed a portion of his arm to radium rays for ten hours, he reported:

> After the action of the rays, the skin became red over a surface of six square centimetres [about one square inch], the appearance was that of a burn, but the skin was not painful, or barely so. At the end of several days, the redness, without growing larger, began to increase in intensity; on the twentieth day, it formed scabs, and then a wound, which was dressed with bandages; on the forty-second day, the epidermis began to form again on the edge, working towards the centre: and fifty-two days after the action of the rays, there is still a surface of one square centimetre in the condition of the wound, which assumes a grayish appearance indicating deeper mortification.

By March 1897, two years after Roentgen's discovery, there were already 69 recorded cases of X rays producing biological damage, and by the end of the century, the number had risen to 170. A German researcher found that mice developed leukemia from constant exposure to X rays, and in 1911 another report listed 94 cases of cancer tumors induced in man by X rays; of these, 50 were among radiologists—the experts themselves.

The early X-ray researchers would later be called "medical martyrs," and their number was to grow to hundreds within two decades. Both the Curies injured themselves severely with their discovery. Marie Curie died at sixty-seven from leukemia caused by overexposure to radioactive radiation. Her daughter, who continued her work, also died of leukemia. Studies would soon show that the incidence of leukemia in radiologists was nine times higher than expected. Because the effects of the X rays were not immediate and because the rays produced by the crude equipment of those early days were of low energy, a casual "I can take it" attitude was adopted by the researchers. The attitude was soon shared by the public. The possibility of removing unwanted hair, warts, acne, and dermatitis was exciting in the venturesome 1920s, and a lucrative X-ray business developed in the beauty parlors of America. Radium itself became the new magic cure, sold with about the same ethics as the old patent medicines.

Doctors readily wrote radium prescriptions for arthritis, gout, hypertension, sciatica, lumbago, and diabetes. Throughout the 1920s, thousands drank radium in solution or had it injected into them as an all-purpose cure. One New York company that produced radium drinking water claimed it had over 150,000 customers. The public's faith in the medical profession never faltered despite the growing body of knowledge about the external hazards of X rays and the equally important fact that no one really knew what radioactive substances would do inside the body.

Radioactive substances are in a state of disintegration or decay. In this unstable state, they give off rays or particles until they finally become stable. There are two different kinds of particles, alpha and beta, and these are distinguished by their ability to penetrate matter. Alpha particles are the weakest; they can be stopped by a piece of paper. Beta particles have a greater energy, but can be stopped by a thin sheet of aluminum. Finally, there are gamma rays, which have the same intensity as X rays. They have the greatest penetrating power and can be stopped only by a very dense element like lead. Collectively, the three, alpha and beta particles and gamma rays, are known as ionizing radiation and are energetic enough to transform anything in their path by tearing a piece off each atom in the material they contact.

At first, only gamma rays were considered harmful. Alpha and beta particles, it was thought, would not harm the human body because they could not penetrate it—at least not beyond the skin. But substances which emit alpha and beta particles can be ingested—eaten or breathed into the body. Damage done by these particles inside the body was first noticed in the case of the women workers who painted the dials of watches with luminous radium paint.

Radium glows in the dark. In 1915, an American medical practitioner and amateur artist, Sabin von Sochocky, devised a formula for luminous radium paint that he thought would look beautiful in landscapes. "This paint," he announced with pride, "would be particularly adaptable for pictures of moonlight or winter scenes, and I have no doubt that some day a fine artist will make a name for himself . . . by painting pictures which will be unique, and particularly beautiful at night in a dark or semidarkened room." Von Sochocky did not live to witness such a spectacle—he died of radiation poisoning in 1928 at the age of forty-five—but before his death he was able to observe the human disaster that flowed from his commercial success.

Von Sochocky founded the New Jersey–based Radium Luminous Material Company in 1915 and established a flourishing business painting luminous dials for wrist watches. His products also included pull switches for electric lights, crucifixes, and instrument panels for World War I aircraft. It was a good trade, and the company employed 250 women in a factory in Orange, New Jersey. They worked in a large open room with high windows, which they called "the Studio." The work was not particularly arduous and the camaraderie was good, especially during the war years. The watches, one dial painter recalled, "were shipped overseas to the doughboys, and we girls would scratch our names and addresses inside the casing. They were cheap watches and when they broke, we girls figured, the men of the American Expeditionary Forces would look inside them. Sure enough, eight or nine months later one of us would get a letter from some young fellow over there saying how lonely he was." Some of the girls found radium so pretty that they used it to paint their teeth, so that they would glow in the dark.

Painting the watch dials was precision work requiring thin fine lines to be drawn on the hour points of the watch. After dipping their camel hair brushes into a cup of canary yellow paint, the women were taught to lick the brush to a point to get the fine line. Each time they did so, they swallowed a tiny piece of radium. By the end of 1924, the local New Jersey medical examiner was convinced that radium was the cause of a growing epidemic of illness among them. Nine were already dead. His studies showed that the radium did not pass through their bodies but instead lodged in their bones and, emitting destructive radioactive alpha particles, gradually ate away the bones.

Five dial painters from the New Jersey plant sued the company for their injuries. Young and crippled, with rotted jaws or spines and shrunken legs, the girls had to be assisted to the witness stand. As the case progressed throughout the spring of 1927, more dial painters became victims. "The Legion of the Doomed," they were called. The case attracted international attention. From Paris, Marie Curie sent her sympathies, and the newly launched *Time* magazine declared: "Newspapers took the five dying women to their ample bosoms. Heartbreaking were the tales of their tortures." They had sued for $1.25 million, but they settled out of court for $10,000 each and a $600 per year pension.

The radium dial painters case was a cruel demonstration of the need for new standards. But one of the basic problems facing the health physicists, as the doctors concerned with the effects of radiation became known, was how to measure radiation. At first, they used a unit called the roentgen, named after the discoverer of X rays. The unit was a measure of the destructive power of gamma rays: how much material, or electrons, could be torn from an atom in a given time by a given concentration of rays.

In the 1920s, British radiation experts attempted to draw up a standard for X-ray and radium workers: a maximum limit of exposure per day. Years went by before the experts felt they had gathered enough data to agree on a standard. In 1925, a body called the International Commission on Radiological Protection (ICRP) was formed, but it took nine years to issue the first maximum-dose rate of one-fifth, or 0.2, roentgens (r) per working day. This represented about 60r a year, depending on the number of days worked. In 1936, the Americans decided that uncertainties about the low-level effects of radiation warranted a lower dose, and they cut the occupational standard in half—to 0.1r a day, or approximately 30r a working year. It was this standard that was used during the Manhattan Project. "If it had not been for those dial painters," one member of the project's fallout monitoring division recalled, "the project's management could have reasonably rejected the extreme precautions that were urged on it—the remote control gadgetry, the dust dispersal systems, the filtering of exhaust air—and thousands of Manhattan Project workers might have been, and might still be, in great danger."

Their diligence about occupational health in the making of the bomb stands in marked contrast to the lax standards allowed when public fallout became a problem. Although General Groves later claimed that he was "much more concerned about this [fallout] than any of the scientific personnel," when it came time to run the Alamogordo test, he allowed the medical division to set an external dose of 50r per week for any nearby community before evacuation was ordered. This figure was more than forty times greater than the weekly maximum (1.2r for a six-day work week) for Manhattan Project workers and one hundred times greater than the level—which would be set a decade later—of maximum exposure to the general public for a whole year.

No one could say definitely that 50r per week was harmful; no one, likewise, could say that it was safe. What could be said was that the internationally accepted levels for occupational exposure did not apply to the unsuspecting citizens in the path of the fallout cloud.

In the days before the Alamogordo test, the scientists of the project's medical division—concerned that sudden wind changes and rain might bring a concentrated dollop of fallout onto a populated area—sent up weather balloons, smoke signals, and an instrument-laden aircraft to try to improve the accuracy of their weather predictions. They hastily assembled a group of mobile monitors who were to follow the radioactive cloud in cars and vans and take readings at prearranged spots.

A thousand sheep and cattle farmers and two hundred members of a tribe of Apache Indians lived in the fallout path. When the Los Alamos medical division suggested that there should be a partial evacuation of these people before the test, General Groves was particularly indignant; the secrecy of the test might be compromised. The evacuation might even get into the newspapers. "What are you?" snapped the general at a medical officer, "a Hearst propagandist?" Groves also vetoed any serious research into the biological aftereffects. Only one experiment was conducted. A string of mice were hung by their tails to a length of rope at a point near the explosion where they could receive large doses of radiation without being fried to death. But the test failed to enlighten radiation scientists; the mice died of thirst before the bomb was detonated.

When the bomb was finally exploded in the half light of dawn on the morning of July 16, 1945, the cloud drifted northeastward at about ten miles an hour, distributing its trail of radioactive products into an oblong area a hundred miles long and thirty miles wide. Cattle were covered with a white mist of particles, and initial reports of the monitoring team showed some readings of 35r an hour. At 4:20 that afternoon, one hundred miles north of the test site, Geiger counters shot off the scale and evacuation plans were triggered, but Groves, ever conscious of security leaks, decided not to move anyone and the cloud passed on. When the monitors themselves returned to the Alamogordo base that evening, at least one of them was so radioactive that he was not allowed to leave the camp.

In the first five years following World War II, the problem of fallout receded as America's tests were conducted in the Pacific. Two postwar series of bomb tests, code-named Crossroads and Sandstone, were exploded in the Bikini Atoll and on the Marshall Islands—part of a tiny group of 2,000 islands known as Micronesia. The problem of the health of the Micronesians was easily solved: with a callous disregard for their homelands, the U.S. government throughout the fifties simply moved the inhabitants to other islands considered a safe distance from the test site.

In the first decade after the war, the radiation protection system followed two broad themes: the first was aimed at improving the standards of occupational exposure, and the second was the gradual extension of those standards to cover public exposure. The days of the single "radium standard" were over; the hundreds of new man-made isotopes behaved in different ways, attacking different organs of the body, and a different standard had to be developed for each of them. The first time many of these new substances were encountered was in the laboratories of the Manhattan Project: products of the nuclear chain reaction, Iodine-131, Strontium-90, Carbon-14, and Plutonium-239, they were later to become catchwords of the nuclear age.

With their newfound experience from the Manhattan Project, the American health physicists soon became the dominant group of the ICRP. They were led by Professor Karl Morgan, who had worked in the wartime Metallurgical Laboratory in Chicago where plutonium was produced. "Radiation need not be feared, but it must be respected," was Morgan's cautious approach. He was put in charge of the new health physics division at the Oak Ridge National Laboratory. He was also made chairman of the working group on internal radiation doses for the ICRP. All calculations were related to a "standard man"—a typical radiation worker. The aim was to arrive at "permissible levels."

New experimental evidence suggested, however, that radiation at any level would affect human genes. It meant, in effect, there was no threshold dose—no minimum amount that was harmless. An irradiated human, it seemed, would sooner or later produce unwanted mutated genes.

Research into the effects of radiation on genes was led by a professor of zoology at the University of Indiana, Herman Muller. A Nobel Prize–winner and the acknowledged father of modern radiation genetics, he had started his work in 1911 as a twenty-year-old graduate student at Columbia University. By irradiating the fruit fly, he had proved that the fly's gene mutations—the chemical changes that can pass on altered characteristics to new generations—increased by one hundred and fifty times.

Each new study confirmed more harmful effects. After World War II, the U.S. Atomic Energy Commission funded a huge research project at Oak Ridge. Eventually known as the "mega-mice" study, it involved tens of thousands of mice. Different parts of the mice were carefully exposed to precise bursts of radiation of different intensity and duration. The first results showed a tenfold increase in the number of mutations over those expected on the basis of Muller's fruit fly experiments. The greater sensitivity of mice implied sensitivity in human beings much greater than earlier genetic studies had indicated, but the results also confirmed that no amount of radiation could be considered safe: even the smallest doses produced mutations. The mutations were proportional to the dose: if a hundred roentgens of exposure created a certain number of mutations, then ten roentgens would cause one-tenth that number.

The geneticists warned that the increase in man-made radiation—through fallout and nuclear power reactors—could have a disastrous effect on the human race. In the past, they argued, the spontaneous mutations of man's evolution had been caused in part by the existence of the natural background radiation from cosmic rays and the naturally occurring radioactive elements, like uranium and thorium. Over the centuries, a genetic equilibrium had evolved. The substantial increase in man-made radiation could now destroy this. Although there was no certain evidence that the old permissible radiation level was too high, the caution of the U.S. health physicists led to another halving of the occupational health standard to .3r a week, or 15.6r a year. This was adopted by the International Commission on Radiological Protection in 1950. The new standard was stated in weekly rather than daily terms to provide greater flexibility for the atomic plant operators. These were the standards in effect when the bomb tests began in Nevada in early 1951, but they were for occupational exposure: there were no national or international standards for public exposure; in particular, there were no standards to protect those people living near the test site.

Postwar testing coupled with increased pressure to have a test site in the continental United States eventually led to a number of fallout studies. Undertaken by the U.S. AEC, they sought to estimate the hazards from the fallout. All the studies stressed the inadequacy of the data and the wide margins of error. The 1949 Oak Ridge study identifying the hazards from fallout was followed by a seminar on radiological hazards. Held at the Los Alamos laboratory in August 1950, its purpose was to review the radiological hazards to citizens who lived in the fallout path of a continental test site. The assembled group of twenty-four scientists, including Edward Teller and Enrico Fermi, considered a rapidly administered dose of 25r as "an emergency acceptable dose, if received only once." A dose of 100r would result in some blood changes, nausea, and vomiting; a lethal dose would be roughly 400r if administered in a period of a few hours. The dose of a normal clinical X ray is less than 1r.

The scientists calculated that the possible gamma doses received downwind from a bomb test could vary from between 6r from a 25-kiloton bomb to 250r from a 250-kiloton bomb at a point 100 miles distant from the test center and two hours after the explosion. At a distance of 300 miles, six hours after detonation, the gamma dose could vary between 0.5r and 5r for the same yield bombs.

The problem was the lack of data. Enrico Fermi stressed the "extreme uncertainty" of 10r, but, he said, people should be warned to stay indoors and take showers. The risk was not that people would be killed, or even hurt, if the yield of the bomb was kept low enough to ensure that the dose remained, say, around 6r. "But," said the report, "it does contain the probability that people will receive perhaps a little more radiation than medical authorities say is absolutely safe."

The AEC health physicists needed greater flexibility for their workers on the test site because the explosions involved short bursts of large amounts of radiation. Their standard, extrapolated from the new international consensus of .3r a week, was set at 3.9r over a thirteen-week period (a quarter of a year). This was adopted as the AEC rule for all scientists and military personnel under AEC control near the test site.*

The key question remained: What standards would be applied to the population that lived near the test site and, in particular—given

*The army, however, decided on almost double these limits, or 6.9r, for thousands of guinea-pig soldiers they positioned seven thousand yards from the blast (as opposed to ten thousand for AEC personnel); the soldiers then were made to maneuver up to 3,500 yards from ground zero after the bomb had exploded. And, in a series of three blasts in 1952, a group of forty-two officer volunteers were placed as close as 2,000 yards from ground zero. Some of them received doses of more than 16r. The intent behind these exercises was to gather information on the possibility of close tactical support by infantry after a nuclear explosion and to get some idea of the psychological reaction of the troops to atomic warfare.

the prevailing winds—to the north and east of it? The AEC experts knew that, short of providing fallout shelters for all of them, it was going to be impossible to ensure that the standard of 3.9r per quarter was kept. Moreover, to monitor 350,000 square miles and cover all points of maximum fallout was something they were simply not going to be able to do.

The most AEC monitors could warn of was a series of average exposures. For public protection, what mattered was an actual maximum exposure not an average. But there was no such thing as a uniform rate of fallout. High concentrations, or "hot spots" of radioactive particles, could occur anywhere in the path of the fallout cloud. Changes in wind speed, changes in the terrain, a sudden fall of rain or snow, could bring the fallout to earth in unpredictable concentrations. And because the AEC chose to keep its monitoring results secret, there was no question of radiation-sensitive film badges being issued to citizens living in the affected areas; the AEC could not risk a public uproar if it was found that one town or another had been exposed to a dose higher than considered acceptable for workers in the atomic factories.

The result was that the monitoring of the external dose rates for unsuspecting citizens was a random affair. More important, during the period of active testing in Nevada when more than one hundred bombs were detonated, there was no systematic effort to measure the amount of radiation—from the alpha and beta particles—ingested by those eating food grown near the test site or drinking milk from cows that had been grazing in irradiated pastures. It was as if the dial painters case, which had demonstrated the appalling effects of the ingestion of radium, had never existed. Before the tests started, the AEC knew that the radioactive isotopes produced by fission bombs—isotopes like Strontium-90—were "bone-seekers" just like radium. They had also learned that another fission product, Iodine-131, when ingested tends to concentrate in the thyroid. In the AEC's atomic factories, these isotopes carried special exposure standards, but no such standards existed for the citizens living near the Nevada test site.

When the first of the new Nevada test series began in January 1951—the one that dropped fallout in the snow over the Kodak factory—the monitoring system around the site was makeshift. It had been thrown together in a greater rush than at the Alamogordo test in 1945. Electrolux vacuum cleaners had been hastily converted into makeshift air sampling devices. As a result, monitoring was perfunctory during the first two years, 1951 and 1952. During this time, about 200 kilotons of explosive power, or the equivalent of ten Hiroshima bombs, were detonated. Even bigger bombs were exploded during the fourth series, code-named Upshot-Knothole, in the spring of 1953. For this series, however, the ground and air monitoring teams were coordinated and the National Public Health Service was included in the off-site monitoring network. The new system, only marginally more efficient, was the first to produce any danger signals.

The fallout from the first shot of this fourth series, a 16-kiloton bomb, came down to earth as predicted, and gamma readings were within the 3.9r standard. The second shot of 24 kilotons, exploded a week later, drifted in the expected northerly direction, but the readings were higher. Fifty miles from ground zero, some two hundred employees of the Lincoln mine were advised by the AEC to stay indoors until the fallout cloud had passed. Eighteen thousand sheep grazing in the vicinity caught the brunt of the fallout, and more than four thousand of them subsequently died from radioactive poisoning. Those that survived had burnlike lesions on the face, neck, and ears. The AEC sent scientists to investigate the deaths. Their findings were suppressed. One of the scientists told a local agricultural official that the radiation levels on the dead sheep were "hotter than a firecracker," but he was ordered to rewrite his report by the AEC, deleting all references to radiation effects.

More trouble came with the seventh shot, code-named Simon. It was a big bomb—some 50 kilotons—or two and a half times the size of the Hiroshima bomb. The last weather report before the test showed the area of maximum expected fallout (estimated to be 8r) right over U.S. Highway 93, running northeast from Las Vegas. When the fallout occurred, however, a maximum reading of 16r was recorded on the highway—twice what had been expected. Even the bland internal AEC report called it an "emergency." Road blocks were hastily set up to catch cars, trucks, and buses and hose them down.

A summary of what happened was later written up by the test site monitoring officer. It was received by the AEC in May 1953, and stamped SECRET SECURITY INFORMATION; it did not become public until April 1979, almost twenty-six years later. The report states that the AEC standard of 3.9r per quarter was exceeded at two hamlets one hundred miles due east of the test site in Nevada. Two readings were taken of 7r and 5.5r equivalent dose per quarter. A total of 264 people were affected. The numbers recorded were small, but the test site report stated, "The level of radiation here is such that if fallout occurred in a populated area immediately adjacent to the Proving Ground, beta burns might be experienced." (They already had been by sheep in the first shots of the series.) "This would be a serious situation indeed since these burns cause hair to fall out and blisters or ulcers to form. This would probably arouse immediate public clamor for the closing of the Proving Ground."

Under a heading: "Effect on the Population," the report went on:

> There are so many unknowns about the biological effects of radiation no one can really say what the effects of this radiation are. Mr. Failla, member of the sub-committee on External Exposure of the National Committee on Radiation Protection [the US branch of the ICRP], and one of our best authorities, is publishing a book which says .3r per week (3.9r per quarter) exposure is acceptable, but that this must be reduced by a factor of ten for minors. This is because of the effect on bone and tissue formation which is going on so rapidly during a child's growth. On the other hand, the Army is allowing a maximum dose of 10r [the dose was sometimes higher for officer volunteers] for soldiers participating in Desert Rock [an Army operation to watch troop reaction to the tests]. Their basis probably is that the danger of injury from radiation even at this level can be accepted as less than danger of injury from more orthodox causes faced during maneuver and combat.

In short, the AEC still had no basis for knowing what effect the fallout would, or indeed could have on citizens living near the test site.

The final paragraph of the test site report stated, "It is apparent from the above that the Test Organization should consider the upper limit of size of tower shots with a view to perhaps holding such shots below some maximum, perhaps 25 kilotons.* The alternates [*sic*] will otherwise almost certainly be over-exposure of nearby populations and conceivably the enforced closure of the Nevada Proving Ground."

A week after the report was received at AEC headquarters in

*This referred to the fact that the test site operators had discovered that high-yield bombs exploded on towers several hundred feet above the ground produced the most fallout. The radioactive isotopes created during the fission explosion were carried on the millions of dust particles sucked up from the ground into the atomic cloud.

Washington, the ninth test in the Upshot-Knothole series was detonated at 5:05 on the morning of May 19, 1953. Code-named Harry, the bomb had a yield of about 32 kilotons and was detonated from a 300-foot tower. It produced fission products that were carried upward in a vaporized mass that was 16,000 feet deep; as it moved away eastward, the base of the cloud was at 27,000 feet and its top at 43,000 feet.

Harry was what the scientists called a "dirty shot"—because of its size and because it was detonated from a 300-foot tower. The following is the report made by the AEC monitor at the town of St. George, population 5,000, 150 miles due east of the test site.

> At about 8:45 AM on the 19th [May, 1953] it was noted that the number of cars requiring decontamination was increasing as well as the readings being obtained. As a result it became necessary to arrange for additional decontamination stations. At about 9:10 readings of .3 to .32 roentgens per hour were being obtained in and out of cars. It then became evident that something was happening, probably fallout. As was suspected the fallout had started at 8:45 AM and at 9:15 the peak had not been reached. Immediately the decontamination was stopped. Word was sent to the people at the road-block to stop monitoring and have the motorists stand by for 20 minutes to an hour. A telephone call was made to the post. At 9:25 instructions were given to the people in St. George to take cover. The sheriff was notified of the situation so the children would not be sent out in the open during the recess period. At 9:40 AM the bulk of the population in St. George was under cover. By this time we had about 200 cars at the roadblock, one hundred from each direction, and about 25 cars and 3 trucks at the service stations in St. George being washed. While everything was at a standstill the monitoring of the area continued. A high reading of .32 roentgens per hour was obtained at 10:15.

The monitoring team was told by worried officials from the test site to stay in St. George. One of them tried to get samples of milk produced in St. George. The report continues:

> I had already contacted the county agent's office to obtain the names of the milk producers in the area. I was unable to get this completely but I got the name of the president of the county dairymen's association . . . it was just as well that neither was available

since I was afraid it would create a disturbance. In view of Tuesday's [the day of the test] episode, everyone in St. George was a little concerned over any unusual incident connected with radioactive fallout and it would not take much to start wild rumors. For this reason it was agreed that the direct approach for the collection of milk samples would not be pursued further at this time. We already had one sample of goat's milk and, if at all possible, I would buy some milk from a local store. That evening I purchased a quart of milk from a local store in the town. I located the producer and in discussing the milk supply in a general way I learned that the milk I had purchased that evening was obtained from the St. George herd on Tuesday evening.

From the monitoring reports, it was later estimated that the external gamma radiation dose to the 10,000 population of Washington County, Utah—half of whom lived in St. George—was between 2.5r and 5r.

If those exposure doses had occurred at one of the AEC's atomic production plants, precise and detailed inspection of those exposed would have been made in order to calculate actual doses received and possible radioactive material inhaled or ingested. Moreover, those workers would have been isolated from any further exposures. No such precautions were taken around the Nevada test site. To have done so would have been to go public with the fact the AEC had taken only the most minimal precautions for the safety of civilians in the area. Rather than telling the truth, the AEC adopted a policy of cover-up—although they preferred to call it a "judicious handling of the public information program."

Whenever the limited monitoring network showed that "permissible levels" had been exceeded, the precautions taken were rudimentary. Residents would be told to stay indoors and take showers. After the cloud passed and the fallout ended, they would be advised to wash the radioactive dust off their cars. There was no program of inspection checks of those exposed—not even for the known high-risk group of small children. The haphazard sampling of the milk in the St. George area after the "Harry" test was typical of the lack of concern in the AEC about the possibly harmful effects of ingested radioactive materials. Several years passed before the AEC was forced to take them more seriously.

13

The Uncertified Possibility

A genuine pragmatist . . . is willing to live on a scheme of uncertified possibilities which he trusts; willing to pay with his own person, if need be, for the realisation of the ideals which he frames.

—WILLIAM JAMES, *Pragmatism*, VIII

Café jokesters in Buenos Aires in the 1950s loved to tell the story of how Ronald Richter lost a thumb when an atom bomb exploded in his hand. Richter, an Austrian-born physicist, was the great hoaxer of atomic energy. He had duped Argentina's President Juan Perón into thinking he knew how to make nuclear power.

After spending the war years in Germany working for Hitler in the Junkers aircraft factory, Richter had arrived in Argentina and, with the help of some of his former Nazi associates, he had managed to persuade President Perón to part with forty-seven million pesos from the government coffers to start an atomic energy project. On a deserted island in the middle of a lake near the Chilean border, a thousand miles from Buenos Aires, Richter and thirty Argentinian technicians started building an atomic laboratory. Perón insisted that the project be kept top secret. He provided a flotilla of patrol boats to keep constant vigil on the island, which at night was bathed in floodlights. Perón told Richter: "If you find an enemy agent, drown him in the lake."

The Argentinian dictator's obsession with security was multiplied severalfold because Richter had told him that his method of making atomic energy was unique. Perón was overjoyed. Ever conscious of his country's inferior industrial status—especially compared with the United States—Perón hoped that Richter's work would enable Argentina to upstage the Americans in the one sector of their technical expertise that the whole world acknowledged and admired before any other: atomic energy.

In February 1951, only eighteen months after work had started on the project, Richter invited members of the recently convened Argentinian Atomic Energy Commission to visit his island laboratory. The commissioners gathered around a contraption, which Richter told them was a reactor, and at the flick of a switch from the Austrian physicist, an array of dials registered instant reactions. Richter claimed that the commissioners had witnessed the world's first controlled thermonuclear reaction.

The experiment as described was way ahead of its time; certainly no one in the United States had done it. American atomic physicists were hard at work in the Los Alamos nuclear laboratories trying to figure out a way of conducting an *uncontrolled* thermonuclear reaction—or, as we came to know it, a hydrogen bomb; they were nowhere near understanding how to *control* one. To this day, physicists have not mastered a controlled thermonuclear reaction.

The Argentinian commissioners were nonetheless convinced by the enthusiastic Richter. So, too, was President Perón. Spurning the counsel of some close advisers who suggested that the most prudent course of action was to organize some peer review of Richter's dazzling claims, Perón decided to announce the experiment to the world at a press conference.

On March 24, 1951, at ten o'clock in the morning, a euphoric Perón, flanked by Richter and members of the Argentine NAEC, called the Buenos Aires press into his office. "First," he said, "I ask your forgiveness for getting you up so early, but I wanted to make this announcement precisely because I want the country to join in the work we are doing . . . what is important is that when I say something, I know what I'm saying. I say it in all sincerity and I check out the truth of what I say before I say it. I have always tried not to lie; I still believe that I have not done so and I will not lie now." Perón then praised the Richter "achievement," emphasizing the revolutionary nature of the project. "The work has been conducted on the basis of thermonuclear reactions, by which atomic energy is freed from the sun," declared Perón. "The problem was obtaining the high temperatures and controlling the reaction." Perón announced proudly that Richter had overcome these problems and Argentina was about to create "artificial suns on earth." Argentina needed atomic energy and would employ it only for peaceful purposes, but "foreign technicians should know that in the course of our experiments the problems of the hydrogen bomb were studied."

These startling words caused a flurry of diplomatic activity in other capitals around the world, particularly in Washington and Moscow. Was Argentina about to become the third atomic power? And with a hydrogen bomb? In Washington, the State Department and the press bombarded the Argentine embassy with demands for clarification of the exact nature of Richter's experiments. In Moscow, Kremlin leaders quizzed German scientists working in Russia about Richter's background. Baron Manfred von Ardenne, who had worked with Richter during the war, told the Russians not to take him seriously. "Fantasy and scientific reality were so confused in him that his work was unreliable," he said.

Western scientists rushed into print. "Perón Atomic Claim Deemed Replete with 'Impossibles,' " said a *New York Times* headline. An American scientist called Richter's claim "the super-duper bull of the Pampas." The chairman of the U.S. Atomic Energy Commission, David Lilienthal, felt there was not the "slightest chance" that Argentina had unlocked a secret of the atom unknown to America, but he greatly admired Perón's capacity for catching the headlines. "They may not know anything about fusion of the lighter elements—and I am persuaded that they do not—but they do know about the methods of American publicity," said Lilienthal.

The scientists had every reason to brand Richter's claims a hoax. As *The New York Times* put it: "The Argentine claim, if true, would require the achievement of at least three miracles—the production of temperatures of millions of degrees, the maintenance of the temperature for longer than a millionth of a second, and the development of materials that would not evaporate long before such a temperature could be attained." When asked how he had reached these millions of degrees, Richter said: "That's the secret; I can only say a new solar physics was born." There was no secret, of course, as the Argentinians soon discovered. They bugged Richter's hotel room in Buenos Aires and overheard him telling close friends how he had managed to fool Perón. Richter was swiftly dismissed from his laboratory, and the jokes about him started to flow.

The Richter affair raised the first, albeit false, alarm of nuclear proliferation—the diplomatic term for the spread of nuclear weapons. After the success of the Manhattan Project, the Americans had expected that the other major powers, the Soviet Union, Britain, and France, would make their own bombs sooner or later. They had also expected pro-bomb lobbies to emerge in the two defeated wartime powers, Germany and Japan. But, until the mid-fifties, the United States had no policy for dealing with the possible spread of nuclear weapons to countries not considered major powers. Officially, the United States remained committed to the 1946 proposals for shared international control as prepared by the Acheson-Lilienthal committee; in reality, the policy had been abandoned. Into this policy void came Eisenhower's Atoms for Peace, a program that opened the way for widespread proliferation of atomic weapons. Promising to supply countries with the "peaceful" atom, the program demanded only a paper declaration from any country receiving American reactors that the "peaceful" atom would not be turned into the "warlike" atom. The central fact that a country which has a power or research reactor producing plutonium can, if it wants, make bombs was conveniently brushed aside. Atoms for Peace started as a simple swords into plowshares idea, converting bomb-grade fissionable material into fuel for power reactors, but it rapidly turned into a weapons proliferation headache. Instead of placing restrictions on bomb-grade material—either as supplied by the United States for fuel or as produced by a country in its new reactors—Eisenhower's generous but simple plan actively encouraged its dispersal. So long as that plan persisted, control of weapons proliferation was a vain hope.

The U.S. State Department policymakers thought the paper declaration rejecting military use could be enforced by on-site inspections of the nuclear facilities of any country that decided to take up the atoms-for-peace offer. It was assumed that, eventually, the politically delicate issue of inspection would be worked out by the newly created International Atomic Energy Agency (IAEA) and this would provide the necessary "safeguard" against any diversion of plutonium or U-235 for making bombs. In the enthusiasm of the moment, however, the Americans ignored a related problem: if an inspection revealed that bomb-grade material had been diverted from a reactor, would there be enough time to bring political and diplomatic pressure—international sanctions of one kind or another—to bear on the country concerned to keep it from turning that material into an actual bomb?

The question of this lag had been addressed as early as 1946. In December of that year, the U.N. Atomic Energy Commission had noted that enriched uranium or plutonium "can be put to use in atomic weapons within days." Four years later, in 1950, a U.S. National Security Memorandum, NSC-68, went further: "In order to assure an appreciable time lag between notice of violation and the time when atomic weapons might be available in quantity, it would be necessary to destroy all plants capable of making large amounts of fissionable material. Such action would, however, require a moratorium on those possible peacetime uses which call for large quantities of fissionable material."

With atoms for peace, such a moratorium was unthinkable if the spirit of the program was to survive. Another option—to insist on a restraining clause limiting the amount of bomb-grade material a country was allowed to have at any one time—would have opened the door to very reasonable charges of "atomic colonialism." So the Americans compromised. They deferred the problem and continued to press for the widest possible powers to inspect all atomic facilities at all times.

The regular presence of foreign inspectors in new nuclear countries was an untried innovation in international relations. It was one that had given the 1946 Acheson-Lilienthal plan, with its hopes of internal inspections of the Soviet Union, a touching naïveté. As the inspection safeguard negotiations progressed, difficulties of persuading countries to accept this infringement of national sovereignty resulted in the inspections themselves taking on an importance of their own. The next step—resolving the question of whether the alarm could be raised soon enough—became moot for the time being, following a natural instinct of backroom diplomacy: cross bridges only when you have to, defer final showdowns, paper over any sticking point with ambiguous wording, and leave it to future diplomats to safeguard the preferred interpretation. This was a policy for the long haul, requiring long-term patience and doggedness of purpose.

Once Atoms for Peace was launched, it was the Department of State that would have to handle these negotiations. As with any other subject to which he gave priority, Secretary of State John Foster Dulles preferred to concentrate the work in his immediate coterie of close advisers. He created a new position of Special Assistant to the Secretary of State for Atomic Affairs and appointed a forty-year-old New York lawyer named Gerard Coad Smith to the post. From 1954 to 1957, Smith was at the center of America's nuclear safeguards diplomacy.

In temperament, physical appearance, and background, Smith radiated solidity. Thin-lipped and square-jawed, even-tempered and serious, Roman Catholic and Republican, a graduate of the exclusive Canterbury School and of Yale: Gerard Smith was sound. A longtime second-string member of the American foreign policy establishment, he was readily accepted in the quiet backroom formulation of American objectives and policies. Smith preferred power to glory; he was hardly known to the American public until 1969, when Richard Nixon appointed him head of the Arms Control and Disarmament Agency and chief negotiator of the SALT negotiations with the Soviet Union. At Smith's confirmation hearings, senators expressed surprise at his previous involvement in these matters: he had not been in the front line. He had written many speeches but never delivered them. Although he had been part of the decision-making process, often at the very center of it, he had never announced the decisions.

From 1954 onward, Secretary of State Dulles gave a number of speeches that had been drafted by Gerard Smith. These warned of the dangers of nuclear proliferation and stressed that, if safeguards on new nuclear facilities were to be effective in halting the spread of nuclear weapons, they had to be applied by all nations acting as suppliers of nuclear materials, including uranium. What was not stated in these speeches was that, to ensure this, Dulles and Smith had organized a secret international cartel of uranium suppliers whose members regularly exchanged information on uranium sales and worked out a common safeguards policy. All major exports of uranium were to be cleared through this secret organization, which became known as the Western Suppliers Group. It was the one part of the American nonproliferation effort that had a measure of success.

The group was a natural extension of another cartel, a wartime agreement between the United States, the United Kingdom, and Canada to buy up all known stocks of uranium in order to prevent other nations from having access to these. The Combined Development Trust, as it was known, would be the only part of this close wartime collaboration to survive the war. Gradually, nations that had significant supplies of uranium were coopted into the new secret Western Suppliers Group.

In May 1954 the United States started secret negotiations among eight uranium-producing nations: the United Kingdom, Canada, South Africa, France, Belgium, Australia, and Portugal. Under Gerard Smith's direction, they began their exchanges as an informal discussion group that met during the drafting of the IAEA treaty; in the early years of the agency's operation, they continued to meet, gradually acquiring a greater degree of formality. The group's meetings were initiated by formal State Department invitations and its delegates were at ambassadorial level. At a time when Atoms for Peace implied no restrictions on exports of nuclear technology, the uranium supplier nations could help to reinforce control over the atomic activities of new nuclear power nations.

But the control was limited. It could not prevent proliferation. The Americans, trapped by their own public relations triumph, could insist only on a policy of detection of new nuclear weapons nations, rather than on prevention.

The Russians were far more successful in enforcing strict controls on the use made of their nuclear exports to Communist Bloc countries; the Kremlin did not have to concern itself with diplomatic niceties, or worry about charges of atomic colonialism; they practiced it openly. In January 1955 the Russians announced that, in return for the continued supply of uranium from Communist countries, they would establish a new program of peaceful nuclear assistance in Eastern Europe and China. But their nuclear export policy was to keep strict control of all weapons-grade material, and they refused to follow the American lead in exporting highly enriched uranium for fueling research or power reactors. Moreover, they always insisted on the return of the spent reactor fuel rods containing plutonium. Detection was nice, but prevention was better.

There was one glaring exception to the Kremlin's rule: China. A faction within the Soviet Politburo convinced itself that a Chinese nuclear weapon would be as subservient to their control as the British "independent deterrent" was to the United States. They would have many occasions to regret this self-deception in the years ahead, as the product of the world's most serious act of conscious proliferation was turned explicitly on the proliferator.

The Soviet Union stopped just short of the final step in proliferation: in 1959 they almost gave China a complete bomb. The transaction was prevented only at the last moment. The sample bomb had been prepared and packed and was ready for dispatch when Yefim Slavsky, the Soviet Minister for Medium Machine Building in charge of nuclear weapons production, approached Nikita Khrushchev. "We've been given instructions to ship an A-bomb prototype to China," he told the Soviet leader. "It's ready to go. What shall we do?" A number of signs of the future conflict between the nations had already emerged, perhaps none more significant than the fact that the Chinese had point-blank rejected a Soviet request to station nuclear weapons under Soviet control on Chinese soil.

"We'd given the Chinese almost everything they asked for," Khrushchev later complained, referring to the exchange of atomic information between the two countries. "We kept no secrets from them. Our nuclear experts cooperated with their engineers and designers who were busy building a bomb. We trained their scientists in our own laboratories." But, Khrushchev added, "We had to draw the line somewhere."

When the dispute between the two Communist giants became public in 1963, the Chinese released a surprising amount of specific information about the ruptured nuclear relationship. They even identified the day—June 20, 1959—on which the split occurred. According to the Chinese, "The Soviet government unilaterally tore up the agreement for national defense concluded between China and the Soviet Union on October 15, 1957, and refused to provide China with a sample of an atomic bomb and technical data concerning its manufacture."

Precisely how much assistance the Russians gave the Chinese is unknown, but there is no doubt it was substantial. A formal agreement between the two nations was signed in the summer of 1955. The Russians promised to train Chinese specialists in Soviet research institutes and universities, to transmit technical data relating to isotope-producing reactors, and to construct a number of reactors in China itself.

Nuclear cooperation between China and the Soviet Union began in 1950 with Russian exploitation of Chinese uranium. Significant deposits had been found in Sinkiang Province. Although the Russians had occupied the province during the war, they had returned it to China when the Communists triumphed late in 1949. On March 27, 1950, a new joint-stock corporation was established, the Sino-Soviet Non-Ferrous and Rare Metals Company. The jointly owned enterprise developed the Sinkiang uranium deposits and established the mining and refining infrastructure exclusively, at first, for Russia's needs.

In February 1953 the first Chinese scientific delegation arrived in Moscow to negotiate details of future cooperation. This was the beginning of Soviet assistance in the basic training of China's nuclear manpower. The leader of the twenty-six-man delegation was a French-trained physicist, Ch'ien San-chang. He had recently been appointed head of Peking Institute of Modern Physics in the newly organized Academy of Science. A few years later, Ch'ien's group would form the Chinese Institute of Atomic Energy in Peking.

As a result of Ch'ien's four-month visit to Russia in the first half of 1953, Soviet aid began to flow later in the year. In July, some 10,000 metric tons of atomic equipment and material were transferred, and Soviet organizers helped the Chinese establish a number of new laboratories. Ch'ien's institute prepared itself for large-scale training of nuclear experts. In China's first Five-Year Plan, announced later that year, eleven priority tasks for science and technology were established: the first was the study of atomic energy, for peaceful purposes.

The next year, 1954, the Russians reinforced their cooperation with other Communist Bloc nuclear programs by creating a Joint Institute of Nuclear Research around a hitherto secret research laboratory in Dubna, on the Volga River north of Moscow. The new facility for basic research and training brought together experts from eleven countries. It was jointly financed; China, as the second largest contributor, paid 20 percent. Eventually, about a thousand Chinese experts received training at the institute. From the start, the Chinese were led by a Western-educated physicist, Wang Kan-chang. The son of one of China's most distinguished physicians, Wang had graduated at the top of his class from Peking's Tsinghua University in 1934. He went on to study at the University of Berlin under Lisa Meitner, the German researcher who made the key scientific interpretation of the Hahn-Strassmann experiments discovering fission. At the end of World War II, he left China again for a period as a research assistant in physics at the University of California. There he was remembered as a "cosmopolite, western in his ways, a natty dresser and an amiable mixer." In the sixties, Western analysts identified him as the scientific director of the Chinese bomb program. Undoubtedly, building a bomb was his top priority as leader of the scientific delegation at Dubna during the two years from mid-1957 to mid-1959.

Information on the Chinese nuclear program is sketchy, more so than even that available about the Soviet Union. There are no memoirs about the early years, such as those that have recently appeared in Russia. Indeed, there are few reports of any kind. Observers can only speculate about the motivations and origins of the bomb program. No one knows, for example, whether an originally peaceful program was later diverted to military use or whether from the outset the Chinese were covertly pursuing the weapons option. And while it appears that Ch'ien San-chang and Wang Kan-chang were the key scientific administrators, that may prove as incorrect as the original Western identification of the physicist Peter Kapitsa as the brains behind the Russian effort. What is clear is that, until the late seventies, the Chinese never tried to extend their substantial base of technical know-how and facilities into the development of reactors for electricity production. For more than two decades, their nuclear effort remained almost exclusively military.

In retrospect, the Chinese nuclear weapons objective seems apparent in the major Twelve-Year Science Plan adopted in early 1956. That contained a huge increase in funds allocated for science, from U.S. $15 million in 1955 to U.S. $100 million in 1956. The first five priorities in the new plan were all directly related to either nuclear materials or missile development, and within the year a new scientific overlord had emerged, Nieh Jung-chen, one of the most powerful political figures in Peking and one of its most senior military men. Nieh became chairman of the Scientific Planning Commission, responsible for implementing the Twelve-Year Science Plan.

In September 1955, when personal military ranks were introduced and decorations given to veterans of the People's Liberation Army, Nieh was awarded the three top decorations and appointed as one of ten Marshals of the Army. He had an impressive curriculum: political commissar turned military commander, veteran of the Long March, top Communist political and military official in the extensive border region plagued by invading Japanese armies in northern China, mayor of Peking when it was triumphantly declared the new

Communist capital, army chief of staff during the Korean War. Nieh was the head of the North China Field Army, one of the five regional power groupings in the new China that were derived from the territorial bases of the victorious Communist civil war armies. His base included the Peking region, and his forces acquired some of the power of a Praetorian Guard. A considerable presence at the center of the new State in the mid-fifties, Nieh went on to launch a career as the nation's top scientific administrator. Technological development emerged as the regime's top priority in a manner not seen again until the late 1970s.

Nieh had come full circle since he had left China to study in Paris. The nineteen-year-old son of a wealthy landowning family in Szechwan Province, he had no firm political views then, but he had an idealistic commitment to work for the independence and modernization of China. During his studies of science and engineering in France and Belgium, Nieh was converted to Marxism and joined the Communist party in 1923. After a year and a half of political and military training in Moscow, he returned to China in 1925. By all accounts, Nieh was a quiet, soft-spoken, scholarly, and cultured man of gentle disposition; he took a serious-minded, judicious approach to even the most staggering problems. For the next decade, he worked as a political organizer for the Communist cause, but with the developing civil war and the Japanese invasion, Nieh became increasingly preoccupied with direct military activity. In 1937 he was given his first independent military command, a small force of two thousand men that was to become the Communists' first guerrilla base behind Japanese lines in Northeast China. With the Japanese in firm control of the cities in his area, including Peking, Nieh developed new guerrilla tactics to isolate them from the countryside. "When they closed in," he told a visiting American reporter, "we moved out. When they stuck out their heads, we chopped off their tails." Visitors to his mountain headquarters in those days saw him as a restless bundle of nerves, a man who could not sit still, his hands always moving, doodling with a Parker pen or playing with a child's Yo-Yo. They also noted his spotless white shirt, his well-tailored uniform, his stylish cavalry boots. "He still takes a Parisian pride in his clothing and is by far the best dressed man in the Eighth Route Army," one visitor reported in 1938. Drawing on the anti-Japanese patriotism of his region, Nieh became the most successful recruiter in the Communist forces; his own army exceeded 100,000 members.

After the defeat of the Japanese, Nieh's battle-trained troops were the key to Communist success in the civil war, and Nieh was one of the first of the top Communists to enter Peking. Although a member of the Communist party's Central Committee, in which capacity he played a central role preparing the new constitution, and despite a period as mayor of Peking, Nieh's primary role continued to be military. He was acting chief of staff of the entire People's Liberation Army from 1949 to 1954 and remained personal commander of the Peking-Tientsin garrison until 1955. By the time he moved on to become overlord of the new drive for scientific development and economic and military modernization, Nieh had had three decades of military command.

In 1956, the new Twelve-Year Science Plan was most actively promoted by Premier Chou En-lai. Nieh had been a close associate of Chou's from their first meeting in Paris more than thirty years before, and immediately on Nieh's return from Europe in 1925, he joined Chou in recruiting for the party at the Whampoa Military Academy. Teng Hsiao P'ing (Dung Xiaoping), who would succeed Chou as promoter of Chinese science, was not only a close personal friend of Nieh's from the Paris days, but they had also been classmates in high school. The appointment of Nieh confirmed the importance the regime attached to scientific research.

Nieh brought to his new responsibilities the same strong commitment to the independence and modernization of China that had first propelled him into the Communist party many years before. He was particularly sensitive to the arrival of the thousands of Russian experts throughout China's military and industrial complex. The Chinese became increasingly apprehensive about this "invasion." Warning of the dangers of having no indigenous engineering industry, one Chinese manager commented in 1957, "Learning from the Soviet Union is a royal road; but some cadre workers do not understand and think it means copying. I say if we copy it will paralyze Chinese engineers."

The caution was particularly apt for the development of military technology, which was Nieh's priority. In August 1958, at the height of Soviet support for the Chinese nuclear program, Nieh warned: "We should and absolutely can master, in not too long a time, the newest technology concerning atomic energy in all fields. . . . There are people who think that as long as we receive assistance from the Soviet Union and other fraternal countries, there is no need for us to carry out more complicated research ourselves. This way of thinking is wrong. . . ."

Two weeks before, in July 1958, the first chain reaction in China had occurred in the 10-Mw Soviet-supplied reactor in Peking. The next Peking research reactor was an all-Chinese model based on Soviet blueprints. After that, even the blueprints were Chinese.

The major engineering feats required for the production of fissile material were, it seems, achieved after the Soviets stopped helping, but it can be assumed that a good deal of design and blueprint information had been handed over. The crucial question that is still unanswered is the extent of Soviet assistance in supplying the new materials and the precision engineering or instrumentation required for the atomic plants. The Chinese bought a lot of precision instrumentation in Europe, ostensibly for use in such things as textile mills but which was diverted to their nuclear and missile programs.

After the Russians left, the Chinese built a number of plutonium-producing reactors, first at Paoxtou in Inner Mongolia and then the Yumen plant in the north central province of Kansu. When the first Chinese explosion occurred at Lop Nor in China's Sinkiang Province in October 1964, Western analysts were in for a surprise. Rather than following the technically easier plutonium route as the British and French had done before them, the first Chinese explosion used U-235, which required the much more sophisticated technical engineering feat of uranium enrichment. The material came from a gaseous-diffusion plant at Lan-chou in the province of Kansu. Powered by a major hydroelectric plant on the Yellow River, the Lan-chou plant was completed after the Soviet aid cutoff. Nieh Jung-chen's policy of self-reliance had paid off. Under his firm military direction, China made rapid progress in nuclear weaponry, and would make a quick transition to hydrogen bombs. At no time, apparently, did Nieh permit his nuclear bureaucracy to divert its attention to the generation of electricity. Despite the broad base of know-how and special facilities, the Chinese showed no interest in nuclear power until the end of the seventies.

When China became the sixth member of the nuclear weapons club in 1964, three other nations had the potential of joining as a result of their peaceful nuclear programs. The first two were the defeated wartime partners, Germany and Japan; the third was Israel. Israel's flirtation with French atomic bomb programs was, even in those early days, well known to Western intelligence. It was always assumed, in Washington at least, that Israel would sooner or later build an atomic bomb, and although it seemed politically impossible for either Germany or Japan to do so, there were men in both those countries who seemed to want to keep the bomb option open.

With the departure of the occupation authorities, the strong sense of patriotism in both Germany and Japan found new expression; nationalism began to reemerge as a basic political value. At a time when the hallmark of national strength and independence had become the possession of nuclear weapons, it was natural that hardheaded practitioners of realpolitik would foresee the day that their nation, through nuclear armament, would be restored to the status of a major power. In both countries, this question was outside the limits of polite debate. For Germany, any step toward acquisition of nuclear arms would mean a preemptive war with Russia. That much was clear. In 1954, Chancellor Konrad Adenauer formally renounced the possession of nuclear weapons, a unilateral declaration made, in the international legal jargon, *rebus sic stantibus*—"until circumstances change." In Japan, the experience of Hiroshima and Nagasaki had left a legacy of domestic hostility to nuclear weapons; legislation for the development of nuclear energy not only renounced weapons development but also sought to enforce the policy by banning any secret research.

No one in either country advocated or, it appears, even wanted the development of weapons. A number of political leaders were, however, conscious that the option to make such a decision in the future could be ensured through the development of "peaceful" uses. They saw it as the responsibility of their generation of politicians to create the circumstances in which the weapons option would at least exist for future generations. No hard evidence is available to show that rapid development of atomic energy in either Germany or Japan was directed toward a bomb program, but the promotional zeal of two prominent politicians in the creation of a nuclear industry in each country deserves a special mention in this respect. They are two maverick politicians, Franz Josef Strauss and Nakasone Yasuhiro. Both men would give hints from time to time that declarations of peaceful intent were made with something less than total conviction. Moreover, each was a vocal critic of his country's ratification of the 1967

Non-Proliferation Treaty, the first international treaty attempting to halt the spread of nuclear weapons from peaceful reactors.

Strauss has been described as "the rogue elephant" of postwar German politics. Nakasone was his Japanese equivalent. Born in 1915 and 1917 respectively, both men were just old enough to have served their countries at the very end of the war, but too young to have had any serious involvement with the discredited regimes. They developed very similar political styles. Energetic, intelligent, autocratic, and ambitious, they shared a mercurial temperament and the impetuous and pugnacious streak of the political maverick. With a flair for showmanship and a capacity for violent rhetoric, they were both members of the "if you see a head, kick it" school of politics. Nevertheless, as well-educated men with quick minds, they were always well briefed and capable of mastering a new field of political relevance with skill and speed. Their capacity and blatant ambition singled them out, and they became the most distrusted, disliked, and feared politicians of their time. Strauss had no time for the mystical ideology of the German race, and Nakasone had no time for the sanctity of the Emperor, but they both had a populist strain that, in the absence of any real political philosophy, achieved coherence through a pugnacious patriotism.

Throughout their political careers, Nakasone and Strauss would each regularly find himself having to explain and qualify some impetuous statement or another. Each was to serve a term in political exile, Strauss after his attempt to suppress criticism by raiding the offices of *Der Spiegel* magazine, Nakasone for falsifying campaign contribution reports and for his involvement with the Lockheed aircraft bribery scandals. The statements that would most frequently get them into trouble, however, were about national assertiveness. Throughout their political careers, they always seemed to be testing the bounds of permissible patriotic sentiment. Despite the limitations of their political bases—Strauss as a leader of a Bavarian political party, Nakasone as leader of a political faction outside the mainstream—each was constantly discussed as a potential national leader. The prospect that either of these might become Chancellor or Prime Minister filled many members of their countries' political elites with deep foreboding. For the political left, they became the objects of special hostility.

The ultimate expression of national manhood is defense policy, and both Strauss and Nakasone came to regard defense matters as their highest political priority. Strauss became the second German postwar Minister for Defense and applied his considerable talents to the reconstruction and expansion of the Bundeswehr during the most important eight years of its development. Nakasone served a term as Minister for the Self-Defense Forces. Even as a parliamentarian he had been a strong advocate of defense independence and was one of the men who drafted the constitutionally controversial laws establishing Japan's paramilitary forces.

In the early fifties, Strauss gave stirring patriotic speeches at meetings of the supermilitant associations of refugees from East Germany and the displaced areas in Poland. Nakasone was one of only two members of the Japanese Diet who dared address the founding meeting of the Congress of War Veterans association, held with conscious insult on the day of the thirteenth anniversary of Pearl Harbor. His speech asserted that the events of Pearl Harbor were, in fact, a conspiracy by Roosevelt and that the United States and Japan were equally to blame for the Pacific war. In 1954, this first Congress of War Veterans adopted a five-point policy: among the usual references to rehabilitation of veterans and widows' pensions was the promotion of nuclear energy. For peaceful purposes.

In 1955, Strauss was appointed the first German Minister for Atomic Questions. "I will make the German people atom-conscious within a few weeks," he declared. He believed nuclear energy would be important for German industry but was cautious on immediate large-scale application. He was more impressed by the slow development of plans for light-water reactors in the United States than the apparent success of the gas-cooled reactor in Britain. Strauss saw the importance of developing the capacity for independent research and of making a priority of training a cadre of specialists, but he did not see the new Ministry of Atomic Questions as of central importance to this goal. He packed the new ministry with second-rate administrators from his home state—they would later be referred to as "the Bavarian Mafia"—and after creating a basic nuclear structure, Strauss moved on to become Minister for Defense. It was in that position that he appeared to display an interest in the weapons option. There is no proof of his interest, just hints and allusions.

In a 1958 interview with the British politician Richard Crossman, Strauss was alleged to have said, "I can guarantee that for three, four or even five years there will be no German nuclear weapons. After that, however, if other states, especially France, produce their own atomic bombs, Germany could also be inspired to do the same." Later he denied the statement, attributing it to "linguistic misunderstanding."

Strauss also became an ardent Europeanist, actively promoting an independent European deterrent based on Franco-German collaboration. As far as is known, Strauss never took any positive steps in the direct promotion of an independent German nuclear capability, but during 1957–58 he did discuss nuclear cooperation with the French Minister of Defense, Jacques Chaban-Delmas. The precise nature of proposals considered at these meetings are still secret, and Charles de Gaulle quickly terminated the discussions on his return to power.

In Japan, the evidence that Nakasone at least shared Strauss's apparent interest in a bomb option is somewhat stronger: he admitted that until 1969 he regarded the Japanese bomb option as open. As a successor to the atomic energy minister Shoriki Matsutaro, Nakasone had proposed expansions in Japan's "peaceful" program that were perfectly in line with a bomb option. But there was no broader vision shaping Nakasone's approach to nuclear bombs, no equivalent to Strauss's European perspective; Nakasone was a good old-fashioned chauvinist with the strong streak of isolationism, racism, and natural arrogance that had always accompanied the more extreme expressions of Japanese patriotism.

Nakasone's most effective means of influence was his close personal relationship with Shoriki. In the nuclear field he was virtually Shoriki's chief of staff and was entrusted with some of the more delicate negotiations leading to the appointment of the first Japanese Atomic Energy Commission. Once the commission was in operation, Nakasone was worried that the bureaucrats would not treat it with the urgency and respect he believed it deserved. As a counterforce to the bureaucrats, he organized a group of parliamentarians; most of them were former technocrats rather than lawyers (all senior positions in Japanese government had been dominated by lawyers) and this group was to have a major influence on Japanese atomic policy.

In 1954, at a time when no one in Japan had begun to think about a coherent nuclear policy, Nakasone moved an amendment to the annual budget providing for the expenditure of 300 million yen for nuclear energy. Of this, 235 million yen were for the construction of a reactor. The figure of 235 was chosen for no better reason than that U-235 was the key fissionable isotope. Reflecting the contemporary enthusiasm, the Diet passed the budget amendment. Nakasone was quoted as saying: "Since the scientists were too immobile I awakened them by slapping them on their cheeks with bundles of yen bills." As he would have to do so many times in the future, he denied the statement in the face of later criticism.

Nakasone led a team of his Diet group to the 1955 Geneva Conference and returned even more enthusiastic. His group immediately pushed for a 10 billion yen budget, and although only 2 billion was allocated, Nakasone influenced the size and distribution of the funds through a behind-the-scene personal contact: the budget officer responsible for nuclear matters in the Ministry of Finance was Hatoyama Iichiro, son of the Prime Minister and a future foreign minister. He and Nakasone had been in the navy together at the end of the war.

In 1959, at the age of forty-one, Nakasone was given a cabinet post; he was its youngest member. His department drew up a long-term nuclear plan in which a definite interest in developing a weapons option was suggested. At the time, Japan had a low-price contract with England for reprocessing spent fuel from its Calder Hall reactor, and there was also a proposal to construct a small Japanese reprocessing plant in Japan. Nakasone decided to double the capacity of the proposed plant, to provide plutonium for "peaceful" breeder reactors. His other major initiative during this term of office was to establish the Japanese space program to launch "peaceful" weather satellites. Read together, these decisions equal fissionable material plus delivery systems, or nuclear missiles.

In 1970, Nakasone firmly declared that he now believed Japan should not develop nuclear weapons, but his statement was treated with some skepticism; it wasn't the first time he had asserted that Japan should renounce the weapons option. Although he began to talk in hardheaded terms of the impossibility of a small, densely populated country's developing a second-strike capability—the key to deterrence capability—it was pointed out that contemporary second-strike weaponry was based on submarines and that he was himself an advocate of the development of a Japanese submarine capability. Nakasone remained silent.

Neither Nakasone nor Strauss represented the majority view in either Japan or Germany; in both nations, most senior decision-makers did not want to create a weapons option. Nevertheless, the option was always there. Inevitably, the plutonium that would accumulate from peaceful reactors would mean that countries like Japan and Germany could become weapons powers almost overnight. For about ten crucial years the Americans, who alone could have stopped this development, did nothing.

American policymakers of the mid-fifties were proud of their pragmatism and reassured by the fact that actual stockpiles of bomb-grade material would not appear in the new nuclear countries for many years. They were confident that their step-by-step backroom diplomacy would work, not realizing that the steps would become harder. American bargaining power was at its height in the mid-fifties, and the policymakers failed to see that it would inevitably decline, that deferral of the key issues would lead to long-term costs. They assigned the problems to the next generation of pragmatic diplomats; by that time, it was too late. The short-term public relations triumph of the Atoms for Peace program would have a dangerous and persistent legacy.

14

The First Ice Age

The screwworm fly was a persistent molester of cattle in the southeastern United States until it encountered the power of the atom. In 1958, some enterprising government officials started a screwworm fly factory in Florida and bred millions of the nasty black insects. They then irradiated them with gamma rays, making them sterile, and dropped them from aircraft on Florida, Alabama, and Georgia.

Over a period of eighteen months, two billion of the sterile flies were allowed to compete with their virile brothers and sisters in the natural breeding process. Sterile males were soon outnumbering virile males by nine to one. By the end of 1959, the screwworm fly was all but eradicated from the three states.

This modest, but clever, triumph represented a high point in the stream of new peaceful uses of the atom that emerged during the 1950s. A host of new uses was found for artificial isotopes in medicine and industry. By simple radiation of existing organisms, for example, the world discovered three new kinds of beans, two kinds of disease-resistant oats, barley that could survive the winter better, peanuts with tougher shells, and a new kind of carnation with fewer but longer-lived petals.

It was found that irradiated strawberries and some types of oranges could be kept longer without going bad. The same went for sweet cherries, prunes, apricots, and nectarines. Potatoes exposed to radiation doses were less prone to sprouting in storage. The exception, for some unexplained reason, was the petulant lemon; it turned spongy and quickly rotted. At one point, the U.S. Joint Committee on Atomic Energy became so excited by the prospect of preserving foods by killing off damaging bacteria and molds with radiation that they started an annual "Irradiated Food Luncheon." They dropped it after the Food and Drug Administration questioned the safety of food prepared in this way.

Each of these new tricks was greeted as a great victory, a portent for the future of the atom in the service of mankind. Success of the peaceful atom in one field became justification for its introduction in another. Promotional speeches from the nuclear advocates would point to the triumph over the screwworm fly as an indication of the potential triumph of nuclear-generated power. The atmosphere was more religious than scientific. An aura of inevitability about the atom and its future grew in the atomic energy commissions that had been specially created to look after it. These new agencies demanded, and received, budgets that were staggering compared with other forms of government-supported science. Private foundations, universities, and corporations with a keen eye on the future expanded the growing stream of nuclear projects—searching for problems the atom might solve.

Among a range of nuclear engineering projects, launched in bizarre proliferation and costing billions of dollars, the nuclear airplane deserves a special mention. The project, started in 1946, was a mirror operation of the atomic submarine, a pet of the Joint Committee on Atomic Energy, and a passion of the air force's strategic bombing school. What could be finer, thought the air chiefs, than to drop atomic bombs from an airplane powered by the atom itself? To run the project, Donald Keirn, an engineer, was given the same two-hat bureaucracy as Rickover; he became the head of the aircraft reactor units in both the air force and the AEC. Although he lacked Rickover's drive and psychic energy, he had the admiral's single-mindedness, and by the end of 1954 the still unproven project was upgraded from "early flight" to "weapons system development." It was blind faith. There was not a penny's worth of justification for it.

The basic design problem was to find a material dense enough to shield the pilots of the plane from the intense radiation produced by the reactor and yet light enough to allow the airplane to take off. No such material could be found; nothing that is dense is light in weight. Undeterred, the imaginative masters of this project actually discussed the possibility of overcoming the problem of radiation by using older rather than younger pilots so that genetic damage to the pilots' offspring would be kept to a minimum.

They also spent a lot of time and money failing to find a solution to the problem of how to guarantee that no radioactive fission products leaked into the reactor's exhaust airstream. But, no one even dared address the other key safety question: What happens to all the radioactive engine parts if the airplane crashes near a populated area?

It was a monster boondoggle, but the air force top brass continued to back the program. Led by the cockpit cowboy, Curtis LeMay, the air force Luddites saw the nuclear bomber as the answer to the threat of the unchivalrous unmanned missiles.

Their enthusiasm was infectious. After the Russians launched Sputnik in 1957, both the AEC and the Joint Committee suggested that a suitable propaganda reply would be to "test fly" immediately a nuclear-powered airplane. The idea, later dropped, was to fool the Russians into believing that they had such a machine, they could put a nuclear reactor into a conventional plane, take off, and claim that it was tested during the flight.

A year later, in 1958, to boost the flagging project, fake "intelligence" reports were leaked, apparently by the airplane's supporters in the military and industry rather than by the administration, saying that the Russians already had a nuclear-powered airplane and that it had been spotted on the ground and in flight. The report created such a stir that, ten days later, Eisenhower, who possessed no such intelligence reports, had to deny it. The nuclear airplane was finally killed by President Kennedy in 1961, after costing the U.S. taxpayer more than a billion dollars.

The Russians themselves were equally enthusiastic about producing new atomic tricks. They led the way with merchant ship propulsion, launching their icebreaker, *Lenin*, in the summer of 1960. It appears to have been very successful, since two more were built. All other ship propulsion prototypes were destined for the physical and intellectual scrapheap. The Americans built the cargo ship *Savan-*

nah, the Germans built its sister ship, *Otto Hahn*, and the Japanese built the *Mutsu*. Glenn Seaborg, chairman of the U.S. AEC at the time, wrote, "I feel that such ships, with smaller but more powerful reactors . . . will inevitably play a major role in worldwide shipping." He was wrong. The *Savannah* and the *Otto Hahn* cruised the world as splendid examples of a brilliant new technology, but they were hugely expensive—the *Savannah* was subsidized to the tune of $3 million a year. They were also quite unnecessary: the job of a cargo ship is to go from one port to another; unlike a nuclear-powered submarine, there is no role for one that can cruise the high seas indefinitely, a real-life *Flying Dutchman*.

Just as ill-fated as the nuclear merchant ships was the American project for nuclear-propelled rockets. Some $2 billion was spent in futile attempts to develop and launch a succession of nuclear rockets. There was Project Pluto, then Project Rover, and, finally, Project Poodle. In the end, the promoters accepted the inevitable: they could not find a way of preventing exhaust gases full of harmful radioactive materials from spreading all over the test site and anyone who happened to be living nearby. The project was canceled, but some still clung to the hope that, one day, they would be able to launch a nuclear rocket from a space platform, out of reach of earthly fears and dangers.

Even the weapons scientists, whose public image was being eroded by the growing concern over fallout, came up with a new concept of "peaceful nuclear explosions." In 1957, Project Plowshare, as it was known in the United States, and its Russian equivalent resulted in a series of experiments designed to prove the "peaceful" uses of the bomb in moving mountains, redirecting rivers, and digging harbors and canals. With overheads already met by the weapons program, nuclear explosions seemed cheaper than the same quantity of TNT. Project Plowshare became part of the American Atoms for Peace drive, and enthusiasm for it would continue for over a decade. At last the bomb could do something useful; that, at least, was the fond hope. When the international treaty of 1963 banned atmospheric nuclear tests, Plowshare enthusiasts focused on underground engineering schemes, such as exploding natural gas trapped in rock. Except in the Soviet Union, the enthusiasm for Plowshare projects gradually waned as more complete cost calculations were demanded—including an account of the residual radioactivity.

In each of these projects, the superiority of the atom was always limited: you could irradiate the screwworm fly and stop its breeding, but the same trick did not work with fruit flies, corn borers, gypsy moths, or the tsetse fly. You could preserve oranges with radiation, but not lemons. Nuclear reactors were useful in submarines and ice breakers, but useless in cargo ships, rockets, and airplanes.

If the atom had more limited application than its visionary supporters claimed, it was in one area vastly inferior when compared with conventional alternatives: the claims made for the superior economic performance of nuclear-generated power were always questionable, if only because of the immaturity of the technology. Unfortunately, no one stopped long enough to question these claims; and so, the progressive image of all things atomic set a herd of white elephants on their path. There was a revolution going on, and no one wanted to miss it.

This progressive environment spawned the greatest hope of the nuclear advocates: electricity production. The promoters had no evidence that it would be cheaper than electricity from conventional means; their extravagant enthusiasm was based on nothing more than the intellectual sex appeal of the new product, an appeal that coal- and oil-fired boilers could not share. But, in the mid- and late fifties, their blind confidence allowed them to predict a set of hopelessly optimistic plans for electricity production from nuclear reactors. Not one of them was fulfilled. The British nuclear power plan of 1955 was trebled in 1957 only to be revoked two years later. The grandiose Russian nuclear estimates in their Five-Year Plan of 1956 simply disappeared. In America, the AEC's twenty-year projection of nuclear generation capacity would be less than half fulfilled. The German plan of 1957 for five reactors would result in one, and the Japanese "Long-Term Plan" of 1957 would not build anything. Even the French, preoccupied with their bomb project, would not achieve their exceedingly modest plans set forth in 1957. The only country that escaped the international downturn in nuclear optimism was Canada. It was preoccupied with developing its own heavy-water technology and had no embarrassing grand targets to meet. For the others, the second half of the decade would be, as the nuclear advocates later dubbed it, "The First Ice Age." It would be a period that would witness the first serious reactor accidents, including the world's worst known nuclear disaster involving radioactive waste. It would emphasize the hazards and the uncertainties of uncontrolled releases of low-level radiation from nuclear power stations. And, finally, it would lay the foundations of the fierce battles for a new worldwide market for reactor sales.

The fanfare that accompanied the Queen's opening of the British power reactor at Calder Hall in 1956 placed great emphasis on the triumphant arrival of the peaceful atom and masked the equally important use of the reactor: to provide plutonium for Britain's bomb project. Indeed, the reactor, cooled by carbon dioxide gas with neutrons moderated by graphite as in the first Fermi pile in Chicago, was primarily designed to boost the supply of plutonium for bombs, not to provide electricity.

As in America, the peaceful atom grew out of the warlike atom. Britain's gas-cooled reactors were siblings of the plutonium-producing graphite piles that had been built from 1946 onward on the site of a disused ordnance factory at Windscale, in northern England. The Calder Hall reactor was built next door. The British were so pleased with the reactor, which had performed extremely well, that they rushed headlong into a series of wildly optimistic forecasts for its future—and for the future of a British nuclear industry.

The first British reactor plan, drawn up in 1955 and based on modifications of the gas-cooled reactor design, proposed a relatively modest program of 2,000 Mw over the next decade. The need for this increased capacity, apart from normal growth, was justified by a predicted energy gap that, it was said, the British coalfields would be unable to fill. At the same time, the British energy planners estimated with great confidence that nuclear power would be competitive with conventional fuels almost from the outset. The government White Paper outlining the plan acknowledged the risk in this capital investment in an untried technology, but declared that "the final reward will be immeasurable."

The Tory government was so sure of a brilliant nuclear future that it made the 1955 decision with a minimum of consultation with the Central Electricity Authority (CEA), the nationalized power industry that would eventually have to use the nuclear plants. The CEA took no part in the detailed preparation of the estimate and was given only just over a month to comment on the draft proposal. When the proposal was made public, criticism was muted; no one thought the program was too large.

More grandiose schemes were to come. Two years later, with the technological breakthrough of Calder Hall behind them, the British were able to put up a bold front against the threats of oil rationing in the wake of the 1956 Suez fiasco and the persistent estimates of a coal shortage. In a dramatic gesture that left many of those involved in the fledgling industry reeling in disbelief, the government trebled the nuclear target from 2,000 Mw to 6,000 Mw. While again acknowledging the risks of the new program, the government said its action was the result of the "tremendous confidence" the country had in its nuclear scientists. This time, the public was not convinced. Atomic energy, said the London *Times* sarcastically, "possessed the power to evoke fantasies." The Atomic Energy Authority, responsible for putting the policy into effect, itself called for a more modest program, but the government was adamant. It put the decision in terms of survival: "The whole national situation depends on it."

The rest of the western world looked on in amazement. But there were some who believed that if the British said the nuclear future was rosy, it must be so. Japan and Italy were so swept up in the boundless enthusiasm that they bought two Calder Hall reactors—the only two ever sold by Britain.

At the center of Britain's ambitious plans—and himself an energetic advocate of nuclear power—was Sir Christopher (later Lord) Hinton, often dubbed the "father" of British nuclear power. As the government decided to treble the program, Hinton was forecasting that the reactors coming into operation at the start of the 1960s would probably be economically competitive with conventional power. But Hinton was also a man of caution. "Like all successes," he had told a congress of electricity producers and distributors in London in 1955, "it brings some danger in its train. The information released [at the Geneva Conference that year] was in the form of papers read mainly by scientists and development engineers. . . . You as practical men know that it is not always reasonably possible to adopt immediately the ideas of scientists; the penalty of proceeding in advance of the reasonable speed can be very severe."

In fact, a private debate had been going on inside Britain's nuclear workshops about the future economic viability of their product. Hinton was one of those who had erred on the side of caution, but his voice was not always heard in the higher echelons of the British government. When the second power program was announced, Hinton was actually away on leave. He and some of his top engineers thought it was crazy: "There was no way in which we could have coped with the demands of the enlarged program and developed the more advanced reactors which the program called for," he said later. It was Hinton who eventually persuaded the British government to curb its enthusiasm and cut back the program: Despite his advocacy of nuclear power, Hinton was a professional and a realist and he always maintained his intellectual honesty.

Hinton's involvement in things atomic began after the war when he was drawn into the ambitious British bomb program. At the time, he was a young and successful engineer with a rewarding career in private industry ahead of him, but he could not turn down the challenge of being part of the atomic revolution, even if it did mean making bombs and taking a cut in his salary to work for the government. He knew the bomb project would be an exacting task, and he was determined to succeed. For ten years, he cajoled and threatened scientists, industrialists, engineers, and civil servants in order to meet deadlines; in the process, he drove many of them—including, in the end, himself—to near-collapse. Later, he would admit that the load had at times been "crippling" on the senior staff of his industrial division at the U.K. Atomic Energy Authority and that they had won through only by being "pig-headed and bloody-minded."

Like most good engineers, Hinton had a greater desire to achieve technical excellence than to hold political power. Born in 1901, he was the son of a country schoolmaster and as a teen-ager he had had to battle his way out of rural obscurity to go to Cambridge University. At the age of sixteen, he was apprenticed to a railroad-yard workshop where he was taught how to make metal parts for locomotive boilers and, as he put it, "learned what boredom is really like—useful to know when you are running a factory." Six years later, he won a scholarship to Cambridge, where he completed the university's three-year engineering course in two years and got a first-class degree. From then on, his rise was rapid. By the age of twenty-nine, he was the chief engineer of the Alkali Division of the British chemical giant ICI. There, Hinton learned the advanced accounting and management practices for which the company was well known (and ahead of its time in forcing its engineers to learn them too).

Hinton spent the war years building the nation's vast network of munitions factories and picking up more valuable managerial experience. Although he knew nothing about atomic energy, he seemed an obvious choice when the war was over to build the huge atomic factories and power plants needed for the British bomb project. In the course of his ten years with that program, he made many enemies with his ruthless, autocratic methods, but he got the job done. Traveling from one end of Britain to the other, this tall, thin man became a familiar figure on the construction sites, dressed in a fawn duffle coat and a trilby hat, striding ahead of his harassed aides. On the job, he was a man of extreme self-confidence, but he was also inflexible, sticking rigidly to his plans. Debate was not one of his strong points, and he would soon become impatient if meetings dragged on too long; when a colleague once commented about the time it had taken to reach a decision, Hinton replied, "That's because you buggers were so slow in agreeing with me."

By 1956, when Hinton escorted the Queen at the opening of Calder Hall, he had joined the select company of Britain's "nuclear knights," a group that included the distinguished physicist James Chadwick, the director of the Harwell research establishment, John Cockroft, and Britain's bomb designer, William Penny.

A year later, Hinton left the Atomic Energy Authority to become chairman of the newly created nationalized power industry, the Central Electricity Generating Board (CEGB), one of the largest industrial enterprises in the world. His opposition to the government's 1957 nuclear plans had led him to resign from the AEA: To Hinton, it had become clear that the economics of nuclear power generation were not quite what they seemed to the Tory government.

Three factors had turned the program sour. The first was the renewed competition from conventional fuels; the second was the fall in the value of the "plutonium credits"—the sale of the by-product plutonium back to the government; and the third was the rise in the interest rates, which adversely affected the large capital costs of building nuclear plants.

Then came a fourth factor: safety. In the early years, it seemed relatively insignificant, but it was to plague the nuclear industry, and because of the short-sightedness of some managers, it was to play a key role in bringing the industry to a virtual standstill in the seventies. Increasing emphasis would be placed on safety after October 1957, when the world's first major reactor accident occurred at the Windscale plutonium-producing plant.

The scenario that was played out at Windscale would come to

have a familiar ring. The core of the reactor—the part containing the uranium fuel rods—caught fire. The only remedy was to douse the core with ordinary water, but the plant engineers feared that this might produce an explosive hydrogen bubble in the core container and that the whole thing would blow up. Careful introduction of the water finally brought things under control without further mishap, but while the fire had burned, large amounts of radioactive isotopes—particularly Iodine-131—had been released into the atmosphere. The British government has never released the full official report on the accident, but it is known that Iodine-131 contaminated an area of five hundred square kilometers surrounding the plant. For the first time, the dangers of this radioactive isotope were publicly recognized: Because the land was used as grazing for cattle, two million liters of milk had to be thrown away—into surrounding rivers and the sea. The Windscale reactors were in due course permanently closed down. It was an ominous portent of things to come.

The Windscale accident caused considerable concern among health physicists in America: Up to that time, the U.S. AEC had taken no precautions to protect the public against the effects of Iodine-131 produced during the bomb tests in Nevada. Windscale also reinforced the fears of some American electric utility companies that the risk of turning to nuclear power was too great.

In 1953, just before the amendments to the McMahon Act made nuclear reactors and atomic fuel available to American private industry, the AEC staff in Washington produced optimistic predictions for the future of the U.S. nuclear industry: America would have between 5,000 and 21,000 Mw of nuclear power by 1975. Three years later, the AEC quadrupled this estimate. Such confidence should have attracted the utility companies to the nuclear revolution in swarms, but it did not. In the event, less than half of the predicted 89,000 Mw would be achieved by 1975; the utilities had never really overcome their reluctance to join the nuclear bandwagon.

Three factors had turned them away: the technical uncertainties; the high capital costs—substantially in excess of conventional plants; and the rising concern over plant safety. In 1957, the AEC produced the first nuclear-plant accident scenarios.

The report, entitled "Theoretical Possibilities and Consequences of Major Accidents in Large Nuclear Power Plants" (also known as the Brookhaven Report because it was prepared at the Brookhaven National Laboratory on Long Island), was intended to allay fears of nuclear power. But the horrific picture it painted had quite the opposite effect. A "worst case" accident—where the core of the reactor melted—would cause the release of radioactive substances that would kill 3,400 people immediately and cause serious injury to a further 43,000, the study concluded. Property damage would amount to $7 billion over an area as great as 150,000 square miles.

The AEC stressed that the report had taken the most pessimistic view and that such an accident was a remote possibility. But the stark figures were enough to turn off even the most faithful commercial investor and did nothing to comfort the insurance companies, which were already balking at the idea of underwriting nuclear power plants.

Even the pro-nuclear activities of Senator Clinton Anderson of New Mexico did not change the outlook of the reluctant utilities. In 1957, Anderson sponsored the Price-Anderson Act, which limited the liability of electric utilities for nuclear accidents and provided substantial taxpayer subsidies for any claims. But the legislation did not immediately overcome the growing resistance to nuclear investments. Outside the handful of self-consciously progressive utility executives who had already become involved, doubts about cost competitiveness continued, and nothing in Price-Anderson spoke to this factor.

Cost was a major factor for the public power organizations—like the Tennessee Valley Authority—but they were in a position to lobby for federal funding. Keen to join the nuclear power club, they sought a massive federal program of reactor construction that would put them in a position to sell power at cost to public companies. But they failed to get the necessary backing from the AEC. The commission's chairman, Lewis Strauss, was a supporter of big business and private enterprise, not public authorities. Despite the support of the Democrats in Congress, the AEC flatly refused to stretch their assistance offers beyond basic research and development at government laboratories, supply of fissionable material for fuel, and a consideration of finance for plants—providing the AEC retained title to the part it had funded.

(What the AEC was reluctant to do at home, it was only too eager to do abroad. Chairman Strauss would preside over generous offers of assistance to nonnuclear nations under the Atoms for Peace program. By 1958, the AEC formally acknowledged that the "urgency" of its reactor program abroad was "dictated primarily by international considerations." The continuing flow of U.S. government subsidies for reactors abroad, running at $2 million a year, was increasingly justified in terms of America's international prestige.)

If the American utilities were cool to nuclear power, their European counterparts, faced with high-cost electricity, were not. Westinghouse and General Electric, undeterred by the poor showing of the utilities at home, joined the Atoms for Peace bandwagon and sought their markets abroad. They quickly developed a new strategy of using the European enthusiasm for nuclear power to develop and market the American light-water reactor technology.

Atoms for Peace had caught Europe in the midst of a burst of enthusiasm not only for nuclear power but also for political and economic unification. Some people, like the president of the French National Railroad Company, Louis Armand, believed that Europe should unify over its nuclear future. In fact, nuclear power seemed to present a much easier platform for unity than coal and steel, which required a delicate balancing act between a host of entrenched interests. Atomic institutions were new, their vested interests hardly established. It seemed a fruitful area to start the first practical supranational organization. In November 1956 the governments of the six European nations involved in the broader unity negotiations appointed three prominent Europeanists, including Louis Armand, to study European energy requirements. They were to work under an umbrella organization, which became known as Euratom.

Armand was a graduate of the Ecole Polytechnique and a member of the elite Corps des Mines. His two colleagues were Franz Etsel, the German vice-president of the European Coal and Steel Community, and Francesco Giordani, former president of the Italian Atomic Energy Commission. Quickly dubbed "The Three Wise Men," the team took as its primary task the need to reduce dependence on oil in the wake of the Suez fiasco and set off for the United States to investigate the prospects of nuclear power. They were given enthusiastic assistance; four senior AEC staff members accompanied them throughout the United States and drafted their final report, "A Target for Euratom." Released in May 1957, it emphasized the advance in American light-water technology and underlined the American willingness to assist the European program. It presented a "target" for nuclear power of 15,000 Mw installed capacity by 1967, about a quarter of all electricity projected for the six European nations for that year.

The team's advocacy of strong links with the American program met fierce opposition. Most opposed were the independent members of France's Commissariat à l'Energie Atomique, the CEA. In particular, the CEA chief, Pierre Guillaumat, completely rejected the idea of having anything to do with the Americans. He was quite satisfied with the CEA's development of the gas-graphite technology, which the British were also developing, and he wanted nothing to do with Yankee technological imperialism.

Armand had expected support from Guillaumat for his grand vision of Euratom; they were, after all, both members of the Corps des Mines; even if they were not close friends, members of "Les X" should stick together. But Guillaumat had no time for Armand's idealistic vision of a united Europe. Euratom's campaign for a unified nuclear-power program using American technology would get nowhere in France.

Already a representative of Gaullist political economy, Guillaumat saw the European Euratom initiative exclusively through French eyes. Euratom was acceptable only if it could expand the scope of French interests and power by reinforcing French military and industrial strength in the nuclear field, especially vis-à-vis the Americans. If it meant that his own institution's objectives had to be compromised in any way, then it was unacceptable.

Euratom's efforts floundered; the original target of 15,000 Mw installed nuclear capacity by 1967—with the United States footing half the bill—was cut by a quarter and, over the next two years, halved again. Only three plants, two in Belgium and one in Italy, with an aggregate of 750 Mw were actually completed. But American light-water technology had a foothold in Europe. As experts have noted: "These three plants are arguably the most commercially important nuclear plants ever built outside the United States. As a direct result of these projects, American manufacturers developed strong working relations with utilities in Germany, Italy and Belgium; and, outside France, the most important potential reactor manufacturers switched development priorities from gas-graphite to light-water systems."

France was not alone in resisting the march of the U.S. light-water reactor salesmen. In Germany, Karl Winnacker, like Guillaumat, was trying to develop a nuclear technology independent of America. He had focused on heavy-water reactors that did not need supplies of enriched uranium from the American enrichment plants. Winnacker spent large sums from the coffers of his chemical company, Hoechst, on projects to produce heavy water and to reprocess plutonium from spent fuel rods for breeder reactors. Both were expensive failures from which Hoechst would only manage to extricate itself completely after Winnacker's retirement. To this day, the Hoechst management are too embarrassed to discuss their postwar nuclear follies.

In June 1957, Winnacker gathered the surviving reactor research leaders, Karl Wirtz and physicist Wolfgang Finkelnburg, at the Hoechst guest house at Eltville on the Rhine. The resulting grandiose 500 Mw program became known as the Eltville Plan, the first reactor program for Germany. Five different industrial consortia were to produce reactors of 100 Mw, each of which would be of a different type. Thus, Winnacker hoped Germany would develop a broadly based indigenous technology. But like all the other grandiose plans of this period, it, too, failed. Only one reactor was built, by Siemens in Bavaria. A heavy-water reactor, it was plagued with difficulties and was finally closed in 1974. The total bill was 350 million Deutschmarks.

For all its isolation, the Soviet Union went through precisely the same cycle as the Western nations during this early period: a burst of optimism about the peaceful applications of the atom that peaked about 1957, followed by a slump. The push came from the scientific community, which, like its counterparts in the United States and England, had acquired an extraordinary amount of political access and influence as a result of weapons work. The Soviet scientists displayed the same yearning as their Western colleagues to see their handiwork applied to something useful and, the wish fathering the thought, became convinced of the imminence of competitive electricity generated from nuclear power.

At the 1955 Geneva Conference, Soviet scientists had presented some of the most optimistic cost projections for nuclear reactors, although thus far, only small experimental units up to about 5 Mw existed. They had talked of building 50-Mw reactors but, reinforced by the enthusiasm of Geneva, they became more ambitious, devising projects up to 400 Mw. The Russian Five-Year Plan of 1956–60 included the world's most ambitious nuclear program: between 2,000 and 2,500 Mw of nuclear capacity by 1960. Only 400 Mw would actually be finished.

The exact turning point is not known, but by 1959 the Minister for Power Stations was giving speeches that casually dismissed the prospects for nuclear power and emphasized the need to reduce costs. The same year, Frol Kozlov, a Soviet Deputy Premier, confirmed the downward trend on a visit to the United States. He specifically complained that the scientists had given misleading cost projections. But there are two other possible reasons for the Soviet downturn: the calamitous accident in the Urals and the loss of power in the state machinery by the nuclear engineer-administrators, the Red Specialists.

There is an ominous silence in the official Soviet literature about standards of radiation protection in the early years of its bomb project. Such evidence as there is, admittedly self-serving to the West, suggests that the crash nuclear program was one of the great public health disasters of the century. An early German refugee drew attention to the primitive protection standards in force at the Saxony uranium mines. American intelligence gathered many uncorroborated horror stories about deaths of sailors in the early Soviet nuclear submarines. It was not until 1968, however, that academician A. P. Alexandrov, Kurchatov's successor as head of the Institute for Atomic Energy, admitted past failures. He emphasized the safety of contemporary reactors in contrast with the past and revealed that, "from 1946 through 1948 some of our employees got radiation cataracts of the eyes." This was a stunning admission: cataracts were the earliest health effect to show up among those who managed to survive the explosions of Hiroshima and Nagasaki. The effect does not generally occur unless the exposure has been as great as 200r or more. For industrial workers to have developed them suggests a huge, continuing health problem with a multiplication of leukemias and cancers in the years ahead.

In February 1957, the Soviet political leadership became acutely aware of the dangers of radioactivity. Vyacheslav Malyshev, who had taken over the top political direction of the nuclear program from Beria in 1953, died from the effects of radiation. Before Malyshev died, a senior German blood specialist had been flown from Cologne to Moscow to treat him; apparently, Malyshev had developed leukemia after exposure to a lethal burst of radiation. But worse was to come. In December 1957 or perhaps the first week of January 1958—the exact date is unknown—what is widely accepted in the West as the world's worst nuclear accident occurred between the major southern Urals cities of Sverdlovsk and Chelyabinsk on the edge of the Siberian plain. The Russians have never admitted the accident occurred, but from Soviet scientific journals, intelligence reports, and refugee accounts it has been possible to put together a reasonable picture of what happened. It is certain that lethal radioactivity spread over hundreds of square miles. The exact cause of the accident has never been released by the Russians, but it seems to have been an explosion of some kind in a radioactive waste dump.

The first reports came from Denmark in April 1958. Journalists there, quoting "diplomatic sources," wrote stories about a catastrophic accident inside the Soviet Union involving radioactive fallout. The United States, through AEC Chairman Lewis Strauss, said it had "no intelligence" of any such event. In May 1958, however, a newsletter published by the Munich-based Institute for the Study of the U.S.S.R., an arm of the U.S. propaganda unit Radio Free Europe, commented on the "unusual amount of attention" being given to radiation sickness in Soviet medical journals and even popular magazines. It said that on January 9, 1958, Radio Moscow had devoted a large segment to radiation sickness, describing in detail a list of possible preventive measures. This strongly suggests that the accident happened around the end of December.

The accident was forgotten for twenty years—until November 1976, when a Russian refugee biochemist, Zhores Medvedev, casually referred to the disaster in an article in the London-based *New Scientist*. To his surprise, he found that it was largely unknown in the West. He suggested that the cause was probably an explosion, more likely chemical than nuclear, in radioactive wastes.

His suggestion produced a hysterical reaction from Western nuclear advocates; at the time, the disposal of reactor wastes had become a major controversy in the emergent debate over nuclear power. It was an especially controversial issue in Britain, where Medvedev's article was published. Medvedev was accused by intemperate officials of the U.K. Atomic Energy Authority of being politically motivated in his explanation of the Soviet accident—especially since he had brought it up so long after the world had forgotten about it.

In fact, Medvedev, who was exiled from the Soviet Union in 1973 and was living in London, is a quiet, mild, unassuming man; he is no militant Russian dissident. He had been asked by the *New Scientist* to write an article about Soviet science to commemorate the twentieth anniversary of Khrushchev's famous party congress speech denouncing Stalinism. Being a biochemist, he had mentioned the Urals disaster as an important historical dateline that brought the atomic physicists together with long-persecuted geneticists.

Experts in Britain, France, and the United States rushed into print denying the story. Most memorable was the response of the chairman of the U.K. Atomic Energy Authority, Sir John Hill, who said that the story was "pure science fiction" and that the idea that a waste dump could explode was "rubbish."

Medvedev's assertion that a large area of land had been contaminated was nonetheless soon confirmed. The CIA itself had documented some kind of nuclear accident in the southern Urals. But the most important evidence was found in Soviet technical publications. Beginning in 1966, first a trickle then a stream of research articles had appeared outlining in great detail the ecological effects of radiation experiments. They were the result, it seemed on the surface, of a planned release of radioactivity for testing purposes. Medvedev's own calculations showed that these reports involved the study of three different lakes, soil samples, and a host of animal and food chains that were consistent only with major contamination over an area of hundreds of square miles. The radioisotopes involved in these "experiments"—especially the large amount of Strontium-90—pointed to nuclear wastes as the most likely source of the contamination. By 1979, American researchers, following up on Medvedev's lead, had identified no fewer than 115 articles in Soviet journals referring to these events.

The most dramatic evidence came from a lake nicknamed "Ilenko's Hot Lake," for the Soviet scientist who had researched its underwater life after the accident. Studies of plankton, water plants, and fish in this lake suggested a high contamination level of radioactive Strontium-90. An American reviewer described it as the most radioactive place on the face of the earth. In an unusual slip in Soviet censorship, one of Ilenko's studies, Medvedev found, said that the samples came from the Chelyabinsk region. The lake in question may have been the one near the town of Kyshtym.

Kyshtym was a special town. It was in a region that had been a center of the Russian armaments industry since the time of Peter the Great and had a relatively high population density, similar to the English midlands or the state of New Jersey. In 1948, long-term residents of Kyshtym were evacuated to make way for a new work force of slave labor. Their task was to construct a secret military plant with a plutonium-producing reactor known as Chelyabinsk-40, or, as the CIA code name called it, Post Box 40. After their work was done, the construction team for Chelyabinsk-40, according to Solzhenitsyn, was "declared to be *a particularly dangerous contingent*" because of their knowledge of the secret plant. When their sentences were over, the prisoners were not allowed to go home, but were transferred to camps on the Kolyma River in the Arctic Circle in the far northeast of Siberia. In view of the later accident, they may have been more fortunate than those that stayed behind.

Medvedev discovered that nuclear wastes in storage could "blow up" under special circumstances: Sir John Hill's outburst was not only intemperate, but ill-informed. It is scientifically possible to have an explosion at a nuclear waste storage site. One such process is described in a 1972 U.S. Atomic Energy Commission report that investigated the accumulation of plutonium in wastes from the Hanford reactors. Low-level radioactive wastes, which had been dumped in unlined trenches in the hope that they would disperse harmlessly in the soil, had in fact produced a layer of highly concentrated plutonium. The layer was removed by the AEC, but the report suggested that a chain reaction could have been set off if water had soaked into plutonium-rich soil. The rapid heating of the water could turn it into steam and the steam could have produced a "mud-volcano type explosion."

Admittedly, this is only one possibility, and in the case of the Kyshtym disaster, it is speculation; the exact cause of the Russian accident may never be known. The number of people who died or suffered radiation damage is not known. Reports by Soviet emigrants suggest that no public protection measures were taken until symptoms of acute radiation sickness were found—days after the accident. The government then ordered a hasty evacuation from the towns where the effects were obvious. As hospitals throughout the region filled up, rest homes, clinics, and hotels were hastily converted into health and evacuation centers. Huge quantities of foodstuffs were seized and destroyed and new supplies were brought in and sold from the backs of trucks, with the queues reminding residents of wartime rationing. The major north-south highway through Kyshtym was closed for nine months. When it reopened, there were signs advising motorists not to stop for twenty miles and to drive at top speed with their windows up.

A Soviet physicist who drove through the area two years after the accident later described the devastation. "As far as I could see was empty land. The land was dead—no villages, no towns, only chimneys of destroyed homes, no cultivated fields or pastures, no herds, no people—nothing. It was like the moon for many hundreds of square kilometers, useless and unproductive for a very long time, maybe hundreds of years."

A decade later, local doctors were still advising pregnant women to have abortions. The region was still dotted with "graveyards of the earth," dumps for heavily irradiated topsoil, with clusters of "giant mushrooms" growing behind the barbed wire. Food was still being checked for signs of radioactivity, and fishing in the lakes was still forbidden.

The Kyshtym catastrophe apparently undermined the reputation of the Soviet atomic scientists: The declared objective of a major expansion of nuclear power plants simply disappeared. Besides the technical, health, and economic reasons for the nuclear pause that followed, there was also a political factor: the declining political influence of the Red Specialists, the managerial elite of the atomic program. They had climbed the Soviet hierarchy under the patronage of Stalin and had reacted adversely to Khrushchev's anti-Stalin campaign in 1956. It was a bad miscalculation. The Red Specialists had come to assume, as a result of Stalin's complete control of the party machine, that the real power lay in the formal organs of the state. The party existed, they thought, to impose only procedural and rhetorical requirements; it did not have power. They saw it as the bureaucratic equivalent of a constitutional monarchy.

At the end of 1956, they moved against Khrushchev by creating a new "Economic Cabinet" within the Politburo; consisting of all the key technocrats, it was to supervise economic decisions. Then, in June 1957, a majority of the Politburo, which included the members of the "Economic Cabinet," tried to dismiss Khrushchev. In an unprecedented move, Khrushchev appealed to the full Central Committee of the Party and, with the logistical support of the military, flew a quorum of the Committee to Moscow. The technocrats and their political supporters were denounced and dismissed as the "Anti-Party Group."

Khrushchev's victory was the equivalent, in terms of Communist ideology, of the reassertion of legislative control over the executive in a Western parliamentary system. Sarcastically denouncing the cadre of heavy-industry specialists as "steel eaters," Khrushchev proceeded to destroy their institutional power base. Twenty-five of the all-powerful central economic ministries in Moscow were abolished and their functions decentralized to regional economic councils. The key members of the managerial elite began a series of steps down the hierarchy of power, and thousands of economic administrators were dispatched to new jobs far from Moscow. Although the atomic umbrella organization, the Ministry of Medium Machine Building, remained in existence, the political power of its administrators was smashed. The Five-Year Plan, with its target of 2,500 Mw by 1960, was abandoned. It was replaced in 1959 with a Seven-Year Plan, which extended goals for most fields but ignored the targets for nuclear capacity. When Khrushchev outlined a new long-term plan for electricity up to 1980, nuclear generation was not mentioned at all. For as long as Khrushchev remained in command, no significant expansion of nuclear power was even discussed. The experimental units under construction were completed, but the Soviet rhetoric of an atomic age was restricted to presentations at international conferences.

The Russian scientists and nuclear engineers joined their colleagues throughout the world—back at the drawing boards. The vision of a nuclear future lingered on, however, and, if anything, adversity made it more intense, more personal.

In both East and West, there is no doubt about the sophistication of the technical achievements that flowed from their work during the next few years. The signs of public anxiety that had already emerged after Windscale and Kyshtym would grow, but the engineers were confident that they could perfect their product and design safety systems even more effective than those required in other industries. Their main objective was to reduce costs. They were convinced that economically competitive nuclear power was achievable; it was only a matter of hard, determined engineering work. The scope and intensity of the technical problems they had to overcome did not daunt them at all. And to the public, the vision of the peaceful atom still had enough appeal to make it possible for the self-consciously progressive image of the scientists and engineers to survive the slump.

The second Geneva Atoms for Peace Conference, held in 1958, reflected the mood of the First Ice Age. The euphoria of the 1955 conference was gone. Few papers dared to give cost estimates; instead, they listed the many technical problems that were now evident. It was a bigger conference—more delegates and more papers—but apart from a passing belief that the problem of controlling fusion could be solved, there was little optimism. Veterans of the frolic of 1955 drew invidious comparisons with the new sense of gloom.

The Americans, however, never short of an atomic gimmick and seeking to regain the public-relations triumph of the first Geneva Conference, had assembled another research reactor. This time, visitors were allowed to "operate" it and, for the pleasure, were awarded with an "Honorary Reactor Driver's License." Even more fun was in store for delegates when, at the third conference in 1964, the Americans brought over their completed nuclear ship, *Savannah*, and flew delegates up for joyrides off the coast of Norway. Throughout the bad years, the enthusiasm of nuclear advocates remained unbowed.

15

The Outcasts

Working out the trick of the hydrogen bomb was not like solving the problem of the uranium and plutonium bombs; it was not a matter of materials, it was a matter of design. The basic ingredients were relatively easy to produce: one atom bomb to trigger a massive explosion in a mixture of hydrogen isotopes. The problem was how to arrange the isotopes in relation to the bomb so that when it went off it produced, for two or three millionths of a second, millions of degrees of heat and thousands of atmospheres of extreme pressure. The shock wave from the fission trigger drives the hydrogen isotopes into a tiny lump of massive density—perhaps 1/10,000 of a millimeter. This is when the heart of the mystery—"the thermonuclear burn wave"—gets going. If the shape of the components and the timing are not absolutely perfect, the fusion will burn but not explode.

Two physicists, one in America and one in Russia, are credited with the single insight that was required to solve these problems. Each made the discovery in the early 1950s. The American was Edward Teller. A Hungarian by birth, Teller had been driven from Nazi Germany. During the war, he worked on the American bomb project, but he had always been much more interested in the more powerful hydrogen device. Much like the public's stereotypical image of a physicist, Teller looks a little mad, his intense blue eyes staring out from under bushy black eyebrows. He was forty-three when he designed the U.S. H-bomb. The Russian was Andrei Sakharov, a tall, thin, blond Muscovite who, at twenty-nine, became the youngest full member of the Soviet Academy of Sciences. It was an honor awarded for his contributions to Russian thermonuclear research and, it seems from the literature, for his contributions to the Russian H-bomb.

Edward Teller came from the brilliant cohort of Hungarian intellectuals, which included Leo Szilard and Eugene Wigner, that made such critical contributions to the U.S. bomb program. John von Neumann, the mathematician whose giant computers at Princeton did the high-speed calculations that helped to reduce the H-bomb to a size small enough to fit on the nose of a missile, was also a member.

Teller was born in Budapest in 1908 to a comfortable Jewish family whose supportive environment was countered by the social turmoil that followed the collapse of the Austro-Hungarian empire. A short-lived and bloody Communist revolution was replaced by an equally bloody and oppressive fascist regime that was also anti-Semitic. Teller left Hungary in the 1920s to join the intellectual whirl of the physics community in Germany. In 1933, he was finally driven out of Europe by the Nazis. He went first to Denmark, then England, and then to the United States, where he arrived in 1935 to take up a teaching position at George Washington University before joining the Manhattan project at Los Alamos.

Teller's self-righteous confidence in his own ideas often seemed like pomposity to others. In his high school days, the intensity of his individualism had marked him as a suitable target for collective scorn and teasing. It would never change: Teller always acted as if he expected to be rejected by others. Even Enrico Fermi, who was able to tolerate Teller's excesses because of intellectual respect, could not resist taunting him: "How come the Hungarians have not invented anything?" he would ask.

During work on the Manhattan Project, the German physicist Hans Bethe, who was head of the Los Alamos Theoretical Division, was often infuriated with Teller's obstreperous behavior. "He did not want to cooperate. He did not want to work on the line of research that everybody else in the laboratory had agreed to as the fruitful line.

He always suggested new things, new deviations . . . so that in the end there was no choice but to relieve him of any work in the general line of the development at Los Alamos and to permit him to pursue his own ideas entirely unrelated to the World War II work." Rejected, Teller sulked and spent his time at Los Alamos working on the H-bomb.

His career was studded with outbursts of petulance; whenever he could not get his way, Edward Teller took his toys and went home. After the war, he refused to stay at Los Alamos because his pet idea, the H-bomb, was regarded as impractical. Even when it was given the go-ahead he refused to work under anyone else and, although he returned to Los Alamos, he left before the project was completed. He hated Los Alamos so much that, together with Ernest Lawrence at Berkeley, he persuaded the U.S. AEC to expand its weapons laboratories and set up the Livermore Radiation Laboratory at the University of California. Although this was originally promoted as "Teller's laboratory," for several years the churlish scientist refused to go there.

Other scientists would chide him. "In my acquaintance," Fermi once told Teller, "you are the only monomaniac with several manias." Two were overriding. Teller had a European's fear of Russia and a deep phobia about Communism. He compared the Soviet Union not just to Nazi Germany, as many people did, but to the Mongol hordes of Genghis Khan. The conflict between the postwar superpowers involved the future of everything he called civilization. It was natural that this became intertwined with his other grand passion: the science of the hydrogen bomb. This was the ultimate weapon, the final defense of the West. Teller's total commitment to the weapon gave him a public image of malevolence: the film character Dr. Strangelove was a conscious mixture of the German Nazi rocket engineer Wernher von Braun and Edward Teller.

This image was not helped by a murky difference of opinion over his actual invention. Stanislaw Ulam, the Polish mathematician, formulated the original design idea that Teller adapted and made into a workable bomb. Ulam described the basic intellectual tool kit that was required for the insight that Teller eventually shared. To unravel the mysteries of the H-bomb, said Ulam, it was necessary to have "a visual and also an almost tactile way to imagine physical situations, rather than a merely logical picture of the problems." He resented Teller's attempt to take all the credit. "He [Teller] either believed he was, or wanted to be known as, not only the main but the sole promoter, defender, and organizer of the work."

Everyone except Teller himself and his closest disciples would accept the fact that Ulam triggered the design breakthrough with an idea Teller would quickly adapt into a more effective alternative. "Ulam triggered nothing," Teller continued to assert many years later. The U.S. government was willing to grant a joint patent on the H-bomb to the two men, but Teller objected. "I found," Teller said, "that under the patent laws I had to make a statement under oath that Ulam and I invented this thing together. I knew that taking this oath would be perjury. I therefore refused to take out the patent and no patent was taken out." This passionate, moralistic sense of injustice stayed with him forever. He thrived on adversity, expected it as his lot in life, and seemed driven to create it whenever it did not exist; a lonely, friendless role, where only disciples could be tolerated. He seemed to relish the path of the prophet in the wilderness. In this regard he is like Andrei Sakharov, though with less justification.

Sakharov, thirteen years Teller's junior, was born in 1921 in Moscow, a city then still recovering from revolution and civil war. He was brought up in an environment permeated by an earnest and dedicated sense of social responsibility. His father, a professor of physics and author of a best-selling textbook, passed on the traditions of the Russian scientific intelligentsia: liberal, selfless, honest, principled. It was "a cultured and close family," Sakharov recalled. "From childhood I lived in an atmosphere of decency, mutual help, and tact, a liking for work, and respect for the mastery of one's chosen profession." It was a comfortable, supportive world. His life was never uprooted like Teller's. He had no experience of physical injury, like the Munich streetcar accident that crippled Teller's foot. The sources of his discontent were neither personal nor social.

Graduating from Moscow State University in 1942, he spent the next three years working in a war production plant. From 1945 to 1947 he did postgraduate work under Igor Tamm, head of the Moscow Lebeder Physics Institute's technical division and the leading Soviet specialist in quantum mechanics. After producing a dissertation on cosmic rays, Sakharov joined Tamm in the new assignment to make the H-bomb.

In 1948, as far as the public was concerned, Sakharov virtually disappeared. "For the next eighteen years I found myself caught up in the routine of a special world of military designers and inventors, special institutes, committees, and learned councils, pilot plants and proving grounds." It was a world of privilege with "special passes" to a secret network of stores with a unique range of luxury goods, chauffeur-driven, curtained limousines, and its own high-quality hospitals and clinics. His salary was frequently supplemented by bonuses: cash from the Stalin Prize and three Hero of Socialist Labor awards. Eventually, he was also given a country house or dacha, the ultimate privilege of the Soviet elite. For most of the time, however, he had to enjoy his wealth in an isolated secret laboratory, which he himself would only describe as "far from Moscow." It is thought to be in Turkmeniya, a southern province of the Soviet Union on the Iranian border. He had no questions about his work at the time. "I had no doubts," he said, "as to the vital importance of creating a super weapon—for our country and for the balance of power throughout the world."

Working separately with research teams in America's atomic laboratory at Los Alamos, and in a secret laboratory in southern Russia near the Iranian border, Teller and Sakharov would each eventually become known in their respective countries as the father of the H-bomb. Both would deny sole rights to this dubious accolade. Sakharov more than Teller would emphasize the supporting work of his colleagues. But neither, at any stage in the development of their discovery, hesitated about going on: both believed deeply in the ultimate value of scientific progress, whatever its unsavory application. They were intense, passionate, brooding, hypersensitive men with a strong sense of their own idealism. They saw themselves as straightforward, open, even innocent people committed to the well-being of mankind and the importance of open inquiry and intellectual freedom—whatever its results. Both were convinced that the future of their countries, as nations, depended on the development of what seemed to all scientists to be the ultimate weapon of mass destruction, for, in theory, there was no limit to the force of the hydrogen explosion.

Each man reacted to his masterful invention in totally opposite directions, however. Teller would stop at nothing to ensure that the discovery was improved and its uses expanded. Sakharov would seek, with equal determination, the abolition of the hideous weapon he had created. For their behavior and their views, both would, in the end, become outcasts, ostracized by the societies that had, at first, hailed their genius. In the eyes of many Americans, Teller became the personification of the evil of the H-bomb. On the campuses, he was branded as a "war criminal." Sakharov, on the other hand, became a powerful critic of the Russian society he had so eagerly sought to protect, and in the West he became a symbol of dissent.

Before this social ostracism set in, however, the praise they earned permitted both men to play significant political roles in the arguments surrounding the first international treaty to limit the harmful radioactive side effects of testing bigger and better atomic bombs. Signed in Moscow in 1963, the Partial Test Ban Treaty banned only atmospheric tests; nations could continue to test their weapons underground, the presumption being that the radioactive products would be kept beneath the earth's surface.

Teller, refusing to accept even such modest limits on scientific progress, made a strident case against any test ban. Playing on the nation's continuing fear of Russia, he stated firmly that any test ban would endanger U.S. security, and a full ban especially because existing means of detection were not good enough to find out if the Russians had cheated. It was this latter argument that made it politically impossible in the end for the United States to accept a total ban. Indeed, it was Sakharov who saved the day. Although he sought precisely the kind of limitations Teller had rejected, Sakharov directed the Soviet Union out of the impasse by arguing the case for an agreement on atmospheric tests only; these could be easily detected.

The movement to ban testing that was to culminate in the 1963 treaty had its origins in America's hydrogen bomb tests. In November 1953, one test, code-named "Mike," had proved the basic principles of the thermonuclear reaction: hydrogen isotopes fused together to give off tremendous energy. The resulting explosion literally blew up the mile-wide Pacific island of Elugelab. The blast was estimated at ten million tons, or ten megatons, of TNT. It was almost a thousand times more powerful than the Hiroshima bomb. But it was not a usable weapon. Mike had required a massive refrigeration plant larger than a two-story house and weighing sixty-five tons to keep the hydrogen fuel in a liquid state prior to detonation. A better—and bigger—bomb was on the way.

On March 1, 1954, using the dry hydrogen isotope Lithium-6 deuteride—which meant the device could function without a refrigeration plant—the test bomb exploded with a force of fifteen megatons, half again as big as Mike. The significance of this bomb, code-named "Bravo," was more than the size of the blast, however. Mike had been a cumbersome device, difficult to adapt to military use. Bravo was a practical weapon in a form that could be dropped by an aircraft or delivered by a missile.

Most important, however, Bravo was the test that made the world conscious of fallout. It did so not because of its awesome size, but because of a slight shift of the wind. Detonated on the surface of Bikini Atoll in the Marshall Islands, it pulverized millions of tons of coral, which were sucked into the rapidly expanding fireball. Each particle of coral became highly radioactive and was carried aloft on a billowing white cloud. As the cloud grew, the high wind changed direction a few degrees to the east. Blown by the wind, the fallout cloud moved quickly over the Pacific Ocean and the radioactive coral particles began to fall by force of gravity over a cigar-shaped area of 7,000 square miles. In its new path, the cloud passed directly over a number of the inhabited Marshall Islands and exposed the islanders to radiation doses of up to 175r.

A panic evacuation was carried out. It was a potential public relations disaster for the test organizers, the U.S. Atomic Energy Commission. Here was the first officially acknowledged fallout catastrophe, but the AEC was confident that the accident was politically manageable: the victims, the Marshall Islanders, were probably far enough away from the United States not to cause an uproar. On March 12, the AEC produced a press release about the Bravo explosion. Several sentences are memorable for their contrived obfuscation and lies. "During the course of a routine atomic test," it began, as if the first deliverable H-bomb could be covered by an assertion of normalcy, "two hundred and thirty-six residents were transported from neighboring atolls . . . according to plans as a precautionary measure," as if the "plans" had existed before the wind shift. "The individuals were unexpectedly exposed to some radiation," it went on, as if the diminutive "some" could cover a very serious dose of 175r. "There were no burns," it stated; but there were, many burns, plus cases of radiation sickness. "All are reported well," it concluded, as if anyone could know at that stage, after the fits of vomiting and diarrhea had stopped, what might follow. In fact, years later, a careful study was conducted of the children exposed to Bravo; it revealed retarded growth, an epidemic of thyroid irregularities, and one leukemia death.

The AEC undoubtedly felt their misleading statements were immune from public scrutiny; if such a disaster had happened in Nevada, it would have been a different story. But, unknown to the AEC, a heavy shower of fallout from Bravo had landed on a Japanese tuna fishing boat, the *Lucky Dragon*. The boat had been to the east of Bikini Island, just outside the proclaimed danger zone, when the bomb was exploded. The shift in the wind had caught the boat in the fallout path. When the *Lucky Dragon* returned to her home port of Yaizu, 120 miles southwest of Tokyo, nearly all the twenty-three-man crew had some form of radiation sickness.

The plight of the fishermen sparked a Japanese national protest against the tests. Then Japanese authorities destroyed huge quantities of contaminated fish, and, six months later, a thirty-nine-year-old crewman of the *Lucky Dragon* died. The Tokyo hospital put the cause of death as radiation sickness. One of the Japanese doctors later said there were three possible causes: (1) serum hepatitis caused by blood transfusion; (2) degeneration of the liver, caused by debris of other radio-sensitive cells destroyed by radiation injury; (3) primary radiation injury. The protests escalated. Accepting the blame for the fisherman's death, the United States, through its ambassador in Japan gave his widow a check for one million yen: about $3,800.

Almost overnight, the Japanese revived a buried interest in their own nuclear victims. For the first time in nearly a decade, the condition of the survivors of Hiroshima became a national preoccupation.

The protest quickly became international, and there followed a burst of worldwide publicity about the awesome potential of hydrogen bombs. The press editorialized at length. Even *Scientific American* concluded: "Just as war is too important to be left to the generals, so the Thermonuclear Bomb has become too big to be entrusted any longer to the executive session of rulers in Washington." Called in by Eisenhower to fend off the protests, AEC chairman Lewis Strauss accused the Japanese fishermen of "inadvertent trespass" into the Bravo test danger area and declared the fallout to be so minor that it would have no harmful effects on any living thing. So confident was Strauss of his ability to ride the storm that he readily consented to answer reporters' questions. He was asked how big the H-bomb could get.

A. Well, the nature of an H-bomb is that, in effect it can be made as large as you wish, as large as the military requirement demands, that is to say, an H-bomb can be made as—large enough to take out a city.

Q. How big a city?

A. Any city.

Q. Any city? New York?

A. The metropolitan area, yes.

Strauss later regretted his remarks, but the damage was done. The headlines the following morning screamed "H-Bomb Can Wipe Out Any City." It was March 31, 1954. The word was out. *Newsweek* warned its readers bluntly, "If you live in a strategically important city, the odds against your survival in an H-bomb war would be a million to one. If you live in the country, your chances would obviously be better. But wherever you live, much of what you live for would be destroyed."

In London, the day before Strauss's remarks, 104 Labour party members of Parliament had signed a motion calling for an end to the tests. In response to the continued testing, a handful of north London housewives started a committee to ban testing; their movement later became the national Campaign for Nuclear Disarmament (CND), and one of the century's great philosophers and mathematicians, Bertrand Russell, became its moving spirit.

At the end of 1954, Russell drafted a manifesto; citing the evidence of the Bikini test, it posed a question he described as "stark and dreadful and inescapable: Shall we put an end to the human race; or shall mankind renounce war?" Russell's manifesto was signed by seven Nobel Prize–winners, among them Frédéric Joliot-Curie, Hermann Muller, and Linus Pauling. Albert Einstein himself, long concerned with the obscene product of his scientific ingenuity and deeply regretting his letter to Roosevelt that sparked the Manhattan Project, signed the manifesto two days before his death in April 1955. It was his last public act. Attracting immediate international attention, the manifesto also led, in 1957, to the creation of the scientists' Pugwash Movement. Twenty-four distinguished scientists from ten countries on both sides of the Iron Curtain met at Pugwash, Nova Scotia, to discuss the risks of bomb tactics and all-out nuclear war. "The principal objective of all nations," they concluded, "must be the abolition of war and the threat of war hanging over mankind." For many years, the movement was the only forum for dialogue between Western and Soviet scientists on problems of the arms race.

Unlike the atomic bomb, this new weapon seemed to take the destiny of the world beyond the scope of human control. The prospect of total annihilation, the destruction of all human life, was no longer science fiction. The scope, range, complexity, and intensity of the Soviet-American rivalry offered no hope that nuclear weapons could somehow be disentangled from the new device. The vision of nuclear holocaust became the stuff of pop culture; in 1957, Nevil Shute's novel *On the Beach*, the story of how fallout from a nuclear war destroyed life on earth, replaced *Peyton Place* on the top of bestseller lists.

Public protests intensified when two of the earliest nuclear critics, Joseph Rotblat, a physics professor at St. Bartholomew's Hospital in London, and Ralph Lapp, an American physicist and author, discovered separately, from examination of the debris from Bravo, that the bomb was not a simple fission-fusion device. The Americans had ingeniously wrapped an additional layer of natural uranium around the fusion fuel, making the total effect a fission-fusion-fission explosion, which meant a much dirtier bomb with more radioactive products. Rotblat was at first advised by a British colleague not to publish his discovery. "There'll be a frightful row with the Americans if it's published by an Englishman," he was warned.

Rotblat held back until the official AEC report on the fallout from Bravo was released in February 1955. It contained still more reassurances: "the radiation exposure received by residents of the United States from all nuclear detonations to date has been about the same as the exposure received from one chest X-ray." Rotblat was appalled at this complacency. If indeed the amount was only equal to one chest X ray, this in itself was dangerous. An X ray is concentrated in one part of the body and does not usually touch the reproductive organs. But radiation from fallout affects the whole body. Serious genetic damage could already have been inflicted on the fallout victims, Rotblat estimated. The time had come, he decided, to make public his discovery of how dirty the Bravo bomb had been. His findings were published in England in the *Journal of Atomic Scientists* and in America in the *Bulletin of the Atomic Scientists*, the organ of the Pugwash movement.

Left largely in the dark by the U.S. AEC and totally in the dark by the British AEA, citizens on both sides of the Atlantic had to make up their own minds about the dangers of fallout. The so-called experts could not agree on the size of the danger. Some said fallout was harmless; others calculated large numbers of people would die and alarming genetic mutations would occur.

The fact that so little was known about chronic low-level radiation permitted wide scope for expert disagreement. The post-Bravo uproar had come at a time when the fears of the geneticists were at their height, and the most recent genetic research had suggested that the damage to human genes from radiation had been underestimated by as much as ten times. The American founder of modern radiation genetics, Hermann Muller, spoke of the "contamination . . . of the pure stream of human heredity." Led by the alarmed Muller, geneticists were pressuring the national and international bodies concerned with radiation protection to accept a basic thesis that "the concept of a safe dose of radiation simply does not make sense." There was no threshold dose, their research suggested, and effects could be found all the way down to the smallest dose. Since the Rotblat-Lapp revelations had also shown that fallout from bomb tests was not confined to areas close to the test site, but could come down all over the world, concern grew. And although Bravo had triggered this concern, the Americans were not the only perpetrators.

On August 8, 1953, seven months before Bravo, the Russians had exploded a thermonuclear device estimated by the United States to be in the order of several hundred kilotons—nothing approaching the fifteen megatons of Bravo, but dry Lithium-6 deuteride had been used as the fuel and this meant that the Russians now had the capability of turning the device into a bomb. And indeed, in November 1955, nine months after Rotblat released his findings on the fallout from Bravo, the Russians dropped an H-bomb from an aircraft. The official U.S. statement acknowledged it was in the "megaton range." Soviet Communist Party Secretary Khrushchev, then on a state visit to India, was jubilant: "I shall not say there has not been such an explosion. It was a terrific explosion. . . . These tests have demonstrated important new achievements by our scientists and engineers." U.S. experts have since interpreted the words "important new achievements" as meaning that this was the first application by the Russians of Sakharov's special configuration for the H-bomb—the same one Teller had worked out in the United States.

In the West, the man who led the public protest against the H-bomb tests was the American chemist Linus Pauling. He is perhaps better known to Americans for his spirited contention that large doses of vitamin C are an effective prevention of the common cold. He is also the only person, other than Marie Curie, to be awarded two Nobel Prizes—the second, in 1963, for peace. The first prize had come in 1954 for his work on molecular structure. The same year saw the beginning of the test-ban movement, and the prize provided the flamboyant, dogmatic, maverick Pauling with a fig leaf of political respectability.

In 1957, Pauling began to organize an international petition of scientists for a test ban. With a core of prominent signatories, including the geneticist Hermann Muller, the petition was eventually signed by more than eleven thousand scientists from forty-eight countries. Notably, there was a comparative absence of physicists: a third of the signatories were from the biological sciences. Three years later, in 1960, Pauling would courageously defend his actions and the integrity of the petitioners before cross examination in a congressional inquiry.

Although there was little hard scientific evidence on the effects of low-level radiation, Pauling produced precise figures for birth defects, leukemias, and reduced life expectancy that would be caused by continuing nuclear testing. He was helped by a report from the British Atomic Scientists Association. Issued in April 1957, on the eve of the first British H-bomb test on Christmas Island in the Indian Ocean, the report estimated that one thousand people would die for every megaton of explosive power released in the test. Pauling expanded this: fallout from the previous tests had already ensured that ten thousand people would die from leukemia, he claimed.

His overweening self-confidence in his own judgment, later reflected in the controversy over the claims he made for the powers of vitamin C, was frequently criticized by other scientists. Those associated with the weapons program were particularly indignant about the precision of his denunciation, especially Pauling's assertion, based on the genetic research, that there was no threshold or "safe dose" and that effects were proportional to dose. Although the basis for his precise calculations of how many would die was shaky, Pauling understood that such precision was politically important: people understood numbers of this kind.

The numbers were challenged by the chief defenders of testing, Edward Teller and U.S. AEC commissioner Willard Libby. Teller was then at the height of his public influence as the father of the U.S. H-bomb. Libby, like Teller, was ideologically committed to the

Cold War. A graduate of the Manhattan Project, he was a chemist by trade and was best known for developing radiocarbon dating techniques for archaeological research. Libby, like Teller, sought to minimize the effects of global fallout. "Generally speaking," he told a conference of the nation's mayors in December 1954, "there is no immediate hazard to the civilian population in this type of fallout."

Libby and Teller firmly believed that Pauling's precise calculations were not based on any real scientific facts. They questioned the absolute numbers—the hundreds of thousands of leukemias and the stated genetic defects. Libby and Teller preferred to talk in terms of percentages and probabilities rather than totals. Their figures looked smaller. They talked in terms of "risks," as if the effects were not definite, and they compared them with other, more familiar risks that the public willingly accepted. "World-wide fallout," Teller asserted, "is as dangerous as being an ounce overweight or smoking one cigarette every two months." Libby said that people were more likely to be hit by lightning than to be harmed by fallout.

Teller and Libby understood that their enemy was what they perceived as the "irrational" fear of the unknown that radiation provoked. The public had to be made to understand that it was a "natural" phenomenon: there had always been background radiation. "The world is radioactive," Libby declared in December 1954, "it always has been and always will be." Ignoring the geneticists' conviction that even natural background radiation was harmful, Libby started to compare fallout doses with "natural" radioactivity. The numbers seemed small and reassuring. Teller pointed out that in Denver, Colorado, the "mile-high city," people received significant doses from cosmic rays, concluding that scientists concerned with minimizing radiation effects would do better to concentrate on moving people out of Denver than arguing against nuclear tests.

The Teller-Libby axis lost some of its momentum, however, when it emerged that the two men, like all the other weapons scientists, had underestimated the danger of one of the man-made radioactive substances: Strontium-90. Like plutonium twenty years later, Strontium-90 was to become the symbol in the 1950s of the nuclear threat, the favorite evil substance of the day.

Produced in large quantities in a fission explosion, the bulk of Strontium-90 shoots up into the stratosphere. Because of its long life, it is still dangerously radioactive as it gradually falls to earth months and even years later. Coming down anywhere in the world, Strontium-90 became the first truly global pollutant. It was an international public health problem for which the traditional techniques of quarantine were irrelevant. Like radium, Strontium-90 is a bone-seeker, and its concentration is particularly dangerous for the smaller growing bones of children.

In order to calculate how much Strontium-90 individuals would absorb from atmospheric testing, a large number of assumptions had to be made. There was great scope for technical disagreement on such questions as how long it would stay in the stratosphere, how evenly it would fall out over the earth's surface, how quickly and completely it would be absorbed in the food chain. Despite the extent of its ignorance on all these matters, the AEC once again asserted its expertise and sought to dismiss the danger. In 1953, its semiannual report stated that the human danger came from animals that fed on grass contaminated by Strontium-90: people might eat "bone splinters which might be intermingled with muscle tissue during butchering," said the report. This was an extremely remote possibility, to say the least, and by no means the obvious hazard; a much more dangerous vehicle to the human body, particularly to children, was cows' milk, but the AEC did not acknowledge this particular hazard until 1956.

Seeking to minimize the scare over Strontium-90, Libby made estimates for how long the substance would stay aloft and how evenly it would fall out. First, he said, it would rise into the stratosphere and take up to ten years to get back to earth. On its way into the human food chain, it would be weathered—by the rain, the soil, and, finally, the cow. Only a minute amount would reach human bones, where it would settle, like radium or calcium. Its beta radiation would be so weak, Libby contended, that it would be harmless.

Ralph Lapp disagreed with Libby. Because of Strontium-90's long life, he said, it could cause bone cancer up to fifty years after the last bomb test. The atmosphere already contained 15 percent of what the world could safely absorb, he warned. Lapp would soon be proved right, Libby wrong.

Teller, meanwhile, was working on his grand obsession, which he saw as the answer to the test-ban lobby: the so-called clean bomb, or what we know today as the "neutron bomb." In the summer of 1957, Eisenhower, still smarting from Adlai Stevenson's success in raising the test-ban issue during the 1956 elections, indicated that he would be "delighted to make some satisfactory arrangement for a temporary suspension of tests." On June 24, Teller met Eisenhower and stressed the prospects of a clean bomb. Not only would this ease the fallout problem but it would enable bombs to be used for "peaceful purposes," like major excavations. The next day, Eisenhower called a special press conference and said that "ending the tests might impede progress on the production of a fallout-free nuclear bomb and on the development of nuclear energy for peaceful purposes."

Pauling countered Teller's clean-bomb idea by introducing the forgotten problem—forgotten, at least, in the West—of the bomb's production of radioactive Carbon-14. When Teller's clean bomb was exploded, it would release millions of neutrons (hence its later name), and these neutrons would combine with nitrogen in the air to produce radioactive Carbon-14. This isotope of carbon has a very long life; in contrast to Strontium-90, it would produce fewer harmful effects over centuries but greater effects over thousands of years. Testing a few neutron bombs probably would not cause harmful effects, but Pauling realized that simply by raising the issue he had tarnished the credibility of the protesting experts by highlighting a danger they had apparently ignored. Teller's clean bomb was no longer clean.

Pauling's protests had a profound effect on Andrei Sakharov. He had already been alarmed by the potential hazards of Carbon-14. He had also published a detailed technical refutation of the "cleanliness" of Teller's bomb. Sakharov's paper calculated the number of leukemias and genetic defects that could result. "The fact that the so-called clean bomb is a radioactive hazard," Sakharov concluded, "takes the wind out of the propaganda sails of those who maintain that this is a qualitatively different weapon of mass extermination." With corroboration from the West in the form of Pauling's work, Sakharov started down the road of dissent. "Beginning in 1957," he wrote later, "I felt myself responsible for the problem of radioactive contamination from nuclear explosions." It was not just Carbon-14 that concerned him: Sakharov had also accepted all of the portentous views presented by geneticists. In 1958, he published a paper on Carbon-14 in which he repeated that there was no radiation "threshold"; effects were proportional to dose for both genetic and immediate injury—and they were cumulative. "The total number of victims," Sakharov declared, "is already approaching one million and each year of continued testing increases this number by 200,000 to 300,000 persons."

Sakharov was straying into troubled waters. Genetics was still a highly politicized subject in the Soviet Union. With the political support of Stalin, and later Khrushchev, the infamous Trofim Lysenko had brutally imposed the reign of his pseudo-science, which proclaimed the possibility of the inheritance of acquired characteristics. Totally rejecting the basic principles of Mendelian genetics—that inherited characteristics were inborn—Lysenko and his supporters purged any independent thinkers; some died in the Gulag.

Orthodox biology and genetics survived only in the atomic energy institutes. With the success of the bomb project, the top nuclear physicists were well connected politically, and their leaders, like Kurchatov and Tamm, created havens for these disciplines in their institutes. They established departments of radiobiology and published basic research through *Atomiztad*, the organ of a specialist publishing house for nuclear energy, rather than in the official organs of established biology. As in the West, it was the geneticists who became most alarmed about the threat of radioactivity; beginning in 1955, they launched a campaign of what they called "underground propaganda" among their politically influential colleagues in the physics establishment. In 1956, Kurchatov himself, the father of the Russian atom bomb, presented Khrushchev with a petition of several hundred signatures calling for the restoration of orthodox genetics, stressing the effects of radiation.

Although scientists and politicians in both East and West were gaining ground in their efforts to bring a halt to the bomb testing, the political process was painfully slow. A few weeks after the Bravo test, Indian Prime Minister Nehru had called for "some sort of . . . standstill agreement in respect to these actual explosions, even if the arrangements about the discontinuance of production and stockpiling must await more substantial agreements among those principally concerned." At the 1955 U.N. General Assembly, the Soviet Union and India proposed an early and separate agreement on the banning of all nuclear tests without supervision, arguing that no significant testing could go undetected. But the U.S., Britain, and France would not consider a test ban unless it was part of an overall disarmament plan.

Two years later, Russia, conscious of the propaganda points to be made, tried again. At the five-power Subcommittee of the Disarmament Commission meeting in London, the Kremlin delegation proposed an agreement on the "immediate cessation of all atomic and hydrogen tests if only for a period of two or three years." They further suggested the establishment of an international commission to supervise the agreement. Still, the Western powers insisted the ban be linked to other disarmament measures, and the Soviet initiative failed a second time. In April 1958, after completing their current test programs, the United States, the United Kingdom, and Russia conceded that it was now possible to detect one another's explosions down to one-kiloton bombs and agreed to a voluntary suspension of tests. It lasted only three years. Meanwhile France, about to become the world's fourth nuclear weapons state, had refused to accept the ban; its first bomb exploded in the Sahara in 1960.

In the summer of 1961, after a failure to formalize the moratorium, Khrushchev called the Russian atomic scientists together to announce his intention to break it. According to Sakharov, breaking the moratorium was part of a desire by the Russians to reinforce their new pugnacious policy in Berlin. The announcement marked a particularly sad moment for Sakharov; he wrote a brief note of protest to Khrushchev, who, in an off-the-cuff dinner speech, publicly rebuked Sakharov for being a meddlesome scientist. Foreign policy and national security were the responsibility of the Soviet premier, not of scientists, said Khrushchev. "I would be a slob and not Chairman of the Council of Ministers if I listened to the likes of Sakharov."

As the new Soviet test series continued, Sakharov's alarm grew. "The Ministry," Sakharov recalled, "acting basically from bureaucratic interests, issued instructions to proceed with a routine test explosion that was actually useless from the technical point of view. The explosion was to be so powerful that the number of anticipated victims was colossal. Realizing the unjustifiable, criminal nature of this plan, I made desperate efforts to stop it. This went on for several weeks—weeks that for me were full of tension."

Sakharov contacted the Minister for Medium Machine Building, Yefim Slavsky, and threatened to resign. "We're not holding you by the throat," the minister replied. In a last desperate attempt to stop the test, Sakharov rang Khrushchev himself, but to no avail. "The feeling of impotence and fright," Sakharov recalled, "that seized me on that day has remained in my memory ever since, and it has worked much change in me."

Sakharov's first open breach with the Kremlin came the following year, 1963. It was not on weapons testing; Sakharov was still part of that secret world. The issue was the continued suppression of genetics, which as far as Sakharov was concerned was directly related to his own work. At an extraordinary meeting of the Academy of Science, Sakharov spoke vociferously against the membership application of one of Lysenko's closest henchmen. In an unprecedented move, the full academy in plenary session overwhelmingly voted down the committee recommendation to admit the candidate. Khrushchev, who actively promoted the Lysenko position, was irked: he threatened retaliation against the academy and was not appeased by Sakharov's letter indicating his personal involvement. For the first time, Sakharov was publicly attacked in the Soviet press.

While he continued his weapons work over the next few years, he became more and more involved with questions of human rights. His course of defiance eventually led to the Nobel Peace Prize in 1973. Propelled by his fear of the effects of weapons testing, Sakharov had entered a new world, spurning the privilege and influence he had known and discovering the oppression and covert activity of a persecuted scientific discipline. The men he met, the things he learned in the next few years, drove Sakharov from a concern with thermonuclear war to a firm conviction that what he called "universal suicide" could be avoided only through the open exercise of intellectual freedom. In mid-1968, his first philosophical tract, "Progress, Coexistence and Intellectual Freedom," was published. Within a month, he was taken off weapons work.

A decade had elapsed since he had published his attack on weapons testing in the Carbon-14 paper of 1958. His concluding words from that paper indicate the origins of his long road to political ostracism.

> Ethically speaking the only peculiarity of this problem is the total impunity of the crime (for in no concrete case can it be proven that the death of a person is caused by radiation) and also the total defenselessness of future generations with respect to our actions. The cessation of test explosions will preserve the lives of hundreds of thousands of people and will have a still greater indirect effect by helping to lessen international tension and to reduce the possibility of nuclear war—the greatest danger of our age.

As Sakharov moved doggedly away from his original commitment to the superbomb, so Edward Teller reinforced his own commitment toward it. In 1961, President Kennedy's inaugural address reflected the new international alarm: "Man holds in his mortal hands the power to destroy all forms of human life." For Teller, this statement became the symbol of his political impotence in the new administration. He firmly believed it was wrong; the world, he thought, would survive a nuclear war.

In fighting what he saw as the unjustified public image of nuclear weapons, Teller helped launch two passing enthusiasms. The first was for nuclear fallout shelters. He argued that for a modest expense, some 90 percent of Americans could be saved even after an all-out attack by the Soviet Union. With great fanfare, Willard Libby, his close associate in the public debate, built a cheap fallout shelter at his home. A few weeks later, it burned down in a brush fire. "This proves," quipped Leo Szilard, "not only that there is a God but that he has a sense of humor." Nuclear critics like Szilard, now a leading force in the Pugwash movement, attacked all fallout shelter proposals as a misleading source of complacency in the face of the unimaginable horror of nuclear war.

Teller's second passing enthusiasm was a more positive public relations tactic: explosions of his "clean bomb" could be "peaceful," he claimed. They could do things that everyone thought desirable: move mountains, dig canals, redirect rivers, build harbors, construct reservoirs, separate oil and minerals from rock. These prospects took him to his most eloquent flights of visionary rhetoric, as he spoke of his dreams for future economic abundance. If the heat from an explosion could somehow be trapped and gradually brought to the surface, it could be used to produce electricity. A single blast could mass-produce diamonds in the right geological strata. A number of blasts could control the weather. "So," Isidor Rabi said wryly, "you want to beat your old atomic bombs into plowshares."

Like all good public relations ploys, the dream was given a catchy title: "Project Plowshare," based on the passage in Isaiah. Had the name-givers read further, they might have been given pause: later, in Joel, the plowshares are easily made back into swords. Yet this dream of Teller's persisted and would bedevil international nuclear discussions for twenty years. The prospect of positive uses of nuclear explosions, without serious consideration of economic costs or side effects, would be invoked against each proposal to conclude a comprehensive test ban and against each attempt to enforce technical barriers to inhibit nuclear weapons proliferation.

For Teller, the development of "clean bombs" for peaceful nuclear explosions fitted exactly his own military objectives. The radiation legacy of Hiroshima had prevented Truman from using small fission bombs in Korea and, as Teller saw it, had created throughout the American political elite a loss of will to use its full strength in any future confrontation with Communism. "Clean bombs" provided the answer.

Edward Teller had done a 180-degree turn. Having invented the most devastating weapon of all time, he was now arguing that "the idea of massive retaliation is impractical and immoral." It was impractical because the Russians had the same weapons and immoral because it slaughtered civilians. The thermonuclear reaction he proposed was best used in the small "clean bomb," the tactical battlefield weapon instead of the strategic blockbuster. The "clean bomb" was perfect for such a limited war strategy.

Teller never confronted the central problem of "limited wars": that no one had ever thought of a way to keep them limited. All the war games, all the analyses, suggested that once the first nuclear weapon was used, other, larger ones would follow automatically. Even a hard-nosed Harvard academic named Henry Kissinger, who first came to national attention with his 1957 book advocating tactical nuclear weapons, had to publish a retraction in 1961. The problem would remain—especially when Teller's clean weapon emerged again in the late seventies as the neutron bomb.

Teller chafed under the security restrictions that prevented him from describing the manifold advantages of the "clean bomb" in the full detail he believed was required. He resented the secrecy. Like most men propelled by a messianic zeal, he was convinced that if only everyone knew as much as he did, they would agree with him. As he had so disastrously done with Robert Oppenheimer years before, he could only understand informed disagreement in terms of malevolence.

His great fear remained the Soviet Union. Succumbing to the temptation of the prophet, he identified total evil with total cunning. For all the reasons he had used to advocate the development of a "clean bomb"—from Plowshare to military efficacy—he believed the Russians would pursue the same objective. America could "fall behind" in the sophistication of its weapons technology. His bitterness grew as the 1958 test-ban moratorium was extended year by year.

With talks continuing about a permanent test ban, Teller alone among the critics was convinced that the Russians were already cheating. On the basis of some valid technical objections, he suggested increasingly bizarre means of how the Russian tests could escape detection: they could conduct them in outer space, behind the sun, or they could dig huge holes underground—thousands of feet in diameter—and explode a bomb in the center that would be detected only as a small earthquake.

His colleagues disagreed. "We are all behaving like a bunch of lunatics to take any such thing as the big hole seriously," said Hans Bethe. He argued that despite the finite but small risk of a failure in the detection system of seismic stations and the limited number of inspections, the Soviet Union could not afford to cheat: their international credibility would be at the mercy of a single defector.

Bethe led that section of the scientific community which saw a test ban as a first step toward reducing the tensions of war. An American representative at the Geneva test-ban talks, Bethe was profoundly disturbed by the irrationality of the position that Teller's vigorous advocacy forced the American negotiators to take. "I had the doubtful honor of presenting the theory of the big hole to the Russians in Geneva in November, 1959. I felt deeply embarrassed in so doing, because it implied that we considered the Russians capable of cheating on a massive scale. I think that they would have been quite justified if they had considered this an insult and had walked out of the negotiations in disgust."

Teller doggedly continued his opposition to a test ban into the early 1960s. But the scientists who had originally supported smaller tactical weapons increasingly were less convinced than he of the importance of further technical advances. Instead, they emphasized the political symbolism of the test ban. They saw a test ban as a tangible "first step" in reducing the levels of mistrust between the two superpowers. They were convinced that a mutual stand-off could soon exist. Their priority was to achieve something, no matter how small, that would get the two sides talking. The horror of the ultimate confrontation, conjured up so vividly by the Cuban missile crisis of 1962, was evidence that a speedy resolution of the problem was desperately needed.

It was Sakharov, according to his later memoirs, who pointed the way to the Kremlin. He showed how it was possible to have a partial treaty: one that banned only atmospheric tests, which were easily detectable, whereas underground tests were not.

It was the much-needed compromise, and it lessened the tensions, certainly, but as an arms control measure, it was almost irrelevant. It was little more than a clean-air act, a symbolic step. There was to be no "second step"; indeed there was not even an intellectual structure or "scenario," as the jargon of the new strategists would call it, for a "last step." Talk of nuclear disarmament, so prominent since the war, virtually disappeared in official circles. Since the mid-fifties, the production of fissile material by both superpowers had reached such levels that it was simply inconceivable that all the fissile material made for bombs could ever be accounted for. Thus, no agreement on nuclear disarmament could ever be completely verified or policed. By the sixties, the entire focus of "disarmament" negotiations had moved to delivery systems, not bombs.

American and Soviet leaders, relying on the comforting doctrine of stability through mutual deterrence, came to believe that their own nuclear arsenals were a positive peacekeeping force. Piling up their own two stockpiles—vertical proliferation—was acceptable. What was unacceptable was horizontal proliferation: the spread of nuclear weapons to other nations. That became the issue. Instead of "disarmament," the superpower leaders talked of "arms control." To appease those who resisted the change, the new U.S. government department set up in 1962 to pursue these issues was called the Arms Control and Disarmament Agency. But when the test-ban treaty of 1963 finally came up for public debate and congressional ratification in America, administration spokesmen stressed its function as a nonproliferation, not a disarmament, measure: Since underground testing was much more difficult and expensive, the treaty would inhibit new nuclear powers. At home, nothing would change. In fact, America's own weapons testing program would be accelerated even if it had to be underground.

PART THREE

The Sixties

16

"Like Butter in the Sun"

"We had a problem like a lump of butter sitting in the sun. . . . Our people understood this was a game of massive stakes, and that if we didn't force the utility industry to put those stations on line, we'd end up with nothing."

—JOHN McKITTERICK, General Electric

In the early 1960s when Alfred Hitchcock selected the tiny Californian fishing village of Bodega and its protected bay as the site for his thriller *The Birds*, he called it "a rare find, remote and unspoiled by man." Today, there is an ugly man-made scar across Bodega's wild landscape. Past the village and the holiday cottages runs a four-lane roadway. It strikes out across the tidal mudflats to the head of the peninsula and ends at a huge hole in the ground, one hundred and forty feet in diameter and seventy feet deep. The hole is all that is left of a nuclear power station site, the first to be successfully opposed by citizen protest.

It was the beauty of the bay that first aroused the protests when the Pacific Gas and Electric Company, America's second-largest private utility, announced its intention to build a nuclear power plant there in 1961. The protests gained momentum when the plant site was found to be only a little more than a mile from the line of the San Andreas fault, the most active earthquake zone in the United States.

The Californian utility engineers were confident they could make the plant, a General Electric boiling-water reactor, earthquake-proof. Their own geologists maintained that the granite bedrock of the land around the bay had held firm during countless previous tremors and would do so again—despite one resident's recollection that in the 1906 quake it had "rolled like the ocean." Independent geologists found a new set of faults, however, and reported that the land immediately under the proposed plant was sixty feet of silt, clay, and sand, not a block of granite, as the utility had said. In 1964, the AEC's division of reactor licensing, which had to approve all plans for new reactors, canceled the project.

The AEC's decision was the kind of setback that any other new industry, trying to get off the ground after a shaky start, might have taken badly. Not so the fledgling U.S. nuclear industry. The confidence of the General Electric engineers in their product was not affected by the Bodega affair; the disappointment was a special case, they thought, not an omen of things to come. Together with the other electric power giant, Westinghouse, they were about to launch a concerted sales attack on the power utilities to make them buy nuclear reactors. The precedent for the future, as far as they were concerned, was occurring on the other side of the United States at Oyster Creek, a river that flows into Barnegat Bay on the New Jersey coastline. They were right: General Electric's boiling-water nuclear reactor at Oyster Creek, which started operation in 1964, was a public relations triumph.

The Jersey Central Power and Light Company, the utility that bought Oyster Creek, was so delighted with its new acquisition that it decided to make a big splash of its arrival in the nuclear age. The company published a detailed cost breakdown of the plant, seeking to "prove" the economic competitiveness of nuclear power. Although GE knew this was no more than propaganda, it did nothing to discourage the wave of international acclaim that followed. Economically competitive nuclear power had finally arrived. The fact that GE, unlike Westinghouse with its Shippingport pressurized-water reactor, could build Oyster Creek without government subsidy was the proof of the new era. Oyster Creek passed into nuclear folklore as the first "commercial plant."

Jersey Central could claim commercial viability because the major costs of building Oyster Creek had been borne by GE. Oyster Creek was based on what came to be known as a turn-key contract—a cheap package deal, including equipment and construction, offered by the electric companies that was so financially attractive American utilities simply could not refuse it. In the five years or so prior to the Oyster Creek "breakthrough," the Atomic Energy Commission had spent more than $1.2 billion to develop reactor technology. Private companies, especially GE, had spent half as much. In the three years following Oyster Creek, GE and Westinghouse sold both fixed-price contracts on equipment and turn-key contracts on which they lost about $1 billion. Such deals were hardly the result of charitable impulses. Of the thirty reactor systems that had originally been supported by the U.S. government through the AEC, two of the light-water type had emerged triumphant: the pressurized-water reactor (PWR) built by Westinghouse and the boiling-water reactor (BWR) built by GE. By offering these turn-key deals, the electric companies not only established a domestic nuclear industry but also, over the next few years, confirmed the American light-water reactor as the primary choice of the emerging nations in the international market.

American utilities, which had so far been reluctant to join the nuclear bandwagon, eagerly reacted to the overoptimistic cost projections promoted by General Electric and Westinghouse. Many would live to regret it. Abroad, less-developed countries that could ill afford nuclear power would be caught up in the wave of infectious optimism, and stubborn, fiercely nationalistic atomic energy bureaucracies, riding the same wave, would boost their own nuclear programs when they would have done better to hold back. For a while, some nations, like France, tried to develop different reactor systems and thus avoid dependence on the United States' monopoly of uranium enrichment services. All but Britain, tenaciously hanging on to its own indigenous gas-cooled technology, would eventually opt for the U.S. light-water system. By the late sixties, the light-water system stood alone as a fully fledged commercial reactor.

The new fixed-price contracts, hopelessly underpriced by the two American electrical companies, helped to put General Electric back into the running with the Westinghouse PWR's, which had been built as a spin-off from Rickover's submarine program. Though the contract losses were larger than either company expected, it seemed to both like money well spent. It was an investment more than a loss, a guarantee of their dominance in a huge new international industry. By the end of 1966, in addition to the twenty-one light-water reactors sold in the United States, another dozen plants were planned in Switzerland, Spain, India, Sweden, Belgium, Japan, and Germany.

Sometimes acting through their local licensee affiliate companies, the two American giants divided the sales more or less equally.

For both GE and Westinghouse, the development of a competitive nuclear industry was a godsend. In 1963, the year of the Oyster Creek order, morale in both of the American electric giants was at a low ebb. Total corporate sales had been stagnant for almost five years; sales at Westinghouse had stayed at $2 billion since 1957, and GE had barely increased its sales of $5 billion. In 1961, for the first time in the seventy-year history of American antitrust legislation, senior executives from both companies had actually been sent to jail for their involvement in a price-rigging conspiracy between electrical manufacturers. Longtime GE executives looked back with nostalgia to an earlier era when the company had been imbued with a missionary zeal to "electrify the world." They remembered the boom years after the war under Charles E. ("Electric Charlie") Wilson. His charismatic personality had symbolized their drive, their excellence, their confidence, and their faith. At one of their annual summer camp meetings in upstate New York, Wilson had warned his team of Westinghouse's plans to take over GE's acknowledged status as the Number One electric company. "They should live so long," Wilson commented. "Now let's all rise and sing 'Onward Christian Soldiers.' " And they did. One GE veteran recalled, "You came out of there with tears in your eyes ready to beat the world for General Electric."

In 1963, both companies went through a change of control at the top. An impetuous former light-bulb salesman, Fred Borch, took over as chief executive of GE's New York headquarters, and Donald Burnham, a long-serving financial executive with Westinghouse, arrived at the top of the Pittsburgh-based company.

Fred Borch was an energetic, cocksure, and impatient man. He had come to the top through the heartland of GE profitability—the Lamp Division. The company traces its origins to Thomas Edison's invention, and it was the Lamp Division that had carried GE through the Depression. The division was an independent empire inside the company. When Borch took over as chief executive, he brought with him a group of his colleagues from the division.

His first job was "to let the outfit know, by God, we were on the move again." He was determined to restore corporate morale by developing a real sense of momentum. Under Borch, the company adopted the advertising slogan "Progress is our most important product" and promoted the "all-electric home." It was Borch who determined to pour GE's substantial cash flow into a series of capital-intensive high-risk ventures, all of which had futuristic appeal: computers, aerospace, and nuclear power.

At Westinghouse, Donald Burnham faced all the same problems, plus an additional one. Despite its initial lead in developing nuclear reactors, Westinghouse was still Number Two in corporate sales. In July 1963, a week after he was named president, Burnham announced, "Our aim is to be first in performance." He immediately set about shaking Westinghouse out of three-quarters of a century of stodgy tradition of pursuing limited aims. The company had never been willing to venture far from its traditional technological base and, even in those fields, had been especially shy about the unknown risks of developing an export market. In contrast with GE's diversification and profitable direct investment abroad, Westinghouse looked dull. As part of the new image, Burnham's top executives were given gold tie pins with a $5 sign emblazoned on them to represent the profit share objective.

With corporate PR handouts proclaiming the "logic of diversity," Burnham drove the company in a number of different directions, all part of his new growth ethic. "Some of our new things," he pointed out, "haven't got a wire connected to them." The new ventures were wide-ranging: car rentals, low-income housing, Swiss watches, a mail-order business, and a record club that distributed the Beatles.

The big ticket growth item was nuclear energy. The early start that Westinghouse had gained from its connection to Rickover's submarine project spurred the company into developing an export potential in nuclear reactors. Burnham wanted to create a new multinational presence, and the reactor business was to be the spearhead.

GE was equally determined to stay ahead of its rival. No one in GE was more conscious of the company's initial failure to become a clear leader in nuclear power than the scientists and engineers at GE's Atomic Power Department in San Jose, California. They knew they were running from behind, that Westinghouse seemed to have an unfair advantage with its navy contracts.

In 1958, the GE engineers had launched an ambitious program of reactor development. Called "Operation Sunshine," it was a phased long-range plan for technical development of the boiling-water reactor. They thought they could produce nuclear plants that were economically competitive with conventional plants and, by the end of the 1960s, a "target plant" that would be cheaper. But they still found it hard to persuade the electric utilities to buy. A key member of Borch's lamp dynasty, John McKitterick, recalled: "We had a problem like a lump of butter sitting in the sun. If we couldn't get orders out of the utility industry, with every tick of the clock it became progressively more likely that some competing technology could be developed that would supersede the economic viability of our own."

The Oyster Creek plant settled the issue. Fred Borch gave GE's Atomic Power Department its head. He decided, as he put it, "to ram this thing right through." The only way to get widespread acceptance by the utilities was to take on the risks that were holding them back, to force-feed them with fixed-price reactors and turn-key contracts where GE would bear all the risks—even in the unfamiliar field of nuclear power plant construction. The GE engineers were convinced that construction costs would fall on a learning curve.

At Westinghouse, the company's Atomic Division was just as determined to maintain its position. They matched every move that GE made. In the post–Oyster Creek enthusiasm, Westinghouse could no longer rely on government subsidy, so they offered fixed-price contracts, too. They also accelerated the development of larger reactors. And they continued to match GE prices and special terms in order to induce utilities to sign up. The nuclear sales started to mount, becoming the biggest contribution to Burnham's growth objective for the company.

Of the two electric giants, it was GE that led the way in the nuclear boom of the mid-sixties. It was the first to publish a formal list of fixed prices and the first to express a willingness to assume all the risks of construction involved in selling plants on a turn-key basis. It also forced the pace in scaling up the size of its plants. At a time when the largest nuclear reactor actually operating was only 200 Mw, General Electric engineers offered bigger and bigger plants, with total confidence in their own ability to overcome any technical problems.

The size of Oyster Creek was just about 500 Mw. In 1965, the turning-point year when the orders escalated, GE offered an 800-Mw model and the next year, a 1,100-Mw, keeping pace with the competitive coal-fired plants that were also being stretched to these sizes. But there was trouble on the way: the capital costs of coal-fired plants were coming down while those of nuclear plants were going up. In its competition with coal, nuclear power had a moving target.

In the middle of 1966, GE made the most dramatic nuclear power deal of the period: it sold a reactor to the publicly owned Tennessee Valley Authority. TVA was right in the heart of coal country and had grown up using coal and hydro power, but, partly for its own progressive image, it decided to "go nuclear." The authority struck a hard bargain; it had made a fetish of its low power costs, and GE had to bid low to win the prize.

In the public tender, GE competed against Westinghouse and the coal companies. GE guaranteed a fixed price of $250 million for each of two 1,000-Mw units to be built at Brown's Ferry, Alabama. The GE bid worked out at $116 per kilowatt of generating capacity compared with $121 for Westinghouse and $117 for the lowest coal bid. GE's advantage was even more decisive in operating costs for fuel: it guaranteed performance of its uranium fuel that brought the total cost of the plant in at 18 percent less than the coal-fired tender. GE was so confident of its cost projections, or at least claimed to be, that it offered to pay TVA a penalty of $1,500 per hour of peak demand that the plant did not function according to contract terms. TVA, proud of its tradition of acting as its own architect-engineer, spurned the GE offer of a turn-key plant in favor of a fixed-price deal on equipment backed by the performance guarantees.

The business press heralded the new age. "Atomic Bomb in the Land of Coal," declared *Fortune*. The new era had arrived: the electric companies not only presumed that costs would come down with each successive plant, they also estimated that the creation of a major market for nuclear fuel would bring down the cost of uranium, too. There were few dissenting voices.

The most eloquent words of caution came from Phillip Sporn, former head of American Electric Power. In 1967, Sporn produced a detailed analysis of the GE bid for Oyster Creek and showed decisively that the claim for the arrival of cost-competitive nuclear power was false. At the same time, he carefully dissected the intense competition between General Electric and Westinghouse and concluded, "These manufacturers, committed to a nuclear future, were determined to maintain strong and closely competitive market positions.

Any pulling ahead or advantage gained by one organization· called forth a challenging competitive response from the other." He warned that long-term decisions should not be based on such short-term public-relations battles. Sporn dubbed the order rush of the mid-sixties "The Great Bandwagon Market."

He was particularly scathing about the TVA contract, dismissing it as a "come-on bid." Events confirmed his analysis. "At the most charitable interpretation," Sporn pointed out in 1967, "the TVA bids obviously were designed to introduce nuclear power to the largest coal burning utility in the United States, and they have not been repeated. Specific requests made of General Electric for a duplicate of the TVA quotation yielded the reply that it was no longer available. As a matter of fact, TVA itself found, in seeking options for a third nuclear unit, that prices had risen considerably."

Phillip Sporn was the doyen of American electricity administrators. A legendary figure, universally acknowledged as an engineering genius, he was sometimes called "the Henry Ford of power." As a teenager, Sporn had worked as a lamplighter for the New York Edison Company. In 1917, he graduated in electrical engineering from Columbia University and joined the American Electric Power Company. He rose to chief engineer in 1933 and chief executive in 1947, a post he retained until 1961. His name was associated with many of the most important technical advances in electricity production and distribution—always pushing the technology to higher efficiencies, bigger sizes, and lower costs. The American Electric Power Company became the largest privately owned utility in the world. Its system extended over seven states of the industrial Midwest, from Lake Michigan to the Tennessee Valley.

Sporn had to be treated with respect: his calculations could not be dismissed. He made three points. The first was that the electric industry—and the government—had been caught up in such a huge wave of nuclear enthusiasm that they had omitted to build into their estimates the fact that the cost of power from conventional fuels was actually going down: the cost of oil or coal was cheaper than it had been a decade before and it was being burned more efficiently. Second, Sporn calculated that the real capital costs of building nuclear plants would be between 10 and 15 percent more than the estimates made by the electric companies. Third, he predicted that the actual cost of producing the electricity once the plant was built would be about 26 percent more.

Sporn's caution did not extend to the question of whether it was truly possible to make reasonably accurate costings. Like the electric companies and the AEC, Sporn thought that, despite the meager amount of reactor operating experience and the small size of existing reactors, reasonable cost estimates for larger plants never before constructed could be made. That was a special kind of confidence.

Sporn was not antinuclear. Indeed, as a leading figure in private power, he had responded to the threat of an "atomic TVA" by participating in a number of nuclear power schemes in the fifties. He was convinced that the technology, if developed slowly, could become cost competitive in some areas. He had devoted his entire life to the careful refinement of electric technology, and the results had been dramatic. But he saw no shortcuts in this slow progress, and he feared that the traditional approach of careful step-by-step engineering development was being subsumed by a passing technical enthusiasm.

Both the electric companies and the AEC objected to Sporn's analysis. GE called it "unduly conservative," and the AEC said, "Mr. Sporn could have been more optimistic." As far as they were concerned, Sporn was being stodgy in face of the exciting prospect of the new wonder power; his hard-headed realism seemed old-fashioned to the nuclear boys. And so the "Great Bandwagon Market" moved on—even to engulf Sporn's old company—American Electric Power.

In 1967, Donald Cook, who had become the president of the coal-rich American Electric Power Company when Sporn retired in 1961, launched into a series of disastrous nuclear investments. With Sporn still issuing words of caution from the wings, Cook would confidently declare, "From the dawn of the nuclear age there has never been a time when the company was not convinced that there was a great future in atomic power."

From the beginning of his presidency of AEP, Cook, an aggressive former chairman of the Securities and Exchange Commission, was determined to combat the increasingly stuffy public image of the utility administrator. Electric companies, he thought, were capable of change and dynamism. "Revolution is the normal thing here," he declared. Building on Sporn's foundation of low electric power costs, Cook spent twice the percentage of revenue on advertising as the industry norm to get his message across. "We are a power supermarket," he asserted. "Cut the price to the bone and the sales will more than match the cut." The financial press called him "restless, demanding, precise . . . tough-minded and tenacious"—the superexecutive. He turned down an offer to become Lyndon Johnson's Treasury Secretary because to him the utility business "still had a lot of challenge."

Despite the huge amount of low-cost coal AEP owned or had access to, it was natural for Cook to be attracted to nuclear power; he was surrounded by nuclear progress. On every side of the AEP service area, utilities were going nuclear. Many of these were headed by the first electric utility nuclear converts from the fifties. In the mid-sixties, these men were all reinforcing their original nuclear commitment, and new companies were hopping on the "Bandwagon." To the west, Harris Ward of Chicago's Commonwealth Edison announced his intent to convert 40 percent of his total capacity to nuclear power. To the north, Walker Cisler at Detroit Edison was promoting the ill-fated breeder program. To the northeast, William Webster, who had put together the Yankee consortium of New England Utilities, was building more nuclear plants. To the southeast, W. B. McGuire of Duke Power was a new convert. Finally, to the south, even the coal-rich TVA had signed up.

Cook was not the sort of man to stay in the background: in 1967 he ordered two reactors; in the company tradition, he named them after himself—the Donald C. Cook One and the Donald C. Cook Two. He would eventually regret the purchase; both lagged hopelessly behind schedule and final costings, raised by having to buy electricity from neighboring utilities, were three times more than originally predicted. In the end, Cook ordered the coal-fired plants that he had hoped the nuclear plants would replace, because, as he admitted in 1970, "We decided we could not count on the nuclear plants coming in on time." AEP's power costs shot up, but Cook never lost his own personal confidence. Five years later, he told an inquiring journalist, "Let me put it this way. We are delighted to have two nuclear power plants. But we are also delighted not to have three."

Donald Cook was unlucky. The utilities that bought their plants before AEP had got bargains. But Cook came into the market after GE and Westinghouse, both realizing how much money they were losing, had stopped offering fixed-price and turn-key contracts. By this time, the two electric giants dominated the world market equally, with Westinghouse still holding the edge in domestic sales. The picture had changed considerably from what it had been in the fifties and early sixties when the two companies had found only a handful of nuclear enthusiasts in the utilities. By 1966, Westinghouse had sold six plants in the United States with a total capacity of 5,000 Mw and had a stunning thirteen additional orders for a total of 11,000 Mw in 1967. GE had sold a total capacity of 7,700 Mw by 1966 and 6,200 Mw in 1967. Moreover, the two giants had been joined by the boilermaking companies, which felt obliged to diversify into reactors to protect themselves. Both Combustion Engineering and Babcock and Wilcox developed a PWR and, in the period 1966–67, sold seven and eight reactors respectively. By 1970, with only 4,200 Mw of nuclear capacity actually operating, a staggering total of 72,000 Mw was under construction or on order.

But if sales were booming, the technical difficulties were only just beginning. All the plants had been ordered at a time when the largest operating plant was only 200 Mw in size. The first of the larger plants, in the 500-Mw range, came on stream in 1969, but the orders had long since jumped over the 1,000-Mw size. Westinghouse would have a lot of trouble with increasing the size of both its conventional and its nuclear plants. GE, more technically successful with its conventional plants, would also run into trouble with its reactors.

The early successes of GE and Westinghouse were repeated with another, ultimately successful, reactor type that was being developed right on America's doorstep. Canadian atomic scientists, aggressively nationalistic in their refusal to take higher-paying jobs south of the border, were busily adapting their wartime experience in heavy-water reactors and creating a new, indigenous nuclear industry. Clarence Howe, the Canadian "Minister of Everything" who had created Atomic Energy of Canada, Ltd. to promote nuclear power, provided the scientists with the political muscle they needed to launch their program. "Howe's boys" held key positions in the AECL and also in the first Canadian utility to take up the nuclear challenge, the Hydro-Electric Power Commission of Ontario. The AECL was only too happy to oblige Ontario Hydro's interest in the heavy-water reactor.

In 1958, a new nuclear power division was created within the AECL and it began designing the scaled-up version of the original Chalk River reactor, built after the war. The pride and confidence of the AECL group was reflected in the name they gave to the commercial reactor: the CANDU, an acronym for Canadian Deuterium (the technical term for heavy-water) Uranium.

The first 200-Mw commercial-size CANDU reactor was ordered in 1959, to be run by Ontario Hydro at Douglas Point on Lake Huron. The program was a technical success as far as it went, and even before Douglas Point was finished in 1964, the British government-owned utility, the Central Electricity Generating Board (CEGB), gave a clear indication that it was considering the CANDU as a possible candidate for a new generation of reactors.

In Europe, the Canadian and especially the American triumphs were major challenges. Both the U.K. Atomic Energy Authority and the French Commissariat à l'Energie Atomique were developing their own gas-cooled reactors. Both these institutions were determined to resist the introduction of other types, particularly the U.S. light-water reactors; their scientists believed they could match anything the Americans could produce. But after the "breakthrough" at Oyster Creek, both the British Central Electricity Generating Board and the French Electricité de France wanted to take a serious look at the competing technology; perhaps the "narrow front" approach of developing only the one line of gas-cooled reactors had been wrong. Fierce battles were fought in both countries between those who wanted to continue with the gas-cooled types and those who wanted to try the U.S. light-water reactors. In the end, France abandoned her gas-cooled program in favor of light-water, but Britain stubbornly refused to make the change—with disastrous results. A decade after the decision to go ahead with the scaled-up versions of gas-cooled reactors, Britain would still not have built one that worked. France, on the other hand, had raced ahead with her light-water technology to become the foremost nuclear-powered nation in Europe.

Christopher Hinton, Britain's leading nuclear engineer, was the key figure in the effort to introduce light-water reactors to the United Kingdom. In 1957, he had moved from the U.K. AEA to head the CEGB, becoming a buyer of reactors rather than a supplier, and he was proving as single-minded in his pursuit of the best economic package as he had been in providing the fissile material on time for Britain's bombs. His former colleague at the AEA, Sir William Penny, the designer of the British bomb, was championing the scaled-up advanced gas reactor (AGR), but even before Oyster Creek, Hinton was not so sure. In 1961 his doubts had burst into the open. In an article in a London banking journal, he reversed his previous public position that the AGR would be the next British reactor and warned that other countries were pursuing different reactor types, ones which might become more successful in the long run. He concluded, "Ultimately everyone connected with the development, design and construction of nuclear power plants must decide his research and development program on the basis of what his customers find most economical and what he can develop and sell to give him a profit and them power at the lowest possible cost."

The implication was clear, and, coming from a man of Hinton's stature and experience, it caused great concern at the Atomic Energy Authority. When Hinton went to Canada a year later to examine the heavy-water CANDU more closely, it simply underlined his doubts and the AEA's fears. The AEA had reason for concern; as a research group, it depended for its existence on the CEGB's buying what it produced. If the CEGB were to choose either the light- or the heavy-water reactor for its next large program, the AEA would effectively be out of business.

Hinton was dogged in his pursuit of efficiency and economic viability, and he refused to accept any research organization as the "ultimate arbiter of commercial policy"; in particular, he refused to accept that the AEA's advanced gas-cooled reactor was the answer to Britain's needs. In government circles, his flirtation with other reactor systems was seen by many as a deeply unpatriotic act.

Yet the British government itself was hardly inspiring confidence in its own product: it was unable to make up its mind which reactor to fund for the next program. British delegates to the fourth Atoms for Peace Conference in 1964 were acutely embarrassed by this delay. After all, at the first two conferences, Britain had been considered a leader, if not the leader, of civilian nuclear power development in the world. The American salesmen of the light-water reactors were most happy, of course. One U.S. nuclear industry journal, *Nucleonics Week*, commented that "the mere fact that LWR's [light-water reactors] are being seriously considered by a nation which has invested heavily in its own commercial power plants enhances the prestige and competitive position of the U.S. nuclear industry everywhere."

Britain's nuclear export potential was never to recover from these internal rows and delayed decisions, and when, in 1965, it finally chose the AGR over the light-water types, it turned out to be one of the great technological catastrophes of the nuclear age. The two 660-Mw reactors that Britain began to build in 1965 were only starting to operate by the end of the seventies. In the interim, the cost of the two had soared from £89 million to £344 million.

There was no inkling of impending disaster at the time, of course. Indeed, the government was absurdly euphoric. In May 1965, the new Labour government's Minister of Power, Fred Lee, told the House of Commons triumphantly: "We have made the greatest breakthrough of all time [in ordering the AGR] . . . we have hit the jackpot." The British press, with rising patriotism, gave Lee's decision a warm, and largely uncritical, welcome.

The basic assumption—which was to prove such a costly error—was that the successful prototype AGR could easily be scaled up into a commercial unit, despite the fact that only sketchy and incomplete designs were available at the time. In 1978, CEGB executives would comment on the still-unfinished AGR reactor, "You don't extrapolate data and jump temperatures by 200 degrees centigrade unless you have done sufficient pre-design and research work before you start on site."

Hints abounded that the decision had been made under political pressure from the government, but no hard evidence emerged to support this thesis. The CEGB, sidestepping questions about political interference, asserted that the decision had been a hard-headed commercial one—that the AGR was quite clearly the most economic of all the types. Whatever the truth behind the AGR decision, it was billed at the time as a setback for the U.S. manufacturers of the LWR's, General Electric and Westinghouse. The setback was only temporary, however. Across the English Channel in France, the giant American corporations were about to make important inroads into an equally stubborn market.

Like their British colleagues, the French atomic scientists in the Commissariat à l'Energie Atomique (CEA) had developed a power reactor directly from their experience in the French atomic weapons program. It was even of the same type, gas-cooled and moderated by graphite. It was self-confidently promoted as a rival to the American light-water technology and was a personal ward of Charles de Gaulle and the fiercely nationalistic CEA. But the French-nationalized electric utility, Electricité de France (EDF), took a much tougher stand against the development of the gas-cooled reactors than its British counterpart, the CEGB. Countering formidable political pressure, EDF pushed France into accepting the U.S. light-water types. Before that policy reversal was made, however, there was a fratricidal battle between the CEA and the EDF, which is best seen in terms of a struggle between two groups of graduates from the engineers school, the Ecole Polytechnique.

It was the elite group of the school's graduates, the Corps des Mines led by Guillaumat, who had taken over the running of the CEA from the scientists after the purging of Joliot-Curie in 1950. Polytechnicians, as the graduates are known, were also in control of the EDF, but they were from a different corps, the Corps des Ponts et Chaussées—bridges and roads.

Ponts et Chaussées was the second choice of the Polytechnique graduates, the top ten traditionally going to the Corps des Mines. Members of Ponts et Chaussées were, therefore, regarded as inferior by members of the Corps des Mines. The battle on reactor choice became much more than a fight over a few centimes per kilowatt hour as produced by reactor types; it invoked one hundred and fifty years of institutional rivalry.

EDF had early shown its determination to keep its options open, and in 1961 it built a small PWR, of Westinghouse design, in a joint venture with a Belgian utility. To plan reactor development, the two French institutions had created a special joint committee, and in 1964, responding to the new wave of nuclear power enthusiasm, the committee planned the first large-scale construction program: one 500-Mw reactor a year. The question was: What type of reactors should they be? Much depended, clearly, on the degree of political clout that the EDF leadership was able to muster.

From 1965 until 1969, the head of EDF was Pierre Massé, one of the foremost economists of postwar France. Together with a small, talented group of engineer-economists, he had refined a system of electricity investment decision-making and tariff-setting that would be universally acknowledged as one of the best in the world. With this success behind him, he was particularly sensitive to the assertions of superiority by the members of the Corps des Mines in the CEA.

Massé had to deal with an engineer in the CEA named Jules Horowitz, who was in charge of the development of the gas-graphite technology. Horowitz conducted what was later described as a form of "intellectual terrorism" over the reactor choice, decreeing certain words taboo and an affront to French patriotism. The term "light-water reactor," for example, was forbidden; the U.S. reactor could only be called a "foreign experimental reactor."

When the plan for the big reactor program began to emerge, the EDF attacked the technical difficulties of the gas reactors and the CEA retaliated with a campaign in political circles and in the press blaming any difficulties on EDF. Massé recalled later that the two sides were irreconcilable. "When the CEA said white, the EDF said black and vice versa," he said. "It did not matter what the subject was. In each establishment, an esprit de corps and a will for power was at play." By the end of 1967, Horowitz and his CEA supporters forced the issue to the top. Putting to good use the political contacts they had made during their weapons work, they asked de Gaulle to call a meeting to settle the matter.

The meeting was held in December 1967. De Gaulle opened it with a full-scale attack on EDF. Massé replied that, as far as he was concerned, there should be "no taboos in this field"—that included the possibility of buying the U.S. technology. Later Massé recalled, "I added that for myself I was not sure on the matter, but that I found it absurd to refuse to undertake an experiment. I have always thought that the person who refuses a priori to undertake an experiment does so because he is wrong."

By the end of the meeting, de Gaulle made it quite clear that he would countenance no large-scale deviation from the gas-cooled reactor development. As a gesture to EDF, and particularly to Massé, however, he agreed to another experiment with light-water technology in a second joint venture with the Belgians. Within a year, new technical difficulties emerged for the gas-graphite reactors. They enabled EDF to delay the expansion of the program until after de Gaulle's departure in April 1969. When the CEA lost its patron, the American light-water system won the day.

De Gaulle's successor, Georges Pompidou, had a healthy suspicion of the tendency of engineers to exaggerate the claims of their pet projects, and he was less concerned about the rhetoric of French independence than he was about comparative economics. Except within the CEA itself, a broad consensus had emerged in the French nuclear industry that light-water reactors were more efficient. In November 1969, the light-water route became official French policy. To rub salt into the wound, Pompidou turned on the CEA, appointed a committee to reorganize it from top to bottom, and detached it from the special position of reporting to the Prime Minister, making it responsible, like EDF, to the Minister for Industry.

It was a considerable triumph for American technology. Westinghouse moved quickly to transform its technical achievement into a long-term industrial power base. It tried to form a new European heavy-industry giant by acquiring electric manufacturers in Belgium, France, Spain, and Italy, but Pompidou vetoed the proposal to acquire the French part of this grand strategy. Some Gaullists were still arguing that France could develop its own light-water reactor on the basis of its research on submarine propulsion. But Pompidou also dismissed this idea; "too late, too expensive," he said. France would have to pay the price of importing technology.

17

Foolish Dreams

Foolish men have foolish dreams.

—W. G. BENHAM, *Proverbs*

On October 5, 1966, the most ambitious of all the American electric utilities' nuclear power projects, Detroit Edison's experimental commercial breeder reactor, named Fermi I, on the shores of Lake Erie, ran into serious trouble. A piece of metal had worked loose at the bottom of the reactor and had partially clogged the flow of the core coolant, the core assembly had become overheated, a tiny part of it had melted, releasing dangerous radioactive gases into the containment structures, and, for all the engineers knew, it seemed as though a major disaster was imminent. At the time, the reactor crew had no idea what had gone wrong, nor how to remedy it. Six months would pass before they could determine the extent of the damage and a full year before they had found the cause. In the early days of the accident, they could only hope the damaged core would stabilize itself without further mishap.

Despite such huge areas of ignorance, in the days following the accident the reactor crew indulged in a display of confidence rare even in the self-assured nuclear business. On the fourth day of the accident, in a conference room only one hundred feet away from the crippled core, the engineers held a sixty-ninth birthday party for their boss, Walker Cisler. For the past fifteen years, Cisler had been the head of Detroit Edison, one of the major utilities in the Midwest; he was one of the most committed of the nuclear zealots. The breeder was his pet adventure.

Cisler's plucky, some would have said foolhardy, group were the early prophets of a sect of nuclear dreamers—scientists and engineers who believed that nature could be tricked into producing reactor fuel for nothing.

By the mid-1960s, the acceptance of the nuclear chain reaction as a means of producing energy had permitted reactor design teams in all advanced industrial countries to make a series of self-serving paper estimates "proving" the need for spending tens of billions of dollars on a new generation of advanced reactors. The most advanced of these types, the breeder reactor, which produces more fissile fuel than it uses, was seen as the key to offsetting the feared shortages of uranium.

The time was right for this kind of big-ticket item. In America, particularly, the triumph of the light-water reactors, along with such other technological achievements as jet aircraft and computers, had decisively overcome the shock of having the Russians steal the space lead with the launching of Sputnik in 1957. Government purse strings were being liberally loosened for anything classed as Big Science.

Political decision-makers were still mesmerized by the brash self-confidence of the international nuclear community and its unshakable belief in the inevitability of nuclear progress. They accepted without demur the logic of progressing from the successes of light-water reactors—now labeled "first-generation" reactors—to new kinds of power plants, which, in unselfconscious bombast, the nuclear scientists had come to call "advanced." No source of independent advice was available to point out the one thing that distinguished each kind of "advanced" reactor from the others: every attempt to build one had failed.

New legislation and funds flowed through the legislatures of every industrial country intent on meeting the challenge. The nuclear research institutes were organized on reactor types and within the institutes each group had its favorite concept: new types of coolants and moderators, but also entire new systems that would make the production of steam by the chain reaction more efficient and less costly.

That meant working on the heart of the nuclear mystery, the core of the reactor itself.

All reactors are "breeders" to a certain extent: the original graphite piles at Hanford in the United States, at Windscale in England, and at Chelyabinsk in Russia bred plutonium for those countries' weapons programs. But the amount of plutonium created was always less than the amount of fissile material used up; those original reactors are sometimes called "burners" as opposed to "breeders." In between the two types there is a group known as "advanced converters"; these can sometimes make as much fissile material as they use.

In the first generation of reactors, the core was fueled with U-238 enriched to 3 or 4 percent of U-235, instead of the less than 1 percent occurring naturally. Natural water was used both to extract heat from the core and to slow down, or moderate, the neutrons whizzing around so that they would be able to find their U-235 atomic targets and sustain the chain reaction. In the breeder, the fuel is U-238 containing between 4 and 6 percent of fissile material, primarily plutonium. Instead of water, a coolant is used that extracts the heat from the core but, rather than slowing down the neutrons, allows them to whiz around striking U-238 atoms and making more plutonium. This sequence involves many more tricks than used in the light-water reactor and some very expensive research and development—as the Cisler group found out to their cost. Before their experiment was finally abandoned in the mid-1970s as too difficult and too expensive, it had in the course of almost two decades cost well over $100 million; the reactor had operated for a total of only 378 hours.

Since the early 1950s, small-scale experimental breeders had been built in the United States (by the AEC), Britain, France, and the Soviet Union. Even during the disastrous history of Detroit Edison's plant, interest in the breeder concept continued to grow. It had become the touchstone of technical achievement in the international nuclear community. It also presented any nation undertaking it with the engaging prospect of energy independence. It was of special interest to countries like Germany and Japan, which had no secure sources of uranium.

But the enthusiasm was based on more than a simple desire for energy independence: research physicists were faced with their own instant obsolescence unless they could come up with a series of "advanced" reactors—ones that would improve on the efficiency of the old types. Like the nuclear weapons designers, who were constantly trying to design more efficient bombs—ones that used less fissile material yet had greater explosive power—the reactor designers were trying to develop reactors that would use less fissile material (or produce more, in the case of the breeder) for a given amount of heat produced.

Prior to Cisler's breeder, General Electric had seriously pursued the possibility of building one, but the technical difficulties proved too great. These two failures, however, did not discourage continued research and funding of new reactor projects in the United States. In the space of two years, from mid-1965 to mid-1967, France, Britain, Germany, Japan, and the Soviet Union all decided to build large-scale prototype breeders. All were bigger than 250 Mw—more than ten times the size of the experimental scale models that had been started in the mid-fifties.

In the march of progress, few bothered to note that a successful breeder would add to the proliferation problem. Four of the countries that launched breeder programs—Britain, France, the United States, and the Soviet Union—were already weapons powers. But the new breeder enthusiasm would involve others, like Germany and Japan, who were not weapons powers and had said they had no intention of joining that club. Their development of the breeder meant they would now be accumulating large quantities of bomb-grade plutonium. Also, some of the advanced reactors would use highly enriched uranium, which any country could turn into a bomb if it so desired. Concern over this potential problem of weapons proliferation would not come for another decade.

Advocates of the breeder seized on the flimsy evidence of dwindling world uranium supplies to justify the development of the new reactors. They measured the highest estimates for uranium demand from the successful light-water reactors against phony figures for world reserves. Great urgency was attached to the breeder programs because of fears of a uranium shortage, yet virtually nothing was known about actual world supplies. Indeed, in the United States, uranium exploration had long been of secondary concern in the nuclear program; searches had been confined to desert regions of Utah and New Mexico where successful strikes had been made in sandstone rocks. Other geological formations that had proved fruitful sources elsewhere in the world were not explored.

The new plan to develop the breeder reactor not only delayed the search for uranium, it also delayed necessary technical developments on existing reactors, first, to increase their safety and, second, to find acceptable ways of disposing of their radioactive waste products. No one personifies these distortions more clearly than the "godfather" of the U.S. breeder program and one of the most talented products of Admiral Rickover's naval reactors' empire, Milton Shaw.

When Milton Shaw was appointed director of Reactor Development of the AEC in November 1964, he faced a battery of different proposals for advanced reactors. The proliferation of ideas bothered him. Technical development, he believed, required clear objectives. Shaw was first and foremost an engineer who wanted total command of his own project. He modeled himself on the Rickover example, and his concept of the engineer's role in technical development was as intensely personal as that of his mentor.

Nothing could appear more anarchic and pointless to a Rickover protégé than the traditional mode of promoting reactor development in the AEC. No one was really responsible for anything. No one set priorities, deadlines, or schedules. There were no hard technical criteria by which to judge proposals. For a man like Milton Shaw, the existing operations were decidedly sloppy.

Over the years, a large number of reactor projects had been subsidized with a minimum of engineering assessment. Worse, they had been allowed to proceed without central direction by government engineers. It was no surprise to Shaw that none of some thirty AEC reactor projects pursued in this fashion had ever succeeded.

Of the two outstanding successes, the Westinghouse PWR reactor had been developed with the support of Rickover's team, while General Electric had produced its BWR alone. At the time Shaw arrived in his new position, this record of failure was about to be repeated on an even grander scale. With the new talk of uranium shortages, the AEC was besieged with proposals for "advanced" reactors, and it looked as if it might mistakenly divide its limited resources by supporting virtually all of them. As far as Shaw was concerned, this meant that none of them would succeed. He had already made up his mind what he wanted to do. He would build the breeder reactor—through all its stages—just as Admiral Rickover had built the pressurized-water reactor.

Shaw's devotion to the breeder was at the expense of time and funds badly needed if the technical difficulties of the light-water reactor were to be overcome; his was a conscious decision not to give these priority. He knew that the technical development of the light-water reactors was far from complete: "We're dealing with an infant technology," he said. "I know the utilities assumed that nuclear power brought with it inherent perfection, but this business is a tough one." Despite this, Shaw diverted funds earmarked for reactor safety to his breeder project and paid little attention to waste disposal. It would not be until 1972 that the first major public hearings on emergency safety systems were held in Washington, virtually destroying Shaw's credibility as an engineer-administrator.

At those hearings, it became clear that during the eight years he was director of Reactor Development and responsible for questions of safety, the construction of facilities to test emergency procedures had been delayed, funds for crucial safety experiments had been cut, unwelcome reports from safety experts had been censored, and internal critics had been intimidated or removed. Even in independent research laboratories, like Oak Ridge, employees who had crossed Shaw and his staff were transferred to jobs out of his jurisdiction.

One focal point in the hearings was the question of the reliability of the emergency core-cooling system. This is designed to pump emergency water into a reactor that has lost its coolant, thus preventing a disastrous meltdown of the nuclear fuel. Documents secured under the Freedom of Information Act revealed that Shaw himself, in a February 1971 internal memorandum, had admitted that "no assurance is yet available" that the safety system would work. Yet Shaw had refused to give the research program any priority.

Despite these revelations, Shaw approached his cross-examination by Harvard economist Daniel Ford with all the self-confidence he had seen Admiral Rickover successfully employ in past public hearings. At one point, Ford pressed Shaw to justify his assertion that engineering conservatism, a basic tenet of the Rickover philosophy, had been applied in the calculation of heat transfers.

FORD: What are the documents you consulted?

SHAW: Mr. Ford, I have been in this business twenty-some-odd years. All right? The information relating to this goes back through those years. My job depends upon this information over these twenty-odd years. I cannot recall every bit of information that I used in this regard nor do I see any good reason to try to do it.

FORD: What documents did you consult?

SHAW: I do not recall.

FORD: Do you not recall any?

SHAW: I do not recall the documents. I am sure I depended a good deal upon my background.

FORD: Well, what is the basic experimental source of information on reflooding heat transfer, Mr. Shaw?

SHAW: Again I believe that is detail, if you don't mind . . .

FORD: Have you ever heard of the FLECHT program?

SHAW: Oh absolutely. In fact I think I initiated it, didn't I?

FORD: But you did not seem to recall that the FLECHT program was the basic source of experimental data on heat transfer in the reflooding period. How in the world do you explain that?

ENGELHARDT (Counsel for the AEC): I object to that, Mr. Chairman. It is argumentative.

CHAIRMAN GOODRICH: I will sustain the objection, much as I would like to hear the answer. (Laughter)

The hearings also revealed that Shaw and his staff had adopted a conscious policy of distancing themselves from the safety issue. As one witness put it, they were "trying to avoid the problem or burden of having to spend a lot of time answering public inquiries that are addressed to them . . . on general questions of nuclear safety."

Shaw's record on waste disposal was equally disastrous, although it must be said that he inherited a huge, unsolved problem from his predecessors. The AEC's program of waste disposal had, to say the least, been haphazard, carried out with unjustified self-confidence and scant regard for long-term safety. Over the years, only one-tenth of one percent of the AEC's budget had been spent on waste disposal. Yet, as early as 1955, a National Academy of Sciences report had warned that "continuing disposal of low-level waste . . . probably involves unacceptable long-term risks." Ten years later, another NAS report on a disposal site in Idaho left the researchers with "two unrelieved major anxieties: (1) that considerations of long-range safety are in some instances subordinated to regard for economy of operation, and (2) that some disposal practices are conditioned on over-confidence in the capacity of local environments to contain vast quantities of radionuclides for indefinite periods without danger to the biosphere." Examples came later: the AEC admitted, in response to a press inquiry, that ducks feeding on algae in the open waste trenches at the Hanford works were riddled with radioactivity. They "would have given a person five times the maximum permissible dosage of radiation if eaten," said the AEC.

In an attempt to overcome the problem of waste disposal, the NAS had for over fifteen years urged the AEC to explore the possibility of storing waste in underground salt formations. Salt beds, usually dry formations, supposedly prevent radioactive leaks from the waste containers from finding their way into surface waters. In January 1971, the AEC announced it had opened an abandoned salt mine near Lyons, Kansas, after spending more than $100 million researching salt-bed suitability. In March 1971, Milton Shaw told the JCAE that "another year's work of research and development in this area on top of fifteen years of work will not be particularly productive. We need the project and are ready to proceed with it. Moreover, we are convinced that the Lyons site is equal or superior to the others." This unusually positive statement was one the AEC would live to regret.

Within six months of Shaw's confident assertion that no more research was needed, the AEC made a startling admission. They had discovered something geologists had known all along: there was water in the salt formations. The company that had been mining salt from the rock had been injecting water into the formation for fifty years as a method of removing the salt, and the mine's tunnels came as close as five hundred yards from the proposed waste dump. The AEC canceled the project. Questions like safety and waste disposal were mere sideshows to Milton Shaw. Jobs to be done, he agreed, but by someone else—someone who was less fitted for the big and difficult tasks.

As a good engineer, Shaw knew the breeder required a large-scale program of basic technical development. Because the breeder has no moderator to slow down neutrons, its core has radiation intensities few materials can withstand. A series of design principles and new materials had to be laboriously tested to determine whether they could stand up to the massive bursts of radiation.

This demanded a considerable effort in research teams and funds. One source for both was the continuing AEC-supported research into other advanced systems. Shaw soon narrowed the AEC's five different systems down to three. By 1965, there was only one, leaving the way open to concentrate on the breeder.

The omens at the time seemed good. General Electric was still vigorously promoting its own breeder project. In September 1964, in the middle of the debate on the advanced reactors, it had stunned breeder teams throughout the world with an announcement that it expected to have a commercial breeder on line within a decade. Only later, as technical difficulties increased, would nothing more be heard of this claim.

Shaw's first priority was to build a huge test reactor that could be used to perfect the new range of materials needed. The basic problem in the design of the breeder concerns neutrons. The unmoderated fast neutrons must avoid colliding with each other and being slowed, and this requires several design tricks in the reactor core.

Apart from the core design, there is also a coolant problem. The common coolant used in breeders is molten sodium. It has a high conductivity and therefore can extract heat from the core with relatively little movement. But sodium also reacts violently with water and with a wide range of other materials. To prevent any reaction inside the core coolant circuit, the sodium has to be covered with an inert gas, such as argon. The gas, however, can get mixed up with the liquid metal and form unwanted bubbles; because liquid sodium is opaque, these are hard to find. Overall, the breeder is one of the most demanding technological challenges ever encountered.

Shaw was undaunted by the challenge, by GE's ultimate failure, by Walker Cisler's accident in the Lake Erie plant, and even by the huge costs. He shared the near-economic illiteracy of the engineering profession, whose members too readily assume that improvements in physical efficiency are simply and directly related to economic efficiency. Shaw thought that the technical marvel of the breeder reactor was its own justification. Like Rickover's naval program, it was to be built to technical standards of perfection and the costs would be of concern only after that had been established.

The only way it could be done was on the Rickover model of absolute control of the private contractors by a disciplined core of determined engineers. Shaw found, as Rickover had before him, that the engineers at GE did not want to be run from the outside, so the prototype breeder was assigned to the always more cooperative Westinghouse.

Shaw's clear priority commitment to breeder development brought international fears of a new American challenge, and nuclear engineers in other countries were determined that the American technological might, already being displayed in the light-water successes, would not be repeated in the future. The key to effective competition against America lay in keeping alive atomic research institutions. In Britain and France, independent research had never stopped. But in Germany and Japan, the research institutions that had been created in the nuclear optimism of the mid-fifties had degenerated over the following decade as their countries had turned to light-water technology. By the mid-sixties, the main hope of restoring some sense of direction within them rested on the breeder reactor.

From its inception, the Japanese Atomic Energy Research Institute had been frustrated in its hope of developing an autonomous national base of independent nuclear technology. The precipitous commitment to import a complete gas-cooled reactor from Britain was the first battle lost by the scientists. In the next few years, they slowly developed a number of other projects, including a small-scale experimental light-water reactor. In 1963, before this project was complete, the Japan Atomic Power Company (JAPCO) decided to import a large-scale light-water reactor from the United States. By 1970, seven imported light-water reactors would be operative in Japan; another five would be completed by the end of the 1970s. GE and Westinghouse had carried on a hard sales drive and shared the spoils equally, six reactors each.

In the early 1960s, Japan's Research Institute had been relegated to minor tasks in support of private industry; its mission to build up Japanese basic research had been almost forgotten. By 1963, leftist scientists, increasingly dissatisfied, had been mobilized by union organizers and a major labor dispute erupted. Officials in Tokyo dismissed the controversy as Communist agitation, but the incident left a legacy of suspicion.

In order to clean up the institute, the government appointed a new head, Niwa Kaneo, who had just retired as chief executive of one of Japan's largest corporations, Mitsubishi Shipbuilding.

Niwa Kaneo brought to his new responsibilities a definite attitude about the promotion of Japanese independence. In the years since the Meiji restoration, which drove Japan into industrial development, the Mitsubishi conglomerate had built a firm foundation of national autarky. The company's slogan had been "To go with the State." In contrast with the more capitalist-orientated Mitsui combine (the largest of the prewar conglomerates), Mitsubishi, the second largest, had eagerly embraced the objectives of the fascist state. The company became known during the war as the "arsenal of Japan," producing every kind of armament, from the "Zero" airplane to the largest battleships. Its growth was based on autonomous technical development, well displayed in the company's shipbuilding operations, notably in the huge Nagasaki shipyards where Niwa worked.

Sweeping aside substantial internal resistance, Niwa totally reorganized the research institute and imposed a system of clearly stated targets for research. With the growing commitment of Japan's electric utilities to import American light-water reactors, Niwa reoriented the research program to advanced reactor systems.

The researchers at the institute, however, no longer accepted uncritically the claims of breeder advocates. Original enthusiasm about the imminence of breeders in Japan's 1957 Long-Term Plan had been abandoned, and the scientists had watched carefully the multiplication of technical problems as breeder research continued elsewhere. Despite this, Niwa forcefully promoted the breeder and had it added to the institute's program.

In late 1964, the Japanese Atomic Energy Commission created a special group called the Power Reactor Development Committee; it was charged with developing a plan for long-term research and development. From his Mitsubishi background, Niwa brought to this committee a formidable model for success. He was the only member of the new group who had definite objectives and a clear structure for implementing them. Earnest, honest, serious, and determined, Niwa was an engineer-administrator of the old school. He even shared the Mitsubishi tradition of being a hard drinker—in Japan, it is said the steel, utility, and Mitsubishi men all drink. Niwa was convinced that with proper direction Japan could leap to the forefront of international breeder development.

At the time the new reactor development committee was formed, Japanese industry was riding a wave of sustained economic success. There was a new spirit of confidence in the nation's ability to compete at the highest levels of modern technology. New long-term projects were being organized in many fields: computers, aircraft, space. It was natural that atomic energy became one of them. In the specific field of the breeder reactor, the large electric manufacturer Toshiba was particularly enthusiastic. The company had links with General Electric through licenses for boiling-water reactors, and felt itself tied to GE's aggressive promotion of the breeder. Indeed, the appointment of the Japanese committee coincided with GE's September 1964 announcement of a commercial breeder within a decade.

Even more significantly, the Japanese electric utilities were enthusiastic about the breeder. Their keenness derived from their strong personal relationship with Walker Cisler of Detroit Edison. Cisler had played host to the first technical teams sent to the United States from Japan to study conventional generating technology in 1950. After that, Cisler regularly visited Japan; he was always welcomed at the top level, which included the Prime Minister. In return, Cisler's Detroit office was the first port of call for any senior Japanese utility man visiting America. They had complete faith in the technical ability of American industry and did not recognize how great the enthusiasm of Walker Cisler had become. He told them that a commercial breeder would be developed, and they believed him.

Technical teams in the Atomic Energy Research Institute were more skeptical about these claims, however. Despite Niwa's commanding position, an interim report of the Power Reactor Development Committee, issued in 1965, recommended only basic research and the construction of a small-scale experimental breeder of the type that already existed in France, Britain, and the United States. Nonetheless, the final report of the committee, released in 1966, recommended not just an experimental reactor of 50 Mw, but also a 300-Mw prototype, to be started well before the experimental reactor was complete. The combined pressure of Niwa's special advocacy, Toshiba's self-interest, and the commitment of the utilities, who were being asked to help finance the new program, had ensured that a more ambitious plan emerged.

Germany had its breeder advocates, too. One was Karl Wirtz, a 1930s graduate of the Kaiser Wilhelm Institute in Berlin who had become a key member of the German wartime bomb project led by Werner Heisenberg. In 1955, Wirtz became a leading figure in the Karlsruhe Nuclear Research Center. He formed a good working relationship with Karl Winnacker, who organized private industrial support for an all-German reactor construction program under Wirtz's direction. Winnacker supported Wirtz's program to build a small natural-uranium heavy-water reactor, but, even before this project was complete, Wirtz was looking for a new program to keep his team together. His attention focused on the breeder.

As the design work on the heavy-water reactor neared completion in late 1957, Wirtz launched his breeder studies. In 1959, he sent the head of his institute's theoretical division, Wolf Hafele, to Oak

Ridge to study breeders for a year. On his return, Hafele was appointed project manager of a breeder team and developed what was, in effect, an institute of his own, moving out of Wirtz's orbit.

Hafele was one of the earliest analysts of the importance of "Big Science" to an industrial nation, and the breeder project, he believed, was one of the most significant for German technological progress. He resisted moves by some Europeanists to combine research with the French—a move the fiercely independent French scientists also rejected. The French were particularly sensitive about German technical prowess and, as in other proposals for European nuclear research, they vetoed any pooling of resources until they were sure they were ahead. Consequently, each of the major European breeder projects would proceed separately.

When General Electric made its claim of building a commercial breeder in a decade, Hafele's team at Karlsruhe panicked, fearing German technology would be left behind. They urgently accelerated their original timetable, devising a new series of cost estimates to support the revision. The estimates were a model of intellectual dishonesty.

Designed to "prove" the economic viability of the breeder, the studies included a farrago of optimism and deceit. Nevertheless, the government agreed to jettison the original carefully planned reactor timetable and approved funds to build two full-scale breeder prototypes at Karlsruhe. One of them was canceled within a few years. Construction of the other, using sodium as a coolant, finally got under way in 1973 in the small town of Kalkar on the Lower Rhine.

In the competitive international fraternity of reactor teams, the sudden emergence of nuclear power in the middle sixties was no surprise; they had taken it for granted years before. When it came, they were not concerned with developing the full range of ancillary services, supplies, and equipment required to construct and operate a large number of reactors of a particular type: the discovery, mining, and milling of uranium; the development of nonnuclear equipment like generators, valves, and pumps; the means of reprocessing spent fuel and storing wastes. Scientists and engineers in each country with a nuclear power program still wanted to devote themselves solely to reactor systems. As a result, when disaster struck, they were totally unprepared.

18 "Transitory, Dangerous and Degrading"

I think that we will not be very successful in discouraging other powers from this course unless we show, by our example and conviction, that we regard nuclear armaments as a transitory, dangerous and degrading phase of the world's history, that before other nations can have a competing armament, there is a good chance that armaments will have become archaic.

—J. ROBERT OPPENHEIMER, 1966

The American government strongly disapproved of Israel's nuclear deal with France: a plutonium-producing reactor in the Middle East could only threaten the region's precarious status quo. Bought from the French in 1957 under a secret agreement, the reactor was built in the heart of the Negev desert at Dimona, halfway between Beersheba and Sodom. At first, the Israelis had said it was a textile plant, a cover story that gave rise to a range of jokes about the rag trade. But in the middle of 1960, U.S. intelligence identified it as a reactor and the Eisenhower administration put pressure on Israel to declare its true identity. David Ben-Gurion finally admitted to the Israeli parliament that it was a reactor; it was being built for "peaceful purposes," he said, under a contract signed with France three years earlier.

The reactor could not produce plutonium on its own: if they wanted to make a bomb, the Israelis had to acquire a special chemical plant, known as a reprocessing plant, to extract the plutonium from the reactor's spent fuel rods. It would all take several years; in the meantime, Israel wanted good relations with the United States, and Ben-Gurion sought a meeting with President Kennedy to patch up the bad feelings caused by the clandestine deal with the French. But during the early days of his administration, Kennedy demonstrated America's serious concern that Israel might be trying to make a bomb by studiously ignoring Ben-Gurion's diplomatic signals, a surprising snub for an American President elected with as small a majority as Kennedy. Finally, Ben-Gurion came to America without an invitation to the White House—ostensibly to accept an honorary degree. He still hoped that Jewish leaders could arrange a meeting, and, in May 1961, the two men did meet, informally, in New York's Waldorf-Astoria Hotel.

Kennedy was stern. Not only had the French deal been struck without consulting Washington, it had also been carried out without any of the standard "safeguard" agreements—such as a prior declaration of peaceful intentions and regular inspections of the reactor—the United States demanded of the countries that bought American reactors. The President asked for and received from Ben-Gurion access for American inspectors to the Dimona reactor site. There were to be spasmodic inspections—not as complete as American inspections of its own reactor exports, but Kennedy's technical advisers believed the inspections would be adequate as long as they knew precisely how much uranium Israel imported. For this information, they thought they could rely on their own covert Western Suppliers Group, which monitored uranium shipments worldwide.

But they were wrong: by 1968, both the inspections and the monitoring of uranium supplies had broken down. The same year, CIA analysts concluded that Israel had succeeded in making nuclear weapons. Kennedy's successor, Lyndon Johnson, attempted no reprimand, however. He would have found it difficult to get tough with an Israeli government still bathing in the glory of the Six-Day War, and a confrontation would have made it almost impossible for the American President to succeed in the last major policy initiative of his administration—a worldwide treaty against the proliferation of nuclear weapons.

After several years of negotiations, the treaty was at a delicate stage in 1968. Many of the countries that the U.S. government hoped would sign had still not made up their minds to do so; it was, then, the worst time to announce that the club of five weapons powers had become the club of six. When the head of the CIA told Johnson that Israel had the bomb, the President decided to do nothing. "Don't tell anyone else," he said.

Another six years went by before the CIA confirmed their earlier estimate. A new report, in 1974, concluded, "We believe that Israel already has produced nuclear weapons. Our judgment is based on Israeli acquisition of large quantities of uranium, partly by clandestine means; the ambiguous nature of Israeli efforts in the field of uranium enrichment; and Israel's large investments in a costly missile system designed to accommodate nuclear warheads."

This statement, explicit in its conclusion yet lacking definitive evidence, sums up the tangled web of information about the Israeli bomb project, a web that included more secret deals with the French for a reprocessing plant, the "theft" by Israeli secret agents of a shipload of uranium in the Mediterranean, and the suspected theft of bomb-grade enriched uranium from a factory in Pennsylvania.

The story shows that the Israeli bomb project never was a mere side effect of a "peaceful" nuclear program. From the beginning, the entire nuclear project had been aimed single-mindedly toward a military option. The French government sat back and watched them do it. With the sole exception of the Soviet nuclear assistance to China, the French involvement with the Israeli bomb project was—and remains—the most serious act of conscious atomic weapons proliferation in nuclear history.

Over the decade of the sixties, the case of Israel's nuclear potential was the most urgent example of the proliferation threat, and it led to a growing concern, first within the United States, and then internationally, about the increasing number of the "near-nuclear" nations, the "*n*th" countries, as they were known. Some countries, India for example, had reason to fear the military intentions of existing weapons powers. In India's case, it was China, which exploded its first atomic bomb in 1964. Others, like South Africa and Israel, could easily foresee circumstances in which their national security would be imperiled unless they had nuclear weapons. Still others, like Pakistan, Brazil, and Argentina, were deeply suspicious of the nuclear intentions of their neighbors.

In Israel's case, the deal with France had started as an alliance of adversity in the immediate wake of the 1956 Suez crisis, in which both countries had been on the "losing" side. It reflected the hostility of the Arab world to both nations: to the Israelis because of their existence and to the French because of their continued occupation of Algeria. Under the personal direction of Ben-Gurion, a series of military and economic ties were formed between the two nations. The secret deal to supply a reactor was signed in 1957.

Reaction in Israel at the time was confined to an inner circle of the Israeli Atomic Energy Commission. Seven of the eight members resigned. The seven were all prominent professors in Israel's leading science departments. No reasons were given for the resignations and no questions were asked in the Israeli Parliament. Only the chairman of the commission, Dr. Ernst Bergmann, an ardent supporter of Israel's pro-bomb lobby, remained at his desk. He held his position until 1966, but for almost nine years he had no commission to chair.

After the deal was made public in 1960, a group of leading Israeli academics and scientists—including two former members of the commission—formed an antibomb faction called the Committee for the Denuclearization of the Israeli-Arab Conflict. The very title of the committee suggests that the region had already been "nuclearized," and it is now known that the nuclear hardware the French gave the Israelis for their Dimona reactor included everything they needed to make a bomb.

The heavy-water reactor at Dimona had a capacity of 24 Mw and a potential plutonium output of 1.2 Hiroshima-sized twenty-kiloton bombs per year. But the crucial question, which remained unanswered until now, was whether the French also gave the Israelis a reprocessing plant capable of extracting the plutonium from the reactor's spent fuel rods. Francis Perrin, the scientific head of the French Commissariat à l'Energie Atomique from 1951 to 1970, has confirmed that a reprocessing plant was an indirect part of the original 1957 deal.

Perrin said, "We refused to sell Israel a chemical extraction plant, but they naturally learned the chemistry of plutonium with us and they asked a French company to help them in building an extraction plant. We just let this company do its work, without specifying exactly what they should have and what they should use. There was undoubtedly some help for the plutonium extraction plant in Israel." Perrin identified the company involved as Saint Gobain, the firm that built reprocessing plants for the French weapons program under license from the CEA. The French company supplied the blueprints and let the Israeli nuclear engineers work out the details for themselves. One reliable source said several engineers died in plutonium-related accidents before the project was complete. The French government simply turned a blind eye to this collaboration

It seems that Israel made the decision to go ahead with the bomb in the aftermath of the Six-Day War in June 1967. The hawks of the original 1957 Israeli bomb lobby—Shimon Peres at the Ministry of Defense, Moshe Dayan, the Chief of Staff, and Dr. Bergmann at the Atomic Energy Commission—had found themselves temporarily leaderless in 1963 when Ben-Gurion was overthrown as Prime Minister and replaced by Levi Eshkol. The bomb project was temporarily frozen under Eshkol, who had made a frustrating and ambiguous statement of intent on the bomb. "Israel," declared Eshkol, "will not be the first nation to introduce nuclear weapons into the Middle East," and this formula remains the official line long after it is generally assumed that Israel has a small arsenal of weapons.

But when Golda Meir became Prime Minister, the old bomb lobby of Dayan and Peres returned from semi-exile on the eve of the war. Victory gave Israel a brief moment when international pressure could be ignored. The bomb option was revived.

In 1968, the Israelis felt they needed to increase their supplies of natural uranium and do so without the U.S. monitors knowing it. Agents of Israel's secret intelligence service, Mossad, conceived a daring plan to steal the cargo of a merchant ship carrying 200 tons of natural uranium. The ship was a modest cargo vessel, the *Sheersberg* A, which had just changed its registration to the Liberian flag. After being loaded with 560 barrels of uranium at Antwerp, in the middle of November 1968, it headed out into the Atlantic bound for Genoa. Then, as far as the rest of the world knew, the ship disappeared. Instead of turning north to Genoa after entering the Mediterranean, it continued to sail east, heading for the waters between Cyprus and Turkey for a quiet rendezvous with an Israeli freighter.

At the beginning of December, several days after it had "disappeared," the ship docked at the Turkish Mediterranean port of Iskenderun. The crew reported that Naples had been the last port of call. Its holds were empty; the monitoring efforts of the secret suppliers' group had been successfully bypassed.

A possible second outside source for the fissile material used to fuel the Israeli bomb has never been proven, but some U.S. decision-makers secretly fear that the Israelis covertly obtained some highly enriched uranium from an atomic plant in the United States. It is known that 200 pounds of highly enriched uranium disappeared from a private American enrichment plant in Apollo, Pennsylvania. The Atomic Energy Commission detected the discrepancy during regular inspections of the plant in 1965, and despite a number of further inquiries by the Commission and the FBI, the missing uranium has never been traced. The suspicion that it somehow found its way to Israel is based on a series of guilt-by-association arguments. The plant was owned by the NUMEC Company, whose chairman is a Zionist; NUMEC had a business relationship with Israel, supplying nonsensitive nuclear materials.

If U.S. intelligence knows where the missing uranium has gone, they have not admitted it. The CIA's intelligence estimate that concluded Israel had the bomb referred only to the "ambiguity" of Israeli activity in uranium enrichment. And in a 1978 briefing, the CIA's top technical analyst downplayed the NUMEC connection. "The question of whether U-235 had been diverted from NUMEC was academic for the CIA because plutonium from the Dimona reactor was believed to be available. Therefore, from the CIA's point of view the diversion did not matter."

Whatever the truth of the NUMEC affair, the Israelis were also capable of making their own fissile material from the Dimona reactor. Since it had started up in 1964, it had been producing enough plutonium for about a bomb a year. All the Israelis needed at Dimona were plentiful supplies of natural uranium, a substance freely available on the international market.

By the mid-1960s the American government decided the time had come to establish a firm policy on the "*n*th" countries. It had taken a decade for the concept of nonproliferation to emerge; indeed, U.S. policy had been lax on the issue. In 1957, the Eisenhower administration had supported the French bomb effort and wanted to hand over weapons information of the kind then being supplied to the United Kingdom. This was stopped by congressional leaders, but the French did receive highly enriched uranium fuel from the United States for their nuclear submarine project.

Kennedy exhibited more concern over the problem, but his method for dealing with it was ineffectual. In the early 1960s, to prevent the emergence of any more independent nuclear forces in Europe, especially in Germany, the most feared proliferant of them all, Kennedy proposed the creation of a European "Multilateral Force," which would place U.S. weapons in Europe under an alliance framework of command. Denounced as an act of proliferation by

Russia and as militarily meaningless by France, the enthusiasm for the MLF was concentrated in the State Department rather than in the capitals of Europe. It was a good example of the confusion over how to tackle the proliferation problem.

After the Chinese explosion, a special Presidential Committee on Nuclear Proliferation was set up in late 1964 under the chairmanship of Roswell Gilpatric, the number-two man in the Pentagon. The report, presented to President Johnson in January 1965, was written in a tone of desperation.

"The spread of nuclear weapons," the report's first conclusion asserted,

> poses an increasingly grave threat to the security of the United States. New nuclear capabilities, however primitive and regardless of whether they are held by nations friendly to the United States, would add complexity and instability to the deterrent balance between the United States and the Soviet Union, aggravate suspicions and hostilities among neighboring states, place wasteful economic burdens on the aspirations of developing nations, impede the vital task of controlling and reducing weapons around the world and eventually constitute direct military threats to the United States.

The report's second conclusion was just as pointed. "The world is fast approaching a point of no return in the prospect of controlling the spread of nuclear weapons. Atomic power programs are placing within the hands of many nations much of the knowledge, equipment and material for making nuclear weapons."

The real turning point came in 1965 when America took the non-proliferation issue out of its general list of arms-control measures and began promoting it on its own. The new mood was made explicit in a dramatic, hard-hitting Senate speech Robert Kennedy gave in June 1965. In the spirit of the Gilpatric committee recommendations, Kennedy stressed the importance and urgency of the issue. "The need to halt the spread of nuclear weapons," he said, "must be the central priority of American policy." He identified Israel and India as the nations of most immediate concern, warning that they "already possess weapons-grade fissionable material and could fabricate an atomic device within a few months."

Kennedy was not alone in his views, although he was one of the few who freely expressed them. Israelis called the speech "The Kennedy Bomb."

Two months later, in August 1965, the United States produced the first draft of the Non-Proliferation Treaty. It called for prohibiting nuclear powers from assisting any nonnuclear state in the manufacture of nuclear weapons and binding every nation other than the five weapons powers to a renunciation of the possession of nuclear weapons.

The proposals were submitted to the Eighteen-Nation Disarmament Conference (ENDC) in Geneva. But because the plan had a provision for the possible future development of a European multilateral force, the Soviets reacted vehemently against it. It was, they believed, nothing more than a loophole for giving atomic bombs to their most feared adversary, Germany. It would take two years for the Americans to back down on this issue. In August 1967, after secret bilateral discussions, America dropped the MLF and the two superpowers presented identical treaty drafts. In the final burst of negotiations, a stern U.S.–U.S.S.R. front brushed aside all the doubts and criticisms of other nations.

But the issue of the multilateral force was not the only problem at the Geneva talks. The nonaligned nations introduced a wide range of proposals to make the treaty less discriminatory against themselves. They sought a credible formula by which the weapons states would guarantee the security of the nations denied the nuclear option. They wanted a binding commitment from the weapons states that they would never use nuclear weapons against a country that did not have them. More generally, they wanted a clause in the treaty that balanced the obligations of the weapons powers and the nonweapons powers. As they saw it, the nonweapons powers were making all the concessions by binding themselves to forgo nuclear weapons. Those countries that already had the bomb were not giving up anything. The dangers involved in the emergence of new weapons states—"horizontal proliferation" as the jargon called it—were not as significant as the intensification of the arms race among countries that already had the bomb—the "vertical proliferation." Therefore, they wanted tangible steps toward nuclear disarmament among the existing powers: an extension of the partial test ban into a comprehensive test ban extending to underground tests, a freeze on further production of fissile material, and a reduction in existing stockpiles. The fears of the discriminatory nature of the treaty would increase as the weapons powers at the Geneva disarmament talks brushed aside all these suggestions.

The first treaty draft that appeared in 1965 spelled out the basic aim in Articles One and Two: nuclear states should not give nonnuclear states the means to make bombs, and nonnuclear states should not accept such technology. But Article Three—the means of monitoring these provisions through safeguards—was left blank. An international safeguards system was still under discussion.

Two nations in particular harbored fears about the implications of any new safeguards system. They were Germany and Japan. Individuals in both countries were reluctant to accept the permanency of a second-rank status implied by the new division of the world into nuclear "haves" and "have nots." By the late sixties, Germany's postwar economic success had made it, once again, a more important power than either France or Britain, and even those Germans who believed their nation should never hold a nuclear arsenal objected to the symbolism inherent in not being allowed to acquire one. But there were other, and more significant, factors in the German and Japanese resistance. The first was the increased cost of implementing the NPT and the second was the treaty's implied threat to the breeder reactor because it used bomb-grade plutonium.

Nuclear power had only just emerged as a significant factor in electricity generation, and it seemed as though it could become even more important in the future. German and Japanese fears grew that the NPT system of safeguards would hurt them economically by increasing the costs of their nuclear industries above those of their competitors in the weapons states. The new safeguards being discussed at Geneva were designed to provide an early warning of a diversion of fissile material from a nuclear plant. Putting them into effect would call for changing the basic design of the plants and would therefore increase costs. The special inspections, cost-conscious German industrialists foresaw, would interrupt production at their plants and add to costs and inefficiencies. It was a problem that carried its own brand of discrimination: Germany's industrial competitors in the United States, Britain, and France would not have to bear these costs because they were already nuclear powers and would, therefore, be exempt from the safeguards system. But beyond concern over costs for existing light-water reactors, the Germans and Japanese were worried about what the NPT would do to their plans to develop energy independence through the breeder reactor. To be at all effective as a timely warning, inspection of the bomb-grade fuel would have to be continuous; there might even be a move to ban breeders altogether.

When, in early 1967, the penultimate draft of the NPT had emerged, alarm bells rang most strongly in Germany. "A second Yalta," declared ex-Chancellor Konrad Adenauer—referring back to the Morgenthau plan to raze German industry. Characteristically, Franz Joseph Strauss, then German Finance Minister, searched for a deeper level of psychosis in the collective folk memory. "A Versailles of cosmic grandeur," he declared. The fears were stimulated by scientists and industrialists closely associated with the German reprocessing and breeder reactor programs. A newspaper headline boldly announced: "The NPT—Against Germany's Breeder."

The NPT debate had erupted in the immediate wake of the German and Japanese decisions to pursue the breeder reactor. Existing safeguards, tied to the Atoms for Peace program, were based on the first generation of reactors—the ones that neither used nor produced, directly, bomb-grade material. Reprocessing plants and breeder reactors did use such material and no one knew how intensive the system of safeguards would be. The economic costs of safeguarding were incalculable; both the Germans and the Japanese in the breeder programs feared that, because of the sensitivity of the material, the safeguards would be especially burdensome.

The Germans, prodded by Karl Winnacker, who was keen to build the German reprocessing plant, and by his close colleague, Professor Karl Wirtz, who established the German breeder project at Karlsruhe, wanted to clarify the new safeguards system immediately. They wanted to use Germany's obvious bargaining power—the threat not to sign the NPT—to ensure that safeguards would have the lowest feasible economic burden, especially on the breeder and reprocessing plants. And they found international support for their position in Japan. Kiyonari Susumu, chairman of the Japanese committee reviewing the NPT draft and the leader of the Japanese team negotiating with the Americans, was also the key executive of the company building both the Japanese reprocessing plant and the breeder reactor. Like his German equivalents, Kiyonari believed that the treaty could make it impossible to develop the breeder. He also emphasized the need to define and simplify the safeguards system.

Before the negotiations were over, the efficacy of the international safeguards system had been compromised precisely because of the adverse cost implications for breeder reactors. Indeed, as later critics would emphasize, the system developed during these crucial years hardly seemed able to safeguard the most sensitive facilities in any meaningful sense of the term. The intervention by the cost-conscious Germans, with the Japanese hiding behind them, had been highly successful.

In the spring of 1967, Karl Wirtz went to Washington to negotiate the new safeguards at a technical level. He took with him a plan that would, if accepted, reduce the number of inspections in any nuclear plant by restricting them to what the Germans called "strategic points." These would be a carefully selected, but limited, number of points in a plant's operation at which the flow of uranium or plutonium could be monitored. If the basic idea was accepted, it would lower the anticipated costs of inspection. Wirtz was able to convince the Americans to write a provision into the NPT that any new system would be focused on certain "strategic points," even though no one had yet worked out where those points would be. That job was given to the project manager of the Karlsruhe breeder program, Wolf Hafele. The German government provided $5 million for the research, and the IAEA agreed to give it their official backing, thus ensuring that the results would eventually be fed into the international decision-making process.

The Germans were not trying to bypass inspections; they just wanted to make sure these were made as cheaply as possible. "The NPT," declared Hafele, "is fine in principle, but the devil is hidden in the detail." Hafele made no pretense of looking for the perfect solution to the problem of inspections. "It is better," he said, "to have an approximate solution to an exact problem than an exact solution to an approximate problem." Hafele began with a careful analysis of the whole fuel cycle—from the crude natural uranium ore to the fissile material produced in the spent fuel rods at the end. By measuring what went into a plant at each point and what came out—a regular bookkeeping exercise—the interference with the plant's operation would be kept to a minimum. The system could work quite easily for some nuclear plants, notably the most common form of electricity-producing plant, the light-water reactor. For reprocessing plants and breeder reactors, however, the system was less effective.

The main problem with the strategic points approach was how to ensure that everything that was supposed to be inside a sensitive plant at any one time was actually there. It was one thing to concentrate on inputs and outputs, but they could still "balance" in the short term even though something inside was missing at the time. Critics stressed the need for a "physical inventory" to compare with the "book inventory"—that is, to count everything inside a plant from time to time. Hafele continued for some years to look for a technical solution to this problem, but eventually he acknowledged that safeguards had to be open-ended; the inspector had to have the right to go into the plant and check. The issue of how often such intensive inspections would occur has never been resolved.

Despite overtones of prevention, "safeguards" have never referred to anything more than an early-warning system. Regular inspections would, in theory, detect missing uranium or plutonium, but the key question was always whether it could be detected in time to bring political pressure on a nation not to make a bomb. In 1970, a specially appointed panel of safeguards specialists at the International Atomic Energy Agency in Vienna believed that if a nation diverted material from a plutonium plant—that is, from either a reprocessing plant or a breeder reactor—it would probably be able to make a bomb within ten days. This period was called the "critical time." In theory, to "safeguard" a plutonium plant against a diversion of bomb-grade material, the plant would have to be inspected at least every ten days.

To cover the problem of how to tell if all the material that was inside the plants was actually there, the IAEA panel suggested what they called regular "wash-outs." All sensitive material would have to be taken out and measured. This, of course, meant production at the plant would have to come to a complete halt. And, according to the "critical time" for plutonium, the "wash-outs" would have to take place every ten days to be effective.

In the end, the IAEA safeguards secretariat—the implementing body—knew that such stringent measures would be politically impossible to adopt because of the economic costs involved, so they suggested that plutonium plants should have a minimum of four "wash-outs" a year. Even this would prove too much for the cost-conscious delegations to the IAEA safeguards committee that met in Vienna during 1970. The system that was eventually chosen would have no references to "wash-outs" or to "critical time." Not even the Americans, who regarded themselves as very tough on safeguards, would support the recommendations of the IAEA safeguards team. Still pursuing the old step-by-step diplomacy of the Atoms for Peace years, all the Americans wanted was confirmation of an old general formula of inspection that they had always asserted was adequate to their masters back in Washington. A safeguards system, they said, had to ensure that the inspectors had "continuity of knowledge" of the plant's operations. They were perfectly content to allow the key point—how many inspections would provide "continuity of knowledge"—to be decided on an ad hoc basis. The essential centerpiece of a system of safeguards—the concept of critical time—was set aside in favor of this loose formula, which amounted to nothing more than an agreement to agree later. It was their old pragmatic step-by-step diplomacy again—deferring the crunch points until they had to be faced. The Americans had never grown out of the promotional zeal of the Atoms for Peace years.

Devising the new safeguards system for the NPT was a marathon process. The IAEA committee met in eighty-two sessions over ten months in the squat, rambling headquarters of the IAEA in Vienna. Throughout the delicate negotiations, with groups of delegates meeting informally in the corridors and lounges of the IAEA building and at a nearby restaurant, the Americans managed to preserve a general reference to "continuity of knowledge" in the system, but there was an overwhelming need for a broad-based consensus. With nuclear power plants escalating throughout the world, all the delegates involved, including the Americans, wanted a definitive final fix by which they could all assert that the new treaty, enforced by safeguards, had solved the problem of proliferation once and for all. As in all such international negotiations, the committee had to act on the lowest common denominator principle; only the matters on which consensus was possible could be codified and anything else had to be papered over by general clauses that effectively deferred them.

In the end, the product of the committee's work—known in the trade as the "Blue Book"—was dominated by Hafele's "strategic points" approach. Yet everyone knew that this type of inspection would only detect "protracted diversions"—where small amounts were directed over a long period. It could never take care of the "abrupt diversions" of large quantities of material. The "Blue Book" talked only in general terms about this problem. It said that a sophisticated system of surveillance and containment would be needed to cover it. New and specialized equipment would be required—like radiation sensors, special locks and seals, and television cameras, all of which would have to be tamperproof. Such technology did not exist when the committee was discussing it; safeguards had, like waste disposal, been given low priority by the nuclear industry.

The Japanese safeguards expert, Imai Ryukichi, summed up the mood after the committee's work was done. "Looking back at the ten tedious months," he said, "what the committee ended up with is almost incredible. The fact that we have finally produced a single document was not necessarily what all of us thought was possible." He concluded, "A great deal was left to subsidiary arrangements for further implementation. . . . As an old expression has it, 'The work of the committee begins with the report and does not end with it.'"

Through it all, the American delegation hid behind their formula of "continuity of knowledge" and failed to press the key issue of "critical time." They assumed, as they had done for almost twenty years, that if a nation was found to have diverted bomb-grade material, U.S. bargaining power would be strong enough to dissuade it from making a bomb.

19

"Listen, World"

In August 1968, a decade after America had tested its early atomic bombs in the Pacific Marshall Islands, President Lyndon Johnson proudly announced that one of the tiny Marshall Islands called Bikini, which had been evacuated to allow bombs to be exploded on its coral atolls and in its shallow lagoons, was fit for human habitation again. Radiation levels, once so high that no human could live there, had dropped below the danger level, or so Johnson said.

The few hundred Bikinians were overjoyed. Since 1946, they and other Marshall Islanders had been shunted from one island to another at the will of the U.S. Atomic Energy Commission and the American military, who thought their beautiful, isolated, and sparsely populated islands were ideal places to let off bombs.

During their twenty-two-year exile, the Bikinians had suffered terrible social and physical hardships. They had been resettled on another island in the Marshall chain; its fishing grounds were inferior and its soil was less fertile, and they had suffered from malnutrition. In 1954, the eleven thousand Marshall Islanders sent joint protests about the tests and the disruptions of their lives to the United Nations. They struck a receptive chord. America governs the Marshall

Islands under a U.N. Trusteeship Agreement of 1947, and each year they are required to report on the welfare of the islanders to a U.N. Trusteeship Council. Responding to the protests, Russia and India proposed a resolution condemning the U.S. tests and calling on America to stop them, but it came to nothing. The council, dominated by nations friendly to America, rejected the resolution and the tests continued for another four years. Nevertheless, the islanders had scored a moral victory. They had revealed that fallout from some of the tests had caused radiation burns, nausea, and loss of hair—sure symptoms of radiation poisoning. They had also made a special plea to get their land back. "Land means a lot to the Marshallese," they had said; "take away their land and their spirits go also." Although the tests continued, the islanders' complaint triggered a widespread international protest that continued to grow, even after the tests ended in 1958.

In 1964, the islanders began to show the latent effects of radiation poisoning, such as thyroid tumors. This alarming development presented the U.S. government with a tricky public relations problem entirely of its own making: the AEC had consistently told the islanders there would be no long-term problems from the fallout. In the wider context, it also posed a serious challenge to the claims of the experts that there were "safe" levels of radiation exposure. As the number of thyroid cases increased over the next four years, it became clearer that the so-called experts had not known what they were talking about. Johnson's announcement that the Bikini Atoll, on which no fewer than twenty-eight nuclear bombs had been exploded, was now safe was precisely the kind of symbolic gesture the experts needed to reaffirm their case. But what appeared to be a public relations triumph soon turned sour.

The Bikinians could not return at once: the island had to be prepared. Topsoil was removed and 50,000 new coconut trees planted. U.S. officials tried hard to restore the island to its previous state, but some things would never be the same. The new vegetation was coarse, marine life in the lagoon had changed, and the undergrowth was littered with the twisted pieces of metal of the destroyed base camp and test facilities.

Slowly over the next seven years, about one hundred Bikinians returned home. In 1975, U.S. officials carried out their first tests to check that the radiation levels remained acceptable. The result was a shock. The radiation report said starkly, "All living patterns involving Bikini Island exceed federal [radiation] guidelines." To prevent adverse reaction, the islanders were not shown the report; in fact, it was not released for another two years. They were only warned to restrict their diet; in particular, they were told not to eat the island's local delicacy, the tree-climbing coconut crab, which had apparently accumulated large quantities of Strontium-90. Over the next two years, the diet restrictions increased, and the Bikinians were advised not to let their children wander into the island's interior, where radiation contamination was found to be even higher. They were told they should eat only one of the island's coconuts a day—although coconuts were the staple diet; and, finally, coconuts were banned, too. Food had to be shipped in.

Far from finding the island fit for habitation, the U.S. researchers discovered that food grown there contained significant quantities of plutonium, a fallout product that, since the tests ended, had been slowly disintegrating through its half-life of twenty-four thousand years. Because they were eating plutonium (which is not as dangerous as breathing it—that is known to cause lung cancer), the Bikinians as a group became a scientific curiosity. In 1976, a U.S. researcher recorded bizarrely that Bikini "is possibly the best available source of data for evaluating the transfer of plutonium across the gut wall after being incorporated into biological systems."

By the beginning of 1978, medical examinations of Bikinians showed them to have internal radiation levels that were nearly twice the U.S. maximum safety standard. In May of that year, the U.S. government was forced to announce that the Bikinians would again be evacuated; the triumphant return, proclaimed by Johnson a decade earlier, had been a dismal failure. The brash radio announcement of the first Bikini test had begun with the words, "Listen, world—this is Crossroads"; twenty-two years later it had a chilling echo.

President Johnson had been able to make his Bikini announcement in 1968 because the self-confidence of the radiation experts—that they did know what they were doing—was still running strong. They had weathered a stormy international debate over fallout in the 1950s; then, scientists of apparently equal qualification had produced opposing views over how much radiation was harmful. A confused public, not knowing whom to believe, had withdrawn, leaving the experts to agree to differ.

By the end of the seventies, however, the public image of the experts had been seriously devalued; it had become abundantly clear that they had not known what they were doing. Citizens affected by radiation because they worked with it, lived near it, or were part of the growing environmental movement were angry that they had been deceived.

The new sense of alarm called for new standards of exposure and intensified the search for a "threshold dose"—some definitive proof that exposure to radiation below a certain level did no harm. In the 1950s, the geneticists, fearing mutations in future generations, had asserted that radiation was dangerous no matter what the dose and they had forced the standards down to a level that many believed too low. During the 1960s, further experiments and observation of those who had been exposed showed the genetic effects were not as great as had been feared. But the equally horrific problem of the latent effects of radiation poisoning received added attention as the number of radiation-linked cancers increased.

No single event caused the erosion of public confidence in the experts. It was the result of several seemingly unrelated incidents that surfaced wherever radioactivity was found. Those affected included the uranium miners, the victims of fallout from weapon testing in the Pacific and near the Nevada test site, hospital patients treated with X rays, and workers in the atomic factories and nuclear power plants.

The nuclear establishment, a closed but visible elite, presented its critics with a single and easily identifiable target. Small groups of concerned citizens, especially scientists, began to question specific aspects of nuclear expertise. At first formed on a local basis, these groups quickly spread nationwide. Before long, there was a worldwide set of antinuclear groups. As these groups successfully prised more information from the secret files of institutions like the U.S. AEC, it became clearer how far the nuclear community had been prepared to go to protect itself—and how far it might be prepared to go in the future.

Nuclear zealots faced a barrage of allegations of cover-up, lies, deceit, and willful suppression of important evidence. Many of these allegations were true. Open hostility broke out between those scientists who worked for the nuclear establishment and those who worked outside it, especially when it was discovered that some on the inside had broken the scientists' tribal code of honesty in research and agreed to suppress data.

America's open government system provided the best examples of this degenerate behavior, but there is every reason to suppose that it happened in each country that was supporting a nuclear program for either warlike or peaceful uses. At the U.S. AEC, the hardening of the institutional arteries gave way to a phase of total intellectual corruption. Expertise was no longer regarded as a process of free inquiry and debate; scientific results were subjected to a test of loyalty to the institution before they were checked for accuracy. Information both inside and outside the AEC became acceptable only if it enhanced the commission's narrow institutional goals—especially the goal of expanding nuclear power. It was only a question of time before some of the scientists on the inside found they could no longer abide the dishonesty and preferred to defect. The decade of the sixties saw the first of the radiation experts become whistleblowers of the nuclear age.

The first of several debates that eventually breached the self-confident image of the nuclear establishment actually started in the late fifties over a question of fallout, not from bombs, but from an accident in a nuclear power plant. The 1957 reactor fire at Windscale had first drawn attention to the radioactive isotope Iodine-131. A short-lived substance with a half-life of eight days, Iodine-131, when ingested, concentrates in the thyroid gland, where it can cause cancers.

The amount of Iodine-131 released from the Windscale plant was tiny compared with the amount released from the bombs exploded at the Nevada test site, but the British found the levels in local milk samples so high that they destroyed two million liters.

Two years after the Windscale accident, in 1959, the U.S. AEC became concerned over a report that significant amounts of Iodine-131 had been found in milk samples in the St. Louis area, over one thousand miles due east of the Nevada test site. One AEC analyst reported, "All things considered, it seems to me more difficult to conclude that levels of Iodine-131 in milk comparable to those measured following Windscale did not occur in many places following several of the early tests than it is to conclude that they did occur."

By 1962, it had become clear that one of the most important dangers of the Nevada bombs—Iodine-131—had been totally ignored in the early years of testing: the AEC monitoring teams around the Nevada test site had not looked for Iodine-131; they had measured the external gamma radiation dose. Armed with the new evidence from Windscale and St. Louis, researchers tried to determine the relationship, if any, between the external gamma dose and the Iodine-131 content. These retrospective calculations were disturbing; they suggested that children who had drunk the milk from cows grazing on the ranches of southern Utah might have received very large doses indeed.

Surprising concentrations of radioactivity were found. For example, fish had Iodine-131 concentrations thousands of times greater than the water they lived in, and some birds had concentrations tens of thousands of times the amount measured in the air they breathed. But calculating average doses was a real problem; radioactive isotopes do not come down to earth in an even pattern from a fallout cloud. The AEC was faced with the same problem as the statistician who drowned in a river of average depth of three feet.

Fearing a public outcry over Iodine-131, the AEC engaged in some of the clearest examples of suppressing information. For instance, Edward Weiss, who in 1967 produced the first study on Iodine-131 and thyroid cancers in the population living around the test site, emphasized that his "most interesting" fact was the sudden appearance of thyroid cancers in children under fifteen. No cancers had previously been recorded in that age group. The AEC subjected the Weiss report to a vigorous in-house sanitizing review, deliberately delaying it; when it was finally published, the unusual upsurge of thyroid cases in children under fifteen was effectively concealed by extending the age bracket to include all those up to the age of twenty. The child thyroids became a statistical blip. During the same period, a study by the U.S. Public Health Service on an increase in leukemia deaths was not published at all.

It was not until the late 1970s, when many of the earlier AEC discussions on the health aspects were declassified, that the actual adverse health effects of the Nevada testing became clear. By that stage, the dishonesty of the official nuclear community was well known, and a second breach in its armor had already been made. This involved the first step in the nuclear fuel cycle: the mining of uranium. In the early 1960s, between 10 and 20 percent of the 6,000 U.S.

miners who had been employed in the postwar American uranium mining boom discovered they were going to die from lung cancer. No one now doubts that the prime cause of their cancers was lack of ventilation in mines, which meant the miners inhaled harmful quantities of the radioactive products called "radon daughters."

In any deposit of uranium in the earth's crust, the natural decay of the mineral through radioactivity produces a radioactive gas called radon. When inhaled, the gas produces its own decay products called "radon daughters." They emit alpha radiation. If these particles become lodged in lung tissue, they can, over a period of several years, cause cancers. The miners were told not to worry unduly about the gas. Their employers advised them, erroneously, that an hour after they had finished work all the radioactivity would have cleared from their lungs. The U.S. government was no more helpful. Regulation of the mines was so lax that surveys of radon gas present in the mines, most of which were located on the Colorado Plateau, did not begin until 1949, and the "radon daughter" particles were not measured at all until 1951. Even then, the surveys were used purely for information purposes, not to devise a measure of health control. That would not emerge until the 1960s when the first of the deaths occurred.

The shocking fact is that the dangers of uranium mining had been known for more than a hundred years before the Colorado mines were opened. Uranium and pitchblende miners in the Erz Mountains on the German-Czech border had a long history of lung complaints. In 1879, lung cancer had been identified as the main cause of their problems and of their subsequent deaths. By the 1920s, the source of the problem, the radon gas, had been identified. By 1939, a German study showed that lung cancers among the miners were thirty times the national average. The disease even had a local name; the miners called it *Bergkrankheit*, or mountain sickness.

In the late 1940s, the European mines operated under enforced standards of mine ventilation; but when the U.S. mines were opened, the AEC ignored the standards—despite studies revealing radon gas levels in uranium mines in Colorado equal to European mines. As a monopoly buyer of uranium in those early days after the war, the AEC had an interest in keeping the costs as low as possible; enforcing radiation standards would have increased them.

Fifteen years after their first exposure to radon gas, some of the miners began to die. Many of them were Navajo Indians, who provided most of the labor in the mines. To medical researchers, the Navajos became a scientific curiosity, because the Indians had no history of lung cancers. Death usually occurred within a year after they had been diagnosed. A local doctor recalled, "It got so that I did not need to wait for the tests. They would come—the wife and six kids were always there, too—and they would say that they had been spitting blood. I would ask and they would answer that they had been in the mines. Then I didn't need to wait to make the diagnosis, really; I already knew."

Imposing meaningful standards in the mines raised the perennial problem of setting radiation standards for any occupation; no one really knew how to estimate the harmful effects of inhaling radioactive material. In the case of radon daughters, it depended on the rate of absorption of the particles by the lung tissue. The health physicists called their new research "lung dynamics." A good example of how complicated the calculations could become was their discovery that the amount of radon daughters inhaled from a given amount of radon gas in the mines could vary by as much as six times.

As the number of deaths continued to grow, the radiation experts who had made the "worst case" assumptions were proved to be right. It was not until 1967, however, that health standards were enforced, although vigorous opposition from the mining industry and, particularly, from the nuclear zealots testifying before hearings of the Joint Committee on Atomic Energy continued.*

Disagreement between the radiation experts would be resolved only with time and prolonged studies of scientifically controlled groups. For the best samples, the researchers looked to the tragedies of the past, like the radium dial painters, the survivors of Hiroshima and Nagasaki, and the British victims of a spinal disease called spondolytis, all of whom had been given large doses of supposedly therapeutic X rays.

*One of the nuclear community's star witnesses at those hearings was Professor Robley Evans, of the Massachusetts Institute of Technology. He was a leading proponent of the "threshold theory"—the view that a level of radiation existed below which no harm could come to the human body. The critics were quick to attack; although there was still plenty of scope for disagreement, the majority of radiation experts no longer subscribed to the "threshold theory." Also, doubts were expressed as to Evans's objectivity. He had received several million dollars from the AEC in support of his research and had also acted as a consultant for a uranium mining company. In the end, instead of protecting the AEC's position, Evans's testimony added to the weight of evidence against it.

Both the A-bomb victims and the spondolytics began to show significant increases in leukemia deaths in the 1950s. During the 1960s, the types of cancer that take longer to form—like those in the lung, breast, stomach, and bone—began to show up. In the period 1965–70, after some largely erratic results, both groups showed a sharp increase in cancers. At the same time, scientifically controlled studies on animals confirmed the human results; they tended to show what most experts had long assumed: the effects of radiation are directly proportional to dose. Exposure to 200r would cause twice as many cancers as exposure to 100r.

As the studies continued, the evidence of some effects at low dose rates increased and convinced the researchers finally to abandon any thought of a "threshold dose," and a new sense of anxiety emerged about the effects of radiation at very low levels. For the first time, the use of clinical X rays became a real focus of concern.

In the past, radiologists had escaped regulation because they were considered sufficiently qualified to be their own judges in the use of X rays. Evidence that X rays might be harmful in small doses had been dismissed out of hand. In 1956, for example, a British epidemiologist named Alice Stewart had studied the effect of X rays on human fetuses. Stewart had found that an apparently harmless X ray examination of a pregnant mother could cause a significant increase in the possibility of leukemia in the child. No one took much notice; the experts preferred to reject Stewart's analysis as statistically irrelevant. But in the early 1960s, her fears were dramatically confirmed in a study of almost three-quarters of a million births in thirty-seven maternity hospitals in the northeastern United States. The children of mothers who had been X-rayed during pregnancy were found to have 40 percent more leukemias and other types of cancer than those who had not been X-rayed.

While there were still some studies that did not support the Stewart findings, a consensus eventually emerged that the fetus was particularly sensitive to radiation. Radiologists stopped using X rays on pregnant women, but clinical X rays remained the largest source of ionizing radiation to the general public, and some health physicists wanted to cut back their use severely—by as much as 90 percent. One of the leaders of this group was an American named Karl Morgan. In 1967, he made the alarming claim that the existing rate of X rays for diagnostic purposes was causing anywhere between 3,500 and

29,000 deaths in the United States each year. His calculation sparked a furious debate among the health physicists.

What began as a scientific disagreement soon turned into a political one. These dramatic assertions about the effects of radiation at low levels hit at the very core of the future of atomic energy: the huge new power reactors of 1,000 Mw then being planned. In operation, each would contain a massive fifteen billion curies of radiation—a curie is a measurement of the amount of radioactivity associated with one gram of radium. If the program of nuclear power was to continue as planned, the AEC knew that the public needed reassurance about radiation, not the alarmist talk that scientists like Morgan were producing. One scientist in particular, Ernest Sternglass, had claimed that 400,000 American babies had already died from the radiation produced from the fallout from the weapons testing in Nevada. The AEC decided to retaliate.

A young Berkeley graduate named Arthur Tamplin, who had been engaged by the AEC to study fallout from the weapons test, was assigned the task of debunking Sternglass. Working out of the Lawrence Livermore Laboratory in California, Tamplin, as expected, produced results that challenged the Sternglass figures—but not in quite the way the AEC had hoped. In a paper entitled "A Criticism of the Sternglass Articles on Fetal and Infant Mortality," Tamplin reported that "due to differing social-economic conditions" ignored by Sternglass, he had made a serious error. "At the same time," said Tamplin, "the existing experimental data indicates that fallout radiation probably did contribute to infant and fetal mortality by way of lethal mutations. . . . The effect is most likely at least a factor of 100 smaller than he proposes." That still left 4,000 dead fetuses and infants—and a huge public relations problem for the AEC.

Tamplin received urgent phone calls and letters from AEC headquarters demanding that he drop the revised estimate of 4,000 deaths from the draft report. Only then would the AEC agree to publication. Tamplin turned for support to his department head at Livermore, John Gofman, who was already working with Tamplin on a long-term study of the possible health effects of a large-scale radiation release from a nationwide nuclear power system.

Gofman had grown increasingly wary of the AEC's willingness to suppress information that was not in its interests. In 1963, he had been appointed as one of the five-member review board to check the disturbing conclusions of Edward Weiss in his study of the effects of Iodine-131 from the Nevada fallout. Gofman had supported Weiss's research—as had all members of the review panel—but when the published work was doctored to make the results look less startling, Gofman was determined never again to be a part of that sort of misrepresentation.

When Tamplin insisted on publishing his own estimate of the 4,000 deaths, Gofman gave him full support, and the two scientists took their step toward excommunication from the nuclear community. The community reacted venomously to their insubordination. As one of their colleagues remarked, "If the Atomic Energy Commission pays to support your research, why do you criticize radiation as a hazard."

The intellectual dishonesty of the AEC led it to become almost totalitarian in its power. Tamplin was subjected to a growing amount of petty, yet significant, harassment. He found his research funds cut off and his staff of twelve reduced to one. He was refused travel allowances to present his case at public meetings—a normal practice in the past—and whenever he did attend such meetings, he had to forfeit a day's pay.

The AEC's campaign of vilification eventually included Gofman as well. The commission was no happier with the results of the Gofman-Tamplin study of the health effects of a nationwide nuclear power system. Gofman and Tamplin had concluded that radiation was a far more serious hazard than had previously been suspected, that the United States could expect twenty times more deaths from radiation-induced cancer and leukemia than had previously been believed, and that genetic damage had also been underestimated. They considered that the existing public radiation protection standard—then 0.17r per year—was too high.

When the two scientists pushed for a lowering of that figure, they ran into vigorous opposition. Gofman wrote later:

> Dr. Michael May, then director of the Lawrence Livermore Laboratory where I worked, visited me in my office. Clearly he had experienced intense pressure from the AEC. In all my experiences with Dr. May I had found him to be a fine person and a first class scientist.
>
> "Jack," he said, "I defend absolutely your right, in fact your

> duty, to calculate that a certain amount of radiation will cause 32,000 extra deaths per year from cancer." But to my disappointment, he then asked, "What makes you think that 32,000 would be too many?" I must presume he was thinking in terms of the hoped-for benefits of nuclear power. . . . technology without a human face.
>
> "Mike," I said, "the reason is very simple. If I find myself thinking that 32,000 cancer deaths per year is not too many, I'll dust off my medical diploma, take it back to the Dean of the Medical School where I graduated, hand the diploma to the dean and say, 'I don't deserve this.' "

Reluctantly, under intense pressure, both Gofman and Tamplin resigned from their posts at the Livermore Laboratory. Neither man was the readily recognizable material of political martyrdom, but both had a strong sense of professional integrity, and both were primarily concerned with the protection of individuals. The intensity of the official hostility toward them, with its strong overtone of political oppression, turned them, in the years that followed, into vocal opponents of the nuclear community.

PART FOUR

The Seventies

20

The Boom Years

The sudden rise in the price of oil during the 1970s presented the world's electric utilities—those gray, monolithic organizations run by faceless men—with a new set of problems for which many were hopelessly unprepared. They had invested heavily in oil-fired generating plants and had been lulled into a sense of false security by the cheapness of the fuel. Some of the utility leaders had already committed themselves to nuclear power in the sixties; at the beginning of the seventies, almost all would turn to it with a new faith and enthusiasm.

Their spending spree was a huge commercial and technical gamble; even with the increases in the price of oil, they had no reason to be so confident that nuclear power would be cheaper than conventional plants. Nor, indeed, did they have any experience of building or operating the huge plants they were ordering. Until 1969, for example, the only operating experience with light-water reactors was on an early 200-Mw model. And while the first reactors of 600-Mw capacity went into operation in 1969, the new orders were for plants of 1,000 Mw and more. It was an extraordinary act of faith in the American reactor system—as if the traditional progressive image of all things nuclear was, in itself, a guarantee of perfection.

Between 1970 and 1973, worldwide reactor orders trebled from a total of 20,000 Mw to 60,000 Mw; it was a signal that two decades of energy abundance based on oil were at an end. During the 1950s and 1960s, huge international reserves of oil, mostly from the Middle East, had flowed to Europe, America, and Japan. During the same period, the international oil companies had produced fuel oil at artificially low costs and succeeded in ousting coal as the dominant fuel in industrial boilers and, especially, in electricity production—the sector that had represented the biggest single market for coal. America had been partially insulated from the oil takeover by U.S. oil import restrictions reserving 88 percent of the world's largest market for domestic production. Utilities elsewhere had ordered oil-fired generating plants under the temporary illusion of low-cost fuel. In some countries, like Germany and Britain, pressure from the coal companies and, particularly, from the miners' unions had slowed down this oil takeover. In others, like France and Japan, the victory for oil had been total.

When the electric utilities turned to nuclear power at the beginning of the seventies, one nation, Japan, took the lead. There were two reasons. First, the country had experienced twenty years of double-digit growth rates. In the early 1950s, its economy was only one-third the size of either the British or the French, but by the late 1970s it was bigger than both combined. Its gross national product passed Britain's in 1962, France's in 1963, and Germany's in 1966. A decade later, its industrial output overtook that of the Soviet Union. The Japanese economy then stood as the second largest in the world, exceeded only by the United States. Electricity demand had grown as rapidly as the economy, and one of the country's nine private electric utilities, the Tokyo Electric Power Company—which serves the capital and its industrial region—was, by 1970, the largest privately owned utility in the world. The doubling of oil prices between 1970 and 1973 and another doubling during October 1973, the month of the Yom Kippur War, made rapid diversification of Japan's sources of energy essential if the country was to ensure the security of its new industrial capacity: 85 percent of its total energy came from abroad.

The second reason for the lead in nuclear power was that Japan's rapid industrial advance had created the worst pollution problems and the most intense urban congestion of any of the advanced industrial nations. Nuclear power, it was hoped, would eliminate the pollution. In the late 1960s the technical men in charge of the utility decisions believed that nuclear power was "clean." Statements to this effect were made by all the utility leaders—from the Soviet Minister of Power Stations to the presidents of American Electric Power and Electricité de France; and in Japan, the belching smokestacks of the fossil fuel–fired generating plants had become the primary targets of the fledgling environmental movement. The utility men, horrified by the cost estimates of the equipment needed to control this pollution, saw nuclear power, despite its huge capital costs, as a panacea.

As it had been in America, the growth of the Japanese nuclear power program owed a great deal to the fears of private utility managers, who wanted to ensure that new technology did not slip into public hands. There was to be no TVA of Japan.

Two utility men supervised the Japanese drive toward nuclear power, and their corporate rivalry added to the surge in nuclear growth. Both were forceful advocates of private ownership. They were Kikawada Kazutaka, who became the president of Tokyo Electric, and Ashihara Yoshishige, who became president of the Kansai Electric Company, the utility that serves the industrial region around Osaka in southern Japan. Both men were hard-driving, dominant executives in a world of low-key collegial decision making. Ashihara was known inside Kansai Electric as "the Emperor"; Kikawada had survived a series of purges and stumbling blocks in his successful drive to reach the top of Tokyo Electric. Each had established special administrative systems that effectively centralized decision making in their hands. Both were self-consciously progressive in their social and political attitudes, believing that big business could be more socially responsible than the survival-of-the-fittest, no-holds-barred tradition of the Japanese corporate world. If they differed in approach, it was only that Ashihara, a sober, cunning merchant typical of the Osaka business community, was less philosophical about his social views than Kikawada. But both felt that scientific progress was one of the major positive social contributions of industry, and they shared the assumption that nuclear power was inherently progressive.

The two men first became associated during the late 1940s when they joined forces to defeat the threat of public power. In 1939, the Japanese militarist state had created a single nationalized power generation company. Hundreds of different distribution companies had been organized into nine regional groups, each distributing electricity for a nationalized power-producing monopoly. At the end of the war, when the American occupation authorities reconstructed Japan, Kikawada urged that the power-generating stations of the state monopoly be allocated to the nine regional distribution companies. Others believed that a single state monopoly on the model of the English Central Electricity Generating Board or the recently created Electricité de France should be set up, but Kikawada formed a strong group of Young Turks in the distribution companies—including Ashihara—to oppose the single corporation.

The political battle on this issue took several years to resolve. The American occupation authorities were convinced that free enterprise could become a pillar of a democratic Japan. Kikawada and his group worked out a detailed plan that called for the creation of a nine-company system to produce and distribute electric power. The plan was readily accepted by the Americans. One part of the structure, however, remained in public hands: the construction of some generating plants. Because the capital costs involved were so high, none of the utilities could bear them on their own. But the construction group remained a potential basis for the spread of public power, and in the late fifties this threat seemed about to be realized as it staked a claim to build the first nuclear plant. In reply, the nine private companies proposed a consortium, the Japanese Atomic Power Company (JAPCO), and with the decisive backing of the wheeler-dealer Minister for Atomic Energy, Shoriki Matsutaro, the government decided that JAPCO could build the plant. The private companies had won. The first contract, in 1957, was with Britain for a gas-cooled reactor—the deal set up almost single-handedly by Shoriki. In the ensuing years, the Japanese experience with their reactor was not good. By 1963, when General Electric made its so-called breakthrough with a BWR at Oyster Creek, JAPCO was faced with institutional lethargy. To overcome this, it needed a new project. The word went out that it was in the market to buy more reactors.

The lobbying from GE and Westinghouse was intense. At the end of 1965, JAPCO opted for what appeared, at the time at least, to be the proven design of the Oyster Creek BWR. It was a choice that, within a year, would lead Kikawada's Tokyo Electric and Ashihara's Kansai Electric to put in their own precipitous and rival orders, launching themselves on a definite nuclear road. Their deals were made for special traditional reasons peculiar to the Japanese utilities; they were made without real economic, or even strategic, assessments.

Japanese electric utilities had always tied themselves to particular manufacturers; they preferred the mutual trust of an established relationship. The tied markets dated back fifty years to the original orders for coal- and oil-fired generators. Tokyo Electric had always bought its power plants from the GE licensee in Japan. Kansai Electric, on the other hand, had always bought its plants from the Westinghouse licensee. The sale to JAPCO, a joint venture of all the utilities, upset this long tradition of individual arrangements.

The fact that JAPCO had bought a BWR plant from GE gave Tokyo Electric an advantage over its rival, Kansai Electric. In undertaking Japan's first light-water reactor, Japanese crews would be trained in the construction and operation of a General Electric BWR. This was the type of plant Tokyo Electric was bound by its traditional corporate ties to purchase if and when it decided to pursue the nuclear route on its own. Kansai Electric, on the other hand, was bound to buy Westinghouse's PWR and would have to struggle to catch up.

Ashihara, all too conscious of the corporate rivalry between his company and Kikawada's, was determined to retain Kansai's reputation for superior technical leadership and effective management. Ashihara simply could not accept the technical back seat that JAPCO's choice of GE's BWR involved. Westinghouse, recently beaten in a string of international competitive bids in Italy, Germany, and India, was ready to offer him generous terms on its PWR. In 1966, Ashihara bought the first PWR in Japan. Not to be outdone, Kikawada immediately ordered a BWR from GE. By the middle of 1966, with these two hasty "prestige" orders, the Japanese utilities had started on a course of nuclear development from which they could not turn back.

The enthusiasm of the Japanese utilities was such that, on their advice, the Japanese Atomic Energy Commission substantially upgraded its Long-Term Plan. In 1961, the plan had predicted an installed capacity of 6,000–8,000 Mw by 1985. In 1967, the new Long-Term Plan projected 30,000–40,000 Mw by that date. Then, in 1970, the forecast was increased again to 60,000 Mw. After the prestige orders of 1966, the utilities were encouraged to press ahead, and the government offered substantial subsidies for them to do so—

including special depreciation allowances and tax cuts for nuclear investments. The nuclear teams in the two big utilities were completely confident that nothing could go wrong. Before even one plant was finished, the origin of the new commitment to nuclear power—a battle for institutional prestige—had been forgotten as a stream of premature self-serving cost estimates and demand projections hyped the advantages of nuclear power even further.

If the Japanese utilities took the early lead, the new wave of infatuation with nuclear power was nowhere more evident than in the French state-owned Electricité de France (EDF). The nationwide monopoly had been one of the great success stories of the postwar nationalizations. EDF was universally respected, indeed admired. It had kept electricity costs down and, over two decades, conducted a large and successful investment program that served as both an example and a spur to the whole of French industry. It was a proud institution, confident of its expertise and of its role. By 1971, the twenty-fifth anniversary of its formation, EDF set about celebrating its past success and reasserting its claims for the future. Promotional pamphlets for the anniversary proclaimed that "the electric revolution was only just beginning," and that "modern man can still discover new reasons to be astounded by the power of electricity." Such an assertive drive quickly produced a single slogan, one that was indelibly impressed on the French public: "Tout électrique! Tout nucléaire!" ("All electric! All nuclear!") became the motto of the full-scale advertising offensive EDF launched in its anniversary year. The technicians of EDF had totally accepted the view that nuclear power was inherently progressive.

The real objective of the campaign was decidedly mundane. At the time, EDF was trying to sell electric home heaters. Fully believing the assumption of the day that nuclear was "cleaner" than coal or oil, they hoped the image of electricity as something modern and clean would cause French consumers to install more home heaters and therefore expand EDF's market. In two decades of success, EDF had learned that the faster the demand for electricity grew, the quicker it could install bigger and more efficient generating equipment. Costs and prices would go down; demand would go up. Like every utility in the world, EDF found that each time it brought a large new plant into operation, it could be made full use of only if demand grew rapidly enough to take up the new capacity. The slogan "all electric" had been used by many utilities in precisely the same way for the same reasons. The phrase "all nuclear" was an EDF addition.

Yet at the time it was first used in France, the slogan "Tout électrique! Tout nucléaire!" was misleading: EDF still had a substantial commitment to oil-fired stations. Two years would pass before a significant nuclear investment program was launched, and another year before EDF really did go "all nuclear." In that time, the desire to maximize electricity demand was transformed from a policy expressing EDF's institutional interests into a national commitment to replace imported fuel with domestic atomic energy. The campaign to get French consumers to do such things as substituting electric heaters for oil was intensified. EDF no longer saw itself as simply a producer of electric power at the lowest feasible cost, but as a major patriotic force: it could help the French nation maximize its flexibility in its relations with other nations and also assist in developing a major new industry that could become a key sector of French industrial strength for a century.

The man who supervised this transition, and its most eloquent public defender, was the director-general of the EDF, Marcel Boiteux. His appointment had been something of an oddity. EDF had always been the preserve of the engineer-administrators of the conservative Ecole Polytechnique, and particularly of the Corps des Ponts et Chaussees. Boiteux was not a Polytechnician; he was a graduate of the Ecole Normale Supérieure, France's greatest seat of liberal education. He had, however, been teaching graduate students from the Corps des Ponts et Chaussées for some years, and was considered an honorary Polytechnician. Certainly, his rigorously mathematical approach to economic decisions gave him the stamp of logic and certainty normally associated with the Polytechnicians.

As a protégé of Pierre Massé, EDF's president, Boiteux was the leading theorist of the talented group of EDF economists. He had won international recognition for his contributions to the rational decision-making techniques used to plan EDF's investments and prices. By the time EDF launched its major nuclear program under Boiteux's direction, it was easy to project future cost increases for oil and coal that "proved" the economic competitiveness of nuclear power—even using the most pessimistic estimates of nuclear costs. Boiteux's total commitment to economic rationality led him inexorably to the conclusion that nuclear power was the answer.

The dramatic successes of the previous two decades of electricity development had resulted in well-defined production and distribution systems: a grid based on large, centralized power-generating stations. This established system made the utility men blind to options of alternative sources of energy: once the hydroelectric sites were exhausted and the fossil fuels became too expensive, only large nuclear power stations fitted into the established systems. For decades EDF, like other utilities, had been interested only in technical developments that could become part of its historically successful approach; other techniques, like small sources of solar power, for example, did not fit into a national grid pattern and were seen not as possible alternatives but as rivals. They were not part of EDF's charter and, like all big institutions, EDF was concerned only with developments that reinforced and expanded its own role.

EDF had not been an early advocate of nuclear power, however. Before the reactor takeoff in the 1970s, EDF had resisted pressure from the Commissariat à l'Energie Atomique (CEA) to launch a large-scale nuclear program. It had also fought a successful battle against the French nuclear bureaucracy over the choice of a reactor system, resisting, on the basis of cost, the CEA's insistence on the indigenous French technology of gas-graphite reactors. Yet even when this battle was won in 1969, EDF did not adopt a major program of nuclear investments, preferring to develop its technical competence with American light-water reactors in a series of scaled-up prototype plants. As late as 1970, EDF still believed that the era of low-cost oil would last. When the oil bubble burst, EDF was left with an urgent need to redirect its entire course and redraft all its carefully worked-out plans. It then adopted nuclear power with an enthusiasm reinforced by its acute embarrassment over the failure of its highly sophisticated forecasting techniques.

Between the years 1971 and 1973, EDF nonetheless instituted only a modest nuclear program, ordering some 3,000–4,000 Mw of nuclear capacity each year. It was not yet an "all-nuclear" program, but rather an insurance policy against the new uncertainty of oil supplies. This gradual program was completely overthrown in the wake of the 1973 Yom Kippur War. At the beginning of 1974, the French government formally adopted a program to switch from oil to nuclear as quickly as possible. EDF was authorized to order 12,000 Mw of nuclear capacity, and in 1976, after a year's hesitation, the new direction was reinforced with orders for 13,000 Mw and options to purchase another 11,000. The commitment was now total. In 1976, France had only ten small-scale reactors with a total capacity of less than 3,000 Mw plus the 870-Mw joint venture in Belgium that had come on stream at the end of 1975. Its new plans required the completion of almost thirty new plants producing almost 30,000 Mw by the end of 1984.

The sheer audacity of the ambitious program appealed to the technical men in EDF; they were exhilarated by the size of the task before them. The renewal of their sense of being in the vanguard of technical progress overcame the embarrassment of their failure to predict the oil crunch. The construction of power stations had now returned to the center of EDF's activity, after years in which distribution and marketing had been more prominent. The self-conscious technical elite inside EDF, the Direction de l'Equipement—the internal architect-engineer department, were known as "the shock troops" of the organization and commanded by a passionate advocate of nuclear power, Michel Hug.

L'Equipement was the section of EDF responsible for planning and constructing all power plants; it was an empire within an empire. It had its own departments for economic studies and public relations and was proud of its heritage of technical achievement. Its members had long been disappointed with the lusterless and even boring technology of oil-fired plants and used to speak of the days of constructing dams in the 1950s as the "heroic age." Now the nuclear drive offered a renewal of the old camaraderie of the dam construction teams, which had worked in isolated and frequently inhospitable surroundings. Nuclear power called for the most sophisticated technical designs and promoted a new sense of grandeur.

Michel Hug, like many French technocrats, was a man of modest origins, the son of a primary-school teacher. He once recalled, with a sense of grievance, how his mother had died young without ever having owned a refrigerator. The nuclear age, he believed, would ensure that everyone's mother had a refrigerator.

Such sentiments never concealed his ambition. He spoke of achievements and objectives in the first person, rather than in the name of his team. It would be *his* Direction de l'Equipement; for the first time, the public knew the name of its director. His ambition to succeed Boiteux as director-general was obvious.

Hug progressively removed any resistance to an all-nuclear program. L'Equipement had traditionally been structured in a series of semiautonomous units, each responsible for the full range of design and construction services in a particular geographical area. The early nuclear decisions, before Hug's accession in 1972, had already required some changes. At l'Equipement's Paris headquarters, a new series of overview sections had been established for hydro, thermal, and nuclear. More significantly, the nationwide requirements of nuclear investment had spawned a strong central assessment group at the head office.

Hug was convinced that a major nuclear commitment demanded a centralized system to plan the entire program and to ensure harmony with the manufacturing industry that had to be put together on a national scale. His attitude was reinforced by the emergence of narrow self-interests in the regions that had been brought into the nuclear field. The region that had been asked to investigate the boiling-water reactor, for example, had already become a complete advocate of the superiority of that system, and the region that had been given responsibility for constructing the first pressurized-water reactor had taken a similar stance on the PWR.

Hug believed that the common features of nuclear plants—their design, materials testing, construction efficiency, and safety regulation—were more significant than regional variations at the sites. The complexity of the problems, and the national urgency of the program, did not permit the luxury of duplicating scarce human resources. For Hug, differences of opinion were signs of inefficiency, and he had a fetish for streamlined organizational structures. Hug totally reorganized the Direction de l'Equipement. The regional units, for so long the very basis of the Direction structure, were completely abolished. There were no longer any separate units concerned with hydro or thermal means of generating electricity. The Direction was organized around the nuclear program and all decisions were centralized. Hug was so convinced that EDF had no other options that he destroyed the institutional basis for even considering them.

By the time the point-of-no-return plan was launched in 1976, there were no dissenters in Hug's division. His commitment was reflected in the words he chose to describe the few internal critics in other departments of EDF. He dismissed them all as "traitors" and called the most persistent doubter, an EDF economist named Louis Puiseux, a "Judas." Hug believed that once a decision was taken, it should be implemented without further debate. Under his strong, authoritative direction—some called it authoritarian—his Direction de l'Equipement was to become the leader of the entire EDF empire.

His fervor did not stop with the existing program of light-water reactors. His commitment to a nuclear age extended to the next "generation" of power-producing equipment—the breeder reactor. Hug turned EDF into an equally passionate partisan of the breeder—even though the economics of breeder reactors were still as dubious and uncertain as any project in the entire history of reactor development.

When the French government made its all-nuclear commitment in the mid-1970s, it was riding the crest of an international wave. In 1974, the International Atomic Energy Agency in Vienna issued what would turn out to be the last of its completely optimistic forecasts of nuclear capacity. It predicted that the existing worldwide capacity of about 55,000 Mw would multiply more than tenfold by 1985 to almost 600,000 Mw. And it would not stop there: by 1990, said IAEA, there would be more than 1 million megawatts and by the year 2000, the figure would be approaching 3 million.

21

"A Shell of the Atom"

We are in the position of Shell in about 1910. It is up to us to build a Shell of the Atom.

—ANDRÉ GIRAUD, Administrator-General, French CEA, 1970

All industrial nations responded to the oil price spiral of the early 1970s with determined efforts to develop sources of energy independent of the Organization of Petroleum Exporting Countries (OPEC) producers. No country resented being subservient to OPEC more than the United States. President Nixon summarized the national mood by launching "Project Independence," a plan to end reliance on foreign sources of energy. Nixon urged Americans to pursue the plan "in the spirit of Apollo and with the determination of the Manhattan Project." Other countries had similar programs for independence, but they also had another kind of freedom in mind: freedom from American domination of the worldwide nuclear industry.

The almost universal triumph of the American light-water reactors had been a challenge to the patriotic instincts of nuclear scientists and engineers throughout the world. Not only was the American reactor design dominant, but, as a result of its weapons program, the United States was also a monopoly supplier of enriched uranium, the fuel used in light-water reactors. The absolute refusal of the British and the strong resistance of the French to adopt light-water technology was due, in large part, to a stubborn, primitive patriotism. But it was also due to their fears about this monopoly. Small countries, like India and Argentina, were willing to pay the higher cost of importing the Canadian heavy-water reactors to escape continued reliance on the United States for enriched fuel.

In Communist Bloc countries, a similar enrichment monopoly existed. Initial attempts to buy heavy-water reactors (which use natural uranium) from Moscow were turned down; in the early seventies, the Warsaw Pact nations unanimously adopted light-water systems and thereby tied themselves to the Soviet Union for the supply of enriched uranium.

Western European nations, in particular, resented U.S. technological imperialism. In the past, U.S. companies had dominated the oil trade; now, the Americans were running away with an expanding nuclear industry that, by the end of 1975, had already become a $50 billion business. The universal faith that nuclear power would become increasingly dominant throughout the world suggested it would also become one of the single largest industries of the next hundred years. No nation serious about remaining an advanced industrial country could afford to stay out of nuclear power. In each developed country, domestic demand was expected to grow rapidly and the international market looked just as promising: it seemed it would become every bit as important as the oil business had been.

Few of the patriots inside the nuclear bureaucracies were more conscious of the potential of the new multibillion-dollar industry than the man who, in 1970, was appointed administrator-general of the French Commissariat à l'Energie Atomique, André Giraud. Giraud's grand vision for France would attack the American monopolies on two fronts: reactors and fuel. Like the oil "majors," there would be nuclear "majors": huge multinational corporations that did everything connected with nuclear power, from mining the uranium to building the reactors, just as the oil majors did everything from drilling for crude to manning the gas pumps. "We are in the position of Shell in about 1910," said Giraud when he took over the CEA. "It is up to us to build a Shell of the Atom."

The analogy with oil came naturally to André Giraud. He had come out of the oil business, rising in the hierarchy of French public power under the patronage of Pierre Guillaumat. Guillaumat had not only established the CEA as a powerful institution driving toward the building of a French bomb; he had also struggled for decades to build a French oil company as fully integrated as any of the oil majors.

Like Guillaumat, Giraud was a Polytechnician and a member of the Corps des Mines. Graduating second in the class of '45, he had spent twenty years working his way up through the special state-owned oil research institutes that Guillaumat had created to serve as the basis for an independent oil company. In the process, he had learned the techniques he later used in creating a nuclear industry in France, an achievement that would earn him the title "The Guillaumat of the Atom."

Giraud was almost forty when, in 1964, Guillaumat took him out of a long career in the backwaters of oil research and made him France's Director of Fuels. It was his first taste of power, and he found that he liked it. He was also good at it. From that time onward, he sought to devour huge lumps of power, pursuing them with an impatient passion. He later recalled the transition with a smile: "However well I understood petrol, my administrative knowledge left a lot to be desired. At that time, I could barely distinguish a decision from an order. That probably allowed me to approach administration from a new perspective—from incompetence." In reality, one thing that was never in question was his administrative competence.

He seemed driven by a persistent need to prove himself; the last victory was never enough. There always had to be a new project, a new battle, a new victory. Under his leadership, the French nuclear program could never stop to digest the last expansion or to reassess the size of the one before that. New projects were swallowed whole—the bigger the better. Nothing was reopened for review once Giraud had put his stamp on it.

His drive and dynamism were frequently compared with that of his mentor, Pierre Guillaumat, but he lacked Guillaumat's quiet self-assurance. The inner core of self-doubt that had kept Giraud in the oil-research institutes for twenty years was suddenly replaced by a brash aggressiveness and, whereas Guillaumat was not very talkative, Giraud was voluble. He was also given to frequent fits of temper, raising his voice to assert his authority. A tall man with silver hair, cold blue eyes, and a thin, well-trimmed mustache, Giraud seemed to have forced himself to adopt the look and air of a man of authority. It did not come naturally: he was a poor boy from Bordeaux, where his father had been a school inspector and his mother a postal clerk. Only his intelligence and energy had enabled him to succeed.

When Giraud took over the CEA at the end of 1970, it was a demoralized and frightened institution. In the wake of its total defeat in the battle with Electricité de France over the choice of a reactor system, it was regarded at the highest levels of the French government as spendthrift, rigid, irresponsible, and inefficient. When President Pompidou shifted ministerial authority for the CEA from the Prime Minister to the Minister of Industry, it was clear it was no longer something special. Worse came as Pompidou appointed an independent Commission of Inquiry to revamp the CEA—the ultimate insult to any bureaucracy. Pompidou's commission recommended a total reorganization of the CEA, breaking down the traditional institutional fiefdoms; everybody's job was on the line. Giraud proved so successful at adapting to the new structure, however, that he accepted full credit for the reorganization, and public reference to the Commission of Inquiry soon disappeared. Within five years, Giraud had created a fully integrated nuclear group on the model of the oil majors. It became a serious challenge to America's leading role in nuclear power.

Giraud knew it was impossible to produce a French reactor as successful as the American light-water types, and he concentrated his efforts on what nuclear technicians call the "nuclear fuel cycle," a rounded, integrative concept based on following uranium through all its stages. The cycle starts with uranium exploration and mining. For light-water reactors, the uranium is then milled and processed into a form suitable for an enrichment plant, where its content of the fissile U-235 is increased. Once enriched, it is fashioned into fuel rods that are put to work inside the nuclear reactors. After they have been "burned up," the "spent" fuel rods—now containing plutonium—are taken out of the reactor. The plutonium is extracted, mixed with more uranium, and "recycled" as fuel in light-water reactors. The plutonium can also be kept separate and used as fuel in breeder reactors. The balance of the radioactive wastes in the fuel rods cannot be used again and has to be stored.

Because of its history, especially its role in the French weapons program, the CEA was already involved in each aspect of the fuel cycle. It explored and mined uranium, it had a small enrichment plant for producing bomb-grade material, it built plutonium and power reactors, and it operated a military reprocessing plant. It was in the vanguard of nuclear research and had a particularly advanced program for breeder reactors. The CEA's advantages were automatically shared by countries that had built the bomb and had developed the disparate parts of the fuel cycle. In Western Europe, this meant the British, who had begun to show a new aggressiveness in their fuel-cycle technology. They had brought everything together under one corporation, British Nuclear Fuels; it was responsible for enrichment and reprocessing. The corporation's ties with the U.K. Atomic Energy Authority were close, and the AEA was in the forefront of international research, particularly in breeders. Unlike the CEA, the AEA did not have direct control of uranium mines, but the huge British mining company, Rio Tinto Zinc (RTZ), had worldwide reserves bigger than those controlled by any single nation. RTZ in tandem with BNF and the AEA equaled a nuclear major.

The Germans were technically strong, but structurally dispersed. Their two electric manufacturers, Siemens and AEG, could build reactors—both light-water and breeders. The chemical giant, Hoechst, was involved in reprocessing, and a consortium of heavy-industry companies dominated uranium enrichment. It would be difficult to pull all these loose strands together. More significantly, the Germans had no secure supplies of uranium and their overseas exploration for it was only just beginning.

Outside Europe, the other potential nuclear major was Japan. Many of the strands of the fuel cycle had been brought together under the government-owned Power and Nuclear Fuel Development Corporation. But the Japanese did not have any supplies of uranium.

Confirmation that Giraud's concept of "nuclear majors" was correct came from the new interest in the nuclear fuel cycle shown by American companies. A number of oil companies had bought up uranium mines and had interests in other parts of the fuel cycle. Getty Oil was developing the reprocessing of spent fuel rods, Exxon was in mining and enrichment, and Gulf Oil—which, with Exxon, had the largest oil-company interest in uranium mining—had bought a reactor manufacturer and formed an alliance with Shell to promote that side of the business. General Electric had ventured into reprocessing, and both GE and Westinghouse had interests in a new enrichment plant. The electric companies were also pursuing uranium exploration.

Giraud never expounded his image of the future of the nuclear majors, but the facts seemed to point toward a regime of "seven brothers": Exxon, Gulf, General Electric, Westinghouse, one British, probably one German—and the CEA. Given France's inexperience in reactor technology—at the time Giraud took over the CEA, it had only just ordered its first light-water reactors—his plan to catch up with these other more experienced giants of the nuclear trade was supremely ambitious, but he threw himself into the challenge with unrestrained vigor and determination.

Within months of assuming office as the head of the CEA, Giraud's grand scheme was on the move. Throughout the decade, Giraud would lead systematic attacks on the monopoly position of the U.S. Atomic Energy Commission in enrichment, as well as on the U.S. potential to dominate reprocessing. He would play a major part in the formation of a special uranium suppliers' club designed to break the U.S. AEC's price control of uranium. By the time he left the CEA at the end of 1978, he had fashioned the only fully integrated nuclear corporation in the world—and this at a time when other corporate structures, with more flexible approaches, found themselves caught in the midst of a floundering industry and were pulling back. General Electric, Westinghouse, and Exxon dropped out of enrichment, at least in the short term, and Gulf closed its reactor business.

Giraud chose the reprocessing part of the fuel cycle as his first target. The French already possessed a small plant near Cherbourg (built as part of their bomb program) for reprocessing the fuel rods from their gas-graphite reactors. The decision to change over to the light-water reactors meant parts of the plant had to be rebuilt, and such reconstruction was quite feasible.

The British were also planning to expand their reprocessing facility, and the Germans were about to build one. All three countries shared a common interest in ensuring that the Americans did not dominate the international market. There was a clear danger that if they did not pool their resources, the three would end up competing against each other and destroy their main purpose. In the short term, there was the added danger of overcapacity: too many reprocessing plants chasing too few spent fuel rods. Eventually, if the projections for nuclear power were correct, there would be plenty of work for everyone. In the meantime, all three decided to collaborate on their reprocessing technology—just as the oil majors had collaborated on supplies and markets. In 1971, France, Britain, and Germany formed a joint company to exchange technical information on reprocessing and to ensure that they did not undercut each other in the international marketplace. No one spoke of a reprocessing cartel; they did not have to—the American threat was enough to justify the arrangement.

The second of Giraud's targets—the mining and marketing of uranium—resulted in a much more aggressive economic group. In 1972, along with other non-U.S. uranium producers, France formed a secret club designed to allocate quotas for the limited uranium market outside America and to set world uranium prices. It is an oversimplification to call this arrangement a cartel: the international uranium market was still a buyer's monopoly effectively dominated by the U.S. AEC. In reality, the club was a "defensive" cartel designed to minimize losses for uranium sellers rather than to maximize profits.

At the time the secret club was formed, the AEC was giving the rest of the world a lesson in how to exploit a monopoly position by making full use of its integration of all aspects of the fuel cycle. Giraud had watched the same power develop in the oil industry: the ability to use strength in one field to overwhelm weakness in another. The AEC was using its monopoly of enrichment plants to hold down the price of natural uranium throughout the world—and not merely for its own purchases. Particularly upset by this were uranium producers like Canada and South Africa who had involved themselves in huge, expensive exploration programs in the 1950s to stave off a feared uranium shortage.

The rich mine in the Belgian Congo that had produced most of the immediate postwar uranium had been supplemented by new Canadian finds, by production of large amounts of uranium as a byproduct of gold mining in South Africa, by the opening of major mines in America's southwestern states, and by a number of new ventures in Australia. In the early sixties, before the nuclear power market was established, all these were cut back as production exceeded U.S. weapons program demands. Dozens of plants were closed in South Africa, Australian production ceased, and, in Canada, only a $100 million government stockpiling program kept a number of mines working at minimum capacity. In 1964, the United States declared an embargo on all uranium imports; it would not be eased until the late seventies. To rub salt into the wounds of the suppliers, the

U.S. AEC declared, in 1967, that it would sell enriched uranium to anybody at a price that effectively fixed the price of natural uranium in the world market at $6 per pound—half the peak price of the mid-fifties.

In 1971, with the glut continuing, market prices outside the United States had fallen to $4.50 per pound—below the cost of production in most mines. Just at this point, the U.S. AEC decided to get rid of its huge stockpile of natural uranium—some 50,000 tons, or ten times the level of international sales per year. Only its monopoly over enrichment enabled the AEC to get away with this: the commission adopted a rule for all its enrichment contracts establishing a fictitious relationship between the amount of uranium that had to be supplied by a customer to get a certain quantity of enriched uranium back. This amounted to about 20 percent less uranium than the actual operating level of the enrichment plants required; the balance came from the AEC stockpile at almost double the market price.

Uranium producers throughout the world were distraught over the continued weakness of their market. The Canadian government, in particular, made vigorous representations on the damage this was doing to the ability to develop mines to fulfill the expected upturn in demand. They found their own customers unsympathetic, and the U.S. government simply ignored the Canadian argument that AEC enrichment practices were inconsistent with America's obligations under international trade treaties.

France was especially irked. After a slow start, she had become one of the major uranium producers, developing its own mines in France and in the former French colonies of Niger and Gabon. Despite gaining political independence, these African nations had become virtual economic colonies of the CEA because of their uranium deposits. During the mid-sixties, huge quantities of natural uranium were stockpiled in France for the CEA's ill-fated gas-graphite program. As a result of the AEC enrichment policy, these would now have to be sold at less than the cost of getting the uranium out of the ground. It was not surprising that the first meetings of the secret uranium suppliers' club were organized under Giraud's empire-building eye, from within the French CEA.

The man in charge of the exercise was Giraud's most trusted lieutenant, Pierre Taranger. As hard-boiled an engineer as L'Ecole Polytechnique had ever produced, Taranger began his career in

Guillaumat's oil empire, and when Guillaumat organized the major French production program for bomb material in the 1950s, he appointed Taranger to run it. He left the CEA after Guillaumat's departure, but was brought back by Giraud in 1971 after a number of the CEA's most senior executives had died in a plane crash. His objective was clear: to overcome the short-term uranium surplus, but to do so with an eye on Giraud's grander scheme of building a "Shell of the Atom."

In the first week of February 1972, representatives of the major non-American uranium suppliers—France, Canada, South Africa, and the British RTZ—met, at Taranger's invitation, in the large, squat headquarters of the CEA on the Rue de la Fédération in Paris. Although it was to be an exploratory meeting, the French quickly established a program. They pointed out that total uranium demand until 1977 would be only 26,000 tons, but production capacity in this period was over 100,000 tons. In addition, there were the stockpiles held by France and Canada. It seemed possible to increase prices as long as production was held down and strict quotas allocated to every producer.

Because of the intense competition caused by the glut, hopes for achieving agreement were not good at first. Furthermore, a number of unusually rich discoveries had just been made in Australia, and these would, because of their richness, probably come in at well below everyone's production costs. Even more advanced than the Australian finds was a major new Canadian mine at Rabbit Lake in Saskatchewan, scheduled for production in 1975. It would become Canada's biggest single producer, but it was not controlled by the Canadian Government. Rabbit Lake was controlled by Gulf Oil, one of the possible rival "nuclear majors" to the CEA.

Taranger had decided initially to restrict the club to existing producers. That meant excluding both Australia and Gulf Oil. But a week after the first meeting, senior executives of Gulf's subsidiary in Canada learned of the Paris gathering. A Gulf internal memorandum later assessed the potential of the new club as follows: "The Club . . . is playing a game in which they hope to a) keep Gulf, as the 'new boy on the block' away from the poker table and b) block the Australians from starting production."

One of Gulf's in-house lawyers summed up the company's need to join the club: "It is at least as important," his internal file memo said, "for Gulf to become a sophisticated and substantial participant in worldwide uranium matters as it was for us to undertake similar efforts with respect to oil and gas thirty or forty years ago." In the course of the next few months, as the club proceeded through tortuous negotiations—first in Paris, and then at a final summit meeting in Johannesburg in May 1972—both Gulf and the Australians forced their way in.

The club was known among its members as the "Club of Five." Those represented were France, Australia, Canada (including Gulf's interests), South Africa, and the British mining corporation Rio Tinto Zinc, which, by the early seventies, looked as though it would eventually control a fifth of the world's uranium deposits outside the United States. It was, therefore, treated as a sovereign state.

The estimated total world uranium market, excluding the United States, for the remainder of the decade was allocated to each of the five groups. Precise percentages were fixed and an elaborate bid-rigging procedure was established. The new cartel price was some 40 percent higher than recent sales; $6.25 a pound became the new floor price below which nobody would bid. It was to escalate gradually on an agreed formula. The club had detailed rules for treating each new contract announced anywhere outside the United States. A central secretariat would decide whose turn it was to get the contract and that company would put in a bid at the club's floor price. Another member of the club would be designated as "runner-up" and would be required to put in a phony higher bid for the contract. In the long run, for deliveries after 1978, the club agreed to end fixed-price bids. If the uranium market took off as expected, the stage would be set for even higher prices.

The Operating Committee of the new club met for the first time at Cannes on July 6, 1972. The committee designated André Petit, an employee of the French CEA, as secretary in charge of coordinating the worldwide bids. The official minutes added, "From the security aspect it was thought convenient that Petit would be 'buried' in the large CEA HQ building."

In a particularly bold step—and in hot pursuit of Giraud's plan for a "Shell of the Atom"—the French used the club to try to keep Westinghouse, a possible rival major, from becoming a force in the uranium market. Special rules adopted by the club were directed against middlemen in the market, and this, for the most part, meant Westinghouse. At the time, Westinghouse was successfully selling its pressurized-water reactors around the world and in most of its deals was offering a "sweetener" of long-term uranium supplies at fixed prices. At the final formation meeting in Johannesburg, the club agreed to a French- and Gulf-sponsored motion to quote higher prices to all "middlemen," such as Westinghouse, in a conscious effort to get them out of the market.

Two members of the club, the CEA and Gulf, joined forces to form the real drive against Westinghouse. Gulf, in particular, wanted to make sure that Westinghouse was left out in the cold, burdened with its special fixed-price uranium deals, because one of Gulf's reactor-producing subsidiaries was in direct competition with Westinghouse at the time for a new high-temperature series of reactors. As a result, Westinghouse was presented with a desperate problem of uranium supply.

In the mid-1960s Westinghouse had begun contracting to supply twenty reactor customers with uranium for the next twenty years at a fixed price of $20 a pound—the U.S. market price at the time was about $8 a pound. But the company never owned anything like the quantity of uranium its guarantees involved. It was playing the futures market, and when the price of uranium began to surge upward, Westinghouse was very badly caught short. By October 1975, the price had already passed $20 a pound and the company saw only one way out: it reneged on its contracts to supply uranium to its reactor customers on the grounds of "commercial impracticability." A group of its customers promptly sued for damages up to $2 billion; and Westinghouse countersued the uranium club for $6 billion.

During the lengthy litigation, the story of Westinghouse's efforts to join the club emerged. Even before the club was formed, a Westinghouse executive had written, "It is anticipated that by the end of the decade an oligopoly of producers will control the uranium market . . . will Westinghouse be a member of the oligopoly?" Excluded from the start by the French and Gulf, Westinghouse had nevertheless watched the formation of the club carefully. It even had a covert source at the first Paris meeting in 1972. Later, it attempted, unsuccessfully as it turned out, to bypass the club's activities by establishing a joint venture with an Australian uranium mining company.

André Giraud had been right. The nuclear industry was developing in the image of the oil business.

Important as uranium mining was in the fuel cycle, however, it was not the centerpiece of Giraud's plan for a "Shell of the Atom";

that was the third target, the scheme for the enrichment of uranium. There was more money in uranium enrichment than in the sale of reactors; estimates projected the enrichment market would grow by the end of the century to a $100 billion business—provided the expansion plans for nuclear power were fulfilled. A fully integrated nuclear major had to have an enrichment plant of its own.

Using the more optimistic projections for nuclear power, a dozen new enrichment plants would probably have to be built before the year 2000 to cope with the increased demand. Giraud was determined that at least one of these plants had to be French; without it, France's claim to energy independence and industrial might could not be realized.

By the late 1960s, the rapid growth of nuclear power in Europe had demonstrated the need for an expansion of the enrichment facilities. The French had already embarked on a new enrichment plant that would use the gaseous diffusion technology they had developed during their bomb program. There was, however, an alternative technology: the centrifuge. This system, abandoned early in the Manhattan Project, was revived in Europe during the 1960s because it was found to have a number of advantages. For one, the diffusion plants have to be built to a minimum size for commercial operation—huge by any industrial standard. The Oak Ridge plant in America spread over an area of 2 million square feet. Not only were centrifuge plants more modest in scale, they could also be gradually increased in size according to the demand for the end product. Even more significant for a European program, however, was the huge amount of power needed to run the diffusion plants: each of the American enrichment plants in the 1960s was consuming as much electricity per year as the city of New York. Given Europe's comparatively high-priced electricity, this was an important consideration. The centrifuge technique, based on spinning metal cylinders rather than on pumping gaseous uranium through miles of pipes, used only 5 percent of the electricity needed for diffusion. It was an attractive proposition.

Despite the promising economics of the centrifuge technology, however, no one had built a commercial plant. Pioneer work on the centrifuge had been done in the German atomic bomb project during World War II, and members of the German centrifuge teams had subsequently worked on the technique in Russia. The Russians had eventually rejected centrifuge as a possible means of producing enriched uranium, preferring the technically proven diffusion route. In

1955, Gernot Zippe, one of the scientists who had worked on the centrifuge in Russia, returned to Germany with a design for a light and durable centrifuge. It sparked considerable interest, and a German company, Degussa, began to develop it. In 1960, the Americans, fearing that the centrifuge could soon be the source of quick, cheap enrichment and, therefore, an easy route to the bomb, insisted that all centrifuge work in Germany be classified. The German company reluctantly agreed to place all centrifuge research in a government-controlled agency on the understanding that it would retain an interest in any future commercial development. The Americans continued to keep the centrifuge under wraps, but in 1968 the governments of Germany, Britain, and the Netherlands entered into secret negotiations on the centrifuge that resulted in a joint research and development group, Urenco.

For André Giraud, the Urenco alliance posed a major challenge: it had serious pretensions to becoming the European response to the American enrichment monopoly. In February 1972, two weeks after the CEA launched the negotiations for the uranium club, Giraud brought all the Europeans together for a study group on enrichment; he called the group Eurodif. His intention, which rapidly became clear to the troika of Urenco, was to drop its centrifuge research and coopt all European members into the French diffusion program. The Urenco members would have none of this. In the end, two separate European enrichment groups emerged: Urenco and Eurodif, the latter consisting of France, Spain, Italy, and Belgium. Eurodif would stick with diffusion.

The French government's substantial commitment to nuclear power and a large contract from the Japanese for enriched uranium persuaded the Eurodif consortium to launch plans to build a full-scale diffusion plant in France. It would be powered by four 900-Mw light-water reactors to be built by France's EDF. It would produce enough fuel for a hundred reactors of that size.

Giraud's brash self-confidence was never better displayed. The Americans had held back on expanding their own enrichment capacity, and this had given Eurodif the opportunity to sign a few short-term contracts, like the one with the Japanese. With an unshakable conviction in the most optimistic assumptions about the future of nuclear power, France—with its revamped reprocessing plant, its secret uranium club, and its enrichment consortium—had launched the most ambitious and comprehensive program of any nation to try to establish its leadership at all points of the fuel cycle.

In 1974, Giraud's plan passed the point of no return and became the prisoner of its own convictions and its own momentum. The orders for nuclear reactors peaked, some were canceled, and it became clear there was an acute danger of a glut of enriched uranium. The Urenco consortium continued to construct its pilot centrifuge plants, secure in the knowledge that it could increase output of enriched uranium to meet demand, if the need arose. Eurodif was stuck with its huge diffusion plant; worse, estimated capital costs had doubled.

Giraud was not, however, diverted from his grand vision. In January 1975, in what was to be his biggest coup, Giraud assigned part of the CEA's shareholding in Eurodif to a new joint French-Iranian company. The Shah promised a major new tied market for Iran's ambitious nuclear plans and presented Eurodif with a billion-dollar loan to bail it out of its financial difficulties. In September 1976, with a renewed sense of self-confidence, Eurodif announced another new diffusion plant to increase its capacity by half in 1985 and to double it within a few more years. Giraud, still determined to build his Shell of the Atom, was not dissuaded by the way international projections of installed nuclear capacity were being lowered everywhere. He was sure this would be temporary.

The reality, however, was somewhat different. The world was facing a substantial oversupply of enrichment capacity and, on the best forecasts, would do so for the rest of the century. The French would find themselves stuck with the high-cost system of gaseous diffusion and no markets. Yet, nothing could curb their enthusiasm. Although the nuclear power programs of the other Eurodif partners were cut back, the French commitment grew. The self-consciously progressive French nuclear industry reacted to growing doubts about a nuclear future with a rigidity that has too often characterized the most regressive parts of French society and industry. André Giraud's conviction in the correctness of his own assumptions was total. Despite the cries of "another Concorde" that greeted each part of the grand strategy as it continued to unfold, the CEA under Giraud had begun to dominate French industrial investment in a manner that would determine the economic welfare of the nation for decades.

22

A Tale of Two Dynasties

It was especially apt that Germany, the country where the fission of uranium had first been described in 1938, should also be the country that made the decisive breach in the American technical monopoly of nuclear power reactors.

For an entrance fee of a mere $2 million—the royalty payment made to Westinghouse for the privilege of using its design for a pressurized-water reactor—and $50 million of its own money, the German engineering firm of Siemens became a leading member of what promised to be one of the most lucrative industries of the 1970s. The company teamed up with its rival German reactor manufacturer, AEG-Telefunken, which made boiling-water systems under license from GE, and formed the only nuclear manufacturing group in the world that was in serious competition with the Americans. It was also the only group anywhere that sold both types of light-water reactors. In 1975, the new German group moved into the international market, winning a massive Brazilian order for eight reactors. A year later, it opened a new market in the Middle East with the sale of two reactors to Iran.

The only other country to make an impact on U.S. reactor sales was France. There, André Giraud's grand design of a "Shell of the Atom" spawned a company called Framatome, which was the Westinghouse licensee in France. French interests in this company gradually increased, and it was encouraged to build production capacity in excess of domestic demand so that it could enter the export market. Unlike the Siemens group, however, Framatome never developed a tradition of technological excellence; its export sales frequently depended on a quid pro quo sale of military equipment by the French government to the nation buying the reactor; for example, Framatome managed to oust Siemens in a reactor sale to South Africa by linking it to a weapons deal. Framatome also secured a contract for two reactors from Iran in 1976. By the mid-1970s, Germany and France each had eight export reactors either completed or already under construction.

Two other reactor producers challenged the Americans, but in a much smaller way. By the mid-1970s, the Canadians had sold six of their heavy-water CANDU reactors, and the Swedes had sold two of their own version of the PWR to Finland.

The British were also potential challengers to the U.S. reactor monopoly, but the U.K. Atomic Energy Authority was still strongly resisting the introduction of the light-water reactor into Britain—even though the failure of their second-generation domestic program based on advanced gas-cooled reactors had made the country a laughingstock in the international marketplace. In 1973, British reactors actually generated less nuclear electricity than they had in 1968, and not one of the five advanced gas reactors was working. One section of the British nuclear industry made a strong bid to develop the American-designed PWR for domestic use and for possible export, but it failed; the patriots in the U.K. AEA were able to persuade the government to go for their own experimental advanced reactor, using heavy water.

Only the Germans were to view the U.S.-dominated light-water market as a wide open challenge. Like GE in the United States, Siemens had started nuclear research because the company could not afford to miss out on a revolution that promised to offer a major energy source. The company's motto was "Only electric, but everything electric." If the source of that electricity in the future was to be nuclear energy, so be it.

As one of the world's largest electric manufacturers, the Siemens corporate style was more rigid than its two American rivals, General Electric and Westinghouse. The German company eschewed the American tradition of rapid mobility and executive headhunting. The Siemens corporate atmosphere was more akin to the Japanese: familial, internally noncompetitive, unthreatening. The primary bond at Siemens, however, was a tradition of intellectual and technical excellence that had been promoted by its paternalistic founder, Werner Siemens. Under his tutelage, the "House of Siemens," as the employees knew the company, had developed a high-priority commitment to original research, irrespective of its immediate commercial application.

Nuclear research had started in secret in 1952—three years before the postwar occupation authorities lifted the ban on German atomic research. Werner Heisenberg himself had suggested that a German theoretical physicist, Wolfgang Finkelnburg, might lead the Siemens group. Finkelnburg, who had joined the Nazi party in 1939 and been the leader of the Nazi University Teachers' Association, had nonetheless won the overall admiration of the German nuclear physicists for his support of Einstein's theories, which were criticized as "Jewish physics" during the Nazi years.

After the war, Finkelnburg emigrated to the United States. His stand against Aryan physics and a year of postgraduate study in the United States before the war made him politically acceptable to the American academic headhunters of Operation Paperclip—a concerted U.S. postwar effort to boost its own pool of scientists by importing as many as possible from war-torn Germany. In the early 1950s, the Siemens executives found Finkelnburg working at the Catholic University in Washington, D.C.

He was perfectly suited to the company's commitment to original research. He came from a family of university professors and, at Siemens, he started an atomic research institute modeled on the distinctive German academic approach; he was the paternalistic professor surrounded by admiring disciples, a father figure for the whole department, a man who knew all his subordinates and their families. Quiet-spoken, eloquent, and flexible, he seemed the prototypical gentleman-scholar. He was too sensitive to be a tough industrial manager, but he created a group with a strong esprit de corps, a sense of purpose and of destiny, one that was stimulated by the ambitious pronouncements of their nuclear colleagues throughout the world.

The team was strongly motivated to move into the forefront of the new international technology.

Unlike many of his wartime colleagues who believed that the atom was the province of the theoretical physicist and that engineering was a secondary function, Finkelnburg emphasized the links between the two disciplines. "The mutual insemination [between the two]," he noted, "results in a tremendous enhancement of efficiency in research." Siemens's research director for the company's heavy-engineering arm, Heinz Goeschel, agreed. "More than ever before," Goeschel declared, "future success depends on close collaboration of the disciplines and the lively exchange of ideas between scientists and engineers." The combination of the two men was to prove particularly effective. Finkelnburg's reactor department became one of the three basic units at the research center that Goeschel created at Erlangen, north of Siemens's Munich headquarters and near its Nuremberg factory. Both Finkelnburg and Goeschel were determined to pursue an independent line of research that could make Siemens—and Germany—a self-sufficient force in the nuclear age. To do so, they had to sever the links to America; that meant building a reactor that did not use enriched uranium.

Throughout the 1950s and 1960s, America's total control of enriched uranium supplies looked as though it would last forever; that, at least, was the view from Europe. Siemens, encouraged by a mixture of German patriotism and corporate and professional pride, therefore embarked on a program of developing the heavy-water reactor, which uses natural uranium as fuel. It was a parallel development to the work being done by their colleagues in Britain, France, Canada, India, Argentina, and Spain; and it was undertaken for the same reason: to ensure national independence in the long term. It soon became clear, however, that the German electric utilities were unanimously in favor of light-water reactors, and so Siemens turned, with equal determination, to its forty-year-old relationship with Westinghouse and sought to develop a PWR.

Since the early days of electric power, Westinghouse and GE had formed relationships with electric manufacturers in every industrial country. In Germany, Westinghouse had linked up with Siemens, and GE had formed an alliance with AEG. The war had interrupted these ties, of course, but they were resumed in the early 1950s.

Siemens's experience with heavy-water reactors—they had already sold one to Argentina—gave them a good start with the light-water types. Many of the most difficult engineering problems of reactor construction were common to both systems; the pressure vessels, the control-rod drives, the pumps and steam generators, were all similar. Siemens did not simply take Westinghouse technical know-how and copy it, as the Japanese would do. The German company used the information as a basis for wholly autonomous technical development.

The licensing agreement between the two companies was twofold. The first part gave Siemens the right to use Westinghouse patents and the second gave Siemens access to nonpatentable know-how. The information was restricted to the state of the art as it was incorporated in Westinghouse's commercial sales, however. Siemens did not have access to Westinghouse research. If the agreement had worked as intended, therefore, it would have severely restricted Siemens's opportunity for competing against Westinghouse; it would always have remained a secondary partner. But Siemens engineers were able to bypass the rules. They went on a series of exchange visits to America and, through the personal contacts they developed, were able to keep up to date on all the Westinghouse developments.

In 1963, the German Atomic Commission announced their first official Five-Year Plan and, sensitive to the still-rival corporate interests between AEG and Siemens, provided substantial government funds for two reactors, one BWR and one PWR. Siemens executives were reluctant to accept sole responsibility for the huge risks inherent in the construction of the large 280-Mw PWR, and so Westinghouse agreed to form a consortium with Siemens. During the negotiations for the consortium, however, Westinghouse displayed a certain inflexibility that annoyed Siemens. It refused to allow Siemens to incorporate the most recent technical advances—which both the Germans and Westinghouse knew could easily be duplicated. Westinghouse also insisted on the American practice of having its own overall architect-engineer for the project. Siemens could not agree to this; it had a tradition of acting as its own architect-engineer.

At the last moment, the negotiations for the consortium fell through, and the previously hesitant Siemens executives decided, for reasons of corporate pride, that they had to pursue the contract on their own: technical development of their PWR would have to proceed in parallel with its construction. In fact, this was exactly the sort of challenge that Finkelnburg's Erlangen research team wanted.

With superefficiency and technical ease, Siemens built the PWR reactor at Obrigheim on the Neckar River and, in the process, demonstrated that German industry possessed the full infrastructure of subcontractors needed to supply all the parts of the plant. (An essential difference between the German reactor industry and the French was that the latter had no such infrastructure.)

In 1967, Siemens negotiated a new license agreement with Westinghouse that recognized the extent of their newfound independence: it provided for a maximum total royalty payment of only $2 million. Considering what Siemens had learned about reactors from Westinghouse, the sum was ridiculously low. It covered the crucial know-how on the chemical and physical properties of nuclear fuel, the design principles for fuel rods, control-rod drives, pumps, and the use of special metals to cover the fuel rods.

AEG had meanwhile built its boiling-water reactor. It had not, however, become so independently expert. It would be several years before it managed to free itself from the GE license arrangement.

Throughout the 1960s, the two German companies competed against each other for light-water orders from the German utilities. Initially, they both priced bids at a loss—exactly as GE had done at Oyster Creek in 1963. The Siemens nuclear department was by now confident that it could knock AEG out of the market altogether, but the move was vetoed by Bernhard Plettner, who had been head of Siemens's heavy-engineering division since 1962. Plettner, a grand conciliator, preferred quiet collaboration to head-on confrontations in the marketplace. From his vast sixty-square-meter office at Siemens's Munich headquarters, a stately mansion that had once belonged to the architect of the Bavarian court, he took the company into a series of joint ventures. He brought Siemens's record subsidiary, Deutsche Grammophon, into an alliance with the Dutch Philips group and the Siemens's household appliance division into a joint company with the Bosch group. In 1969, as makers of electricity-generating equipment everywhere came under pressure to amalgamate because of increased costs, Plettner brought Siemens together with its rival, AEG, into a new company called Kraftwerkunion (KWU). This was immediately recognized as a real challenge to America. Westinghouse precipitously canceled its license agreement with Siemens; it had become painfully obvious that the German company was too good at using its access to Westinghouse research to produce its own improved products. At Siemens, they were unconcerned. "No one shed a tear," said one top executive.

At first, because of their separate license agreements with GE and Westinghouse, the two companies had to keep their nuclear reactor divisions separate. But they shared research and development problems, produced a single new type of turbine generator, and pooled their marketing efforts. With a domestic monopoly in a major market, the merger also promised to become a powerful alliance in the international market.

Within Germany, the AEG boiling-water reactor initially proved the most successful. Six orders were placed for it by German utilities between December 1970 and January 1972. The Siemens PWR was still in the proving stage. Meanwhile, using the PWR technology, Siemens was engaged in building the largest reactor in the world: 1,200 Mw, at Biblis on the Rhine. It had been ordered by the German utility RWE. With absolute faith in their previous experience, which had shown larger plant sizes meant lower costs, RWE forced Siemens to stretch the reactor to a size requiring a turbine that had not yet been built, not even for coal-fired plants. Siemens accepted the challenge and succeeded. With the Biblis order, they finally completed their independence from Westinghouse. They were even able to pay their royalty payment of $2 million out of this single contract. From that time on, sales of the PRW from Kraftwerkunion continued at a steady pace and soon overtook the sales of the company's BWR. In 1975, the Siemens side of the business became predominant: six orders from Germany and eight orders from abroad. Plagued with cost overruns and technical problems, AEG decided to cut its losses in 1976 and get out. It sold its half of KWU to Siemens, and for a few years the company went from strength to strength. Its first setback was the loss to France's Framatome of the sale of two reactors to South Africa. Faced with that loss, a Siemens executive commented dryly, "We wish that competition would concentrate on technological and economic aspects." But Framatome had inherited a somewhat different approach to international business: it was an approach in which politics and family dynasties played as important a part as technical expertise.

It was under the presidency of Valéry Giscard d'Estaing that Framatome became the leading French nuclear reactor company. Its corporate rise was not an easy one. Framatome was part of the sprawling investment empire of the third Baron Empain, a man regarded with some suspicion by leading French businessmen. For a start, he did not have a French pedigree; he was of Belgian stock. Even though the Empain family had managed, at the beginning of the 1960s, to acquire controlling stock in the French Schneider empire—the major French heavy-industry and armaments group, which included Framatome and was known as the "French Krupp"—that was not enough to convince French industrialists that the company was "French." The checkered history of the Empain family itself did not help matters.

The first Baron Empain, a Belgian and an engineer, had created a financial empire through a series of audacious investments at the turn of the century. These had endeared him to the newly aggressive King of the Belgians, Leopold II. The Baron helped the King develop the Belgian colony in the Congo, became a general responsible for supplies to the Belgian Army during World War I, and was ennobled by the grateful monarch. He developed a powerful presence in the electrical industry through the construction of electric railways, including the first Paris Métro, and the ownership of a string of utilities in Belgium, northern France, and Paris. His son and heir, Jean, the second Baron, was more of a playboy than a businessman but had enough dynastic pretensions to fret over the failure of his wife to produce a son.

The problem was solved one fateful night in London in 1935 when Jean was holding one of his dissolute parties at which the main attraction was the American beauty "Goldy" Rowland, star dancer of the Ziegfeld Follies. She appeared wearing nothing but gold paint, and within the year she had produced a son, the third Baron. Jean did the respectable thing: divorced his first wife and married Goldy.

In 1960, Charles Schneider, the patriarch of the "French Krupp," died without a son and heir. To the dismay of the rest of the Schneider family, his widow, the film actress Lillian Constantini, decided to exercise her rights under Charles's will and attempt to run the Schneider empire. Dissident members of the Schneider family promptly sold their interest in the company to the Belgian Empain group, enabling it to build up its share on the market to a controlling 25 percent. Outraged by this affront to her and to France, the widow Constantini sought, and received, a personal audience with de

Gaulle, entreating him to beat back the Belgian intruders. She obviously put on a magnificent performance because de Gaulle introduced into the Schneider board a number of high-ranking French civil servants, and it was not until de Gaulle's departure in 1969 that the third Baron Empain, Goldy's son, was able to get rid of the Frenchmen. Gradually, the subsidiary Framatome emerged as one of the most thriving sections of the empire.

The Baron was not yet committed to its success, however. As always, how and where he spent or made his money was largely a matter of indifference to him. When Westinghouse was desperately trying to establish a strong united presence in Europe at the end of the sixties, the Baron willingly sold the Americans his Belgian electric subsidiary and attempted to sell them another electric subsidiary of the Schneider group. President Pompidou vetoed the plan and put pressure on the Baron to sell the subsidiary to a wholly French manufacturer named CEG, a consortium with a license from General Electric to build light-water reactors. The Baron resisted Pompidou's approach, thereby making some important enemies in Gaullist circles.

When the French electric utility finally opted for light-water reactors instead of the indigenously built gas-cooled types, the Baron realized that his intransigence would cause a problem. The Gaullists in the French government would prefer that the orders for the light-water reactors go to the French company, CEG, rather than to his own Framatome. He was forced to follow the path of the electric manufacturers in America and Germany and price his initial bids at a loss. He was successful: he sold one PWR to EDF in 1970 and three more in 1971. The rival French company, CEG, was content to watch Framatome lose its money, believing that EDF would surely want to try both types of light-water reactors. But, as Siemens discovered, factors other than economics and technical expertise were at work.

The political position of the Empain group was transformed with the death of President Pompidou early in 1974. The new President, Giscard d'Estaing, was more open-minded and, in any case, was a family friend. Giscard had married a Schneider, from the "aristocratic" branch of the family, which had led the sale to the Empain interests in 1963. More significantly, he did not share the primitive Gaullist prejudice against Belgians. The third Baron Empain was Giscard's kind of businessman. Aggressive, tall, handsome, with a beautiful wife—the Baron had style. When the Baron was kidnapped in 1978 (by ransom hunters without a political motive), the police received their orders directly from the Presidential Palace in the successful two-month manhunt that followed.

Giscard's determination to build Framatome as *the* French company for nuclear reactors was apparent. The imprimatur of the President led to the total acceptance of the Empain group at the top of the French business community. The Baron himself was admitted to one of the most exclusive "clubs" in France: he became a member of the council of the powerful French employers' association Le Patronat Français; he was the first "foreigner" ever to be included. This was much more than a naturalization certificate; it was acceptance at the center of French power and wealth. Goldy Rowland's boy had come a long way.

Framatome was ready to fill the gap in André Giraud's plans to build a "Shell of the Atom"; it would become the French company that could produce light-water reactors. Giraud would make sure of that.

Giraud had his own access to Giscard d'Estaing. He had known Giscard for years, having been his senior by one year at L'Ecole Polytechnique. Furthermore, Giscard's first cousin and boyhood friend, Jacques Giscard d'Estaing, had been appointed head of the Financial Division of the CEA. Most significantly, Giraud's success in turning the CEA from a lethargic research institution into a flourishing nuclear cartel had established his reputation and his power at the top of the French administrative elite. As Marcel Boiteux, director general of Electricité de France, said: "A member of the Cabinet of the Minister of Industry told me one day . . . 'We have the agreement of the President of the Republic. We now have to get Giraud's.' "

In August 1975, a new deal was announced. First, Framatome was to have a monopoly of the domestic market and develop a base for exports; the EDF would simply cancel the orders for BWR's it had already placed with CGE. Second, the CEA was to buy out the major part of the Westinghouse 45 percent interest in Framatome. (In fact, it would get only 30 percent—less than the crucial blocking minority interest—with the remaining 15 percent retained by Westinghouse until the overall license agreement expired in 1982.) Third, a major research effort would be mounted, as a joint CEA-Westinghouse exercise, to develop French technology so that the license agreement would not have to be extended beyond 1982. France would then be independent, no longer having to pay royalties—estimated at 5 percent of the cost of each reactor sold. EDF stepped up its orders to ensure that the massive new investments were justified. It was a crash program for national independence, a program brimming with self-confidence.

In Germany, the Siemens company had kept a watchful eye on the whole event, of course, but without alarm. Despite Framatome's obvious political clout, Siemens was convinced that only political accidents would enable Framatome to beat them in export markets. Framatome, after all, had no tradition of technological excellence, no clarity in its mission, no coherence in its organization or investment plan. Siemens could afford to sit back and wait for the Baron Empain to get bored with the nuclear business.

23

"A Certain Event"

The cumulative effect of all defects and deficiencies in the design, construction, and operation of nuclear power plants makes a nuclear power plant accident, in our opinion, a certain event.

—Three senior General Electric engineers, February 1976

Nineteen seventy-six was a major turning point for the antinuclear movement. In that year, a handful of nuclear engineers resigned from responsible positions in the industry, declaring that nuclear plants were unsafe and a threat to man's existence. The stalwarts of the industry had never before publicly voiced such criticism. Prior to these resignations, nuclear industry leaders had found it relatively easy to dismiss antinuclear groups as emotional, campus-style upstarts. Such labels could not be pinned on these dissident engineers. Like many intelligent, earnest scientists of the mid-fifties and early sixties, they had been attracted to the progressive image of atomic energy. In the beginning, they had shared the conviction of the profession: that the technical problems affecting the safety of reactors could be solved. In theory, it seemed so. But they came to believe that the degree of precision needed both to build and to operate the reactors was unattainable in practice.

The first to resign, in January 1976, was Robert Pollard, a thirty-five-year-old electrical engineer with the U.S. Nuclear Regulatory Commission, the federal agency that had replaced the AEC as regulator of nuclear plants. Pollard, one of forty-eight project managers on the NRC, was in charge of the safety review of several nuclear plants, including the controversial Indian Point plant, twenty-six miles north of New York City. In a deeply disturbing resignation letter, Pollard said "sheer good luck" was the only basis for thinking that a very serious accident could be avoided given the inadequate design of the Indian Point reactors.

Pollard's departure was followed in February by the resignation of three senior engineers from General Electric's nuclear energy division in California. They immediately announced their commitment to the antinuclear cause. One of them, Dale Bridenbaugh, said he had grown "increasingly alarmed at the shallowness of understanding which has formed the basis for many of the current designs." Nuclear power, he said, had become a "technological monster and it is not clear who, if anyone, is in control."

Eight months later, in October 1976, another NRC engineer, Ronald Fluegge, resigned. The NRC had "covered up or brushed aside nuclear safety problems of far-reaching significance," he said. "We are allowing dozens of large nuclear plants to operate in populated areas, despite known safety deficiencies that could result in very damaging accidents."

These defections were a terrible blow to the nuclear advocates. They had become accustomed to the criticism of experts from the biological sciences and they had also taken in their stride the significant minority of physicists who had joined the band of nuclear critics, primarily because of fears of nuclear proliferation. The defection of engineers was more serious; this was the heartland of the nuclear community, the center of its self-confidence and the foundation of its public acceptability.

Even before the fifth resignation *Nucleonics Week*, an industry journal, declared: "Four Resignations Radically Change Complexion of Nuclear Fight." Suddenly a new credibility had been given to the criticism of nuclear safety. If, as the engineers said, reactors could never be made safe enough, then the nuclear industry appeared doomed; its already shaky public image might never recover if that opinion became accepted.

The three GE engineers produced a litany of defects in GE's boiling-water reactor: pipes that cracked, seals that leaked, valves that stuck open, and unpredictable vibrations. GE and the NRC issued a point-by-point rebuttal showing that some of the problems had already been solved and the others were being studied. But the dissident engineers' case did not rely on single defects; it was the cumulative impact that mattered. The fact that so many different things could go wrong in so many different places—in design, construction, and operation—had convinced the engineers that they should give up their life's work. After resigning, they submitted testimony to the Joint Committee on Atomic Energy that concluded: "The cumulative effect of all defects and deficiencies in the design, construction, and operation of nuclear power plants makes a nuclear power plant accident, in our opinion, a certain event. The only question is when and where?"

For the first time, the public had conflicting evidence from the experts, the nuclear engineers themselves. If no plant could be guaranteed totally safe, then the nuclear industry had been making its design, construction, and operating decisions on the basis of "safe enough." The question of how safe is safe enough was, however, a political decision, not a technical one. From this point on, nuclear engineers, government-agency watchdogs, and the electric utilities would increasingly be called upon to justify the social and political implications of their actions. If engineers asserted a right to decide degrees of acceptable risk, they would be treated like politicians and not like impartial experts.

As the flow of technical criticism increased—especially from the United States—the expressions of overall hostility to nuclear power exploded throughout the world with a uniformity and spontaneity that was unprecedented. No environmental issue became so international so quickly, and the question of the safety of nuclear reactors was the focal point. The belief that the public had been denied information about nuclear power was quickly transformed into an assumption that the nuclear community had something to hide. The arrogance of the nuclear community added to this belief; in every industrial nation, the nuclear communities had regarded themselves as a special technical elite. Their confidence and their optimism had left the public totally unprepared for the possibility of accidents. So when accidents began to occur, such as the Hanford nuclear waste tanks leak in 1973; or were revealed as having occurred—such as the Russian catastrophe at Kyshtym, which became known in 1976—no reassurances from the experts could restore public confidence. Even a series of relatively minor reactor problems was received with alarm.

To the antinuclear groups, each incident—valves sticking open, pipes cracking, control rods being inserted upside down, defective welding, fuel rods shrinking, loose switches—became magnified, sometimes out of proportion to its significance. No matter how significant or insignificant any one of the incidents may actually have been, they all sounded serious, both because they were part of a nuclear power plant and because the experts were no longer credible. The feeling grew that, whatever the perfection of the system in theory, the practical prospects of ensuring that every component was manufactured with the requisite precision and that every plant was built and operated as perfectly as the textbooks said it could be was too much to ask for in a system reliant on human beings. Similarly, each release of radioactivity to the environment came to be regarded as a "near-miss." And such releases were universal. They had happened at the Dresden nuclear station near Chicago, at Turkey Point in Florida, at the Millstone plant in Connecticut, at Wurgassen and Brunsbuttel in Germany, at Tsuruga and Fukushima in central Japan. No matter how small these emissions were, it was easy to conceive of circumstances that could have made them more severe. As the nuclear critics developed their technical competence, they analyzed each mishap to show exactly how things had gone wrong, often suggesting that if something else had happened at the same time there would have been a catastrophe.

A large number of unrelated errors at each point of the fuel cycle added up to an overriding public doubt about the capacity of the "experts" to control nuclear technology. There were failures in proposed reprocessing schemes—as often technical as economic—in West Valley, New York; Barnwell, South Carolina; and Karlsruhe, Germany. In Japan, leaks were found at the Tokai plant just after it started up. Technical difficulties at the French La Hague plant near Cherbourg—most of which were smothered in official secrecy—resulted in a buildup of levels of radioactivity in marine life off the Brittany coast. Technical difficulties associated with the weapons program also undermined the image of professional expertise. The most significant was the fire at the Rocky Flats plutonium bomb factory in Colorado in 1969, which caused plutonium contamination in the surrounding countryside. Disregard for adequate safety procedures in uranium mining and tailings storage in the U.S. mines plus the failure to deal adequately with weapons wastes from the plutonium-producing reactor caused a similar lack of confidence. Other nations were equally cavalier in their disregard for safety standards in uranium production for bombs. For example, the failure to protect the stocks of uranium tailings at Rum Jungle in Australia's Northern Territory or in the Manhattan Project uranium-processing plant in Port Hope, Ontario, led to contamination of neighboring land, beaches, and towns; even of schools. There were also legacies of the early radium industry from an even more distant past, when safety standards were almost nonexistent, but they came back to haunt the nuclear advocates of the seventies. As if that were not enough, a number of bizarre incidents added to the growing doubts about technical expertise in the handling of radioactivity. In January 1978, the Soviet satellite Cosmos 954 disintegrated over Canada with its highly enriched uranium power pack spreading radioactive particles over the Arctic snow; and in the same year came the revelation of the 1969 incident in the Himalayas when the CIA had to abandon a nuclear-powered device spying on China. The device was thought to have broken up and polluted the source of India's Ganges River.

The nuclear critics quickly learned that the interdependence of the uranium fuel cycle meant that nuclear power could be challenged at each point: in mining, in transporting, in the power reactor, in storage, in reprocessing, in waste disposal—huge quantities of harmful radioactivity were produced throughout the cycle. For their part, the nuclear advocates responded by emphasizing how each incident had been successfully contained; the safety systems had worked. The nuclear engineers had always known, of course, that things could go wrong: even though each failure was, by definition, unexpected, they had devised backup safety systems—multiple lines of defense with emergency systems in reserve. So far, the advocates said, everything had worked as well as they had expected. And, they repeatedly stressed, there had been no fatalities at nuclear reactors. (This was true as far as the general public was concerned, but three workers had, in fact, been killed at an AEC experimental plant in Idaho, and in accidents in Yugoslavia and Czechoslovakia. In 1976 two workers were killed at the Jaslovske Bohunice nuclear plant in Czechoslovakia. A safety valve failed and they were doused with radioactive coolant. The Czech authorities refused to acknowledge the deaths until 1980.)

Despite the assurances, public doubts grew. The primary thrust of many expressions of public concern was the demand for more information before any irrevocable commitments were made. But the nuclear community had always confused public information with public relations: the official attitude—almost instinctive in the traditionally secretive nuclear world—had been to regard all public information mechanisms as a means of selling its own views. Such sham campaigns simply reinforced the public's opinion that the community had something to hide. And indeed, as the antinuclear movement penetrated the official handouts and obtained archival material, it would find that all too often they had.

Unlike the antiwar movement, which had been waged largely on political, moral, and emotional grounds, the antinuclear movement's greatest strength would come from its ability to disseminate facts about the decline and fall of nuclear power: the cost overruns, the health hazards, the safety hazards, and the apparent insolubility of the waste disposal problem. Hence, the movement's initial thrust was aimed at forcing the release of information. Everywhere, they sought to provide access to divergent views, to establish sources of independent advice, to test the unanimity of the official party line. In the United States, critics sought legislation requiring full public hearings on new nuclear construction. In England, Canada, and Australia, they fought for special ad hoc commissions of inquiry. In Sweden, Austria, and Switzerland, the issue became powerful enough to require resort to public referenda on the future of nuclear power, leading, in the Austrian case, to the mothballing of a reactor project.

By 1976, critics in the United States had forced the issue of nuclear power onto the ballots of six states. These referenda were all beaten by two-to-one majorities, allowing the advocates to claim a great victory over what they saw as the public's irrational fears of nuclear power. Increasingly, however, it became clear that an intense, active, hostile group, even if a minority, could not be ignored.

The extent of the feeling against nuclear power became apparent when action went beyond petitioning and picketing to a number of sporadic acts of civil disobedience occasionally leading to violence. Bomb explosions and other acts of sabotage occurred at nuclear power plants in France, Spain, and the United States.

Most of the opposition centered in local groups intent on blocking construction of nuclear plants in their communities. Yet the proliferation of such groups ultimately made them a national force. In the sixties, the focal point of the early antinuclear groups had been the environmental issues. Concern centered on the nuclear plant's cooling system, which was based on water drawn from nearby rivers or lakes. The water was heated to intense temperatures before being discharged back into these rivers and lakes, and it was known to have disastrous effects on the local environment. Protests were heard from the German wine growers on the Rhine, the fishermen in Japan's coastal waters, and the anglers of middle America.

Some early protestors also thought of nuclear power in terms of Hiroshima, believing that a power plant could explode like a bomb. But even those who understood that such was not the case came to realize the dangers of a catastrophic accident. Antinuclear groups that had initially organized because of a threat to their rural lifestyle or to the local ecology soon came to apprehend that their very lives were at stake. If the radioactive poisons that were continuously being produced inside a nuclear reactor should ever escape into the environment, everything in their path would be threatened. These anxieties would be reinforced in the ensuing decade as scientists came to challenge the prevailing view that power plants were safe.

Environmentalist groups had been slow to see the dangers of nuclear power. Initially, they had shared the image projected by the nuclear industry that electricity generated from the atom was especially "clean." During the sixties, utility executives throughout the world—from the head of New York's Consolidated Edison to the Soviet Minister for Power Stations—had justified their decisions to build nuclear plants in terms of reducing air pollution. The fledgling environmental movement, concentrating on the more visible pollutants from coal- and oil-fired plants, started to challenge this view only as their analysis of the invisible potency of radioactive substances became more sophisticated.

The nuclear fraternity was ready to meet the attacks of all these groups: its members were proud of the detailed attention that had been given to ensuring the safety of the reactors. The engineers especially were acutely conscious of how many apparently insuperable barriers they had overcome in devising special materials, designs, and operating procedures. They had approached their task with a commitment to "defense in depth"—a series of barriers designed to contain the dangerous radioactive substances such as uranium and plutonium. First, the dangerous substances in the fuel rods were enclosed in a special metal alloy cladding. Second, they were trapped by the reactor pressure vessel surrounding the fuel rods. Third, if the first two failed, the radioactive poisons would be caught by the special containment built around the reactor pressure vessel. In addition, special provision had been made for a series of emergency backup systems—some of them duplicated to ensure reliability. Nuclear engineers were supremely confident that all these precautions were adequate.

Despite its overwhelming self-confidence, the nuclear industry had been apprehensive about adverse public reactions. Radioactivity, after all, was a special kind of poison: it could not be seen, touched, or smelled. To most people, it was a total mystery and they were afraid of it. The nuclear advocates labeled such a response irrational, but their awareness of it reinforced their normal penchant for concealment. This, combined with their unshakable self-confidence, turned them into an arrogant, secretive elite.

As long as the public remained dependent on the experts themselves for information, the nuclear community could continue to act independent of real restrictions. Their virtual monopoly of information and the arrogance with which they treated any questioning of their preserve are perhaps nowhere better illustrated than in the case of the nuclear submarine *Thresher*.

On April 10, 1963, *Thresher* sank while on a diving exercise in the Atlantic. All 129 aboard died. Twenty-four hours later, without any conclusive evidence, Admiral Hyman Rickover dismissed out of hand any suggestion that the accident could have resulted from a malfunctioning of the submarine's nuclear power plant. Throughout the naval and congressional inquiries that followed, Rickover—scarcely a disinterested party—remained the sole source of expertise on the nuclear aspects of the accident, and his word was accepted without question. To this day, the full proceedings of the inquiry remain secret, but the available record indicates an alarming degree of public innocence as it unquestioningly accepted Rickover's assurances.

At a Pentagon press conference held the morning after the submarine disappeared, the Chief of Naval Operations conveyed Rickover's assurances that there was "no radioactive hazard." Later in the day, Rickover himself issued a statement detailing the "many protective devices and self-regulating features [that] are designed to prevent automatically any melting of the fuel elements." It was an extraordinary statement to make at a time when no one could have had any idea what had actually happened to the *Thresher*; the first small pieces of debris had only just been collected, and none had yet been analyzed. Nevertheless, the press readily praised the "inherent safety" of the Rickover reactors and ran articles that ruled out the possibility of a nuclear malfunction. Once excluded, it was never seriously considered again.

When Rickover appeared before the Naval Court of Inquiry at Portsmouth, New Hampshire, on April 29, he was quick to bolster his original assertions with evidence. Samples taken from the ocean bed where the *Thresher* had disappeared showed no signs of increased radioactivity, he reported. By offering the information, Rickover implied that it was important: that if the reactor had caused the accident, there would have been signs of radioactivity. In fact, the information was largely irrelevant: when the submarine sank, for whatever reason, it would have imploded under the pressure of the water, scattering pieces of the reactor over a wide area. Some of these pieces would have been in touch with the reactor core and would have been intensely radioactive. Even if some of them had been found, it would not necessarily indicate the cause of the accident. But the court of inquiry was not to know that: Rickover was the court's only source of expert advice on reactor physics and, according to the limited proceedings made public, the court never got around to asking how accidents might happen in a submarine reactor. Rickover had said that they could not happen, and that was good enough. When the court of inquiry reported on June 20, it issued a brief three-page statement from twelve volumes of secret testimony and stressed that no radioactivity had been found "in the search area."

Rickover assiduously steered the court's deliberations away from his reactor by introducing evidence that indicated some defects in the submarine's piping systems in an area unrelated to the reactor. Theories other than Rickover's had suggested that the *Thresher* sank because it lost power—that, for some reason, the reactor failed. Rickover dismissed such talk, and the court finally adopted his theory of a burst pipe as the "most probable cause" of the tragedy. This effectively exonerated the nuclear part of the submarine.

Questions about the submarine's reactor remain, however. One, in particular, concerns the last intelligible message that *Thresher* sent to its accompanying surface ship. The message said simply that the submarine was "experiencing minor problems." It was delivered in a voice later said to be "very relaxed." It seems most unlikely that a pipe failure with a consequent loss of oxygen or, perhaps, intake of water could be described by a submariner as a "minor problem." On the other hand, an unexplained loss of power could be described in that way. Another question, still unanswered in a satisfactory way, concerns the evidence given to the inquiry by a chemist from the Portsmouth Naval Shipyard. He reported that pieces of plastic of the kind used in reactor shielding and recovered in the search area showed signs of having been scorched by a "rush of flame." He also found that some of the plastic fragments had jagged edges with metal fragments embedded in them, suggesting some kind of explosion in or near the reactor. If a reactor loses its cooling water and the emergency cooling system fails, then the core of the reactor will continue to heat up, becoming unstable—a condition that can lead to a core meltdown. The facts brought to the inquiry by the Portsmouth chemist were consistent with a core meltdown, but Rickover did not mention that possibility to the court. Instead, a few days later, one of Rickover's lieutenants appeared before the inquiry and dismissed the "rush of flame" theory. He said he had tested one of the largest fragments and the discoloration that looked like a burn "apparently" came from the lubricant of a drill that had been used in the construction of the vessel. He did not refer to other smaller pieces of plastic that showed similar burnlike damage. The inquiry accepted his evidence; the other burn evidence from the Portsmouth chemist was ignored. The possibility of a core meltdown was never pursued again, either during the court proceedings or during the later congressional inquiry.

At the time of the *Thresher* incident, the concept of a core meltdown was not widely understood. It was to be several years before the general public became aware of the possibility of such a meltdown and what it entailed. During normal operations, the core of the reactor, which contains the fuel elements, is cooled usually by ordinary water to around 550 degrees Fahrenheit. If the flow of water is interrupted—because of a burst pipe or a stuck valve—the fuel elements can melt. In theory they could become a molten mass that could sink through the reactor containment and down through the earth's crust, coming out on the other side. The American nuclear engineers dubbed the phenomenon the "China Syndrome," assuming the molten mass from a U.S. reactor would emerge somewhere in China. To prevent this type of accident, the engineers devised a scheme called the Emergency Core Cooling System—ECCS—a separate cooling system that would trip in when the reactor overheated.

In the early 1970s, as international orders for the light-water reactor reached their highest levels, the effectiveness of the ECCS became the crucial point in debates over reactor safety. It soon emerged that even the experts could not agree whether ECCS was a viable safety procedure.

In July 1971, a Boston-based group called the Union of Concerned Scientists revealed that the AEC had just completed some small-scale tests on the ECCS; each one of the official tests had failed. The system had not worked in the way that the computer studies had predicted. The tests had been run at the AEC's research reactor at Idaho Falls, Idaho. Only negligible amounts of emergency core cooling water reached the core in any of the tests. For example, Idaho Nuclear, the company responsible for the experiments, gave these results:

> Test 845: Early analysis of test data indicates that essentially no emergency core coolant reached the core.
>
> Preliminary analysis for Test 846 indicates little or no core cooling by the emergency coolant.
>
> Tests 847 and 848: Preliminary analysis of the results of these tests indicates little or no core cooling by the emergency coolant.
>
> ECC liquid was ejected from the system in Test 849, as in previous tests with accumulator ECC, and at no time did ECC liquid reach the core.

The revelation was given wide coverage in newspapers and by two network television news programs, badly shaking the credibility of the nuclear community. The secrecy and banal assurances they had indulged in to protect the public from its own "irrationality" were about to rebound on them.

In the wake of the new revelations, the AEC held a series of public hearings on the ECCS. They were aimed, in part, at stalling the growing criticism of the ECCS at license hearings for new reactors. Between 1966 and 1971, only twenty-four out of the seventy-four license hearings were contested, but the rate was climbing rapidly. Rather than have each hearing delayed by a new critique of the

ECCS, the AEC decided to have the issue aired in special hearings. It had not bargained for the response from the Boston scientists, however; hearings that were originally scheduled to last only six weeks continued for almost two years, included one hundred and twenty-five days of public testimony, and produced twenty-two thousand pages of verbal evidence and one thousand supporting documents.

Ironically, the reason the hearings took so long was disagreement over the ECCS within the nuclear community itself. Henry Kendall, a physics professor at MIT, and a young economist, Daniel Ford, spoke for the Union of Concerned Scientists. But they were helped considerably by a flood of leaks about the ECCS from inside the AEC and from key AEC-funded laboratories.

To collect the evidence, Ford recalled, "We met in hotel rooms, in people's houses at 4 a.m. . . . one man proposed leaving documents at a gasoline station for us to pick up." Ford and Kendall received thousands of pages of internal AEC documents, many mailed in plain brown envelopes by unknown AEC employees. They learned of dozens of internal AEC safety experts who had questioned the reliability of the ECCS. Altogether, Ford and Kendall found some two dozen scientists inside the industry who had misgivings about the ECCS. They pieced together almost a decade of secret internal controversy within the AEC during which many suggestions for improving the safety of power reactors had been canceled, deferred, delayed, and rejected.

Useful as these leaks and contacts were, the depth and scope of the internal hesitancy over the effectiveness of the ECCS came out only under persistent cross-examination from the Union of Concerned Scientists. For example, in August 1971, George Brockett, a leading ECCS researcher, identified by the AEC regulatory staff as one of the country's leading experts in reactor safety technology, had written that present AEC reactor safety analysis was "unverified," "inadequate," "incomplete," and "uncertain."

During the hearings, it became clear that there had been no such thing as free discussion of reactor safety inside the nuclear community. On the contrary, it emerged that any nuclear scientist who spoke out of turn would face serious retaliation.

The hearings passed into the rapidly growing body of antinuclear folklore—not only in the United States, but throughout the world. Controversy about the effectiveness of the emergency system would continue, and the detailed revelations and point-by-point challenges of the marathon hearings were to have a lasting impact on every aspect of reactor safety. Writing about the hearings, even the industry journal, *Nucleonics Week*, would note: "They have opened up a Pandora's box of scientific doubts and bureaucratic heavyhandedness." Yet the overwhelming majority of nuclear engineers remained confident that the ECCS would work. At their most pessimistic, they saw it as the last line of defense against an improbable event. Not to the nuclear critics; they required a more detailed and comprehensive response than the simple assertion that nuclear accidents were "highly unlikely."

The most complete reply to the critics came in the United States in the summer of 1974 with the publication of the *Reactor Safety Study*. Professor Norman Rasmussen, dean of engineering at MIT, and a staff of fifty had poured $4 million over a three-year period into the study, which was funded by the AEC. Its basic conclusion was that the consequences of a reactor accident were no larger, and, in some cases, much smaller than the consequences of other everyday accidents to which the public is exposed. Accidents involving a meltdown were more likely than had previously been imagined, but the consequences were less than expected. It was the most intensive effort ever made to identify the precise scope of the dangers from nuclear reactor accidents, and its results were enthusiastically adopted and promulgated by nuclear advocates throughout the world.

The Rasmussen study was aimed at producing a consensus from the collective wisdom of nuclear engineers that nuclear power was safe—or at least that the risk of a major accident was low enough to be acceptable to the public. Specifically, it was aimed at helping gain renewal of the Price-Anderson Act—the legislation limiting the liability of the utility companies in the event of a catastrophic accident. More generally, the AEC desperately needed a favorable statement about reactor safety to help restore flagging public enthusiasm for nuclear power. Instead of simply analyzing the worst possible accident scenario, as had been done in earlier studies, the Rasmussen study group was directed to determine precise probabilities.

When a draft of the report became available in the summer of 1974, initial news stories emphasized one dramatic conclusion: that a disastrous nuclear accident was about as likely as a meteorite crashing into a city, or a chance of one in the course of a million years. Later, a more critical look at the document demonstrated that the "Executive Summary," from which the first press reports had been prepared, passed over the considerable detail in the twelve volumes of the report itself that, taken, in its entirety, gave a slightly different picture.

The Summary, for example, spoke only of "prompt deaths": one major reactor accident causes, say, ten deaths, which is the same number that would be caused by a meteorite shower falling on a heavily populated area. What the Summary omitted, however, was that the estimates in the appended reports showed that the reactor accident that caused the ten "prompt deaths" would eventually cause seven thousand cancer deaths, four thousand genetic defects, and sixty thousand thyroid abnormalities as well as contaminating some three thousand square miles of land. Overall, the comparison with the meteor looked very deceptive.

One critic of the report, the Union of Concerned Scientists, felt that the figures for the deaths and injuries were sixteen times too low. Another, the American Physical Society, in a study on light-water reactor safety, concluded that Rasmussen had underestimated by a factor of fifty the numbers for long-term cancers and genetic defects.

It was not the figures, however, that became the centerpiece of the attacks on the report; even after adjustment, the risks still looked low in comparison with car accidents, dam failures, and cigarette smoking. The critics plowed into the methodology of the report and tore it apart. The Rasmussen team had used computer analysis to go through a wide range of possible defects or component failures, assigning precise probabilities to each of them. The critics pointed out that a number of the "probabilities" were no better than guesses multiplied by guesses. They showed how the report had underestimated the possibility of a single accident, like a fire or an earthquake, resulting in multiple failures to a number of backup safety systems. They stressed that the averages produced by the team for weather patterns at reactor sites were meaningless; they could vary by up to one thousand times in the rate of dispersal of a radioactive cloud, for example. They pointed out that the report made no allowance for the unexpected or the unforeseen—like new materials defects or new chemical reactions between components—even though the entire history of reactor safety was littered with such examples. They criticized the way the report refused to adopt the consensus of the radiation specialists on the health effects of radioactivity, producing, instead, figures about half the size of those that had been widely accepted. They disputed the figures on how quickly evacuation and decontamination measures could be implemented in case of an accident. Finally, they criticized the misleading way the report, by limiting its analysis to a five-year reactor lifespan, had glossed over the fact that the component parts of nuclear plants become less reliable with time. Moreover, even within the AEC, reactor safety experts pointed out how the Rasmussen staff, claiming to have studied all possible accidents, had not mentioned a large number of accidents and component failures that had happened.

With a host of technical experts now available to the antinuclear cause, the Rasmussen report was put through a process of appraisal that was more intensive than any other document in nuclear history. The casualty at the end was not so much the precise assertion that nuclear power was safe as the ability to make any such assertion whatsoever, in view of the extraordinary range of gaps in knowledge. Once the dust had settled on the debate, one thing was clear: no one knew precisely how safe a nuclear reactor was.

By the time the full Rasmussen report appeared in 1975, nuclear power had become a much more highly politicized issue than it had been even in 1972 when the committee was appointed. Within those three years, the number of technical experts critical of nuclear power had grown; each small blemish in the Rasmussen study was categorized as part of the overall conspiracy of a scheming official nuclear community willing to distort the truth for reasons of institutional or personal interest. By 1979, the weight of technical criticism was such that the Nuclear Regulatory Commission—successor to the AEC—formally disowned the report. It rejected many of the calculations and was particularly scathing about the politically motivated promotional aspects of the report, notably in the misleading "Executive Summary." The NRC's reassessment was not, however, that nuclear plants were less safe than Rasmussen's team had said, but that the experts were less sure of the conclusions. In technical terms, the reassessment conceded that the range of uncertainty was greater or, in layman's terms, that no one really knew precisely how safe a nuclear reactor was. For a public that had shown itself increasingly unwilling to accept the assertions of expertise from the nuclear community, there was little reassurance in the limits of expertise being underlined in this official way.

As the technical competence of the antinuclear groups increased and they became a direct challenge to the nuclear community, their political protests progressed from small demonstrations of local groups into national and sometimes international campaigns of civil disobedience. The first large-scale operation—with more than twenty thousand people taking part—took place in Germany, a nation with no tradition of political action of this kind.

In February 1975, three hundred residents of Whyl, a small village in the southwestern Rhineland well known for its Riesling and Liebfraumilch wines, occupied the construction site of a 1,350-Mw nuclear power station. They sat down in front of the bulldozers and halted work. The German police responded with water cannon and made some arrests, but the Whyl residents persisted and were later joined by others from throughout Germany and from across the border in France.

The Whyl residents had begun the demonstration out of fear they might lose their livelihood—that the extra humidity caused by the plant's cooling towers would increase the fog and frost in the area and damage their vines. But their concern quickly expanded into a full-scale attack on the safety of the proposed reactor. Like residents near other reactor sites throughout the world, the people of Whyl and neighboring villages became conscious of the fact that a full-scale reactor would contain huge amounts of lethal radioactivity; the health of everyone in the region depended on containing the entire mass of radioactivity inside the plant with a perfection that some scientists said was beyond technical capability. None of the residents was willing to accept talk of risks and probabilities; only the size of the consequences of something going wrong seemed important.

With both national and international support and intense local organization, the Whyl residents were able to occupy the site for an entire year; they even built a makeshift village. Their petition against the proposed reactor attracted 90,000 signatures. In the end they won: construction was halted. The demonstration became an important precedent for future confrontations in Germany, France, and the United States; and in many there would be violent clashes and mass arrests.

On April 30, 1977, some two thousand Americans occupied the parking lot of a nuclear plant construction site at the quiet seaside town of Seabrook, New Hampshire. For almost ten years, the state's utility, the Public Service Company, had been planning to build twin 1,150-Mw reactors on a 715-acre site near a marshy estuary that is a breeding ground for a variety of birds and marine life. It was estimated that the cooling system for the plant would use a billion gallons of sea water a day and return it to the ocean after raising its temperature by 30 degrees Fahrenheit. As elsewhere, the first protests came from local residents afraid for their jobs—particularly the clam diggers and lobstermen, who feared that the warmer waters would affect the marine life. Brushing aside both these objections along with the fact that the site of the reactors was on a geological fault line and that the original price tag for the plant had more than doubled (to almost $2.5 million), the Public Service Company went ahead with its plans. In 1976 it was granted a license by the NRC.

A local antinuclear group, the Clamshell Alliance, was inspired by the citizens of Whyl to organize a civil disobedience campaign. The "Clams," as they were called, soon became the best-known of similar alliances—with names such as "Crabshell," "Oystershell," and "Abalone"—that sprang up all over the United States. Composed mostly of young people, but also attracting the support of local residents, fishermen, and farmers, the movement produced widespread civil disobedience campaigns.

The April 1977 occupation of the Seabrook site's parking lot was a huge success, largely due to the hard-line reaction of New Hampshire's conservative governor, Meldrim Thomson. An enthusiastic supporter of the Public Service Company's nuclear ambitions, Thomson ordered a mass arrest of demonstrators, thus assuring nationwide media coverage of the event. Altogether 1,414 people were interned. More than half of them refused to pay bail and were held for two weeks in national guard armories, thus providing daily news stories about their welfare.

Seabrook was quickly assured of a place on the antinuclear protest calendar as the movement grew in America and Europe. Although the political power of the movement remained locally based, the issue had, by the end of the seventies, received so much attention that it was a national debate.

24

"The Buddha Is Smiling"

The coded message flashed from Pokaran in the Thar Desert to government offices in New Delhi on May 18, 1974: "The Buddha is smiling." India had successfully exploded an atomic bomb—the sixth nation to become a member of the nuclear weapons club, and the seventh if Israel's untested bomb was included. The event was to have the greatest impact on the course of world nuclear development since the first Soviet test of August 1949. Within months, active concern with the issue of nuclear proliferation, which had peaked with the Non-Proliferation Treaty in 1970 and then receded into backroom dickering, became a new and pressing matter on the international diplomatic agenda.

A psychological barrier had been broken. The Indian test appeared to be the first step toward a new age of nuclear chaos. It now seemed that anyone might build a bomb: desperately poor nations, mad dictators, even political terrorists.

International opinion had long since reached the conclusion that there was no difference between a "peaceful" and a "military" explosion; the 1970 Non-Proliferation Treaty had specifically emphasized the inseparable link between the two. Despite this and in the face of strong international criticism, the Indian government claimed the explosion was "peaceful"; it was to be the first of many explosions that would be used for such peaceful pursuits as changing the course of rivers and building dams and harbors. Indian citizens, greeting the test with a great outpouring of national pride, dismissed the criticism as hypocrisy: if developed nations were allowed to explode nuclear devices, why not underdeveloped countries?

Few believed the Indian claims. Many physicists saw atoms for peace and atoms for war as Siamese twins. To the growing body of critics of the nuclear power business, nothing had proved this point better than the Indian explosion. It had taken only a change of political will for India to cross the line from making electricity to making bombs.

Who was to blame for this blatant act of proliferation? The plutonium used by India came from nuclear plants that had originally been wholly peaceful in design and intent. Two countries, Canada and America, had provided India with the technology and the materials. Both were acutely embarrassed at the outcome, though they reacted in different ways. The Canadians had supplied the Indians with a heavy-water reactor and they immediately cut off further nuclear assistance for the power reactor program. As supplier of the heavy water, the Americans had also contributed to the final product, but they tried to shed their responsibilities.

Spending much time blaming the Canadians, Secretary of State Henry Kissinger chose to deny that the United States had contributed any nuclear material for the bomb. He also declined to pressure the Indians to abandon their bomb program. Kissinger accepted proliferation as a fact of life and, under his foreign policy leadership, the Nixon administration downgraded the priority of nonproliferation policies that had been developed under Kennedy and Johnson.

Two years would pass before the Nixon administration admitted publicly that it had supplied the heavy water to India—and had done so without placing any restrictions on its use. At the same time, congressional inquiries would disclose that American nuclear assistance to India had included training no fewer than 1,367 Indian nuclear physicists in various aspects of atomic know-how. Some of them had been taught the secrets of plutonium, the rationale being that they would eventually use their knowledge to build breeder reactors. At Hanford, where the American reprocessing plants were located, special classes for foreign students had been held from 1958 until 1972. Of the 169 students who had attended, 14 had come from India. In fact, all the nations that would become potential nuclear weapons states had been represented: 7 Brazilians, 12 Germans, 5 Israelis, 29 Japanese, 11 Pakistanis, 8 South Africans, and 9 Taiwanese.

As the congressional inquiries progressed—particularly those of the Government Operations Committee led by Senator Abraham Ribicoff of Connecticut—the extent to which American policy had in fact allowed proliferation in the name of the "peaceful atom" became much clearer. For example, at least four of the Indians at Hanford had been partly funded by the long-term, virtually interest-free loan America had given India to purchase a power plant from the United States. The U.S. nuclear assistance programs had not been designed to give other nations the capacity to make bombs, but that is effectively what had happened.

These congressional inquiries eventually identified two critical areas that required an immediate change in American policy. The first was the "backyard bomb" with its potential terrorist connection. The second was the "overnight bomb": the fact that any nation possessing separated plutonium could build a bomb virtually within hours. Although the hearings began as an outgrowth of congressional concern over the laxness of conditions on the export of nuclear materials, these volatile scenarios led them to undertake a full-fledged review of the entire posture of American nonproliferation policy.

The basic proliferation issue that returned to center stage was the one of "critical time": the time it would take a nation to make a bomb, providing it had enough fissile material. The 1970 NPT safeguards had ducked the primary question: What good were inspections if a nation could build a bomb almost overnight? Even if the early-warning system provided by NPT inspections worked, so that the international community learned in advance of a country's intent to build a bomb, did it provide enough time to bring political pressure to bear to force that country to drop its plans? And what if it did not work? What sanctions could then be brought against a nation that already had a small arsenal of atomic bombs?

Congress took the first step. As an outgrowth of the post-Indian bomb hearings, it proposed cutting off foreign aid to any nation that built bomb-sensitive facilities—such as a reprocessing or an enrichment plant—that would give that nation access to bomb-grade fissile material. By the end of the seventies, this action would lead to a dramatic new turning point in nuclear history as the United States sought a total halt to the development of the breeder reactor and its associated plutonium fuel cycle. It was not a popular move among America's allies; those already committed to a breeder policy, such as Britain, France, Germany, and Japan, resisted it strongly. But it was the heart of the problem: India's development of a reprocessing plant to extract plutonium had, after all, originally been based on a desire to build a breeder reactor.

At the time, nothing had seemed more progressive or benign. The 1956 agreement with the Canadians that gave India a plutonium-producing heavy-water reactor was signed during the first ebullient days of the Atoms for Peace program. It came in the immediate wake of the personal triumph of Homi Bhabha, founder of India's nuclear program, as president of the 1955 Geneva Conference. There was no American policy requiring "safeguards" at the time; the word of a nation like India, which promised to use the reactor for peaceful purposes only, seemed enough. No warning bells were sounded even when, two years later, Bhabha announced his intention to build a reprocessing plant to extract plutonium from spent fuel rods. The plant was duly constructed at Trombay, 25 miles from Bombay; everything seemed open and aboveboard. There was no secrecy about it. Like his colleagues elsewhere, Bhabha assumed that sooner or later electricity from nuclear energy would be produced by breeder reactors. Indeed, Indian "progress" was internationally acclaimed.

Only in the years that followed, when nuclear exporters like the Canadians and Americans began to insist on the idea of safeguards, did the Canadians express concern about their absence in the case of the Indian reactor. In 1963, when India sought a new contract with the Canadians for a bigger power-producing reactor, the Canadians insisted it should be safeguarded by inspection. Some Canadians wanted to backdate the agreement to include the first reactor. "We knew that reactor was naked," a Canadian negotiator recalled. "Here was a chance to do something about it. But the commercial people kept saying that if we didn't give the Indians what they wanted, they'd get it elsewhere." So the first reactor remained without safeguards.

Bhabha had known from the beginning that the reactor and the Trombay reprocessing plant gave India a weapons option. In October 1964 his sense of urgency about that option increased. The Chinese, with whom India had fought a border war only two years before, exploded their first bomb.

The Chinese bomb hurt Bhabha's pride as much as his patriotism. For years, the Indians had been boasting about how much more advanced they were in nuclear science than their larger neighbor. Within weeks Bhabha was saying publicly that a nuclear deterrent seemed to be the only way to balance China's nuclear threat. "The only defense," he said, "against such an attack seems to be the capability and threat of retaliation." But he added, "We are still eighteen months away from exploding either a bomb or a device for peaceful purposes, and we are doing nothing to reduce that period." Within a few months, however, he received approval from the Indian Prime Minister: Lal Bahadur Shastri gave Bhabha the signal to prepare a test site. As a signatory to the partial test ban treaty, India was prevented from testing in the atmosphere, so an underground test site had to be constructed. In 1966, land was acquired—ostensibly to provide a new "artillery testing range"—at Pokaran. Preparations required the relocation of twenty-five villages. Some 200,000 people were affected, and the construction aroused considerable local protest, despite the compensation payments made.

The test site preparations were a major step toward a bomb, but the Indian military opposed any major diversion of resources from conventional arms. Even Bhabha himself did not advocate full-scale development of a nuclear arsenal. All he wanted was a single explosion to prove India's nuclear capacity. His sensitivity to matters of status and reputation had been nurtured in the artificial world of an international community in which symbols were regarded as being as good as the real thing. In Bhabha's mind, reputation and reality had become hopelessly confused; he believed that the single explosion was actually relevant, indeed important, to national security. Ironically, he never lived to witness the actual event; at the end of 1966, Bhabha died in a plane crash.

With Bhabha's death, the momentum of the Indian bomb program was temporarily lost, and his successor as chairman of the Indian Atomic Energy Commission, Vikram Sarabhai, did not pursue the project. Indeed, during his term of office, Sarabhai accepted much tougher safeguards on India's new nuclear plants.

Within the Indian AEC, however, a pro-bomb faction, centering on Bhabha's key lieutenants, remained. They were able to get Sarabhai's permission to build a plutonium-fueled research reactor—ostensibly to pursue breeder reactor research. In fact, the pro-bomb lobby was handling—and producing—significant quantities of plutonium. They even called the research reactor, code-named CIRUS (Canadian-Indian Reactor, United States), their "bomb." All they were waiting for was a change of political will. It came soon after Sarabhai's death in 1971, under the new leadership of Indira Gandhi.

Gandhi was ready to revive the bomb project. Her father, Jawaharlal Nehru, had been the patron of science and, especially, of Homi Bhabha. He had isolated nuclear policy from the usual checks and balances of the Indian democratic system and allowed decisions to be made by direct contact between the Prime Minister and the Atomic Energy Commission. No other ministers were involved, and such other sections of the bureaucracy as were needed—defense, foreign affairs, or finance—were brought in only *after* decisions had been made. Within this closed world, the option to test a bomb had been around for so long that everyone assumed it was just a matter of time before it would be exercised. It was easy for Indira Gandhi to convince herself that this was the stuff of patriotic immortality. Her own self-image of personal strength and her sense of mission would be reinforced if she could invoke the full strength of Indian patriotic sentiment with a nuclear explosion. After the transient glory of the 1971 victory in the war with Pakistan and the creation of Bangladesh, Gandhi's domestic political fortunes had begun a rapid downward spiral, which would culminate in the State of Emergency in 1973. A dramatic act of national assertiveness could reverse the trend. So, at the beginning of 1973, Gandhi ordered the Pokaran test. In the end, rather than the decision of a leader strong enough to take it, it was the decision of a leader weak enough to need it.

In the month following the test, the sense of nuclear anarchy intensified. President Nixon, on a tour in the Middle East, offered power reactors to both Israel and Egypt. Such nuclear largess in that unstable region seemed especially threatening. This was followed by a series of international nuclear deals which only served to confirm the feeling that the spread of nuclear weapons was now out of control. The momentum of the worldwide nuclear industry seemed unstoppable as it dispersed facilities that would produce bomb-grade material. Germany signed a massive contract with Brazil that included every aspect of the nuclear fuel cycle, including a reprocessing plant and an enrichment plant. France signed contracts with Pakistan and South Korea to supply reprocessing plants. The French also agreed to supply Iraq with research reactors that used highly enriched uranium. The Taiwanese were caught (by the Americans) in the act of building a clandestine reprocessing plant of their own. The South Africans announced they were going to proceed with an indigenous enrichment plant. In the midst of a rapidly multiplying series of international sales of power reactors, it was this proliferation of technology using bomb-grade material that was most disturbing.

Smarting from the embarrassment of the Indian test, Canadian diplomats urged the reconvening of the old Western Suppliers Group, no longer merely to monitor the supply of uranium but to control the supply of reactor technology. As yet, there were no international restrictions on technology transfers; indeed, the International Atomic Energy Authority safeguards negotiators had explicitly decided not to require NPT safeguards on technology.

In the fall of 1974, six months after the Indian test, the secret suppliers' group reconvened at the U.S. embassy in London. The first meetings were attended by the United States, Russia, Britain, France, Canada, Germany, and Japan. They drew up a list of sensitive technology and agreed that items on it should be restricted from export unless they were properly safeguarded. The list became known as the "trigger list"—a phrase implying that any item on the list would "trigger" the need for safeguards. Another seven nations—including Belgium, Holland, Sweden, and Switzerland—were asked to join the group. In 1975 its existence became public.

Two important proposals failed to get the approval of the fourteen member nations. The Canadians, who had been pressing for the toughest safeguards in the wake of the Indian test, suggested that no nuclear material—either fuel or equipment—should be sent to any country that had an unsafeguarded plant of any description. This would have included the plutonium-producing reactors in India and Israel, plus South Africa with its unsafeguarded enrichment plant. The fourteen-nation group could not reach agreement on this proposal; toward the end of 1976, first the Canadians and then the Americans imposed the condition on themselves. They also proposed a ban on the export of reprocessing and enrichment plants, but the French and the Germans, who had just completed sales of these items, refused to go along. The group managed to come up with only a vague agreement to exercise restraint on sales. Dissatisfied with this loose arrangement, the Americans worked toward rolling back the sales that had already been made. These included four French deals with South Korea, Taiwan, Pakistan, and Iraq, and the German deal with Brazil. Under strong U.S. pressure, South Korea canceled its contract with France to buy a reprocessing plant and the Taiwanese dismantled their plant, which was already under construction. This left the French deals with Pakistan and Iraq and the German deal with Brazil. While the Germans eventually agreed to halt future exports, they refused to pressure Brazil to end the agreement for its enrichment and reprocessing plants.

The French deals, however, had been made in 1975 under the premiership of Jacques Chirac. A leader of the Gaullist party, Chirac was the special custodian of the Gaullist tradition that saw any act of agreement with American nuclear policy as an affront to French patriotism. Tweaking American noses in the nuclear field—even if it meant exporting sensitive materials—had long been a touchstone of the Gaullists. In August 1976, Chirac resigned, and under the new premiership of Raymond Barre, French policy changed. President Giscard d'Estaing undertook unilaterally to prevent any further exports of reprocessing plants and sought to overturn the agreements already made with Pakistan and Iraq. France eventually pulled out of its contract to supply a reprocessing plant to Pakistan.

The United States and Canada slowly led other nations toward a new policy on proliferation. Mere detection of a diversion of bomb-grade material, the United States and Canada maintained, was no longer enough. The central issue was not, as the NPT signatories had thought, the act of exploding an atomic device; it was having the capability of exploding one. Goaded by the North Americans, the Western nations and Japan were coming around to adopting a nuclear export policy that had always been followed by the Soviet Union—except for their Chinese adventure. The Russians had never allowed their customers to keep any fissile material. All spent reactor fuel rods had to be returned. It was too late for the West to impose that restriction, but their new policy was heading in that direction. Canada and America were now proposing that enrichment and reprocessing plants—the producers of the "inherently dangerous" bomb-grade materials—be built only on an international basis. It was, in fact, a throwback to the conclusions of the Acheson-Lilienthal plan, which had said explicitly that no workable safeguards could exist as long as "inherently dangerous" materials were allowed to remain under national control.

In 1976, the outgoing Ford administration shifted its policy in the direction of a tougher stand on nuclear proliferation. Before leaving office, Ford declared that the long-standing assumption the breeder would be the reactor of the future and everyone would have a reprocessing plant should no longer be taken for granted. He urged a policy of restraint on its development. And, at the beginning of 1977, the newly elected U.S. President, Jimmy Carter, took this policy realignment to its logical conclusion and stunned the nuclear nations by announcing that the United States hoped to ban reprocessing and breeder reactors, or at least halt their development, until energy needs grew to such an extent there was no alternative.

In taking Ford's policy a major step further, Carter declared that the projected demand for world electricity did not justify the building of the breeder. Instead of a uranium shortage—the key to making the plutonium breeder reactor an attractive proposition—there was in fact, a glut. Not only had demand for known sources dropped with the cutback in nuclear power plans, new sources had been discovered that had yet to be exploited. With more than enough uranium to meet forecasted needs, the breeder was unnecessary. Moreover, the cost estimates of the full-scale breeders were still rising rapidly, and many plans to build commercial reprocessing plants, needed for the proposed breeder programs, had collapsed. Now was the time, said President Carter, before any irrevocable commitment to the plutonium economy was made, to develop alternative nuclear power technologies that would not involve the use of bomb-grade material.

In the name of a crusade against the proliferation of nuclear weapons, the entire thrust of a thirty-year technical assumption that breeders were the ultimate reactor types was suddenly to be halted. All the long-term plans for the industry would have to be rewritten. For his part, Carter showed the strength of his conviction by announcing that the United States would set an example to the rest of the world by delaying its own plans for reprocessing and breeder plants.

Carter had to wage his crusade on several fronts. At home, the Congress, refusing to accept that America should have to set an example to the rest of the world by canceling its own projects, proceeded to vote funds for the technically obsolescent breeder reactor on the

Clinch River in Tennessee. America's allies in Europe and Asia were even more reluctant to agree with the President. With the oil embargo still fresh in their minds, they did not believe it was time to cut off any energy option, however bad its economics appeared in the short term or whatever the proliferation risks. The Carter policy became a major source of friction between the United States and its closest allies, leading to a prolonged stand-off.

France, the United Kingdom, Germany, and Japan all had commitments to proceed down the breeder path; the momentum in their nuclear bureaucracies and research institutes was overwhelming. None of them felt that they had the flexibility to consider alternative energy systems in the medium term, the way the resource-rich United States did. Whatever the risks, they were all willing to pay penalties for a degree of energy independence.

The French, in particular, had gone further than anybody else in committing themselves to breeder development. They believed their technology was more advanced than America's and that the United States was cynically stalling for time because of its technical lag. More than anything, however, they bitterly resented the President of the United States—and a host of American experts in his wake—publicly challenging all the assumptions of their domestic nuclear policy, a policy that was already controversial enough at home. The emergent trend of greater French cooperation with America's nonproliferation drive promptly ended.

Like the French, the British still believed the breeder reactor was inevitable. They also shared the French determination to supply a major slice of the international market with reprocessing services. It quickly became clear that the United States could not prevent reprocessing development in the existing weapons states. The key question became how to control those suppliers who were not already weapons powers. That meant Germany and Japan. But Carter encountered equally fierce opposition from them. A Japanese diplomat expressed the sense of frustration with the U.S. proposal when he said: "For twenty years we have followed U.S. guidelines on nuclear policy. Now you are saying you made a complete mistake. But it is too late."

Perhaps more significant, however, than the drive for long-term energy independence, the momentum of nuclear bureaucracies and ongoing research, and the potential international market was the unintended threat embodied in the Carter proposal to the whole nuclear industry. At the time, the industry was under heavy criticism for its failure to do any serious technical work on how to dispose of highly radioactive spent fuel rods. The assumption had been, in countries like Germany and Japan, that reprocessing would take care of this. In both those countries, no new reactors were being licensed unless the operators had adequate waste-disposal plans, and in many cases, these plans were based on the fact that the spent fuel rods would be reprocessed. Further technical advance might eventually show that wastes could be disposed of in other ways, but for the moment reprocessing was a technical imperative.

The American policy had to be taken seriously in Europe and Japan, however. The United States still held a nuclear trump card as the largest supplier of enriched uranium for the light-water reactors, and the Europeans and the Japanese—both in government and in private industry—were well aware that future supplies might be conditional on a country's renouncing the breeder. In the long term, this threat would carry less weight because the four enrichment plants being built in Europe—two in France, one in Holland, and one in Britain—would soon overturn the American monopoly. But in the short term, the Carter policy could not easily be swept aside. Three uranium suppliers—the United States, Canada, and Australia—controlled more than half of the Western world's reserves, and each was taking a hard line on nonproliferation. All were insisting that any country that bought uranium for fuel rods must obtain their consent before reprocessing the spent rods. As yet, only America was willing to support a total ban on reprocessing: Canada and Australia talked instead about "improving" and "intensifying" the existing system of safeguards.

By the end of the 1970s, most American safeguard experts, agreeing that bombs could be made in a matter of days from fissile materials diverted from a reprocessing plant, wanted to revive the safeguards concept known as "critical time." This required linking the frequency of inspections of sensitive facilities, like reprocessing plants, to the amount of time it would take a country to make a bomb; if the time was, say, ten days, the concept called for an inspection every ten days. The first round of safeguards meetings in the late 1960s had focused on critical time as the centerpiece of a safeguards system, but it had been abandoned as economically unacceptable. The frequency of the inspections was considered too great an interruption to the running of the plants.

Germany and Japan, both nonweapons powers and both planning breeder programs, were at the center of resistance to critical-time inspections, which would make their breeder programs economically unworkable. They nevertheless understood the problem well. The Japanese safeguards expert, the soft-spoken and amiable Imai Ryukichi, admitted, "It is doubtful that any safeguards system can cope effectively both in quantity and in quality with the anticipated growth of the nuclear industry . . . when fast breeder reactors dominate the scene, plutonium will be consumed at an annual rate of hundreds of tons. Effective inspection will prove impossible unless a dramatically different concept of safeguards is introduced."

Under new political pressure from the United States, the IAEA's advisory group on safeguards met in Vienna in January 1978. The minutes of the meeting, couched in the ponderous bureaucratic language of all previous safeguards debate, nonetheless recorded a new sense of urgency. "The continued unavailability of a credible set of values for detection time might raise questions as to the inherent safeguardability of certain types of nuclear facilities, particularly reprocessing plants . . . an internationally accepted set of values for detection time must be established at an early date."

The group agreed that it would take only ten days to convert the fissilc plutonium in reprocessing plants into metal and as little as two days to make the separated plutonium metal into a weapon. They went on to recommend that the "critical time" should be calculated as the same number of days. Not surprisingly, the German and Japanese delegates objected and the meeting ended with the safeguards matter unresolved. The fears of uncontrolled proliferation that had started with the Indian bomb test remained.

After more than twenty years of active promotion of nuclear power through the international program of Atoms for Peace, the U.S. policy of dealing with the spread of weapons on a step-by-step basis—facing the problems only when they could no longer be deferred—had failed to produce any international consensus about how to handle proliferation. At the same time, along with other nuclear weapons powers, the U.S. commitment to large nuclear arsenals had intensified. Stockpiles had grown at alarming rates, and nonweapons states had become frustrated and angry. Not only were they reluctant to accept what they saw as unjustified economic hardships in the name of nonproliferation, they were also unwilling to accept the basic premise of America's nonproliferation crusade: that they should not be allowed to handle bomb-grade fissile material because they were, somehow, less responsible.

They began to point to the provision in the 1970 Non-Proliferation Treaty that required the weapons powers to negotiate in good faith both for the "cessation of the nuclear arms race" and for "nuclear disarmament." The SALT process had partially fulfilled the arms race obligation, but there had been no negotiations on disarmament at all. As the nonweapons states saw it, the superpowers were asking them to accept restrictions that went well beyond the letter of the treaty while they themselves were not adhering to the treaty as signed. More than that: the nonweapons states were being asked to do these things in a way that cut off possible energy sources at a time when energy independence had become a universal basic policy objective. It was hardly a propitious moment for the United States to decide that it had to face the implications of its policy of deferral.

Like the safeguards question, every aspect of the new American stance became a matter of technical controversy: whether enough uranium existed to defer breeder reactors, whether new kinds of reactors or entire fuel cycles that did not use bomb-grade material were technically or economically practical. Developing new rules and a new consensus on ways to halt proliferation would take time—a lot of time. The Carter administration seemed determined that no irrevocable commitments should be made anywhere until a completely new direction had been found. During the interim, American vigilance would have to be directed not just at "sensitive technologies" but also at "sensitive countries."

In some of the "near-nuclear" countries, the American policy of deferral worked. South Korea and Taiwan dropped their plans for reprocessing. The ambitions of the Shah of Iran disappeared with his regime in 1979, and the large Iranian reactor-building program was canceled. All the steps taken by Arab nations—from Libya to Iraq—to acquire a balance to the bomb they assumed the Israelis possessed were frustrated. Even the German deal with Brazil to supply an untried enrichment plant and a reprocessing plant in an eight-reactor package seemed to be likely to fall under its own weight. Hailed as the largest commercial contract ever when signed at $4 billion, there were no similar cries of acclaim when it became clear the cost would be more than $20 billion. With such cost escalation, the availability of low-cost undeveloped hydroelectric sites, and growing criticism about nuclear safety in Europe and the United States, it seemed likely that most of the Brazilian package would be delayed for many years.

There were some countries, however, where the United States found itself already bereft of bargaining power. The most significant was South Africa, which had its own substantial uranium reserves and an enrichment plant built in part with German-supplied equipment. This powerful combination would allow it to develop highly enriched uranium as bomb-grade material. No one had any leverage over the independent South African program—not as long as it remained a covert proliferation.

In August 1977, the Soviet Union publicly announced that its satellite observation of South Africa had revealed preparations for a nuclear test site in the Kalahari Desert. American surveillance systems, alerted by the Russians, soon confirmed the allegation. A flurry of international threats from the United States, the Soviets, France, and the U.K. followed. If a test had been planned, it did not take place. Nothing more was said about the alleged test-site preparation. Then, in September 1979, an American Vela satellite, designed to spot the characteristic double flash of atomic explosions, detected what appeared to be a bomb test somewhere in the southern hemisphere.

The United States said it was unable to corroborate the satellite picture with any other evidence, such as telltale fallout. The nearest land mass to the flash appeared to be South Africa, and the South Africans immediately became a prime suspect; another was Israel. There was some speculation that it might have been a joint South African–Israeli effort. Both governments denied any involvement, and the United States backed off. The consensus among scientists who examined the satellite picture was that a nuclear. explosion had, in fact, taken place, but the source of the explosion remains a mystery.

India was another problem country for America. Despite explicit American displeasure and notwithstanding efforts within India itself by Prime Minister Moraji Desai to restrain the country's nuclear hawks, it continued to increase its supply of unsafeguarded plutonium. Meanwhile, India's neighbor, Pakistan, became the focus of international attention in 1979, when it was discovered covertly buying parts for a centrifuge uranium-enrichment plant. In his final testament from his death cell in 1979, Prime Minister Ali Bhutto had given the first hint: "Christian, Jewish, and Hindu civilizations have this [nuclear] capability," he said. "The Communist powers also possess it. Only the Islamic civilization is without it, but that position is about to change." It was Bhutto who, taking over ministerial responsibilities for the Pakistani nuclear program in 1965, had established it as a serious enterprise. He said at the time, "If India builds the bomb, we will eat grass or leaves, even go hungry, but we will get one of our own."

His plans to buy a reprocessing plant in 1966 had been overruled. But after Bhutto himself took control of the country in 1971, they were revived, and a contract signed with France. After the French policy switch at the end of 1976, it became clear that they would renege on the contract, and Pakistan turned to uranium enrichment. A Pakistani physicist living in Holland had been given access to crucial classified information at the Dutch centrifuge plant in Almelo, and he had been able to put together a shopping list of all the different materials and equipment required for an enrichment plant, together with the companies that could supply them. The Pakistan Ordnance Service set up a "Special Works Organization" that created a covert network of dummy corporations and began buying all of the guarded materials for "textile mills." Western intelligence belatedly put together the full ramifications of the Pakistani buying spree and cut off supplies. No statement was made about how much had already been delivered.

The shadow of an "Islamic bomb" that Libya's Colonel Muammar al-Qaddafi had tried to buy in 1970 seemed a real prospect by 1980 as the fourth decade of nuclear history ended with no international consensus on how the dangers of nuclear proliferation would be overcome.

25

Pride and Prejudice

There are few more ridiculous documents in the history of nuclear overstatement than the one published in 1974 by the Vienna-based International Atomic Energy Agency. Entitled "Market Survey for Nuclear Power in Developing Countries," the survey projected a potential demand for nuclear power plants in the world's less-developed countries that was hopelessly overoptimistic. It said 140 plants of between 500 and 600 Mw would be built, but the national grid systems of these poorer nations were simply not equipped to take such high input from a single plant. In Bangladesh, for example, the IAEA survey said there was a market for six reactors of up to 400 Mw capacity and another four of up to 600 Mw, yet at the time the entire Bangladesh electricity system had only just exceeded 200 Mw at peak demand. Bangladesh faced the same problem as other developing nations: no one power plant on a grid system can exceed 10 to 15 percent of the total grid capacity or the breakdown of one plant will disrupt the entire system.

Within months, the bizarre distortions in the report caused it to be shelved. Yet, in the underdeveloped countries, the idea persisted that the market it had forecast existed. Three years later, a Bangladesh atomic energy official recalled nostalgically that, although the report had been based on unreliable data, it had given "renewed hope to the interested developing countries that reactor manufacturers would now be willing to supply small and medium reactors on a commercial basis." In the past, the Bangladesh AEC, like its counterparts in other developing states, had pleaded in vain to the reactor makers to produce smaller units.

Despite the obvious flaws in the report, nuclear advocates from the less-developed countries continued to use their numerical voting power at the IAEA to maintain the flow of aid for their nuclear power programs. Agency funds were still available—a hangover from the era when nuclear power was regarded as something special, a panacea for the world's energy problems. Unfortunately, the nuclear aid was totally unrelated to the general aid needs of the recipient nations.

None of this, of course, was new. During the age of optimism, the mid-fifties, several of the less-developed countries—India, Taiwan, South Korea, Argentina—had created atomic energy commissions. The political elites in those countries, convinced that science and technology and, especially, prestige programs like nuclear power held the key to their future prosperity, were only too happy to accept development funds; that those were often more readily available for nuclear research than for any other field of science posed no problem for them.

Yet nuclear power was far too costly for these nations, and other natural sources, like coal and hydropower, were almost always more readily available. None of these considerations diminished their enthusiasm, however. Even the poorest diverted a part of their limited pool of trained manpower into nuclear energy. And their plans were constantly reinforced by the international nuclear community's endless round of ego-massaging seminars and conferences.

The Indian subcontinent provides some of the most poignant examples of this nuclear madness. At the beginning of 1970, India had a ten-year plan calling for the construction of one heavy-water producing plant per year and the installation of a massive 2,700 Mw of nuclear reactor capacity by 1980. Yet the Indian electrical grids were simply not big enough to take such loads and the industrial capacity was not available to build them. The plan was abandoned within two years as far too ambitious, but that did not stop the nuclear bureaucracy. Only in 1978, after the expenditure of hundreds of millions of dollars, would the Indians admit that nuclear power was, for them, a high-cost option that would not play a major role in their power program.

Even more premature than India's nuclear program was the insistence of the Bangladesh government that nuclear power was relevant to their energy plans. The country's nuclear advocates were driven to extraordinary economic distortions in their special pleading. The Bangladesh AEC continued to proclaim the cost advantages of small reactors that were simply not on the market. In 1977, a government study, comparing different energy sources on the basis of capital and operations costs, concluded that the time had not yet come when it was necessary to introduce nuclear power into Bangladesh, but this did nothing to deter the zealots in the country's AEC. At a nuclear power conference in Salzburg the same year, they attacked the report: "Neither the fact that Bangladesh will need to depend upon nuclear power in the long run and, as such, should learn and absorb the technology as soon as possible, nor the removal of regional imbalance in economic development, nor the credit due to a nuclear station in a developing country for its unquantifiable but genuine social benefits, etc., featured in their economic assessment." The audience of nuclear power advocates was sympathetic, of course.

Some developing nations, like South Korea and Brazil, did have power grids big enough to accommodate large nuclear units. Others, like Taiwan and the Philippines, proceeded to order nuclear plants in the hope that their grids would grow. In each case, however, the continued faith in nuclear power as a future technology blinded them to any serious cost comparisons with other energy sources. Aggressively patriotic and self-consciously progressive, the nuclear bureaucracies of the less-developed nations were still the prisoners of a passing Western fad.

The reactor manufacturers responded eagerly to this continued nuclear enthusiasm. At a time when nuclear orders had dried up in the industrialized nations, each new sale in an underdeveloped country acquired an added significance. In some cases, the nuclear salesmen of these multimillion-dollar deals, like their colleagues in the aircraft industry, were unreasonably eager to make a sale.

In the 1970s, the international bribery attached to aircraft sales by Lockheed and Northrop to Europe and Japan grew out of a corporate scramble for massive amounts of money. Nuclear power plants were in exactly that league: single orders worth hundreds of millions of dollars. Rumors spread of payoffs as part of some nuclear sales, with reactor salesmen asserting that bribery by their rivals had lost them a contract. Hard evidence was difficult to find: payments of this kind in international commerce are lost in the balance sheets under such convenient euphemisms as "agent's fees" and "finder's fees." But in 1976, two contracts won by the state-owned Canadian maker of heavy-water reactors, Atomic Energy of Canada, Ltd., came under the close scrutiny of the Canadian Auditor General, the country's public expenditure watchdog, and a number of strange transactions were revealed.

In the mid-1970s, AECL was still struggling to find a place in a market dominated by American light-water reactors. It had managed to sell two of its heavy-water kind to Argentina and Korea. They were important sales because Canada's new, tough nonproliferation stance, taken after the Indian bomb in 1974, had cut off its old markets in both India and Pakistan. In order to clinch the Argentine sale, AECL had followed the by now well-trodden path of selling the plant at a loss. "Call it a loss leader if you like," said one Canadian official. But the Auditor General's report suggested that more than a simple loss leader had been involved. The government watchdog discovered that AECL had deposited a sum of $2.4 million in a Lichtenstein bank account for an unnamed agent who had helped the sale. A second sale by AECL to Korea was also deeply suspect. This time, AECL had paid a Tokyo-based middleman $8.3 million for "expenses" incurred in "public relations" connected with the deal.

More significant than the percentages creamed off by unscrupulous agents, however, was the fact, discovered in 1977 by the nuclear safety section of the IAEA in Vienna, that sales by Western reactor companies to less-developed nations were frequently of a less-safe design and had less-rigorous operating specifications than were required in the exporting countries. The IAEA report, compiled by an American, Dr. Morris Rosen, focused particularly on special models that had been made to fit specific requirements and had never been subject to regulatory review. Rosen was well known in reactor safety. He had made a name for himself in the early seventies when he objected to the dominant view of the U.S. AEC that U.S. domestic safety standards were sufficient, an objection that led to his "promotion" out of safety inspection. In his 1977 report on developing countries,

Rosen highlighted the sale of a reactor by Westinghouse to the Philippines, a sale that became one of the most notorious nuclear power deals in an underdeveloped country.

The central character in the deal was a Filipino of Italian descent named Herminio Disini, a high-flying self-made businessman and close friend of Philippine dictator Ferdinand Marcos. Disini started his own business, making cigarette filters, in 1970. The company consisted of himself and two employees in a one-room rented office. Within five years he controlled a wide-ranging conglomerate of thirty-five companies with assets in excess of $200 million. His empire extended into textiles, bricks, cellophane, paper, detergents, shampoos, oil exploration, airline chartering, engineering construction, telecommunications, and petrochemicals. His biggest contract, however, was related to a billion-dollar Westinghouse nuclear power plant, the first in the Philippines.

Nuclear power fitted exactly into the self-image of the Disini group: aggressive, progressive, dynamic. It also matched the sense of self-importance of President Marcos, who assumed dictatorial powers by declaring martial law in 1972. The next year, Marcos gave the go-ahead to the ambitions of the Philippine nuclear advocates by announcing with considerable fanfare that the first two nuclear power plants—to be followed by ten more—would be constructed on the outer edge of the Bataan Peninsula of Manila Bay. It was just the kind of farsighted project that Marcos believed justified his autocratic rule. Like his counterparts in many other developing nations, Marcos was convinced that nuclear power was something progressive, something that could enhance the international image of his country. The new contract for two power plants would be the largest construction order in the nation's industrial history. A lot of money was involved—to be made and to be lost.

Marcos's martial-law regime was guided by a strange mixture of high-level technocratic advice and the personal bickering of a feudal court. Of the two, Marcos and his powerful wife, Imelda, preferred the courtiers. The style of official life involved a pomp and extravagance that the world had not seen since the fall of the Winter Palace—indeed, perhaps since the fall of the Bastille. It was an endless whirl of parties and meetings, of dinners at the presidential palace and cruises on the presidential yacht—all replete with court jesters, favorite entertainers, and a troupe of dancing girls. With opposition political leaders in jail, the press tightly censored, and the oligarchical family structure—the traditional Philippines power base—in a state of flux, it was a great opportunity for a man with a courtier's touch.

Herminio Disini prospered at court. Under Marcos, the basis of both political power and financial success was access to the first couple, the President and his wife, and Hermie, as he was known, started with distinct advantages. His wife, a doctor, was a cousin of Imelda Marcos, a former governess to her children, and her personal physician. Hermie himself became a regular companion and golfing partner of the President's. (As one of his associates put it: "We leave it to Hermie to play golf with Marcos. That's his job.") There was no better business contact: Marcos could issue government contracts without review and change the tax laws by decree. Hermie's new empire was built on privileged access to government loans, special taxes that wiped out his competition, an endless stream of government contracts, and the acquisition of assets at bargain prices after their previous owners had been subject to government prohibitions.

In 1974 the Philippine technocrats were coming to the end of a year of detailed negotiations for the nuclear contract. Although the project was criticized by the government's senior economics minister as an extravagance, the presidential decree kept the wheels moving—indeed, Marcos even dedicated the second week of December as "Atomic Energy Week." By this time, it looked probable that the contract would be issued to GE. The company had completed a detailed study and, in a four-volume presentation of costs and specifications, had offered two units for a total of $700 million. According to one report, Westinghouse had submitted "nothing more than its standard advertising brochures . . . [and] merely quoted a price of $500 million for the two reactors." They had done no detailed study of the suitability of their reactors to the Philippine case. It came as a considerable surprise, therefore, when Marcos overruled his advisers and issued the contract to Westinghouse. An unnamed spokesman for the Philippine National Power Company noted, "The decision to choose Westinghouse was a political decision." Westinghouse negotiators openly admitted that their success was attributable to the influence of Herminio Disini, whom they had retained as a sales representative. Later, Westinghouse further admitted that they had paid Disini a "normal" commercial fee for his services. Some sources said it was "four million dollars"; others suggested the fee was a percentage of the contract price, which could have made it tens of millions of dollars.

The $500 million quoted by Westinghouse was ludicrously low. Even the U.S. embassy in Manila, which was actively promoting the sale as the "Filipino Aswan dam," commented that the Westinghouse bid had been too low. And indeed, by the time the company completed its detailed submission, the contract price had leaped from $500 million for two plants to over $1.1 billion for one and showed every sign of heading toward $2 billion.

The economics of the nuclear plant were dubious even before Westinghouse submitted its real cost estimates. The Philippines was still developing substantial low-cost hydroelectric and geothermal resources. And as was the case in other less-developed nations, the large size of the nuclear plant—600 Mw—would threaten the stability of the small grid when, and if, it came into operation.

The first sign of trouble came from Dr. Morris Rosen's 1977 reactor safety report. Rosen pointed out that the safety of its design had been "referenced" to a sale to Yugoslavia that was in turn "referenced" to a Westinghouse sale to Brazil. The Brazilian sale was "referenced" back to a proposed Westinghouse project in Puerto Rico. There was one problem in the chain. The Puerto Rico reactor had been canceled, so that the Westinghouse system had never been subject to independent regulatory review at all. As far as Rosen was concerned, it was as safe as a house of cards. Rosen went on to point out that the regulatory capacity of the nations that had bought this system was, frankly, "sub-minimal." The Philippine Atomic Energy Commission was scarcely an independent body and, in any case, hardly had a licensing system worth its name.

There was something even more disturbing, however. The Philippines are situated in a strong earthquake zone and also contain a number of volcanoes. In 1978, a special IAEA safety mission reviewed the design of the plant and identified some major problems with the site on the Bataan Peninsula. Most dramatic was the physical proximity of the site to several volcanoes, the closest being only ten miles away. It had earlier been dismissed as "extinct," but the IAEA report stressed that a volcanic eruption had to be considered as a "credible event" and the reactor license should be changed with this in mind. The IAEA review also identified shortcomings in the earthquake protection specifications of the original license and concluded that they should be tightened considerably. In mid-1970, these new concerns, together with a growing understanding of the poor economics of the project, eventually persuaded President Marcos to suspend construction on the site. As in other developing countries, the progressive dream of nuclear power had become a nightmare.

No developing nation, however, suffered more from its nuclear investments, both in economic terms and in the misuse of its trained scientists, than India—the only developing country with a substantial nuclear program. Ironically, the burst of enthusiasm displayed by other developing countries for the potential of nuclear-generated electricity coincided with a major about-turn in the ambitious Indian project: the high-flown rhetoric that had accompanied the program in the 1950s and 1960s disappeared as it became increasingly clear that nuclear power was an unacceptable high-cost option. It was also a national disaster: the nuclear program had undermined efforts to overcome the electricity shortages that were a major obstruction to Indian economic development. For decades, nuclear reactors had been the only type of power-generating plants to receive financing from the federal government; other energy sources, particularly underdeveloped low-cost hydroelectric schemes, were left to depend on the state governments and had thus been starved of funds. It was only in the 1970s that these hydro projects began to receive federal financing.

Until his death in 1966, the founding father of the Indian nuclear program, Homi Bhabha, kept alive the vision of low-cost electricity from nuclear power. Under the banner of energy self-reliance, he concentrated on the development of heavy-water reactors—the type that would keep India independent of America and its monopoly control over the supply of enriched uranium used in light-water reactors. Even when much wealthier nations, like Germany and Japan, conceded the superior economic potential of the light-water reactors, Bhabha committed the country to a major program of building Indian versions of the Canadian heavy-water CANDU reactors. These reactors produced a high proportion of plutonium, and Bhabha thought that would eventually fuel the breeder reactors of the second phase of his plan.

In the wake of the growing enthusiasm for light-water reactors after the Oyster Creek "breakthrough," however, Bhabha decided to make a single exception to his grand strategy. Conscious that other segments of the Indian bureaucracy doubted his grandiose claims for the economics of nuclear power, he ordered a single boiling-water reactor from General Electric. It was to be built at Tarapur, 250 miles north of Calcutta. The post–Oyster Creek rhetoric about the arrival of commercial nuclear power would help to silence the doubts—or so he hoped.

With Nehru's death, Bhabha lost the absolute political patronage he had enjoyed, and doubts about his program grew. When he died in 1966, the doubts were temporarily silenced by yet another grandiose plan, devised by his successor as chairman of the Indian AEC, Vikram Sarabhai, The new plan called for an ambitious ten-year strategy. Sarabhai shared Bhabha's faith in the importance of scientific development in India, but he was more sensitive to the industrial and economic requirements of practical application. Unlike Bhabha, who was allergic to economic analysis, Sarabhai regarded it as central to scientific decision making. Unfortunately, his own knowledge of economics was inadequate for him to insist on a fundamental review of the dubious basis of calculating costs and benefits that had been developed in the Indian AEC for purposes of nuclear advocacy.

Sarabhai, who had helped to found the Indian Institute of Management, brought to his new position a faith in the rationality of management that he had displayed in modernizing his family's textile industry. He was particularly attracted to operations research, the dominant management tool of the fifties and sixties. The failing of this tool (later replaced by the practice known as "systems analysis") was that it did not interrelate the needs of one sphere of investment decision making with others. By restricting his analysis to the bounded nuclear field, Sarabhai was able to devise a long-term plan for India's nuclear program that gave the appearance of rationality. His Ten-Year Plan of 1970 reflected the optimistic projections of the nuclear industry throughout the world at that time. It called for more heavy-water reactors that would bring an installed nuclear capacity of 2,700 Mw by the end of the decade, including an advanced reactor and a breeder program already under construction. The operations-research approach simply assumed that the Indian construction industry would be able to build the units to this timetable. It also ignored the fact that Indian electricity grids were simply not large enough to absorb 200 Mw units, let alone the 500-Mw units proposed. As for the level of investment required for the capital costs of the program—that was someone else's business. Within two years, the entire plan was abandoned as far too ambitious.

At the end of 1971, Sarabhai died, leaving the practical engineering problems of the nuclear program to escalate year by year. The construction program for heavy-water reactors and for plants to produce heavy water fell further and further behind. Costs increased from one estimate to another. The single light-water reactor at Tarapur, plagued with difficulties from the outset due to a range of equipment and material failures, was subject to frequent shutdowns. The heavy-water reactors ran into a series of construction delays and, when finished, were also subject to major shutdowns. And the heavy-water-producing plants were struck by one disaster after another, requiring the importation of heavy water to operate the completed reactors.

Although the Tarapur plant eventually became a vital part of the electricity supply, it also demonstrated the inappropriateness of nuclear power for developing countries. Electricity load and voltage fluctuations in the inadequate grid system caused fuel damage and frequent shutdowns. The Indians, however, could not afford to shut down the reactor for long periods for fear of disrupting the electricity supply. As a result, maintenance of the plant was minimal and the residual radioactivity, caused by the initial operating defects, was never cleared up. As one Indian journal reported in 1979, "Tarapur is so heavily contaminated . . . that it is impossible for maintenance jobs to be done without the personnel exceeding the fortnightly dose of 0.4r in a matter of minutes. Thus the maintenance worker . . . holding a spanner in one hand and a pencil dosimeter in the other, turning a nut two or three rotations and rushing out of the work area, is a common phenomenon at Tarapur."

The problems of Tarapur were unprecedented and attracted criticism within India itself. To counter the critics, the Indian AEC pointed out that nuclear plants throughout the world were facing unexpected technical failures and low-capacity factors—as if that made it all right, as if the only thing that mattered was that their own technical expertise should not be questioned. The Indian AEC demanded to be judged by international standards—as if India was wealthy enough to indulge in such technical experimentation.

As India's nuclear bureaucrats massaged their status anxieties, they blamed others for the delays and cost overruns: foreigners above all, but also Indian manufacturers and the deficiencies of the electricity distribution system. The Indian bureaucracy, however, was becoming increasingly restless with the political requirement that they simply accept AEC cost estimates and projections without review. Bodies like the Indian Planning Commission began giving the nuclear bureaucracy a hard time. They sought more detailed cost estimates and, particularly, projections of future costs. When they were met with silence, they began to question why so many related costs were never included in the estimates and sought justification of the optimistic assumptions about operating availability. When these were not produced either, Indian industrial planners began to turn to the huge untapped resources of hydro and coal. Whatever the economic merits of coal or nuclear power, hydro was definitely the lowest cost option. Although many of the hydro sites were not conveniently located, there was still major extra capacity available in the medium term. To counter this, nuclear advocates continued to use out-of-date estimates showing that the total hydro capacity was only 4,100 Mw when all the experts knew that the actual potential was twice that.

Finally, in 1978, the draft Five-Year Plan acknowledged that nuclear was a high-cost option that would not play a major role in the nation's power program. Yet few dared ask how the hundreds of millions of dollars still in the program would be spent or what the thousands of scientists would do. Instead, Indian officials dismissed the relevance of nuclear power to the overall nuclear program and talked about the possibility of sideshows like irradiating agricultural material, making so-called peaceful nuclear explosions for large earth-moving projects, and, more candidly, of the ultimate excuse for the economic expenditure—national security. Perhaps the thought that the whole effort had been a waste of time as well as money was too awful to contemplate.

26

The Second Ice Age

The summer of 1977 found the nuclear industry in the midst of what its supporters called the Second Ice Age. New orders for nuclear plants were about as rare, said an industry journal, as the appearance of a comet.

By the end of the decade, the optimistic projections of nuclear generating capacity were demonstrably and hopelessly out of line in several countries. The U.S. Atomic Energy Commission had estimated a staggering 300,000 Mw by the end of 1985; the fledgling nuclear industries in Japan and Germany had targets of 60,000 and 50,000 Mw by the mid-1980s. By 1980, it was clear that little more than a third of any of these plans would be achieved.

Not only was there a decline in new orders, but utilities also started to defer old ones. Between 1974 and 1975, American utilities actually canceled or deferred 130,000 Mw of nuclear generating capacity.

Only France and the Soviet Union continued to develop their nuclear programs. In the wake of the 1973 Yom Kippur War, the French government had doubled its 1985 nuclear power goals to 25 percent of electricity needs and, at the end of the seventies, raised it again 50 percent. In the Soviet Union's tenth Five-Year Plan (1976–80), nuclear power stations accounted for more than a third of new electrical production capacity, and by 1980 they were scheduled to supply 10 percent of all needs in the western part of the country. This compared to 3.2 percent in 1975.

The underlying cause of the downturn elsewhere was the economic recession of the mid-seventies. All Western nations experienced a dramatic reduction in the growth of electric consumption, and the electric utilities were suddenly faced with excess production capacity. Up to this time, utilities had regarded an annual growth rate of 7 percent (in effect, doubling electricity use each decade) as a law of nature. In the two-year period 1974–75, however, U.S. utilities experienced virtually no growth; over the long term it seemed, the growth rate had dropped to less than 5 percent.

Even had a need for new power stations existed, the utilities would, by the end of the decade, have been reluctant to choose nuclear power. The hoped-for economic advantages of nuclear power over coal had not materialized. Part of the problem was that the actual cost of the reactors had doubled and showed every sign of doubling again. All types of reactors had always been at a disadvantage to coal in terms of the capital costs: 70 to 90 percent of the cost of electricity generated by a light-water reactor comes from the capital charges associated with its construction; in a coal-fired plant, the figure is between 30 and 65 percent. Experience had also shown that the American light-water reactors, the ones that were supposed to have been the best and the cheapest, were operating less efficiently than originally estimated. Generating costs of the light-water reactor had been based on the assumption that the plant would operate at about 80 percent of its capacity, but the U.S. plants had been operating at around 60 percent. Their failure to perform at planned capacity significantly increased the cost of the electricity they produced.

Beyond these hard operating realities about nuclear power, those utility executives who had been trapped into going nuclear by a seductive combination of manufacturer's propaganda, fascination with the new technology, and questions of corporate prestige now faced a new range of uncertainties: these included questions about reactor safety, a sudden increase in the cost of uranium, the future of the high technology of the breeder reactor, and an increasingly sophisticated antinuclear movement that challenged the nuclear industry to produce hard evidence it should be allowed to exist—that mankind needed nuclear power and that it could be made safe.

For almost a decade, the nuclear community had asserted that no significant changes were required in the light-water reactor system to make it safe. They had done so at a time when they had virtually no experience of operating reactors. As the larger models, which had been ordered in the mid-sixties, came on stream, the problems began to show up. They occurred at different points inside the reactor. Under the intense radiation and the high pressures and temperatures, pipes, valves, and nozzles cracked and malfunctioned, tubes corroded, fuel rods bent, and a variety of new ways of breaching the so-called containment systems—the ones designed to trap the radiation in the event of an accident—were discovered.

All this gave rise to a public debate on the ability of the system to continue its excellent record of safe operations. It also meant a considerable increase in the costs of the system, each new problem requiring expensive redesign of the plant.

Regulatory agencies throughout the world poured out an endless stream of new design requirements: concrete containment walls had to be made thicker with more steel reinforcement, and new equipment had to be installed to improve existing safety systems. Reactors everywhere were shut down for long periods while these improvements were effected, increasing significantly the average costs of producing electricity. In 1970, for example, a nuclear plant in the United States could be built at a cost of about $200 for each kilowatt of electricity produced. By the end of the decade, this figure had risen to $1,000 and showed no signs of easing. It was a much faster increase than the rate of inflation.

Nuclear power continued to score against its conventional rivals because the uranium fuel cost less. But in the mid-seventies, the world price of uranium increased suddenly. Uranium prices doubled during 1974, doubled again in 1975, and stabilized only after a further increase in 1976. The price had gone from under $5 per pound to over $45 per pound.

A number of factors, including the formation of the cartel of uranium producers outside America, combined to bring about this sudden price rise. But the most important single event was the change in U.S. policy concerning sales of its enriched uranium. America was still a monopoly supplier of the enriched uranium fuel required by light-water reactors. Until 1973, electric utilities had been able to order supplies of enriched uranium more or less when they needed them; at most, they had to give less than six months' notice. But during 1973, the year of the oil crisis, the U.S. AEC adopted a series of new sales rules designed to avoid the feared shortage of enrichment capacity that was expected to result from an increase in the number of nuclear plants. The new rules required two things. First, the utilities had to give eight *years'* notice of their requirements instead of six months. Second, they had to supply an increased amount of raw uranium to get the same amount of enriched uranium. The battered utilities, facing rising interest rates as a result of inflation, were forced to buy long-term contracts for uranium supplies. But these contracts were hard to find: potentially large production from the new uranium mines in Canada and Australia had been delayed due to lengthy public inquiries forced by the growing antinuclear movement.

In 1975, the utilities received an even greater shock when Westinghouse announced its intention to renege on past contracts that had guaranteed uranium supplies at fixed prices with the purchase of its pressurized-water reactors. During the boom years, Westinghouse, as part of their sales drive for the PWR's, had contracted to supply some 80 million pounds of uranium to twenty-seven utilities in the United States and Sweden. The average price of this uranium was $9.50 a pound and, although twice the market price in 1970, that looked good as a long shot. By 1975, Westinghouse had bought only 15 million of the 80 million pounds, and the price of uranium had shot up to more than $40 a pound. Westinghouse faced a loss of $2 billion if it covered the shortfall. At Westinghouse headquarters in Pittsburgh, it became known as "The Uranium Thing," and a company consultant admitted, "It is the most stupid performance in the history of American commercial life."

Bob Kirby, who had succeeded Donald Burnham as president of Westinghouse in 1975, made the potential $2 billion loss his top priority. Kirby had come to Westinghouse with a glowing reputation for troubleshooting. As one admirer put it, "He is one of the world's leading toy puzzle solvers, particularly those involving intertwining loops." Kirby took the desperate decision to renege on the contracts and risk the possibility of a major legal action by the electric utilities. The step was in sharp contrast with the company's old advertising campaign, which had proclaimed, "A Westinghouse nuclear power plant provides utilities protection against rising fuel costs." It had been a successful campaign and had certainly helped the company to sell reactors, but it now threatened the economic stability of the entire corporation. Company lawyers came up with what seemed like a plausible legal excuse. The American Uniform Commercial Code contained a little-used provision that exempts compliance with contracts on the basis of "commercial impracticability."

When the twenty-seven utilities sued Westinghouse for breach of contract—the majority combining in a joint action in the courts of Richmond, Virginia—Westinghouse fought the issue each step of the way. The legal costs soon mounted into tens of millions of dollars, making it the most expensive private litigation of all time.

In the middle of 1976, the battle took a new twist. The Australian branch of the environmentalist group Friends of the Earth secured an enormous pile of internal documents from an Australian subsidiary of the British mining conglomerate Rio Tinto Zinc (RTZ). These documents established for the first time the existence of the uranium cartel formed by the governments and uranium exporters of Canada, France, Australia, South Africa, and the "sovereign state" of RTZ. Until then, although the existence of some kind of club was known in the industry, the extent and operating procedure was still secret. Westinghouse was given a new defense and basis for legal attack—ironically, thanks to the Australian environmentalists.

Westinghouse attributed the full increase in uranium prices over the previous years to the cartel. But Westinghouse's own actions in offering guaranteed uranium supplies that it did not own had been as, if not more, significant. The apparent security of these contracts had depressed uranium prices by keeping demand off the market. In 1970, Westinghouse had rejected substantial offers of uranium at low prices. As the cartel pushed the price up, the company decided to renege on its contracts; the utilities reacted with panic to cover themselves, and the price jumped.

The legal quagmire of the Westinghouse contracts reinforced the anxieties about nuclear power among electric utility decision-makers. It indicated just how different the uranium market was in comparison with coal; for instability, it even matched the new turbulence in oil markets. For the first time, it seemed impossible to make the long-term plans necessary in electricity production; not only were new uranium prices high, future availability of the ore was apparently subject to massive fluctuations of a scale that made the normally conservative utility executives very apprehensive. Although the new reduced estimates of installed nuclear capacity would, in theory, increase the life of the known uranium deposits, the utility men concluded it was still possible to project a demand for uranium that would exceed supply, at least in the medium term before unexplored deposits were worked. Uranium prices would therefore continue to be high, they feared, and could go even higher.

These fears were combined with growing doubts about the prospects for reprocessing and breeder technology. For decades it had been assumed that the used uranium fuel rods from the light-water reactors would be reprocessed and the plutonium produced during the chain reaction extracted and put to use—either by being mixed with uranium and used as fuel for light-water reactors or by being burned directly in breeders. Now, the new American stance on nonproliferation sought to exclude both the reprocessing and the plutonium economy—and therefore increased the likelihood of short-term uranium shortages and high prices.

The new American policy apart, serious doubts still surrounded the commercial viability of the plutonium economy. In particular, there were doubts as to whether reprocessing plants could be built at a cost that would produce plutonium at reasonable prices. Of the three planned U.S. reprocessing plants, one had closed because of technical failure and the other two had never opened because of cost overruns. In Germany, a consortium of companies that had planned to build a reprocessing plant abandoned the idea after they determined it was not a commercial proposition. The British and the French had each developed reprocessing capacity at the beginning of the seventies with plants, one at Windscale and another at Cap La Hague. Both British and French plans to expand the capacity of the plants met opposition, however. In France it came from the trade unions, and in Britain from the environmental groups. The Windscale reprocessing plant was temporarily delayed by a public inquiry. By 1980, the excess in reprocessing capacity so feared at the beginning of the decade had turned into a major shortage, and the cost of reprocessing had increased tenfold.

The new cloud hanging over the future of breeders and reprocessing plants hid another uncertainty. The international nuclear community had relied so completely on the concept of reprocessing that they had ignored the problem of what to do with spent fuel rods if they could not be reprocessed. The antinuclear movement seized on this appalling deficiency in the plans of the advocates, and it rapidly became a major item in the nuclear power debate.

When the fuel rods are removed from a reactor core, they are intensely radioactive. To "cool them down," they are first stored in a pool of ordinary water at the reactor site. This allows the substances with a short radioactive half-life to decay: Iodine-131, for example, one of the most dangerous of the fission products, has a half-life of only eight days. Until reprocessing and the breeder came under attack, the assumption had always been that, after storage in the water pools for one to three years, the rods would be packed up and shipped to a reprocessing plant. There they would be dissolved in nitric acid and the usable uranium and plutonium separated out.

Even with reprocessing, there are waste-disposal problems. The leftovers, still containing a large variety of radioactive material, are known in the trade as "high-level wastes" and have to be stored, somewhere, away from contact with humans for thousands of years. Countries with a nuclear bomb program had all along been operating reprocessing plants and had already accumulated significant quantities of these "high-level wastes." In America, 100 million gallons of these wastes generated by the weapons program—including the naval nuclear submarine program—had been buried in giant steel tanks on military reservations. The waste from power plants was now increasing at such a rate, however, that by the mid-eighties, without some kind of solution to the problem of how to dispose of it, countries would be running out of storage space for the spent fuel rods.

The problem had always been there, of course, and members of the nuclear community had remarked on it from time to time. In 1963, David Lilienthal, the former chairman of the U.S. AEC, wrote, "To project as a goal of our government, as the AEC does, a program for the construction of ten to twelve full-scale [nuclear] power plants during the next dozen years *until* a safe method to meet this problem of waste disposal has been *demonstrated* seems to me irresponsible in dealing with public and private funds; more important, it reveals a disturbing attitude about the importance of protecting public health and safety against catastrophic mistakes that might occur."

In the short term, at least theoretically, the wastes can always be stored in new tanks away from the reactor site and away from large population centers—except that no one wants a nuclear garbage dump in his backyard. In the long term, some place—proposals range from deep underground to beneath the ocean floor—has to be found for them. And they have to be stored so that they do not leak into underground water and so that natural groundwaters cannot corrode the containers holding them; there have already been leaks from the American tanks holding military high-level waste. The nuclear industry is replete with suggestions about suitable geological formations that could hold wastes and of ways to surround them with glass or ceramic, which cannot be corroded. But the fact remains that, more than thirty years since the first nuclear power stations left the drawing board, no agreement has been reached on which disposal method is the best.

It was not just the legacy of years of neglect of waste management by the nuclear industry nor the faltering economics of the technology itself that made the utilities nervous about the future. It was also the growing sophistication of the antinuclear movement.

At first, nuclear power had prospered because the environmentalist movement had slowed the progress of alternative means of generating electricity. Hydroelectric schemes—on the Hudson River in New York and in Sweden's northern rivers for example—had been canceled under conservationist pressure. Moreover, the new concern over air pollution had boosted coal costs significantly. New, expensive equipment had to be installed to trap the releases of sulfur and nitrogen into the atmosphere, adding significantly to the capital costs of coal plants. Environmental objections had also hampered the development of low-cost open-cast coal mines—strip mining—adding to the price of the fuel. One successful environmental movement in Australia had prevented the development of low-cost coal-carrying supertankers that could have revived the international coal trade. In the sixties, Daniel Ludwig, a billionaire recluse and shipping magnate, had proposed a major development of coal deposits south of Sydney. He hoped to build a new superport to cater to huge coal supertankers, but the scheme was canceled after intense environmentalist objections. When the oil crisis came, the technology for coal supertankers did not exist.

Environmentalists were also concerned about nuclear, though, initially, to a lesser degree. When they began to focus on it, eventually forcing stricter regulations, the environmentalist movement added significantly more to the costs of nuclear power than it had done to the cost of the alternatives. By the end of the 1970s, the impact of the movement on the cost of coal-fired plants seemed to have leveled out, but the impact on the costs of nuclear plants showed every sign of increasing indefinitely.

The antinuclear movement also began to point to the hidden government subsidies that nuclear power enjoyed—subsidies that distorted electric utility decision making in favor of a nuclear choice and without which, the critics argued, there was no economic justification for nuclear power. The "unfair" advantages that nuclear power had received from these subsidies became a significant debating point in the antinuclear cause.

The important question to the utilities was how this line of argument might affect future cost comparisons. Much of the subsidy was so deeply embedded in the history of nuclear energy that it could not be removed. Nothing could take away the massive advantages to nuclear technology that came from its close connection with military applications, both bombs and submarines. Moreover, the decades of enthusiasm about the "peaceful atom" had resulted in billions of dollars' worth of subsidies from government sources.

Looking for some way to justify their existence in the midst of the downturn, nuclear advocates seized on the growing concern over the "energy crisis." They proclaimed that a shortage of oil was the same thing as a shortage of "energy." But the term "energy" was a political buzz word, a metaphor that obscured the variety of things for which it was used and overrode the technical and economic difficulties of substituting one source of energy for another. In fact, nuclear power could substitute for oil in only about 10 percent of its use; and even that 10 percent could not be changed overnight. Some uses of oil—in home heating, for example—could be replaced by electricity, and therefore nuclear power, but this would take time. And even if totally successful, it would not represent a substitute for the major use of oil in industry and motor vehicles.

The simple, indeed simplistic, calculations of the energy planners and forecasters assumed that the different types of energy could be easily interchanged, and, as the political preoccupation with energy totals had given rise to the false conception that an energy shortage was imminent, it seemed reasonable that the electric utilities should build more nuclear plants. But the utilities had become unwilling to commit themselves to the uncertain future of nuclear power. When the pressure came to order new plants to fill the so-called energy gap, the utilities, at least in the United States, would respond by ordering coal rather than nuclear and thus would extend the second ice age.

Two nations would stand out in this period for refusing to compromise; one was France, the other was the Soviet Union. Rather than trim their nuclear programs, they expanded them. In both nations the nationalized electric utility monopolies were driven by government orders to expand nuclear capacity in what was seen to be the national interest.

At Electricité de France, Marcel Boiteux and Michel Hug changed the direction of their institution's goals from the provision of low-cost electricity to the achievement of energy independence for France.

Hug was a Polytechnician of the Corps des Ponts et Chaussées. He never displayed the glimpses of doubt or of reasonableness about the nuclear program that Marcel Boiteux increasingly revealed. By the mid-1970s, debate about the forced pace of the program was growing inside France. Responding to this, Hug tried to turn the old slogan "Tout électrique! Tout nucléaire" into a battle cry. At the same time, Boiteux was calling it a public relations disaster, arguing that the nuclear program was not an EDF initiative and trying to shift the onus—to the government, to the laws of economics, to the CEA. Unlike Hug, who relished his authority and his power, Boiteux became increasingly uneasy as EDF began to be called "The State Within the State." Boiteux rejected this imagery of independent strength.

As public criticism of nuclear power intensified, the charming and reserved Boiteux simply could not comprehend how he could be called "Monsieur Tout Nucléaire" or "Public Enemy No. 1." He did not understand why people painted Hitlerite mustaches on his pictures or why anyone made the threatening phone calls he received or planted the bomb that exploded (without injuring anyone) in his Paris apartment. Throughout, he remained calm and rational. He did not like power, he did not seek it, and, most of the time, he was not conscious that he had it. The power to persuade was something else; he believed in persuasion, not orders. If this was "power," then he willingly accepted responsibility for its exercise, and he put his eloquence and debating skills—his sentences always perfectly formed, his ideas precisely expressed—to good effect. Inside EDF, his diplomatic astuteness and intellectual charisma were appreciated as the public debate on nuclear energy escalated.

Much more the university professor than the hard-nosed chief executive, Boiteux carefully explained the economic facts of life to the French people. He presented the detailed case for the nuclear choice in terms of precise costs and benefits. The figures, he believed, spoke for themselves. His personal commitment to economic rationality was as total as Hug's technical drive. Both men believed that their techniques were politically neutral; neither really understood the political or social dimensions of decisions. In Boiteux's elegant economic models, social issues were suppressed as unmeasurable and figured in the equations with the customary range of *ceteris paribus* (other things being equal) assumptions. Something that could not be measured, no matter how sophisticated the technique, was, by definition, a constraint and not something to either maximize or minimize. Boiteux had surrendered to the ultimate trap of his approach to decision making: he had come to believe that the things that could be measured were in some way more important than the things that could not. He would never admit it, but he had forgotten that in a democracy it is the public, and not the experts, that defines the issues of what is politically relevant.

Russia's revival of interest in nuclear power began in the mid-sixties, but it was a more gradual process than the self-proclaimed boom in the West. Nothing had been said about nuclear plans under Khrushchev's rule, although in his later years, the old heavy-industry group he had suppressed began reappearing at the center of the Soviet elite. They were among the most obvious beneficiaries of Khrushchev's overthrow in October 1964; the following year, talk of nuclear power began to build.

In 1965, the first articles about a new commitment to nuclear electricity generation appeared in the Soviet press. Past failures were admitted. The grandiose plan of the mid-fifties that had come to nothing and the incorrect cost estimates were explicitly acknowledged. The campaign for nuclear power was in full swing within a year, with a series of articles on nuclear power, new cost estimates, and a highly publicized launch of a biography of Igor Kurchatov, father of the bomb and proponent of peaceful uses.

The key to this revival, however, remained the attitude of the Soviet Ministry of Power Stations, the nationalized electric utility monopoly. Like their counterparts throughout the world, the engineer-administrators in this ministry believed the only question was cost. In the previous decade, the Soviet program had built a series of smaller prototype plants. Although the Russians have never admitted it, the cost estimates for scaling up those plants must have involved the same uncertainties faced by administrators in the West. And within the utility monopoly, there were differences of opinion on costs. Georgi Yermakov, who had been a senior officer in the special Atomic Power Division of the Ministry for Power Stations created during the 1950s boom, displayed all the exasperation of his Western colleagues when faced with doubters: "Now the 'opponents' of the construction of atomic electric power stations (and there are those among the electric power specialists) raise new objections: atomic electric energy is too expensive and therefore, such power stations cannot play an essential role in the development of our power system." Yermakov conceded that earlier cost estimates had proven optimistic, but claimed that years of struggle by scientists and engineers had now improved the efficiency of nuclear power. As in the West, the only way to prove this claim was to build the plants.

By 1968, the Soviet electric utility planners were sufficiently convinced of the economics to begin a modest expansion in nuclear power, and the special Atomic Power Division of the Ministry of Power was revived. In 1971, the 24th Party Congress inserted a new target into the ninth Five-Year Plan: it called for an expansion of the existing capacity of 1,600 Mw by adding 6,000–8,000 Mw in five years. The resolution even specified that the extra capacity should be in the new 1,000-Mw units.

The ninth Five-Year Plan did not meet its target, but the planners were just as ambitious in stating their targets for the tenth plan, which was to cover the 1976–80 period. They hoped to bring into operation 13,000–15,000 Mw in this time. Furthermore, they announced their intention to construct new facilities that would enable them to build 10,000 Mw every year in the 1980s. Questions of safety were by and large suppressed or ignored. The Western press reported occasional incidents of local hostility to large-scale nuclear construction, and in 1976 the celebrated Russian physicist Peter Kapitsa stressed the dangers of nuclear power in an address at the celebrations of the Academy of Sciences two-hundred-fiftieth anniversary. His speech was not reported in the Soviet press. By then, nuclear advocates in the West were commenting wistfully on the smooth efficiency of the Russian program.

27

In the Spirit of Paracelsus

Toward the middle of the sixteenth century, a Swiss physician and alchemist called Paracelsus, who is acknowledged as the father of medical chemistry, was sacked from his job as the town doctor of Basel because of his unorthodox practices. Banished from Switzerland, he sought refuge in the small Austrian town of Salzburg and continued his work. His basic problem was that his ideas were ahead of his time: while his colleagues were trying to turn base metals into gold, Paracelsus was testing minerals in medicinal preparations. His forceful denunciations both of the work of his predecessors and of those contemporaries who disagreed with him made him many enemies. Isaac Asimov describes his life as one "marked by eccentricity, quarrelsomeness and a vast army of enemies lovingly manufactured by himself." He was also a great self-promoter. His "loud mouth . . . did most to bring [his work] to general notice [but] he did not always achieve happy results, for his almost psychotic cocksureness led him to use medicines such as compounds of mercury and antimony even after practice had shown them to be toxic." His determination to prove himself right never waned, however, and he met criticism and banishment with an extraordinary measure of self-confidence and even arrogance.

In a fit of high conceit, he discarded his real name, Theophrastus Bombastus von Hohenheim, and adopted the name Paracelsus, meaning "better than Celsus." (The Roman physician Celsus, whose works had only recently been translated from Latin, had made a great impression on the alchemists and physicians of the period.) Despite his radical approach to alchemy, Paracelsus always deluded himself about one thing: he never excluded the concept of the philosophers' stone, the elixir of life that would, he believed, allow him to live forever. Indeed, late in his life he claimed to have found it. Nonetheless, in 1541, at the age of fifty, he died.

One part genius and one part charlatan, Paracelsus is the personification of the nuclear community of the late 1970s. They thought their marvel machine, the breeder reactor, would enable their enterprise to live forever; they believed, like Paracelsus, they had found the "philosophers' stone." And, as Paracelsus had done before them, they met their critics around the world with gritty determination and self-assurance. It was in such a spirit of defiance that, in the spring of 1977, the international nuclear community sought refuge in Salzburg for the fifth of their Atoms for Peace conferences.

The nuclear business was in disarray. Antinuclear groups around the world, citing the hazards of radiation from reactor accidents and waste disposal, had launched a series of vigorous, sometimes violent, demonstrations in their fight to halt nuclear power. Electric utilities, faced with an excess of capacity, had stopped ordering new plants. The future of nuclear power, as the title of one of the conference papers suggested, was at a turning point. In fact, energy specialists, including some from inside the industry itself, had identified a period of zero energy growth: a time when the United States, and possibly the rest of the world, could do without more nuclear power plants. Some forecasters had even gone so far as proposing a nuclear moratorium. For the first time in the history of nuclear power, some of its advocates were acknowledging that there was a choice between nuclear and alternative sources of energy.

At the same time, the antinuclear movement was growing in strength and sophistication. Even as the two thousand pro-nuclear delegates arrived at Salzburg, a counterconvention of antinuclear groups had just ended in the city. The message from their week-long meeting, held under the uncompromising banner "Conference for a

Non-Nuclear Future," was clear enough: They warned they were prepared to take action to secure a future without atomic energy. One German delegate, referring to the successful occupation of the nuclear site at Whyl in 1975, declared at a press conference, "The tractors of the farmers are the panzers of the people's movement."

Convening at Salzburg, the nuclear zealots were unmoved, their certainty in the future of nuclear power unshaken. André Giraud, the French head of the CEA and one of the community's leading advocates, commented: "Renounce nuclear energy? That leads inevitably to penury." While the commitment to nuclear power remained firm, other, more grandiose plans for the atom had been quietly dropped. Even the city of Salzburg had been carefully chosen in an effort to avoid any reminders of the optimistic Geneva conferences, when nuclear power was a surging success story and an unquestioned panacea. The nuclear advocates had even chosen a new theme: abandoning the all-embracing "Atoms for Peace," they had settled for the dry and quite specific "Nuclear Power and Its Fuel Cycle." For the first time, the entire convention would be devoted solely to questions relating to the generation of electricity from the atom. There would be no papers on the use of radioisotopes in medicine, the excavation of harbors by "peaceful" nuclear explosions, the extermination of insects by irradiation, the nuclear propulsion of cargo ships.

The nuclear community was a group under siege; yet, like Paracelsus, the zealots were, if anything, more confident than ever. They still saw themselves as a scientific elite, something special and apart. In their overriding arrogance, they quickly turned the nuclear fuel cycle—from the mining of uranium to the reprocessing of spent fuel rods—into a majestic, totally self-contained grand scheme that took on an organic life of its own. The concept was banal. Many industrial processes find uses for by-products; the nuclear fuel cycle was certainly no grander conceptually than the petroleum cycle. Yet this fact was conveniently lost in the fancy graphic displays on the walls of the conference room in Salzburg, where multicolored arrows representing the integrated concept of the nuclear fuel cycle swept around in cyclical harmony.

Anyone daring to challenge this group's sense of eternity was committing an act of calumny. The nuclear zealots had come to Salzburg to close ranks, to seek justification of the present and reassurance for the future. In their midst, however, was someone who did not entirely subscribe to this mood; nor was he intimidated by it. He was Alvin Weinberg, the nuclear community's iconoclast-in-residence, as he was once dubbed by the journal *Science*.

Weinberg had as impressive a set of credentials as any nuclear advocate in the world. He was one of the most senior scientific administrators in the international community. He had been at the University of Chicago during the Manhattan Project, was the originator of the pressurized-water reactor, and had become director of the Oak Ridge atomic laboratory in Tennessee. Unlike most advocates, whose approach was narrow-minded, Weinberg was always trying to broaden the scope of the laboratory's research and, most important, to take account of the social implications of that research.

Under his leadership, the Oak Ridge scientists extended their horizons far beyond their primary nuclear tasks, studying such secondary areas as the purification of sea water and the biological effects of radiation. They had carried out extensive programs of genetic research, explored sources of nonnuclear energy, and looked at the environmental effects of all energy sources. Such a perspective had been ahead of its time.

Although Weinberg accepted, with almost religious faith, the ability of science to solve any technical problem, he also cautioned against the pitfalls of "Big Science"—the worship of high technology without due regard for its social consequences. In the early sixties, he had tried to get his colleagues to assess the social purposes of their research. "Those cultures," he said, "which have devoted too much of their talents to monuments which had nothing to do with the real issues of human well being have fallen on bad days. We must not allow ourselves, by short-sighted seeking after fragile monuments of Big Science, to be diverted from our real purpose, which is the enriching and broadening of human life."

Weinberg recognized the nuclear power option was a Faustian bargain: the promise offered to mankind by the atom was hedged with risks. He believed that the risks—such as a human error causing a reactor meltdown or bomb-grade fissile material being stolen by terrorists—could be controlled by a "nuclear priesthood," an elite cadre of specialists whose professionalism alone would ensure that these things did not happen. "We nuclear people," Weinberg wrote in 1972, "offer—in the catalytic nuclear burner—an inexhaustible supply of energy . . . but the price we demand of society for this magic energy source is both a vigilance and longevity of our social institutions that we are quite unaccustomed to."

Five years later at Salzburg, the slender, gray-haired Weinberg warned his audience that if nuclear energy became the dominant energy system, it could require the dedication of a nuclear cadre that "goes beyond what most other techniques demand." He added, "We are unaccustomed, perhaps unwilling in a sense, to face up to the consequences of complete success . . . one cannot avoid recognition of the special social problems posed by our technology." As if the title of his address, "Nuclear Power at the Turning Point," was not enough, Weinberg concluded, "Our enterprise is not merely on the defensive; in some quarters it is in danger of extinction." It was not what the delegates had come to Salzburg to hear. In contrast to the lengthy applause that greeted each bombastic statement about the importance of nuclear power, Weinberg's tendency to bare his tortured soul in public was greeted with murmurs of disapproval. The delegates were too conscious that any dissenting view, any gap in the ranks, would be swiftly exploited by the antinuclear groups hovering in the wings.

Weinberg had two candid messages, neither of which the nuclear advocates wanted to discuss. The first concerned core meltdowns. The unpalatable fact was that a fully developed world nuclear economy would, on present estimates, result in one complete core meltdown every four years. The second message was that by the year 2050 there would be 5,000 breeder reactors in operation, for which 100 tons of plutonium would have to be reprocessed each day. The nuclear industry had to face up to these facts by creating a "nuclear priesthood" to run it, and the public would have to begin to accept the risks. "The public will eventually accept radiation as part of life's hazards," declared Weinberg. But, he warned, "Our technology is very different . . . we would do well to admit this instead of denying it." There was no point in denying that processing 100 tons of bomb-grade plutonium each day was not a problem.

Weinberg's audience listened solemnly. A number of antinuclear activists present eagerly, but selectively, adopted his words of warning. The German author Robert Jungk, one of the earliest and most persistent of the chroniclers of atomic matters, recalled with satisfaction, "I have seldom seen an audience so disconcerted as that which listened to the lecture he gave." Antinuclear activists could not, however, embrace Weinberg's position because, as he assured his audience, in the end there was no real alternative to nuclear energy. According to estimates, he said, by the year 2050 the world would be producing six to nine times the amount of energy it did in 1977. If the bulk of this was provided by fossil fuels, the associated problems of increased carbon dioxide in the atmosphere could reach catastrophic proportions. If the world decided to reject the nuclear alternative, an all-solar world might eventually be possible, but it could not provide nine times the current production and, in any case, the world would have to pay much more for a unit of energy than it paid at present. Here, then, was the "energy gap." "If the world forswears nuclear energy, from what we now know of solar energy the world would have to adjust permanently either to much more expensive energy than we now enjoy, or much less energy than the developed countries use, or both."

A conscious challenge to this thesis had come the week before from the Conference for a Non-Nuclear Future. To date, the antinuclear movement had in the main been characterized by its negative approach: hostility to the dangers of nuclear power in terms of reactor safety, waste disposal, and weapons proliferation. Though elements in the movement were concerned with alternative energy sources, they tended to be a less visible group. The Salzburg counterconference marked the maturity of this new and more positive theme. All of the delegates stressed the range of alternatives to nuclear power, underlining the fact that it was not a necessity in any absolute sense.

One young scientist stood out as the intellectual leader of this new approach. He was Amory Lovins, an American energy expert working for the British branch of Friends of the Earth. Slim, ashen-faced, and rather disheveled in appearance, Lovins is an energetic and earnest man with an extraordinary capacity to organize the jumble of interrelated facts required for energy decision making.

Lovins told the counterconference, "Nuclear energy is a future technology whose time has passed." The alternatives, he said, were a host of different ways of producing and using energy—some technically new, many quite old. Nuclear advocates were saying that the world had no choice, that nuclear power on a large scale was inevitable because of the demands for energy throughout the world. Lovins argued that this basic assumption was wrong. At a time when more and more impartial observers were resigned to accepting nuclear power as no better than a necessary evil, Lovins argued that it was definitely more evil than necessary.

The assertion that there was a choice had received its most complete elaboration two years earlier when the Ford Foundation Energy Project published *A Time to Choose.* The experience of the 1973 oil embargo had brought to the fore the prospect of an actual physical shortage, and the idea that the world would "run out" of oil had made "energy" into the most powerful political metaphor of the seventies. There was a general consensus (though Lovins dissented) that a physical shortage might be a reality in the short term, but for a decade at most. Beyond that, however, no absolute shortage need exist. The real question was simply one of cost. Alternatives to nuclear were technically feasible, but they could be more expensive.

With the release of the Ford report, the antinuclear community had begun to emphasize the range of possible alternatives: windmills and solar heaters; conversion of garbage and vegetation into liquid alcohol fuels; new, cleaner forms of burning coal in a system known as fluidized bed combustion. There were schemes to use the heat in the bowels of the earth, called geothermal energy, and schemes to use temperature differences in the oceans to generate electricity. New emphasis was given to the extraordinary variety of ways in which existing forms of using energy could become more efficient—by changing the way homes are built or cars and industrial machines are made. The nuclear critics also stressed the institutional barriers imposed by electric utility companies when they discouraged the use of waste heat in a myriad of existing factories that, by "cogeneration," could produce electricity as well. Solar energy in all its forms was a renewable resource; unlike oil, coal, or uranium, the world's reserves were replenished daily. As the list of alternatives grew in the mid-seventies, the nuclear advocates were forced to respond. All of this speculation, they argued, was nice but irrelevant: such methods could never produce "enough" energy.

Lovin's major contribution was to refute this position. Adding up all the possibilities, he argued that, in total, these alternative sources could be sufficient in the long term. In October 1976, six months before the Salzburg conference, he had published his conclusions in the American journal *Foreign Affairs*.

Lovins was an eloquent proponent of the "small is beautiful" philosophy. For political and ethical reasons, he advocated the development of technologies that could be used on a small scale in a decentralized manner. Beyond the ideological questions, however, his analysis struck at the heart of the case as put by the nuclear advocates: that there was no alternative. Increasingly, it appeared that, in the medium term, alternatives did exist. The reason it would take time to implement them was because of the lost decades in which scientific and engineering talent throughout the world had been preoccupied with nuclear energy. If there was no alternative, even if only for a period, it was the fault of the very people who asserted the fact. Lovins was vigorously assailed by the nuclear advocates. He was accused of underestimating both the costs and the difficulties of technical development. He was attacked for overstating the maturity of his preferred technologies and the ease or speed with which they could be introduced on a large scale. He was criticized for failing to take account of the full range of side effects his proposals might cause. In short, Lovins and his supporters were attacked for all the things that had characterized the nuclear community for decades. The irony passed unnoticed.

For the nuclear advocates, Lovins was a good target, an "outsider." Weinberg was a member of "the group." But what really upset the nuclear advocates about Weinberg was that he agreed with Lovins on the basic point: long-term energy planning must envisage a transition to a new energy system not linked to exhaustible resources, such as coal, oil, and uranium. Both also agreed that, in most countries, it was possible to defer at least temporarily the further development of nuclear power. Both agreed that coal could fill the transition period; huge reserves were available throughout the world, and in some countries in Africa and South America many are still unexplored. In the long term, of course, the reserves of coal would be limited—less so than oil, but still limited. Moreover, there would be considerable pressure to conserve the limited supplies: the use of coal to make petrochemicals would become more valuable than burning it to make electricity. But coal could fill the gap in the short term. The essential difference between the two men was that Lovins believed his wide variety of renewable resources—especially solar energy in all its forms—could prove adequate to the world's needs. Weinberg disagreed, arguing that the cost of developing the renewable resources would be prohibitively high and that, although uranium is not a renewable resource, its life could be extended indefinitely by the introduction of the breeder reactor.

Seeking to deemphasize coal as a possible alternative even in the transition period, nuclear advocates pointed to the dangers of accidents in the coal mines and to the adverse health effects of burning coal. They needed only to cite another Ford Foundation study, *Nuclear Power: Issues and Choices*, completed in 1977. It recorded that, in normal operation, a 1,000-Mw nuclear power plant produced roughly one fatality a year from occupational accidents and radiation risks to workers and to the public. A comparable new coal plant, meeting the then current standards, was estimated to produce from two to twenty-five fatalities a year—two from mining and transportation accidents and the rest from the adverse health effects of sulfur-related pollutants.

Weinberg himself had emphasized another coal-burning hazard in his Salzburg address when he warned of the increase in carbon dioxide levels it would cause. An increased use of coal could produce a blanket of carbon dioxide around the earth that would act as a sort of insulation layer, raising the overall temperatures beneath it and causing drastic climatic changes. In reply, Lovins and his supporters had pointed out that new coal-building technology was much cleaner, but also that they advocated its use only for the transition period to totally renewable resources. Other papers at the official Salzburg conference showed analagous environmental problems with the expansion of nuclear power—especially if breeders were used on a large scale. There would be a significant increase in the release of long-lived radioactivity products, particularly from the reprocessing plants, and, as yet, there was no national or international agreement to limit these "normal" emissions. Reprocessing plants were a source of potential global pollutants of even greater magnitude than those released in the decades of atmospheric nuclear testing. Like the fallout from those tests, reprocessing emissions represented an international public health problem for which the traditional methods of quarantine were irrelevant. Moreover, although much was known about the global environmental pathways of weapons fallout, virtually nothing was known about the pathways of the different types of radioactive substances that would be routinely released in a fully deployed nuclear economy. Nor was it likely that any nation would have the incentive to pay the economic price of limiting its releases: most releases would "fall out" over other nations.

In accordance with the forecasting tradition, in which confidence in projections tends to increase the further a forecaster looks into the future, Alvin Weinberg had assumed a range of figures for global energy demand that indicated that sooner or later everyone would have to use nuclear power. The problem was, his forecasts came at a time when the economics of the nuclear industry were subject to an even greater uncertainty than most of the alternative sources of energy.

Weinberg was candid about the limitations of his analysis. "I put the previous scenarios forward with much diffidence," he said, "especially since events proved me so poor a prophet when . . . ten years ago I estimated nuclear energy would be ten times less expensive than it has turned out to be." Yet, at Salzburg, projecting forward almost a hundred years, Weinberg seemed oblivious of the fact that the things he did not know were more important than the things he could predict. The most basic assumption in his forecast, shared by other nuclear oracles, was that the technical and social environment a century from now will bear some relationship to the environment existing today. For all the difficulties of social and technological forecasting, the assumption that things will be related in any way whatsoever to the present is the *only* assumption that we can say, with absolute certainty, is wrong. It is a tenuous basis for making major commitments, as distinct from keeping options open.

28

"We All Live in Pennsylvania Now"

If I had a friend in Harrisburg, I guess I'd—I don't think I'd tell him to move, I'd tell him to keep close to his radio, something, if you had somebody really close in, you might tell him, if he didn't have to stick around, why maybe he oughtn't be there.

—COMMISSIONER VICTOR GILINSKY, Nuclear Regulatory Commission, March 31, 1979

On March 16, 1979 a Hollywood thriller called *The China Syndrome* opened in New York and swiftly became a box-office hit. In the film, a nuclear reactor goes berserk and is only just brought under control in time to avoid a meltdown. As a result of the film's success, "meltdown" and "China Syndrome" became household terms, preparing the world for the worst accident in the West's twenty-three-year history of nuclear power.

On March 28, 1979, at Three Mile Island, a thin strip of land in the middle of the Susquehanna River in eastern Pennsylvania, a pressurized-water reactor suffered a loss of coolant, and for five days the world watched in suspense as the crippled plant was coaxed back from the brink of disaster.

Within hours of the news of the accident, the media descended on the small town of Middletown across the river from Three Mile Island. Their coverage, filled with alarming stories, swamped newspapers and radio and television broadcasts. Headlines raged about a possible meltdown, dangerous radiation leaks, and an explosive hydrogen bubble that might blow up in the reactor core.

In the end, there was no meltdown and there was no explosion.

The experts from the nuclear industry were able to proclaim there were no deaths, no injuries; there was a smoldering reactor under control. The emergency safety systems had worked, or so it seemed. As the full picture of the accident emerged during the coming days and weeks, however, the case was not so simple. Where the experts had admitted to only minor difficulties, there had been, from the start, evidence of a major accident. Where they had claimed that the accident was under control, there had been total confusion over how to prevent its becoming more serious. Where the safety backup systems had been available, they had not been used correctly, and serious questions had arisen for which all the safety manuals and previous operating experience could provide no ready answers. In fact, at no stage during the first four critical days of the accident were any of the experts—the operators of the plant or the Nuclear Regulatory Commission—in a position to reassure the public of anything. Yet, as they had done many times before, the experts confidently provided answers that were at best evasive and, at worst, lies. Over more than two decades of trying to establish a credible relationship with the public, a relationship that would dispel the criticisms raised by the antinuclear movement, the experts had, apparently, learned nothing.

Reporters, their imaginations fired by *The China Syndrome*, believed—as it turned out, correctly—they were not getting the whole truth and wrote their stories and filmed the event accordingly. The result was not helpful. In the course of a background interview, one hapless member of the NRC was explaining to a reporter what would happen if the accident got worse and mentioned the word "meltdown." Within hours, the world news organizations were carrying the doomwatch word in their headlines. When pregnant women and young children were evacuated from Middletown on the third day, one television crew, seeking to present its audience with a ghost town—where, in fact, there wasn't one—asked people to stay off the streets while they filmed. The people obliged. Powerful emotional dramas were built around the ignorance and very real fears of the residents of Middletown. One woman believed that she could protect her baby from radiation fallout from the plant if she put a napkin over its head. Farmers believed radiation was making their livestock sick; so did a woman with a hemorrhaging poodle. A dairy farmer refused to drink his own animals' milk even though he had been advised by health authorities it contained no radiation. The overall ignorance and fear that was transposed into such headlines as the New York *Post*'s "Race with Nuclear Disaster" outweighed the more sober—and equally misleading—headline in the Manchester (New Hampshire) *Union Leader*, "No Injuries Reported in Nuke Mishap."

Some of the reporting was balanced and accurate, some of it cheap sensationalism. Although there was nothing to see at Three Mile Island's Unit 2, one reporter vividly described the "evil" steam dripping "like candle wax" down the four huge cooling towers—a report clearly designed to suggest the steam was a radioactive leak.

An angry reaction spread quickly across America and then Europe; people who had never seriously questioned nuclear power felt they had been duped by the industry's propaganda about "safe" electricity, and the antinuclear movement rejoiced in their free publicity with a chorus of "I told you so." Many American newspapers were quick to call for a moratorium on nuclear power development until a full inquiry could show what had really happened and until the nuclear industry could prove that it would not happen again. "We all live in Pennsylvania now," was the cry in Europe. In a matter of days, Three Mile Island became a public relations disaster from which, it seemed at the time, the nuclear industry might never recover. Containing their own uncertainties, the industry remained publicly buoyant with assertions of confidence that, in the end, the technology had proved itself to be safe. People's irrational fears would soon be overcome, or so they hoped. The industry's lobby, the Atomic Industrial Forum, cautiously forecast that the accident had set the nuclear industry back from one to three years. A year later, despite new concern over rising oil prices and instability of oil supplies, the actual amount of electricity being produced by nuclear power in America had dropped from 12.9 percent to 10.6 percent of the total. There were no new orders for reactors, and five electric utilities discarded plans to build nine new plants. Other utilities were considering whether reactors under construction should be converted to coal.

Doubts about the future of nuclear power also cropped up in Europe. In Germany, Sweden, and Austria, a de facto moratorium on new reactor orders looked as though it might be continued into the mid-1980s. Only in France and the Communist Bloc countries was the enthusiasm for nuclear power unabated, and Russia went ahead with plans to increase nuclear output ten times over the next decade. Canada, Japan, and Britain lay somewhere in between the two extremes.

The instant, and perhaps, in the long run, the most important legacy of Three Mile Island was an increased public awareness of the possible effects of low-level radiation. Before the accident, many people had believed the industry propaganda that nuclear power was "clean"; they had not understood, however, that radiation in small amounts is often vented from nuclear power plants. Although fallout from atomic clouds was well known, fallout from the fission reaction of a nuclear power plant was virtually unheard of. The public soon learned some scary figures. For example, on March 28, the inventory of fission products in the Three Mile Island Unit 2 included about seven hundred Hiroshima bombs' worth of Strontium-90 and Cesium-137 and fifty Hiroshima bombs' worth of Iodine-131. The public also learned that nuclear power plants regularly release tiny amounts of relatively harmless radioactivity, and when something goes wrong they can release quantities that can be dangerous. Although the amounts released during the accident at Three Mile Island were small—and calculated by the experts to be negligible in terms of the extra cancers they might produce—the accident occurred at a time when the debate about low-level radiation effects had still reached no satisfactory conclusion. Experts were continuing to argue about a "threshold dose." In fact, in the last few years of atomic energy's fourth decade, a determined minority of specialists working on the effects of radiation had begun to assert that the harm radiation could do to human beings, even at low levels, was significantly greater than anyone had ever assumed.

What happened at Three Mile Island, often known as TMI, was a mixture of mechanical failure and human error. Pumps feeding water to steam generators of the plant's Unit 2 reactor (TMI-2) stopped working shortly after 4:00 A.M. on March 28. The flow of water to the generators ceased and no more steam was produced by them. The plant's safety system automatically shut down the steam turbine and the electric generator it powered. This apparently simple event was much more serious than it sounds. In a nuclear power plant, the production of steam is used not only to power the turbines but also to remove heat from the water that is cooling the reactor core, the coolant, which in a PWR is ordinary water. Without steam in the steam generators, the coolant heated up, expanded, and created an abnormal pressure in the core. At this point, the first safety system, a relief valve at the top of the core, opened automatically, releasing excess pressure. The valve was only a small one, however, and could not release enough pressure to stop the heating up of the coolant. A second automatic safety device, known as a "scram," dropped the control rods into the reactor core to stop the fission, but decay products from previous fissions continued to produce heat—though it was only 6 percent of the total fission heat. In theory, the overheated coolant would cool down, if slowly.

The relief valve at the top of the core coolant system should have closed thirteen seconds into the accident when the pressure returned to normal. This was to prevent any excessive loss of the core coolant. A light on the control panel indicated that the electric power that had opened the valve had gone off, leading the operators to assume that the valve had shut. But the valve had stuck open and would remain so for two hours and twenty-two minutes, draining vital coolant water. In the first hundred minutes of the accident, about thirty-two thousand gallons of coolant would escape this way. Had the valve been closed or had the operators realized that it was stuck open and operated a secondary valve to block the flow of coolant, the accident would have remained minor.

A second error concerned the temperatures being created at the top of the relief valve. An emergency operating procedure states that a pipe temperature of 200 degrees Fahrenheit indicates an open valve. However, the operators failed to notice the temperatures, which recorded data show reached 285 degrees Fahrenheit. They said later that the temperatures regularly registered high because either the relief valve or some other valve would always be leaking slightly.

A combination of these and other errors led to the top part of the core being uncovered and heated to the point where the zirconium alloy of the fuel rod cladding—the coat surrounding the uranium—reacted with steam to produce hydrogen. Some of this escaped through the open relief valve into the containment building, where it caused a minor gas, not nuclear, explosion on the first afternoon of the accident. The rest of the hydrogen remained in the reactor itself and formed a hydrogen bubble that was to cause the experts anxious moments in the days ahead. The first evidence of the uncovering of the core and the rupture of the fuel claddings was at approximately 6:00 A.M. on March 28—two hours after the feedwater pump to the steam generator had failed.

Almost immediately, the operators of the plant, the Metropolitan

Edison Company (Met Ed), began a public relations campaign to reassure the public either that there was no crisis or that the crisis would soon be over. Neither was true.

Appearing on morning television on March 29, the second day of the accident, the utility's president, Walter Creitz, confidently asserted that the plant soon would be safely closed down without injury to anyone. Later the same morning, at a press conference, the company's top technical official, John Herbein, acknowledged for the first time that serious things had gone wrong: radioactive gas and steam were building up to potentially explosive levels in the plant's auxiliary building. Middletown's mayor, Robert Reid, demanded to know why the company had delayed telling his community of the dangers. A reporter, who lived a mile from the plant, yelled, "What are you going to do to protect my family?" Another, summing up the public's total confusion over what had really gone wrong, asked, "Mr. Herbein, is your plant a lemon?"

With a host of television lights and cameras focused on his perspiring face, Herbein lost his composure. "We didn't injure anybody with this accident," he stammered. "We didn't seriously contaminate anybody and we certainly didn't kill anybody."

Actually the operators knew little of the extent of the accident at this stage—and even less of what to do about it. There had been a loss of the water coolant surrounding the reactor core, the core was overheating, and water pressure had been building up in one of the tanks holding coolant leaking from the reactor; moreover, that coolant was highly radioactive. On the third morning of the accident, March 30, a helicopter hovering over the crippled plant picked up a sudden large increase in radioactivity. The data was flashed simultaneously to the Washington headquarters of the NRC and the office of Pennsylvania's governor, Richard Thornburgh. The governor wanted to know whether to evacuate the local population. The radioactive gas released from the plant had formed a plume of radioactivity that was drifting with the wind to the northeast of the plant—directly over Middletown. The NRC advised against evacuation—yet. So far, the NRC had calculated that the radiation dose to the population at one mile from the plant was about .017r per hour and about half that amount at two miles. In a lung X ray, a patient receives about .05r over his chest, and even though the people of Middletown were being exposed over their entire bodies, exposure at that level was not cause for immediate concern, they believed. The question was: How much more radiation was Met Ed planning to release, and when?

The public knew nothing of the radiation release. Middletown citizens had been informed by Met Ed that an accident had occurred, but they had no idea of the total confusion among the experts at Met Ed and the NRC headquarters. This picture is painted in all its frightening absurdity in transcripts of taped meetings of the NRC in Washington. The transcripts show, in vivid detail, that forty-eight hours into the accident the agency still lacked any clear idea of how to deal with the problem.

In the edited extracts that follow, these characters appear: the NRC chairman, Joseph Hendrie; commissioners Victor Gilinsky, Peter Bradford, Richard Kennedy, and John Ahearne; the NRC's nuclear reactor regulation director, Harold Denton; and the NRC's public affairs director, Joe Fouchard. They are discussing the release of radioactivity from the plant and whether to evacuate people living nearby. The wind is blowing radiation from the plant to the northeast—over Middletown.

DENTON: Yes, I think the important thing for evacuation to get ahead of the plume is to get a start rather than sitting here waiting to die. Even if we can't minimize the individual dose, there might still be a chance to limit the population dose.

BRADFORD: It ought to be made clear that you are not talking about lethal doses.

HENDRIE: Harold, the recommendation on evacuation was for that direction, right? Northeast?

DENTON: But the people at the site are obviously much better [able] to direct and run emergency plans than we are, and I would hope the plant people and our people are really monitoring what is going on in there and acting on it from moment to moment. . . . It just seems like we are always second, third-hand; second guessing them [Met Ed]. We almost ought to consider the Chairman [Hendrie] talking to the owner of the shop up there and get somebody from the company who is going to inform us about these things in advance if he can. We seem not to have that contact.

GILINSKY: Well, it seems to me we better think about getting better data.

FOUCHARD: Well, the Governor is waiting on it, and Mr. Chairman, I think you should call Governor Thornburgh and tell him what we know.

I don't know whether you are prepared at the present time to make a commission recommendation or not. The Civil Defense people up there say that our State Programs people have advised evacuation out to five miles in the direction of the plume. I believe that the commission has to communicate with that governor and do it very promptly. . . .

GILINSKY: Well, one thing we have got to do is get better data. Get a link established with that helicopter to make sure that from now on we get reasonable data quickly.

FOUCHARD: But it does seem to me you have to make a judgment promptly.

GILINSKY: Well, that's right, but also it doesn't look like this thing is going to be over with that judgment call.

DENTON: It just seems to me we are going to have to operate on the basis primary coolant has very high dissolved gas levels, it's a five-day half-life and it is going to persist like this for a long time until they really get to a situation where they don't have to get any of the primary coolants out of the containment area. As long as they keep piddling around with pressurized levels, blow-down and let-down they're going to give a continual occasional case like this. I sure wish I had better data.

HENDRIE: Yes. Your current link out to the site is just not operating or what is the situation? Who are you talking to out there, and is there an open line out there?

DENTON: I'm not sure. Let me ask. What is our line at the site? We talk to our guy in the control room who bends over and asks the questions while we are talking to him and gets back on the phone. So we do have our people in the control room who search out the answers. But with regard to any actual or hard numbers for release rate, curies, quantities, doses off-site, that process seems to take hours.

FOUCHARD: Don't you think as a precautionary measure there should be some evacuation?

HENDRIE: Probably, but I must say, it is operating totally in the blind and I don't have any confidence at all that if we order an evacuation of people from a place where they have already gotten a piece of the dose they are going to get into an area where they will have had .0 of what they were going to get and now they move some place else and get 1.0. [He is referring to the movement of the radiation plume away from the site.]

GILINSKY: Does it make sense that they have to continue recurrent releases at this time?

DENTON: I don't have any basis for believing that it might not happen—

is not likely to happen again. I don't understand the reason for this one yet.

FOUCHARD: I believe as a precautionary measure—

HENDRIE: I think we had better get—Harold, see if you can get some sort of a better link established.

DENTON: We need to have some way so that when we ask a question we get an answer eventually . . . but everybody who is up there in a supervisory role is doing something else. And we don't know what they are doing, they may be taking just the proper action. . . . People who go up there fall into a morass, it seems like they are never heard from. It seems like you might want to consider having something like rotating shifts through senior people there in the control room or in a room off the control room that we could communicate with about these kinds of things directly. I would be happy to volunteer and see how things go along for a while.

HENDRIE: You decide whether you ought to be one, Harold. . . . This could go around the clock for the next couple of days. I don't know what you can do to improve the communication situation, but it is certainly lousy. Now, Joe, it seems to me I have got to call the Governor—

FOUCHARD: I do. I think you have got to talk to him immediately.

HENDRIE:—to do it immediately. We are operating almost totally in the blind, his information is ambiguous, mine is nonexistent and—I don't know, it's like a couple of blind men staggering around making decisions.

Hendrie reached the governor in his offices in Harrisburg. They discussed the difficulties in getting precise information from the site about the extent of the radiation plume. Hendrie explained that one release of radioactive gas from the plant produced 1.2r per hour, giving .12r on the ground. "That is still below the EPA evacuation trigger levels; on the other hand, it certainly is a pretty husky dose rate to be having off-site," commented Hendrie.

THORNBURGH: Do we have any assurances that there is not going to be any more of these releases?

HENDRIE: No, and that's a particularly important aspect I want to talk to you about. As best I can judge from the kind of information coming from the plant, it is not clear that they won't get into this kind of situation again. I trust not again without all of us knowing in advance. . . .

Hendrie rang off, and there was a discussion at NRC headquarters about how to improve communications with the plant. Hendrie concluded, "I mistrust our ability to establish any direct communication with the site, and that is precisely what we haven't been able to do for 24-plus hours." This information gap continued, the NRC officials at the site becoming exasperated. It meant the NRC was unable to calculate the extent of the damage. Roger Mattson, an NRC systems safety director, explained the problem:

MATTSON: Now, B and W [Babcock and Wilcox, the reactor designers] and we have both concluded, some hours ago, now—I don't know, sometime close after midnight—that we have extensive damage to this core. That corroborates with the release that we are seeing. . . . My best guess is that the core uncovered, stayed uncovered for a long period of time, we saw failure modes the likes of which has never been analyzed. It isn't like a LOCA [Loss of Coolant Action]. Some kind of swelling, rupture, oxidation near the top of the quarter of the assembly.

We just learned . . . three hours ago, that on the afternoon of the first day, some 10 hours into the transient, there was a 28-pound containment pressure spike [explosion]. We are guessing that may have been a hydrogen explosion. They [Met Ed], for some reason, never reported it here until this morning. That would have given us a clue hours ago that the thermo-couples [temperature readings] were right and we had a partially disassembled core.

Mattson then mentioned that a gas bubble had been forming in the top of the core; it was apparently made up of noncondensable gases, mostly hydrogen. These are gases that, under the temperature and pressure conditions in the core, would not liquefy but stay as gases and therefore form a potentially explosive mixture. "We have got every systems engineer we can find, except the ones we put on the helicopter, thinking the problem, how the hell we get the noncondensables out," said Mattson, concluding that the situation was like a horse race:

MATTSON: Do we win the horse race or do we lose the horse race? If you are lucky and you have overestimated the noncondensables, you might win it. If you are not lucky and you have got the right number on the noncondensables, you might lose it. . . . The latest burst didn't hurt many people. I'm not sure why you are not moving people. Got to say it. I have been saying it down here. I don't know what we are protecting at this point. I think we ought to be moving people.

HENDRIE: How far out?

MATTSON: I would get them downwind, and unfortunately the wind is still meandering, but at these dose levels that is probably not bad. . . . I might add you aren't going to kill any people out to ten miles. There aren't that many people and they have had two days to get ready and prepare. . . . It's too little information too late unfortunately and it is the same way every partial core meltdown has gone. People haven't believed the instrumentation as they went along. It took us until midnight last night to convince anybody that those God-damn temperature measurements meant something. . . . My principal concern is that we got an accident that we have never been designed to accommodate, and it's, in the best estimate, deteriorating slowly, and the most pessimistic estimate, it is on the threshold of turning bad. And I don't have a reason for not moving people. I don't know what you are protecting by not moving people.

KENNEDY: It's going to be in the newspapers this evening at 5:00, "NRC contemplating evacuation." If that's what you want, all right.

After conferring with the White House, Governor Thornburgh had, in fact, already advised pregnant women and preschool children to leave the area within a five-mile radius of the plant until further notice. Thornburgh was at pains to stress there was no reason for panic, but the sight of a convoy of yellow school buses followed by cars packed with bedding leaving Middletown was ominous to the local inhabitants.

The White House and the President's press secretary, Jody Powell, talked to Hendrie about coordinating press releases. Governor Thornburgh also asked Hendrie for the worst possible scenarios that he should prepare for in the next couple of days.

HENDRIE: I think it would be prudent to have the evacuation plan and emergency people go on alert status . . . if we suspect a major release, I think we probably ought to talk about going out to 20 miles.

THORNBURGH: Is there anyone in the country who has experience with the health consequences of such a release?

HENDRIE: Ah—not in the sense that it's been studied and understood in any real way. There were back in the days when they were doing bomb testing, they managed to give groups of soldiers and occasionally a few civilians doses in the low r range—a subject of discussion these days—but that's about the only comparable experience that occurs to me. You are talking now about a major release, not about the small releases that have occurred thus far.

THORNBURGH: What is your opinion now of a chance of a meltdown?

HENDRIE: I think pretty small.

Sustaining themselves with hamburgers and chocolate milkshakes, the commissioners continued their marathon session into Saturday, March 31, the fourth day of the crisis. They were still considering whether to evacuate people from the immediate vicinity of the site.

As the plant's condition improved over Sunday and Monday, no evacuation orders were issued. By Tuesday, the sixth day of the crisis, the crippled plant was in a relatively stable condition and the temperatures were coming down, albeit slowly.

Thus Three Mile Island passed into nuclear folklore: the worst single reactor accident the West had ever known, producing the largest amount of radioactivity released in the West since the peaceful nuclear power program began. The antinuclear movement could not have wished for better publicity. A month later, on May 6, more than 100,000 demonstrated in Europe over the next several weeks. At one extreme, the critics wanted a total shutdown of all plants immediately; at the other, they wanted at least a moratorium until the industry could show that the problems were solved.

The nuclear industry, facing its worst crisis, was as stubborn as it had ever been, confidently predicting that America's energy could not do without nuclear power now or in the future; existing plants were not enough—more nuclear power was needed to keep the country from running into a serious power shortage. Utilities in 20 states were in the process of building 92 reactors—in addition to the more than 140 already operating. On the table at the NRC were requests for permission to build another 28 reactors. "The unfortunate events at Three Mile Island should not distract us from the fact that nuclear power must play a significant role in the nation's search for energy," declared Thomas Ayers, chairman of Chicago's Commonwealth Edison Company. The utility, one of the first in America to go nuclear, had seven nuclear power plants operating, six under construction, and three more awaiting construction permits.

At the U.S. Department of Energy, the forecasters were not so bullish. They were swiftly revising their estimates downward. By 1985, they thought, nuclear energy would account for 118,000 Mw. In 1977, the projected figure had been 140,000; in 1976 it had been

245,000, and in 1970, the old Atomic Energy Commission had forecast 300,000 Mw by the mid-1980s.

At the NRC, a de facto moratorium on licensing new plants was in effect.

At the White House, President Carter said it was out of the question to shut down all nuclear reactors immediately, as the antinuclear movement wanted, but he appointed a blue-ribbon commission to study the Three Mile incident and to come up with recommendations about the future of nuclear power in America. The commission was to be chaired by John Kemmeny, a mathematician who had worked at Los Alamos after the war. Working with him were a host of consultants and eleven commissioners, including a forty-three-year-old Middletown housewife, Anne Trunk, who was chosen as a representative of the Three Mile Island community. "I'm going to be the average consumer," Trunk told one news magazine. "They're going to have to make me understand. If I can understand it, everyone in Middletown can." The commission was asked to report by the end of October.

Throughout the summer, the crippled Unit 2 reactor at TMI lay sealed and silent, still causing the nation's best scientists and technicians to scratch their heads over what to do about it. Inside the containment vessel, half a million gallons of spilled coolant had created a dank, radioactive atmosphere, sealed from the outside by steel and concrete walls. The radioactivity was still too high for anyone to enter, and most estimates guessed the levels would stay too high until the end of the year. Plans were being drawn up to clean the containment building and the connected auxiliary building where another half a million gallons of radioactive water was sitting. Finally, there was the reactor itself. "That could be the biggest problem of all if the damage has fused things together," said an engineer from the NRC. In July, experts from the owners, the General Public Utilities Corp., estimated that it would take four years and cost up to $430 million to recommission the plant; that figure would soon rise over the $1 billion mark. If the company included the electricity needed to replace the amount that would have been produced by TMI-2, the cost would run over $2 billion.

At the end of October 1979, the Kemmeny Commission reported. Based on more than 150 formal depositions and on collected material that fills more than 300 feet of library shelf space, the report is an extraordinarily lucid and easy-to-read document. Its main conclusion came as a surprise: instead of focusing on the breakdown of the technology (the commission agreed that an equipment failure initiated the event), the report concluded that the fundamental problems were people-related.

"The most serious 'mindset' is the pre-occupation of everyone with the safety of equipment, resulting in the down-playing of the importance of the human element in nuclear power generation," said the commission. "We are tempted to say that while an enormous effort was expended to assure that safety-related equipment functioned as well as possible, and that there was back-up equipment in depth, what the NRC and the industry have failed to recognize sufficiently is that the human beings who manage and operate the plants constitute an important safety system."

The commission's strongest criticism was reserved for the Nuclear Regulatory Commission: "Fundamental changes will be necessary in the organization, procedures, and practices—and above all—in the attitudes of the NRC and, to the extent that the institutions we investigated are typical, of the nuclear industry."

Of its forty-four recommendations, only two were concerned with technical aspects. The rest were aimed at reform of the institutions charged with running nuclear power. The NRC, said one commissioner, was "at present a headless agency that lacks the direction and vitality needed to police the nuclear power industry on a day-to-day basis." The utility and its suppliers were in need of stricter standards to guarantee safety. Better training of operating personnel was required; so was more concern for the health and safety of plant workers and the public. There should be improved emergency-response plans by utilities in case of future accidents and establishment of these plans should be contingent on granting licenses. Finally, the commission called for greater public access to technical information and greater education of reporters in handling complicated news stories.

Both the owners and the operators of the TMI plant were openly criticized. Met Ed "did not have sufficient knowledge and was not sufficiently prepared" to deal with the accident; its procedures "were inadequate and, in some cases, hopelessly confusing. The control room where the accident was managed is greatly inadequate for managing an accident." More than one hundred alarms had gone off during the first few minutes of the accident "with no way of suppressing the unimportant ones and identifying the important ones." One control room operator told the commission, "I would have liked to have thrown away the alarm panel. It wasn't giving us any useful information."

The commission, however, decided not to recommend any delay or halt to nuclear power plant construction while its safety recommendations were being implemented. This single conclusion dominated the debate on the report and to a large extent obscured the value of its recommendations. Protagonists of a moratorium need not have been concerned: in the coming year, there were no orders and eleven cancellations of earlier orders. The de facto moratorium looked to be around for some time. "The almost universal faith and optimism of earlier years have been superseded by concern, pessimism, and not infrequently hostility," commented a report by a special team of energy experts done jointly for the Rockefeller Foundation in New York and the Institute for International Affairs in London.

The industry reacted by spending $22 million on two new organizations aimed at improving the technical designs and operating procedures of the reactors and the training of those who run them. The first was called the Nuclear Analysis Center and the second the Institute for Nuclear Power Operations.

While industry accepted, as a rule of thumb, that the moratorium would probably last for three or four years, they also still firmly believed in a projected "energy gap" that would require more nuclear plants to fill it. Some, indeed, believed the nuclear capacity of the United States would have to be doubled by the year 2000. Robert Kirby, the Westinghouse chief executive, was one of them. Bertram Wolfe, the General Electric vice-president in charge of the nuclear fuel and services division agreed with James Schlesinger (the former Secretary of Energy) that, "If we don't substitute nuclear for other energy forms in a realistic fashion the United States just might not make it."

Outside America, the TMI-2 accident was headline news for more than a week. There were also huge antinuclear demonstrations, some of them violent. In Germany, even the conservative news magazine *Der Spiegel* devoted twenty-five pages of its April 9 issue to the nuclear debate, and its lead editorial called for a fundamental reappraisal of the country's nuclear plans. Four days later, two bombs destroyed a pylon linking the newly completed Esenshamm nuclear plant near Hamburg to the grid. At the beginning of May, Chancellor Helmut Schmidt called for an intergovernmental conference on the safety of nuclear reactors.

In France, the April 14 issue of the conservative news weekly *L'Express* ran a twelve-page cover feature on TMI-2 entitled "The Nuclear Risk." It concluded, "There never has been, even in the National Assembly, the least debate [or] vote on civil electronuclear plans . . . the French have the right to demand, first, a commission of inquiry into nuclear safety, independent of the Executive; and thereafter, a true Parliamentary debate based on the conclusion of this commission." *L'Express* was too late: the week before the article appeared, André Giraud, now Industry Minister, had given the government's answer: it would press ahead with its plans for nine additional nuclear plants in the next five years. "France has no choice," declared Giraud. "It's either nuclear energy or economic recessions."

Even in those countries with less-developed nuclear plans, demonstrations and protests followed Three Mile Island. In early April, there were antinuclear demonstrations in Switzerland. A typical slogan read: "Today Harrisburg—tomorrow Mühleberg," a reference to the Swiss nuclear plant near Bern. In Spain, on June 13, a bomb exploded at a nuclear plant site. In Holland at the end of June, the Dutch Parliament approved a motion calling on the government to be "extremely cautious" over nuclear energy.

In some European countries, politicians knew their attitude to nuclear power could be the difference between victory and defeat at the polls. In Germany, on May 16, in a half-hour live TV broadcast, the Minister-President of Lower Saxony said he would refuse permission for a reprocessing plant at Gorleben, not because it would be unsafe, but because it would be politically unacceptable. In Sweden, where the civil nuclear power program had become the single most contentious political issue and had been blamed for the fall of two governments before the TMI accident, the former Prime Minister, Thorbjoern Faelldin, on March 31 called for the shutdown of Sweden's only operating PWR. All parties agreed eventually to hold a referendum on nuclear power. A year later, the Swedes gave nuclear power a limited "Yes"—allowing the government to expand its reactor plans from six to twelve over the next decade, but endorsing a principle of phasing out nuclear power over the next twenty-five years.

Often in Europe, the industry's reply to the TMI accident was to say "It couldn't happen here, we have different"—or "better-designed"—reactors. Such was the case in Germany, for example; at a press conference on April 10, Klaus Barthelt, executive director of Kraftwerk Union, insisted that extra safety features of the German-designed PWR meant the same accident could not happen there. Britain, of course, was the one country where its nuclear advocates could boldly declare, as did Prime Minister James Callaghan, "I can assure the country that the incident which took place in Harrisburg could not take place here because of the different types of reactors." He was referring to the fact Britain's is a gas-cooled type. "We have been very wise," Callaghan continued, "in concentrating on a safer type of reactor." (A year later, the Conservative government of Margaret Thatcher approved plans to build a Westinghouse PWR in Britain.) In none of their responses did the European advocates speak to the human factor the Kemmeny Commission would highlight six months later.

Before the Three Mile Island accident, the debate over low-level radiation had focused on effects on three groups: guinea-pig soldiers caught in the fallout from the Nevada tests, residents living near the test sites, and workers in the atomic factories. The month before the accident, a *New York Times* article had listed some of the more disturbing questions still being debated:

> Could it be that radioactive fallout from the nuclear weapons tests of the 1950's and the 1960's maimed the brains of millions of children and lowered their IQ's? Could it be that several hundred thousand ship-yard workers servicing nuclear-powered vessels have been exposed to too much radioactivity over the past 20 years, causing leukemia in some of them? Could it be that thousands of workers in the nation's nuclear weapons plants have been exposed to levels of radiation that induced cancer?

The issue of the safe threshold dose was never more alive, however, than at Three Mile Island. A year after the accident, Middletown residents were suffering from what one of the local doctors called "nuclear neurosis"; they were ready to blame anything that went on in their lives, from colds to inflation, on Three Mile Island.

Pregnant women, who were evacuated briefly from Middletown during the accident, had listened, bewildered, to experts discussing whether a tiny increase in the number of infant deaths and child thyroid cases could be the result of radiation poisoning. The residents had watched, equally confused, a year of newspaper stories of "strange goings-on" with farm animals in the area: spontaneous abortions, stillbirths, sterility, blindness, and sudden death. The farmers and local vets said these did not occur before the accident. Local health authorities said they had nothing to do with radiation or any other kind of poisoning from the nuclear plant.

In March 1980, a former Pennsylvania state health chief disclosed an alarming 50 percent increase in the number of infant deaths within ten miles of the reactor during the six months following the accident compared with the same period the year before. Two days later, an epidemiologist from the state health department said that recent studies contradicted that information: the studies found "no significant changes" in the infant death rates before and after the accident. The epidemiologist produced figures to show a *drop* of 13 percent in infant deaths for the six-month period after the accident compared with the six months immediately before. He said it was "very unfortunate" that the earlier "raw data" had been disclosed "because the sensational numbers, by themselves, could cause unnecessary fear."

A local vet, Robert Weber, who had practiced in the Middletown area for thirty-two years, disclosed that his farmers had been experiencing widespread bone problems with their cows ever since the first Three Mile Island reactor started up in 1976. After giving birth, said Weber, "the cows go down and they can't go back up." In 1979, farmers reported birth problems with their pigs; sows were unable to give birth normally despite massive hormone injections. After the accident, Weber was performing one cesarean section a week whereas previously he had done only one or two a year. But the Pennsylvania agriculture department said they had collected tissue samples from dead animals on one hundred farms in the area since the accident and had found no evidence of any radiation damage. Immediately after the accident, Iodine-131 had been found in milk samples, but the levels were thirty times lower than federal standards and ten times lower than Iodine-131 levels found at U.S. monitoring stations around the country after a Chinese nuclear test in 1976. "The problem," said Joseph Leaser, a local GP who also owns livestock, "is that no one really knows what has caused these upsets. All the evidence so far is empirical. You hear old timers say they had four barren mares [since the accident] and last year, for the first time, I had trouble breeding my own animals. But these are off the top of the head observations; there are no hard statistics."

The lack of statistics about radiation damage to Middletown's population before the accident has also made it hard to produce scientifically sound comments on post-accident medical research. In a few of his patients, Dr. Leaser found an increase in eosinophils; white blood corpuscles associated with the body's immune system, eosinophils increase in number during allergic reactions, but also during exposure to radiation. "We have found them in patients who don't have allergies and we don't know the cause," said Leaser. The state health department said his results were probably due to early spring pollens in the air, but Leaser also found the increase last winter. "Even so," he said, "the evidence is only just enough to whet the appetite; there's no proof."

The illness for which there was plenty of proof, however, was "nuclear neurosis." "The emotional trauma is ten times as important as the physical effects [of Three Mile Island] and people are not addressing that problem; it's like a powder keg waiting to explode," Leaser said.

Behind these very real fears, a dramatic rethinking about public radiation protection standards was taking place. Over the years, the health physicists had generally agreed that effects detected at high levels were proportionally true all the way down the line to the lowest doses. If 100r of exposure causes 100 cancers, then 10r would cause 10 cancers. A minority had argued that this view was too conservative—one-tenth the dose would cause less than one-tenth the effect—and some believed that there was a "threshold" below which radiation had no effect at all. In the late seventies, a new group of alarmed experts emerged, warning that the old assumptions were not conservative enough. They claimed new evidence suggested that at low level the effects were more than proportionate: one-tenth the dose would have *more* than one-tenth the effect.

The new approach would require a basic rethinking of all existing evidence on radiation effects and all existing social uses of radiation sources: not just nuclear power but X rays, building construction materials, color television sets, false teeth, jet travel, and the full range of trivial uses of material that is radioactive. The new thinking would require a complete reconsideration of the effects of "natural" background radiation—the kind that affects everyone and that comes from cosmic rays or materials in the earth. For years, the figures had suggested that only 1 percent of all cancers that occur naturally could be attributed to this background radiation. Indeed, nuclear power advocates had adopted the practice of justifying nuclear plants by asserting that their operation would give the public "only" a small percentage of the background radiation dose. If the new figures were right, then even the "natural" background radiation would become grounds for concern, for rather than the 1 percent that had been assumed, background radiation could explain more than half of all the cancers that occur.

Nuclear advocates scorned the new hypothesis; they knew that it would be almost impossible to disprove but equally as difficult to prove; the times, however, no longer gave the nuclear community the benefit of the doubt. At the center of the new controversy was the small, quiet-spoken, gray-haired English epidemiologist, Alice Stewart. It was she who had detected, as long ago as 1956, the first scientific indication that low levels of radiation had any effects at all. Stewart had worked in Oxford University's Institute of Social Medicine and had participated in one of the world's biggest systematic research studies of children's health, known as the Oxford Child Health Survey. This study would provide a huge amount of information in the field, but one of the earliest results—or rather indications—would prove the most controversial. By 1956, the Oxford Child Health Survey figures were already suggesting that children whose mothers had received pelvic X rays during pregnancy were more likely to get cancer. That was all the survey could show: the only information available was whether an X ray had been administered. The study could not be of very much use to the health physicists because the amount of radiation actually received in an X ray, though generally low, could still vary greatly and, without figures giving the size of the dose, effects could not be correlated.

That no one knew the precise levels was a general defect shared by the stream of subsequent research confirming the effects—both in Oxford and in the United States. As the evidence accumulated, however, and some new studies indicating other effects at low doses were published, the consensus in the radiation community began to shift. In 1976, when Alice Stewart was seventy, the final seal of approval was given. Sir Edward Pochin, head of the British National Radiological Protection Board and chairman of the International Commission on Radiation Protection—the official organ of orthodoxy for government agencies throughout the world—gave official blessing to Stewart's twenty-year-old study. "It now appears likely therefore," he intoned, "that absorbed doses of only a few rads in the fetus may induce malignancies of various types." It was the first time that the international hierarchy of official radiation experts had acknowledged that effects at such low levels were real and not simply hypotheses.

Just over a month before Pochin's acknowledgment, Alice Stewart had taken a new course of action that would lead her to assert that even this official blessing was no longer sufficient. In May 1976, Alice Stewart was asked by Professor Thomas Mancuso, of the University of Pittsburgh, to help with an epidemiological study of a sample of individuals who had worked at the plutonium-producing factories at Hanford. Stewart and British statistician George Kneale quickly set up a statistical method—one that was to become very controversial—to work out the Hanford figures. They concluded that, even at the very low levels of exposure involved (and in the case of Hanford, precise dose figures were available), there was a definite increase, as against the general population, in the rate of death from cancer among Hanford workers. The only other large-scale sample that had been exposed to low doses—some of the Hiroshima survivors—had not shown any effects. But Stewart seized on new evidence that people who had come into Hiroshima after the bomb went off did show a higher rate of cancer. Stewart was convinced she had shown for the first time a precise relationship between doses and effects at the very lowest dose rates. The disturbing thing was that the relationship was much larger than everyone had always assumed would be the worst case. The effects were *more* than proportional.

Just after Stewart announced her new results, other researchers in a number of institutes came out with similar kinds of figures. A study of workers who maintained nuclear submarines at Portsmouth, New Hampshire, showed higher than expected cancer rates, indeed twice the national average. A look at a group of a few thousand soldiers who had been marched into the Nevada testing grounds after a bomb went off concluded they had twice the number of expected leukemia deaths. Another, a study of residents living near the test site in southern Utah, showed that the leukemia rate was two and a half times what it had been before the tests.

All of these studies were challenged on their statistical perfection; only time would prove whether the low levels produced at Three Mile Island were, as some experts claimed, harmless and insignificant: for the moment, the old standards remained. For a public increasingly concerned with the effects of nuclear power, it was disturbing to discover how wide the difference among the experts had become.

Epilogue

As the atom enters its fifth decade, the nuclear barons are still considering the theme that has repeated itself throughout this book: whether the application of nuclear knowledge should be expanded.

The overwhelming issue, against which others seem almost trivial, is the continued presence of the huge stockpiles of nuclear weapons. They represent the threat of annihilation for millions of people and must never be accepted as a fact of life; the concept of deterrence through "mutually assured destruction" is truly insane, a manifestation of intellectual barbarism.

Today there are signs of renewed citizen action against the threat. Mainly in Europe, but also to some extent in the United States, public consciousness about a nuclear holocaust has been reawakened by the cooling of the superpowers' relationship and by the threat of even bigger stockpiles. If we can persuade governments to agree that the bombing of civilians has no place in war—as sooner or later we must if we are to survive—there will be no reason to develop new generations of strategic nuclear weapons and existing stockpiles will become outmoded.

The process of this disentanglement cannot stop with the superpowers' arsenals: the problem of the spread of nuclear weapons to other countries remains. Nations without nuclear weapons will always want them if others have them, and the easiest way to acquire them will continue to be through nuclear power programs. Thus, the core of the proliferation problem is the expansion of nuclear-generated electricity: more nuclear power plants mean more fissile material available for diversion into bomb projects. There is no way of preventing this at present: the nuclear materials market is too large and too diverse to be controlled through export sanctions, and the current early-warning system of a diversion of nuclear material, known as safeguards, is inherently ineffective.

Even without the link to nuclear proliferation, nuclear power carries dangers of a magnitude that we ought not to accept. There is something profoundly stupid about continuing to multiply a series of engineering marvels that contain fifteen billion curies of radiation. We do not know enough about radiation and cannot be sure enough of our technical prowess to allow this system to dominate our energy supply. Moreover, the instinctive human fear of radioactivity is not irrational, as the nuclear advocates assert; it is also so universal and so enduring that it is a political fact of life.

This does not mean that all nuclear power plants should be closed down and no more built: there may be no practical alternative in the medium term for some nations where the cost and energy-security penalties of other sources would be considered too high.

Partly for this reason, there is no consensus about what to do about nuclear power. Beyond the minority who totally oppose it, however, there is a widely held view that, although nuclear power may be necessary, its use should be kept to a minimum. The problem, as the entire nuclear history illustrates only too forcefully, is that we have never been able to devise human institutions dedicated to minimizing their own activities. We must continue, therefore, to treat the claims of those in charge of the nuclear complex with suspicion; by their past actions they have not earned our trust or our confidence. Their almost compulsive sense of self-importance has been given a serious blow by the strength of the citizen reaction to events like Three Mile Island, but it has not been defeated. Those who want to expand nuclear power continue to repeat the old refrain that we have no choice, yet that has always been wrong and it is wrong today.

The nuclear story so far has been about a scientific and technological revolution that transformed a yearning to understand and control the powers of nature from fiction to fact. It was a stunning achievement of many clever men. What is in question now is not their cleverness but their wisdom.

Bibliography
Notes
Acknowledgments
Index

Bibliography

Acheson, Dean. *Present at the Creation*. New York: W. W. Norton, 1969.

Alexandrov, A. P. "Nuclear Physics and the Development of Nuclear Technology in the USSR," *Oktyabri Nauchnyi Progress*. Moscow: Academy of Sciences, 1967.

Allard, P., et al. *Dictionnaire des Groupes Industriels et Financiers en France*. Paris: Seuil, 1978.

Allardice, C., and E. Trapnell. *The Atomic Energy Commission*. New York: Praeger, 1974.

Amann, Ronald, et al. (eds.). *The Technological Level of Soviet Industry*. New Haven: Yale University Press, 1977.

American Assembly. *Atoms for Power: United States Policy in Atomic Energy Development*. New York: Columbia University Press, 1957.

Anderson, Clinton P., and Milton Viorst. *Outsider in the Senate*. New York: World Publishing, 1970.

Anderson, Robert S. *Building Scientific Institutions in India: Saha and Bhabha*. Montreal: Centre for Developing Area Studies, McGill University, Occasional Paper Series, no. 11, 1975.

Anderson, William R., with Clay Blair, Jr. *Nautilus 90 North*. Cleveland: World Publishing, 1959.

Ardagh, John. *The New France: De Gaulle and After*. London: Penguin, 1970.

Ardenne, Manfred von. *Ein Glückliches Leben für Technik und Forschung*. München: Kindler Verlag, 1972.

Asimov, Isaac. *Biographical Encyclopedia of Science and Technology*. New York: Doubleday, 1964, 1972.

Astashenkov, P. N. *Kurchatov*. Moscow: Molodaia Guardiia, 1968.

Astley, Joan Bright. *The Inner Circle: A View of War at the Top*. Boston: Little, Brown, 1971.

Atlantic Council of the United States. *Nuclear Power and Nuclear Weapons Proliferation*. Vols. 1 and 2. Boulder, Colo.: Westview Press, 1976.

Atomic Energy Commission of the United States. *In the Matter of J. Robert Oppenheimer*. Washington, D.C.: U.S. GPO, 1954.

Azrael, Jeremy. *Managerial Power and Soviet Politics*. Cambridge, Mass.: Harvard University Press, 1966.

Bader, William B. *The United States and the Spread of Nuclear Weapons*. New York: Western Publishing, 1968.

Bailes, Kendall Eugene. "Stalin and Revolution from Above: The Formation of the Soviet Technical Intelligenstia, 1928–34." Doctoral dissertation, Columbia University, unpublished, 1971.

Barwich, Heinz. *Das Rote Atome*. München: Scherz, 1967. (Translated as *L'Atome Rouge*. Paris: Laffont, 1969.)

Batchelder, Robert C. *The Irreversible Decision, 1939–50*. Boston: Houghton Mifflin, 1962.

Bechhoefer, Bernhard G. *Postwar Negotiations for Arms Control*. Washington, D.C.: Brookings, 1961.

Berger, John. *Nuclear Power, The Unviable Option*. Palo Alto, Calif.: Ramparts Press, 1976.

Bernstein, Barton J. "American Foreign Policy and the Origins of the Cold War." In Barton J. Bernstein and Allen Matusow (eds.), *Twentieth Century America*, 2nd ed. New York: Harcourt Brace Jovanovich, 1972.

———. "The Quest for Security: American Foreign Policy and International Control of Atomic Energy 1942–6." *Journal of American History*, vol. 60, Mar. 1974.

Bernstein, Barton J. (ed.). *The Atomic Bomb: The Critical Issues*. Boston: Little, Brown, 1976.

Bertin, Leonard. *Atomic Harvest: A British View of Atomic Energy*. London: Secker and Warburg, 1955.

Beyerchen, Alan D. *Scientists Under Hitler*. New Haven: Yale University Press, 1977.

Bhabha, Jamshed (ed.). *Homi Bhabha as Artist*. Bombay: A Marg Publication, 1968.

Bialer, Severyn (ed.) *Stalin and His Generals*. New York: Pegasus, 1969.

Bieber, Hans Joachim. *Zur Politischen Geschichte der Friedlichen Kernenergienutzung in der Bundesrepublik Deutschland*. Heidelberg: Evangelischen Studiengemeinschaft, 1977.

Birkenhead, Frederick, 2nd Earl of. *The Prof in Two Worlds: The Official Life of Professor F. A. Lindemann, Viscount Cherwell*. London: Collins, 1961.

Blackett, P. M. S. *Fear, War and the Bomb: The Military and Political Consequences of Atomic Energy*. New York: McGraw-Hill, 1949.

Blair, Clay, Jr. *Admiral Rickover and the Atomic Submarine*. New York: Henry Holt, 1954.

Blum, John M. (ed.). *The Price of Vision: The Diary of Henry A. Wallace 1942–46*. Boston: Houghton Mifflin, 1973.

Blumberg, Stanley A., and Gwinn Owens. *Energy and Conflict: The Life and Times of Edward Teller*. New York: G. P. Putnam's Sons, 1976.

Boffey, Philip M. *The Brain Bank of America: An Inquiry into the Politics of Science*. New York: McGraw-Hill, 1975.
Borkin, Joseph. *The Crime and Punishment of I. G. Farben*. New York: Free Press, 1978.
Brodie, Bernard. *Strategy in the Missile Age*. Princeton, N.J.: Princeton University Press, 1957.
Brodine, Virginia. *Radioactive Contamination*. New York: Harcourt Brace Jovanovich, 1975.
Bupp, Irvin C., and Jean-Claude Derian. *Light Water: How the Nuclear Dream Dissolved*. New York: Basic Books, 1978.
Burn, Duncan. *Nuclear Power and the Energy Crisis: Politics and the Atom Industry*. London: Macmillan, 1978.
Burns, E. L. M. *Megamurder*. New York: Pantheon, 1967.
Calvert, James F. *Surface at the Pole*. New York: McGraw-Hill, 1960.
Cameron, James. *Point of Departure: An Attempt at Autobiography*. New York: McGraw-Hill, 1967.
Cartellieri, Wolfgang, A. Hocker, and W. Schnurr (eds.). *Taschenbuch für Atomfragen*. Bonn: Festland Verlag, 1959, 1961, 1964, 1968.
Carter, Jimmy. *Why Not the Best?* New York: Bantam, 1976.
Caute, David. *The Great Fear*. New York: Simon & Schuster, 1978.
Chaban-Delmas, Jacques. *L'Ardeur*. Paris: Stock, 1975.
Chevalier, Haakon. *Oppenheimer: The Story of a Friendship*. New York: Braziller, 1965.
Childs, Herbert. *American Genius: The Life of Ernest Orlando Lawrence*. New York: E. P. Dutton, 1968.
Chuikov, Vasilii. *The Battle for Stalingrad*. New York: Holt, Rinehart and Winston, 1964.
Clark, Ronald. *The Birth of the Bomb*. London: Phoenix House, 1961.
———. *The Life of Bertrand Russell*. New York: Knopf, 1976.
———. *Tizard*. Cambridge, Mass.: M.I.T. Press, 1965.
Cochran, Thomas B. *The Liquid Metal Fast Breeder Reactor*. Washington, D.C.: Resources for the Future, 1971.
Cohen, Bernard L. *Nuclear Science and Society*. New York: Anchor Press, 1974.
Committee on Government Operations, U.S. Senate. *Peaceful Nuclear Exports and Weapons Proliferation: A Compendium*. Washington, D.C., U.S. GPO, 1975.
Compton, Arthur Holly. *Atomic Quest*. New York: Oxford University Press, 1956.
Congressional Research Service, Library of Congress. *Commercial Nuclear Power in Europe: The Interaction of American Diplomacy with a New Technology*. Washington, D.C.: U.S. GPO, 1972.
———. *Nuclear Proliferation Factbook*. Washington, D.C.: U.S. GPO, 1977.

———. *Nuclear Weapons Proliferation and the International Atomic Energy Agency*. Washington, D.C.: U.S. GPO, 1976.

———. *Project Interdependence: U.S. and World Energy Outlook Through 1990*. Washington, D.C.: U.S. GPO, 1977.

———. *Reader on Nonproliferation*. Washington, D.C.: U.S. GPO, 1978.

———. *United States Agreements for Cooperation in Atomic Energy*. Washington, D.C.: U.S. GPO, 1976.

Couve de Murville, Maurice. *Une Politique Etrangère 1958–59*. Paris: Plon, 1971.

Craven, Wesley F., and James L. Cate (eds.). *The Army Air Forces in World War II*, 7 vols. Chicago: University of Chicago Press, 1948.

Curtis, Richard, and Elizabeth Hogan. *Nuclear Lessons: An Examination of Nuclear Power, Safety, Economic and Political Record*. Harrisburg, Pa.: Stackpole Books, 1980.

Dalberg, Thomas. *Franz Josef Strauss: Porträt Eines Politikers*. München, Gütersloh: Bertelsmann Sachbuchverlag, 1968.

Danielsson, Bengt, and Marie-Therese Danielsson. *Moruroa Mon Amour: The French Nuclear Tests in the Pacific*. Melbourne: Penguin, 1977.

Davenport, Elaine, Paul Eddy, and Peter Gillman. *The Plumbat Affair*. New York: J. B. Lippincott, 1978.

Davis, Nuel Pharr. *Lawrence and Oppenheimer*. New York: Simon & Schuster, 1968.

Dawson, Frank G. *Nuclear Power: Development and Management of a Technology*. Seattle: University of Washington Press, 1976.

Dean, Gordon. *Report on the Atom*. New York: Knopf, 1957.

Denison, Merrill. *The People's Power: The History of Ontario Hydro*. Toronto: McClelland & Stewart, 1960.

Divine, David. *The Broken Wing: A Study in the British Exercise of Air Power*. London: Hutchinson, 1966.

Divine, Robert A. *Blowing on the Wind: The Nuclear Test Ban Debate 1954–60*. New York: Oxford University Press, 1978.

Doern, G. Bruce. *Science and Politics in Canada*. Montreal: McGill University Press, 1972.

Donnelly, Warren H., et al. *Western European Nuclear Energy Development*. Congressional Research Service, Library of Congress. Washington, D.C.: U.S. GPO, 1979.

Donovan, Robert J. *Conflict and Crisis*. New York: W. W. Norton, 1977.

———. *Eisenhower: The Inside Story*. New York: Harper & Bros., 1956.

Dyson, Freeman. *Disturbing the Universe*. New York: Harper & Row, 1979.

Eggleston, Wilfrid. *Canada's Nuclear Story*. Toronto: Clarke, Irwin, 1963.

Elasser, Walter. *Memoirs of a Physicist in the Atomic Age*. New York: Science History Publications, 1978.

Emmerson, John K. *Arms, Yen and Power: The Japanese Dilemma*. Tokyo: Tuttle, 1972.

Endicott, John. *Japan's Nuclear Option*. New York: Praeger, 1975.
Epstein, Barbara. *Politics of Trade in Power Plants*. London: Trade Policy Research Centre, 1971.
Epstein, William. *The Last Chance: Nuclear Proliferation and Arms Control*. New York: Free Press, 1976.
Erikson, Erik H. *Gandhi's Truth*. London: Faber & Faber, 1970.
Erikson, John. *The Road to Stalingrad*. London: Weidenfeld & Nicolson, 1975.
Faulkner, Peter (ed.). *The Silent Bomb: A Guide to the Nuclear Energy Controversy*. New York: Vintage, 1977.
Feis, Herbert. *The Atomic Bomb and the End of World War II*. Princeton, N.J.: Princeton University Press, 1966.
Fermi, Laura. *Atoms for the World: United States Participation in the Conference on the Peaceful Uses of Atomic Energy*. Chicago: University of Chicago Press, 1957.
Fleming, Donald, and Bernard Bailyn (eds.). *The Intellectual Migration: Europe and America 1930–1960*. Cambridge, Mass.: Harvard University Press, 1969.
Forman, Harrison. *Report from Red China*. New York: Henry Holt, 1945.
Francis-Williams, Edward. *A Prime Minister Remembers*. London: Heinemann, 1961.
Franks, C. E. S. "Parliamentary Control of Atomic Energy Corporations." Paper delivered at the Conference on Legislative Studies, Canada, Simon Fraser University, February 1979.
Fuller, John G. *We Almost Lost Detroit*. New York: Reader's Digest Press, 1975.
Gaudy, René. *Et la Lumière Fut Nationalisée*. Paris: Editions Sociales, 1978.
Gilpin, Robert. *American Scientists and Nuclear Weapons Policy*. Princeton, N.J.: Princeton University Press, 1962.
———. *France in the Age of the Scientific State*. Princeton, N.J.: Princeton University Press, 1968.
Glasstone, Samuel. *Sourcebook on Atomic Energy*, 3rd ed. New York: Van Nostrand Reinhold, 1967.
Glasstone, Samuel, and Philip J. Dolan. *The Effects of Nuclear Weapons*. Washington, D.C.: U.S. Department of Defense and U.S. Department of Energy, 1977.
Gofman, John W. *Irrevy: An Irreverent, Illustrated View of Nuclear Power*. San Francisco: Committee for Nuclear Responsibility, 1979.
Gofman, John W., and Arthur R. Tamplin. *Poisoned Power: The Case Against Nuclear Power Plants*. Emmaus, Pa.: Rodale Press, 1971.
Goldschmidt, Bertrand. *The Atomic Adventure: Its Political and Technical Aspects*. London: Pergamon, 1964.
———. *Les Rivalités Atomiques 1939–66*. Paris: Fayard, 1967.

Goldsmith, Maurice. *Frédéric Joliot-Curie: A Biography*. London: Lawrence and Wishart, 1976.

Golovin, Igor I. *I. V. Kurchatov: A Socialist-Realist Biography of the Soviet Nuclear Scientist*. Moscow: Mir Publishers, 1969. (Also translated by Joint Publications Research Service [JPRS 37 1804], Washington, D.C.)

Gowing, Margaret. *Britain and Atomic Energy, 1939–45*. London: Macmillan, 1964.

Gowing, Margaret, with Lora Arnold. *Independence and Deterrence: Britain and Atomic Energy, 1945–52*. 2 vols. London: Macmillan, 1974.

Gravelaine, Frédérique de, and Sylvie O'Dy. *L'Etat E.D.F.* Paris: A. Moreau, 1978.

Green, Harold. "The Oppenheimer Case: A Study in the Abuse of the Law," *Bulletin of the Atomic Scientists*, Sept. 1977, pp. 12–16, 56–61.

Greenwood, Ronald G. *Managerial Decentralization: A Study of the General Electric Philosophy*. Lexington, Mass.: Lexington Books, 1974.

Greenwood, Ted, Harold A. Feivson, and Theodore B. Taylor. *Nuclear Proliferation Motivations, Capabilities and Strategies for Control*. New York: McGraw-Hill, 1977.

Grodzins, Morton, and Eugene Rabinowitch (eds.). *The Atomic Age: Scientists in National and World Affairs*. New York: Basic Books, 1963.

Groom, A. J. R. *British Thinking About Nuclear Weapons*. London: Pinter, 1974.

Groueff, Stéphane. *Manhattan Project: The Untold Story of the Making of the Atomic Bomb*. Boston: Little, Brown, 1967.

Groves, Leslie. *Now It Can Be Told: The Story of the Manhattan Project*. New York: Harper & Bros., 1962.

Gyorgy, Anna. *No Nukes: Everyone's Guide to Nuclear Power*. Boston: South End Press, 1979.

Halperin, Morton H. *China and the Bomb*. New York: Praeger, 1965.

Hanson, Haldore. *Humane Endeavour*. New York: Farrar & Rinehart, 1939.

Harkavy, Robert E. *Spectre of a Middle East Holocaust: The Strategic and Diplomatic Implications of the Israeli Nuclear Weapons Program*. Monograph Series in World Affairs, University of Denver, 1977.

Harrod, Sir Roy. *The Prof: A Personal Memoir of Lord Cherwell*. London: Macmillan, 1959.

Heisenberg, Werner. *Physics and Beyond: Encounters and Conversations*. New York: Harper & Row, 1971.

Heller, Robert, and Norris Willatt. *The European Revenge*. London: Barrie & Jenkins, 1976.

Herblay, Michel. *Les Hommes du Fleuve et de l'Atome*. Paris: Pensée Universelle, 1977.

Herken, Gregory F. "American Diplomacy and the Atomic Bomb, 1945–47." Unpublished doctoral dissertation, Princeton University, 1974.

Hewins, Ralph. *The Japanese Miracle Men*. London: Secker & Warburg, 1967.
Hewlett, Richard G., and Oscar E. Anderson Jr. *The New World, 1939–46*. Washington, D.C.: U.S. Atomic Energy Commission, 1972.
Hewlett, Richard G., and Francis Duncan. *Atomic Shield, 1947–52*. Washington, D.C.: U.S. Atomic Energy Commission, 1972.
———. *Nuclear Navy 1946–62*. Chicago: University of Chicago Press, 1974.
Hodes, Aubrey. *Dialogue with Ishmael*. New York: Funk & Wagnalls, 1968.
Hohenemser, Christoph, Roger Kasperson, and Robert Kates, "The Distrust of Nuclear Power." *Science*, vol. 196, Apr. 1, 1977.
Holloway, David. *Technology, Management and the Soviet Military Establishment*. London: Institute for Strategic Studies, 1971.
Hsieh, Alice Langley. *Communist China's Strategy in the Nuclear Era*. Englewood Cliffs, N.J.: Prentice-Hall, 1962.
Huff, Rodney Louis. "Political Decisionmaking in the Japanese Civilian Atomic Energy Program." Unpublished doctoral dissertation, George Washington University, 1973.
Huntington, Samuel P. *The Common Defense: Strategic Programs in National Politics*. New York: Columbia University Press, 1961.
Imai Ryukichi. *Kaku to Gendai No Kokusai Seiji* (Nuclear Energy and Contemporary International Politics). Tokyo: Nihon Kokusai Mondo Kekyu-Jo, 1977.
———. "Nuclear Safeguards," in *Adelphi Papers*, no. 86. London: International Institute for Strategic Studies, 1972.
International Atomic Energy Agency. *A Study of the Nuclear Controversy in the United States 1954–1971*. Vienna: IAEA, n.d.
Irving, David. *The Destruction of Dresden*. London: Kimber, 1963.
———. *The Virus House: Germany's Atomic Research and Allied Countermeasures*. London: Kimber, 1967.
Jabber, Fuad. *Israel and Nuclear Weapons: Present Options and Future Strategies*. London: Chatto & Windus, 1971.
Jacchia, Enrico. *L'Affaire "Plumbat."* Paris: Seuil, 1978.
Jacobson, Harold Karan, and Eric Stein. *Diplomats, Scientists and Politicians: The United States and the Nuclear Test Ban Negotiations*. Ann Arbor: University of Michigan Press, 1966.
Jain, J. P. *India and Disarmament: The Nehru Era*. New Delhi: Radiant, 1974.
———. *Nuclear India*. 2 vols. New Delhi: Radiant, 1975.
Japan Atomic Industrial Forum. *Nihon No Genshiryoku* (Japanese Atomic Energy). 3 vols. Tokyo: JAIF, 1971.
Jenkins, Roy. *Mr. Attlee*. London: Heinemann, 1948.
Johnson, Brian. *Whose Power to Choose? International Institutions and the Control of Nuclear Energy*. London: International Institute for Environment and Development, 1977.

Johnson, Chalmers. *Japan's Public Policy Companies*. Washington, D.C.: American Enterprise Institute, 1978.

Joint Committee on Atomic Energy. *Fallout, Radiation Standards and Countermeasures*. Hearings, 88th Congress, 1st Session, Washington, D.C., 1963.

———. *The Nature of Radioactive Fallout and Its Effects on Man*. Hearings, Washington, D.C., 1957.

———. *Radiation Protection Criteria and Standards: Their Basis and Use*. Hearings before the Special Subcommittee on Radiation, 86th Congress, 2nd Session, Washington, D.C., 1960.

———. *Selected Materials on Radiation Protection Criteria*. Committee Print, 86th Congress, 2nd Session, Washington, D.C., 1960.

Jones, R. V. *Most Secret War*. London: Hamish Hamilton, 1978.

Jungk, Robert. *Brighter Than a Thousand Suns*. New York: Harcourt, Brace, 1958.

———. *The New Tyranny*. New York: Warner Books, 1977 (with "Aftermath of Three Mile Island" added in 1979).

Kaku Kenryo. *Cycle Mondai Kenkyukai* (Research Group on Problems of the Nuclear Fuel Cycle), *Nippon No Chosen* (Japanese Challenge). Tokyo: Nikkan Kohgyo Shinbunsha, 1978.

Kamakura Tano. *Denryoku Sangoku-shi* (History of Three Countries in Electricity). Tokyo: Seikai Sha, 1967.

Kapur, Ashok. *India's Nuclear Option: Atomic Diplomacy and Decision Making*. New York: Praeger, 1976.

———. *International Nuclear Proliferation: Multinational Diplomacy and Regional Aspects*. New York: Praeger, 1979.

Keck, Otto. "Fast Breeder Reactor Development in West Germany: An Analysis of Government Policy." Unpublished doctoral dissertation, University of Sussex, 1976.

Keeny, Spurgeon M., et al. *Nuclear Power Issues and Choices: Report of the Nuclear Energy Policy Study Group*. Cambridge, Mass.: Ballinger, 1977.

Kelleher, Catherine M. *Germany and the Politics of Nuclear Weapons*. New York: Columbia University Press, 1975.

Kenton, John E. "Building the Nuclear Navy." *Nucleonics*, vol. 17, Sept. 1959.

Kevles, Daniel J. *The Physicists: The History of a Scientific Community in Modern America*. New York: Knopf, 1978.

Khrushchev, Nikita. *Khrushchev Remembers: The Last Testament*. Boston: Little, Brown, 1974.

Kilbourn, William. *Pipeline: Transcanada and the Great Debate*. Toronto: Clarke, Irwin, 1970.

Kistiakowsky, George B. *A Scientist at the White House: The Private Diary of President Eisenhower's Special Assistant for Science and Technology.* Cambridge, Mass.: Harvard University Press, 1976.

Klein, Donald W., and Anne B. Clark. *Biographical Dictionary of Chinese Communism, 1921–1965.* Cambridge, Mass.: Harvard University Press, 1971.

Kohl, Wilfred L. *French Nuclear Diplomacy.* Princeton, N.J.: Princeton University Press, 1971.

Komanoff, Charles. *Power Plant Performance.* New York: Council on Economic Priorities, 1976.

Kondo Kan'ichi and Osani Hiroshi (eds.). *Sengo Sangyo-shi E No Shogen* (Testimony to Postwar Industrial History). Tokyo: Mainichi Shimbun, 1978.

Kosciusko-Morizet, Jacques A. *La "Mafia" Polytechnicienne.* Paris: Seuil, 1973.

Kowarski, Lew. *Réflexions sur la Science.* Geneva: Institut Universitaire de Hautes Etudes Internationales, 1978.

Kramish, Arnold. *Atomic Energy in the Soviet Union.* Stanford, Calif.: Stanford University Press, 1959.

———. *The Peaceful Atom in Foreign Policy.* New York: Harper & Row, 1963.

Kuby, Erich, et al. *Franz Josef Strauss: Ein Typus Unserer Zeit.* Wein: K. Desch, 1963.

Kulkarni. R. P., and V. Sarma. *Homi Bhabha: Father of Nuclear Science in India.* Bombay: Popular Prakashan, 1969.

Lambright, W. Henry. *Shooting Down the Nuclear Plane.* Inter-University Case Program No. 104. Indianapolis, Ind.: Bobbs-Merrill, 1967.

Lamont, Lansing. *Day of Trinity.* New York: Atheneum, 1965.

Lang, Daniel. *From Hiroshima to the Moon: Chronicles of Life in the Atomic Age.* New York: Simon & Schuster, 1959.

Lapp, Ralph. *The Voyage of the Lucky Dragon.* New York: Harper & Bros., 1957.

Law, Charles, and Ron Glenn. *Critical Choice: Nuclear Power in Canada.* Toronto: Corpus, 1978.

Lawrence, Robert M., and Joel Larus (eds.). *Nuclear Proliferation: Phase Two.* Lawrence, Kans.: Regents Press of Kansas, 1974.

Lear, John. "Ike and the Peaceful Atom." *Reporter,* Feb. 12, 1956.

Lebedinsky, A. V. (ed.). *Soviet Scientists on the Danger of Nuclear Tests.* Washington, D.C.: U.S. Joint Publications Research Service, U.S. Dept. of Commerce, 1959.

Lenoir, Yves. *Technocratie Française: La Démarche Technocratique de Louis XIV à l'Atome.* Paris: Pauvert, 1977.

Leonard, Wolfgang. *The Kremlin Since Stalin.* London: Oxford University Press, 1962.

Lewis, Richard S. *The Nuclear Power Rebellion: Citizens vs. the Atomic Industrial Establishment*. New York: Viking, 1972.

Lewis, Richard S., Jane Wilson, and Eugene Rabinowitch (eds.). *Alamogordo Plus Twenty-Five Years: The Impact of Atomic Energy on Science, Technology and World Politics*. New York: Viking, 1975.

Lieberman, Joseph. *The Scorpion and the Tarantula: The Struggle to Control Atomic Weapons 1945–49*. Boston: Houghton Mifflin, 1970.

Lilienthal, David. *Atomic Energy: A New Start*. New York: Harper & Row, 1980.

———. *Change, Hope and the Bomb*. Princeton, N.J.: Princeton University Press, 1963.

———. *Journals: The Atomic Energy Years, 1945–50*. New York: Harper & Row, 1964.

———. *Journals: Venturesome Years, 1950–55*. New York: Harper & Row, 1966.

———. *TVA Democracy on the March*. New York: Harper & Row, 1953.

Lindberg, Leon N. (ed.). *The Energy Syndrome: Comparing National Responses to the Energy Crisis*. Lexington, Mass.: Lexington, 1977.

Liu, Leo Yueh-Yun. *China as a Nuclear Power in World Politics*. New York: Taplinger, 1972.

Lovins, Amory. *Is Nuclear Power Necessary?* London: Friends of the Earth, 1979.

———. *Soft Energy Paths: Towards a Durable Peace*. London: Penguin, 1972.

Lowrance, William W. *Of Acceptable Risk: Science and the Determination of Safety*. Los Altos, Calif.: Kauffman, 1976.

MccGwire, Michael (ed.). *Soviet Naval Influence*. New York: Praeger, 1977.

MccGwire, Michael, et al. (eds.). *Soviet Naval Policy: Objectives and Constraints*. New York: Praeger, 1975.

McKnight, Allan. *Atomic Safeguards: A Study in International Verification*. New York: United Nations Institute for Training and Research, 1971.

McPhee, John. *The Curve of Binding Energy*. New York: Farrar, Straus & Giroux, 1974.

Major, John. *The Oppenheimer Hearing*. London: Batsford, 1971.

Markin, A. *Power Galore: Soviet Power Industry*. Moscow: Progress Publishers (no date).

Martin, Daniel. *Three Mile Island: Prologue or Epilogue*. New York: Ballinger, 1980.

Medvedev, Roy A., and Zhores Medvedev. *Khrushchev: The Years in Power*. New York: W. W. Norton, 1978.

Medvedev, Zhores A. *Nuclear Disaster in the Urals*. New York: W. W. Norton, 1979.

———. *Soviet Science*. New York: W. W. Norton, 1978.

Messer, Robert L. "The Making of a Cold Warrior: James F. Brynes and American-Soviet Relations, 1945–46." Unpublished doctoral dissertation, University of California, Berkeley, 1975.

Metzger, H. Peter. *The Atomic Establishment*. New York: Simon & Schuster, 1972.

Michelmore, Peter. *The Swift Years: The Robert Oppenheimer Story*. New York: Dodd, Mead, 1969.

Modelski, George. *Atomic Energy in the Communist Bloc*. Melbourne: Melbourne University Press, 1959.

Moran, Lord. *Winston Churchill: The Struggle for Survival, 1940–1965*. Boston: Houghton Mifflin, 1966.

Morely, James (ed.). *Dilemmas of Growth in Pre-War Japan*. Princeton, N.J.: Princeton University Press, 1971.

Morgan, Karl Z., and J. E. Turner (eds.). *Principles of Radiation Protection*. New York: John Wiley, 1967.

Moss, Norman. *Men Who Play God*. New York: Harper & Row, 1968.

Mullenbach, Philip. *Civilian Nuclear Power*. New York: Twentieth Century Fund, 1963.

Murphy, Arthur W. (ed.). *The Nuclear Power Controversy*. Englewood Cliffs, N.J.: Prentice-Hall, 1976.

Myrdal, Alva. *The Game of Disarmament: How the United States and Russia Run the Arms Race*. New York: Atheneum, 1978.

Nader, Ralph, and John Abbotts. *The Menace of Atomic Energy*. New York: W. W. Norton, 1977.

National Academy of Sciences. *The Biological Effects of Atomic Radiation*. Washington, D.C.: National Academy of Sciences, 1956.

Nau, Henry R. *National Politics and International Technology: Nuclear Reactor Development in Western Europe*. Baltimore: Johns Hopkins University, 1974.

Nehrt, Lee C. *International Marketing of Nuclear Power Plants*. Bloomington, Ind.: Indiana University Press, 1966.

Nelkin, Dorothy. *Technological Decisions and Democracy*. London: Sage, 1977.

Nero, Anthony. *A Guidebook to Nuclear Reactors*. Berkeley: University of California Press, 1979.

Neustadt, Richard. *Presidential Power*. New York: Signet, 1964.

Newman, Peter C. *Renegade in Power: The Diefenbaker Years*. Toronto: McClelland & Stewart, 1963.

Nieburg, Harold L. *In the Name of Science*. Chicago: Quadrangle, 1970.

———. *Nuclear Secrecy and Foreign Policy*. Washington, D.C.: Public Affairs Press, 1964.

Nogee, Joseph. *Soviet Policy Toward International Control of Atomic Energy*. South Bend, Ind.: University of Notre Dame Press, 1961.

Noorani, A. G. *Aspects of India's Foreign Policy*. Bombay: Jaico, 1970.

Novick, Sheldon. *The Careless Atom*. Boston: Houghton Mifflin, 1964.

Nuclear Energy Policy Study Group (report sponsored by the Ford Foundation). *Nuclear Power Issues and Choices*. Cambridge, Mass.: Ballinger, 1977.

Office of Technology Assessment, Congress of the United States. *Nuclear Proliferation and Safeguards*. New York: Praeger, 1977.

Ohtani Ken. *Denryoku o Meguru Seiji to Keizai* (Politics and Economics of Electricity). Tokyo: Sangyo Mohritsu Tanki, 1978.

Oppenheimer, Robert. *The Open Mind*. New York: Simon & Schuster, 1955.

Orlans, Harold. *Contracting for Atoms*. Washington, D.C.: Brookings, 1967.

Palit, D. K. and P. K. S. Namboodiri. *Pakistan's Islamic Bomb*. New Delhi: Vikas, 1979.

Parmet, Herbert S. *Eisenhower and the American Crusades*. New York: Macmillan, 1972.

Patterson, Walter. *The Fissile Society*. London: Earth Resources, 1977.

———. *Nuclear Power*. London: Penguin, 1976.

Pearson, Lester. *Mike: The Memoirs of Lester B. Pearson*, 3 vols. New York: Quadrangle, 1972–75.

Pesch, Jurgen Peter. "Staatliche Forschungs und Entwicklungspolitik in Spannungsfeld zwischen Regierung, Parlament und privaten Experten untersucht am Beispeil der Deutschen Atompolitik." Unpublished doctoral dissertation, Rechtwissenschaftlichen Fakultat, Albert-Ludwigs Universität, 1975.

Petrosyants, A. M. *From Scientific Search to Atomic Industry: Modern Problems of Atomic Science and Technology in the USSR*. Danville, Ill.: Interstate Printers and Publishers, 1975.

Peyrefitte, Alain. *Le Mal Français*. Paris: Plon, 1976.

Pickersgill, J. W., and D. F. Forster. *The Mackenzie King Record*, 4 vols. Toronto: University of Toronto Press, 1960–70.

Pocock, Roland F. *Nuclear Power: Its Development in the U.K.* London: George Allen & Unwin, 1977.

———. *Nuclear Ship Propulsion*. London: Ian Allen, 1970.

Polmar, Norman. *Death of the Thresher*. New York: Clinton Books, 1964.

Porter, John. *The Vertical Mosaic*. Toronto: University of Toronto Press, 1965.

Poulose, T. T. (ed.). *Perspectives of India's Nuclear Policy*. New Delhi: Young Asia, 1978.

President's Commission on the Accident at Three Mile Island. *Report*. Washington, D.C.: U.S. GPO, 1979.

Primack, Joel, and Frank Von Hippel. *Advice and Dissent: Scientists in the Political Arena*. New York: Basic Books, 1974.

Pruss, K. *Kernforschungspolitik in der Bundesrepublik Deutschland.* Frankfurt: Suhrkamp, 1974.

Puiseux, Louis. *La Babel Nucléaire.* Paris: Galilée, 1977.

Quester, George. *Nuclear Diplomacy: The First Twenty-Five Years.* New York: Dunellen, 1973.

———. *The Politics of Non-Proliferation.* Baltimore: Johns Hopkins University Press, 1973.

Rabi, Isidor. "My Life and Times as a Physicist." Two lectures, Claremont College, 1960.

Rapoport, Roger. *The Great American Bomb Machine.* New York: E. P. Dutton, 1971.

Richards, Denis. *Portal of Hungerford.* London: Heinemann, 1977.

Rickover, H. G. *How the Battleship Maine Was Destroyed.* Washington, D.C.: Department of the Navy, 1976.

Robert, Chalmers M. *The Nuclear Years: The Arms Race and Arms Control 1945–70.* New York: Praeger, 1970.

Roberts, Leslie. *C.D.: The Life and Times of Clarence Decatur Howe.* Toronto: Clarke, Irwin & Co., 1957.

Rochlin, Gene I. *Plutonium, Power and Politics: International Arrangements for the Disposition of Spent Fuel.* Berkeley: University of California Press, 1979.

Rosenberg, David Alan. "American Atomic Strategy and the Hydrogen Bomb Decision." *The Journal of American History,* vol. 66, no. 1, June 1979.

Roskill, Stephen W. *Hankey: Man of Secrets.* 3 vols. London: Collins. 1970–74.

Rotblat, Joseph. *Scientists in the Quest for Peace—A History of the Pugwash Conference.* Cambridge, Mass.: M.I.T. Press, 1972.

Ryan, William L. *The China Cloud: America's Tragic Blunder and China's Rise to Nuclear Power.* Boston: Little, Brown, 1968.

Sabotta, J. (cd.). *State, Science and Economy as Partners.* Berlin: Verlag A. F. Koska (no date).

Sakaguchi, Akira. *Ishikawa Ichiro.* Tokyo: Kashima Shuppan-kai, 1972.

Sakharov, Andrei D. *Alarm and Hope.* New York: Knopf, 1978.

———. *Sakharov Speaks.* London: Collins, 1974.

Sampson, Anthony. *The New Europeans.* London: Hodder & Stoughton, 1968.

Scheinman, Lawrence. *Atomic Energy Policy in France Under the Fourth Republic.* Princeton, N.J.: Princeton University Press, 1965.

Schilling, Warner R., Paul Hammond, and Glenn Snyder (eds.). *Strategy, Politics and Defense Budgets.* New York: Columbia University Press, 1962.

Schlesinger, Arthur, Jr. *A Thousand Days.* Boston: Houghton Mifflin, 1965.

Schubert, Jack, and Ralph E. Lapp. *Radiation: What It Is and How It Affects You.* New York: Viking, 1957.

Scott, John. *Behind the Urals.* Boston: Houghton Mifflin, 1942.

———. *Heavy Industry in the Soviet Union East of the Volga: A Report Prepared for the Board of Economics Warfare*. Washington, D.C., 1943.

Seaborg, Glenn T. *Science, Men and Change*. Washington, D.C.: U.S. Atomic Energy Commission, July 1968.

Shapley, Deborah. "Nuclear Navy." *Science*, July 18, 1976.

Shepley, James, and Clay Blair, Jr. *The Hydrogen Bomb: The Men, the Menace, the Mechanism*. New York: David McKay, 1954.

Sherwin, Martin. *A World Destroyed: The Atomic Bomb and the Grand Alliance*. New York: Knopf, 1975.

Shoriki Matsutaro. *Fifty Years of Light and Dark: The Hirohito Era*. Tokyo: Mainichi Shimbun, 1975.

Simonot, Philippe. *Le Complot Pétrolière*. Paris: Moreau, 1976.

———. *Les Nucléocrates*. Grenoble: Presses Universitaires, 1978.

Slusser, Robert (ed.). *Soviet Economic Policy in Postwar Germany*. New York: New York Research Program on the U.S.S.R., 1955.

Smith, Alice Kimball. *A Peril and a Hope: The Scientists' Movement in America, 1945–47*. Chicago: University of Chicago Press, 1965.

Smith, Perry M. *The Air Force Plans for Peace*. Baltimore: Johns Hopkins University Press, 1970.

Snow, C. P. *Science and Government*. Cambridge, Mass.: Harvard University Press, 1961.

Snow, Edgar. *Red Star Over China*. New York: Grove Press, 1968.

Solzhenitsyn, Alexander I. *The Gulag Archipelago 1918–1956*. New York: Harper & Row, 1975.

Sporn, Philip. *Vistas in Electric Power*. 3 vols. New York: Pergamon, 1968.

Steenbeck, Max. *Impulse and Wirkungen: Schritte auf Meinen Lebensweg*. Berlin: Verlag der Nation, 1977.

Stern, Philip M., with Harold P. Green. *The Oppenheimer Case: Security on Trial*. New York: Harper & Row, 1969.

Stobaugh, Robert, and Daniel Yergin (eds.). *Energy Future: Report of the Energy Project at the Harvard Business School*. New York: Random House, 1979.

Stockholm International Peace Research Institute. *The Near Nuclear Countries and the NPT*. Stockholm: Almqvist & Wiksell, 1972.

———. *The Nuclear Age*. Cambridge, Mass.: M.I.T. Press, 1974.

———. *Nuclear Proliferation Problems*. Cambridge, Mass.: M.I.T. Press, 1974.

———. *Safeguards Against Nuclear Proliferation*. Stockholm: Almqvist & Wiksell, 1975.

Strauss, Lewis. *Men and Decisions*. Garden City, N.Y.: Doubleday, 1962.

Suleiman, Ezra N. *Elites in French Society: The Politics of Survival*. Princeton, N.J.: Princeton University Press, 1978.

———. *Politics, Power and Bureaucracy in France: The Administrative Elite.* Princeton, N.J.: Princeton University Press, 1974.

Suttmeier, Richard P. *Research and Revolution: Science Policy and Societal Change in China.* Lexington, Mass.: Lexington Books, 1974.

Sutton, Anthony C. *Western Technology and Soviet Economic Development 1945–65.* Stanford, Calif.: Stanford University Press, 1976.

Szasz, Paul C. *The Law and Practices of the International Atomic Energy Agency.* Vienna: IAEA, 1970.

Takamiya Shin. *Kikawada Kazutaka No Kein Rinen* (Kikawada's Management Idea). Tokyo: Cenryoku Shinpo Sha, 1978.

Taylor, A. J. P. *English History, 1914–45.* London: Oxford University Press, 1965.

Taylor, Lauriston S. *The Origins and Significance of Radiation Dose Limits for the Population.* Washington, D.C.: U.S. Atomic Energy Commission, 1972.

Teller, Edward, and Allen Brown. *The Legacy of Hiroshima.* Garden City, N.Y.: Doubleday, 1962.

Thoenig, Jean-Claude. *L'Ere des Technocrates: Le Cas des Ponts et Chausées.* Paris: Organisation, 1973.

Thomas, John R., and Ursula M. Kruse-Vaucienne (eds.). *Soviet Science and Technology: Domestic and Foreign Aspects.* Washington, D.C.: National Science Foundation, 1977.

Truman, Harry S. *Memoirs: Years of Decisions.* Garden City, N.Y.: Doubleday, 1955.

———. *Memoirs: Years of Trial and Hope.* Garden City, N.Y.: Doubleday, 1956.

Uhlan, Edward, and Dana L. Thomas. *Shoriki: Miracle Man of Japan.* New York: Exposition Press, 1957.

Ulam, Stanislaw M. *Adventures of a Mathematician.* New York: Charles Scribner's Sons, 1976.

Union of Concerned Scientists. *The Nuclear Fuel Cycle.* Cambridge, Mass.: M.I.T. Press, 1975.

———. *The Nugget File.* Cambridge, Mass.: M.I.T. Press, 1979.

Verrier, Anthony. *The Bomber Offensive.* London: Pan, 1974.

Weart, Spencer R. *Scientists in Power.* Cambridge, Mass.: Harvard University Press, 1979.

Weart, Spencer R., and Gertrud Weiss Szilard (eds.). *Leo Szilard: His Version of the Facts.* Cambridge, Mass.: M.I.T. Press, 1978.

Webster, Charles, and Noble Frankland. *The Strategic Air Offensive Against Germany, 1939–46.* Vol. 1. London: H.M.S.O., 1961.

Weiher, Siegfried von, and Herbert Goetzler. *The Siemens Company: Its Historical Role in the Progress of Electrical Engineering.* Munich: Siemens, 1977.

Wheeler-Bennett, Sir John W. *Sir John Anderson, Viscount Waverly*. New York: St. Martin's Press, 1962.

Whitaker, Donald P., et al. *Area Handbook for the People's Republic of China*. Washington, D.C.: U.S. GPO, 1972.

Whitson, William H. *The Chinese High Command*. New York: Praeger, 1973.

Wilczynski, Jozef. *Atomic Energy for Peaceful Purposes in the Warsaw Pact Countries*. Bloomington: Indiana University Press, 1974.

Wildavsky, Aaron. *Dixon-Yates: A Study in Power Politics*. New Haven: Yale University Press, 1962.

Wilensky, Harold. *Organizational Intelligence*. New York: Basic Books, 1967.

Williams, Roger. *The Nuclear Power Decisions*. London: Croom Helm, 1980.

Willrich, Mason. *Global Politics of Nuclear Energy*. New York: Praeger, 1971.

Willrich, Mason (ed.). *International Safeguards and Nuclear Industry*. Baltimore: Johns Hopkins, 1973.

Willrich, Mason, and Theodore B. Taylor. *Nuclear Theft: Risks and Safeguards*. Cambridge, Mass.: Ballinger, 1974.

Wilson, Jane (ed.). *All in Our Time: The Reminiscences of Twelve Nuclear Pioneers*. Chicago: Bulletin of the Atomic Scientists, 1975.

Wilson, Richard. *Anatomy of China: An Introduction to One Quarter of Mankind*. Rev. and updated ed. New York: Weybright & Talley, 1968.

Winnacker, Karl. *Challenging Years: My Life in Chemistry*. London: Sidgwick & Jackson, 1972.

Winnacker, Karl, and Karl Wirtz. *Nuclear Energy in Germany*. La Grange Park, Ill.: American Nuclear Society, 1979. (Translation of *Das Unverstundene Wunder: Kernenergie in Deutschland*. Düsseldorf: Econ Verlag, 1975. Published in French as *Atome, Illusion ou Miracle?* Paris: Presses Universitaires, 1977.)

Wohlstetter, Albert, et al. *Moving Towards Life in a Nuclear Armed Crowd*. Los Angeles: PanHeuristics, 1976.

———. *Swords from Plowshares: The Military Potential of Civilian Nuclear Energy*. Chicago: University of Chicago Press, 1979.

Wonder, Edward F. *Nuclear Fuel and American Policy*. Atlantic Council Policy Series, 1977.

Yanaga, Chitoshi. *Big Business in Japanese Politics*. New Haven: Yale University Press, 1968.

Yergin, Daniel. *Shattered Peace: The Origins of the Cold War and the National Security State*. London: André Deutsch, 1978.

York, Herbert. *The Advisors: Oppenheimer, Teller and the Superbomb*. San Francisco: W. H. Freeman, 1976.

———. *Race to Oblivion*. New York: Simon & Schuster, 1970.
Young, Elizabeth. A *Farewell to Arms Control*. London: Penguin, 1972.
Zhukov, G. K. *The Memoirs of Marshal Zhukov*. London: Jonathan Cape, 1971.
Zumwalt, Elmo R., Jr. *On Watch: A Memoir*. New York: Quadrangle, 1976.

Notes

CHAPTER 1

The basic sources that proved particularly useful in compiling this summary of the early years of atomic physics include: Spencer Weart and Gertrud Weiss Szilard (eds.): *Leo Szilard: His Version of the Facts*; Spencer Weart: *Scientists in Power*; Daniel Kevles, *The Physicists*; Lew Kowarski, *Réflexions sur la Science*; Alan Beyerchen, *Scientists Under Hitler*; Walter Elasser, *Memoirs of a Physicist*; Margaret Gowing, *Britain and Atomic Energy*; Werner Heisenberg, *Physics and Beyond*; Maurice Goldsmith, *Frédéric Joliot-Curie*; Donald Fleming and Bernard Bailyn (eds.), *The Intellectual Migration*; Igor Golovin, *I. V. Kurchatov*; Arnold Kramish, *Atomic Energy in the Soviet Union*; George Modelski, *Atomic Energy in the Communist Bloc*; Martin Sherwin, *A World Destroyed*; Robert Jungk, *Brighter Than a Thousand Suns*. Interviews with the authors and Lew Kowarski, Nov. 1978, Boston, Mass.; Gertrud Weiss Szilard, Sept. 1978, La Jolla, California.

3 Szilard's time in London comes from Spencer Weart and Gertrud Weiss Szilard: *Leo Szilard: His Version of the Facts*. The letter of rejection of Szilard's patent is mentioned on p. 18. Copy of the letter, dated Oct. 8, 1935, is in the Szilard papers at the University of California, San Diego.

5 "the merest moonshine": Appears in the summary of a speech by Ernest Rutherford, British physicist, *Nature*, vol. 132, Sept. 11, 1933, pp. 432–33. The full sentence was "One timely word of warning was issued to those who look for sources of power in the atomic transmutations—such expectations are the merest moonshine."

6 The best account of Joliot-Curie's prewar experiments appears in Spencer Weart, *Scientists in Power*.

6 "atomic poppycock": David Irving, *The Virus House*, p. 42.

6 ". . . sleep fairly comfortably in our beds": Lord Hankey, Minister Without Portfolio in the War Cabinet, quoted in Margaret Gowing, *Britain and Atomic Energy*, p. 39.

7 "Almost overnight, physicists have been promoted from semi-obscuri-

ty . . .": J. Hammon McMillen, Kansas State College, quoted in Daniel Kevles, *The Physicists*, p. 320.

8 "like a stirred up ant heap": Letter from Leo Szilard to Lewis Strauss, Jan. 25, 1939, quoted in Weart and Szilard, op. cit., p. 62.

8 "France's reputation was for women's fashions . . .": Interview Lew Kowarski, Nov. 1978.

8 Szilard's description of his uranium fission experiment: Weart and Szilard, op. cit., p. 115.

9 "We take the liberty . . .": Joint letter to German War Office, Apr. 26, 1939, from the young Hamburg physics professor Paul Harteck and his assistant, Dr. William Groth, quoted in David Irving, *The Virus House*, p. 36.

9 "It means that a bomb can be built . . .": Private communication, quoted in Herbert York, *The Advisors*, p. 29.

9 For a brief discussion of the motives of the German physicists during the war, see R. V. Jones, *Most Secret War*, p. 483, notes; also, the letter of General Leslie Groves to Sir William Penny, Dec. 13, 1965, Groves Papers, West Point. Groves discusses the importance of the so-called Farm Hall tapes. At the end of the war, the German scientists, including Heisenberg, were taken to an English country house called Farm Hall. Unknown to the scientists, the rooms of the house were bugged. The British have never released the tapes: unofficially saying they might be embarrassing. Heisenberg died in 1976, but Carl Friederich von Weisacker, a pupil of Heisenberg's, is still alive. In his letter to Penny, Groves discusses a request from David Irving to look at the tapes (Groves had a copy). The general denied the request. Groves's own conclusion from the tapes, clearly, is that Germans had tried to conceal the fact that they had willingly aided the bomb project.

10 An excellent account of the French heavy-water adventure is in Spencer Weart, *Scientists in Power*, pp. 135–37.

11 British plans for acquiring Belgian uranium: See Gowing, *Britain and Atomic Energy*, p. 35.

13 Frisch and Peierls memorandum: Ibid., pp. 40–43.

13 Letter from Einstein to Roosevelt appears in full in Weart and Szilard, op. cit., pp. 94–95. Letter from Szilard to Sachs: Ibid., p. 97.

14 "swimming in syrup": Appears in Eugene Wigner, "Are We Making the Transition Wisely," *Saturday Review of Literature*, vol. 28, Nov. 17, 1945, quoted in Kevles, op. cit., p. 324.

14 Letter from Einstein to Sachs, Mar. 1940: Quoted in Weart and Szilard, op. cit., p. 120.

CHAPTER 2

The basic sources on the Manhattan Project are the two official histories: Richard Hewlett and Oscar Anderson, *The New World*, and Margaret Gowing, *Britain and Atomic Energy*. Other source books used include Ronald Clark, *Birth of the Bomb*; Arthur Compton, *Atomic Quest*; Nuel Pharr Davis, *Lawrence and Oppenheimer*; Stéphane Groueff, *Manhattan Project*; Leslie Groves, *Now It Can Be Told*; Robert Jungk, *Brighter Than a Thousand Suns*; Lansing Lamont, *Day of Trinity*, Martin Sherwin, *A World Destroyed*; Jane Wilson (ed.), *All in Our Time*. Private collections of papers that were particularly useful were the Ernest Lawrence Papers in the Bancroft Library of the University of California, Berkeley, and the Leslie Groves Papers at West Point.

18 British intelligence had established by the beginning of 1944 that the Nazis were not seriously pursuing a bomb. See Margaret Gowing, *Britain and Atomic Energy*, p. 367. The scientists, so highly motivated by a fear of a German bomb, were not told.

18 ". . . technically sweet . . .": Gowing, ibid., p. 368.

19 The sketch of Leslie Groves draws on all the basic Manhattan Project references listed above and on his own papers at West Point. In addition, biographical material from *Current Biography*, 1945, and *Collier's*, Oct. 13, 1945, was also used.

19 "The Secretary of War has selected you . . .": Stéphane Groueff, *Manhattan Project*, p. 5.

20 Mrs. Groves never knew . . . : Leslie Groves, *Now It Can Be Told*, p. 21.

20 ". . . I fear we are in the soup": Groueff, op. cit., p. 9.

21 For more on Eger Murphree see *Biographical Memoirs*, National Academy of Science, vol. 40, Columbia University Press, 1969; *New York Times*, Mar. 28, 1956, and Apr. 2, 1956; Standard Oil Company press release, May 5, 1950; *Current Biography*, 1956. Murphree had been the key technical contact man with I. G. Farben in the global cartel arrangement by which Farben agreed to stay out of the oil business and Standard Oil agreed to stay out of chemicals. During 1942 a series of congressional inquiries exposed the deal, and eventually Standard Oil's largest shareholder, John D. Rockefeller, Jr., intervened and the two Standard Oil senior executives with whom Murphree had climbed the corporate ladder were removed. At the time the centrifuge project fell behind, Murphree was preoccupied with his own corporate power base which must have been in doubt for some time. See Joseph Borkin, *Crime and Punishment of I. G. Farben*, pp. 93–94. In view of the public controversy, Murphree's illness, which allegedly prevented his participation in the atomic project, was conveniently timed.

22 John Dunning's basic instincts were toward engineering rather than pure science, but he was attracted by the new frontier between the disciplines that Ernest Lawrence had established. Like Lawrence he was a boyhood radio buff. He went to Columbia University to pursue electrical engineering and after the war became dean of Columbia's engineering faculty. The official historians of the Manhattan Project, Richard Hewlett and Oscar Anderson, noted, "Dunning seems to many to be more an engineer than a physicist. He had been known to suggest that possibly a scientist could have too much knowledge, that too many facts would make him overly sensitive to the obstacles in the path of technical advances." See Hewlett and Anderson, *The New World*, p. 122. On Dunning, see *Current Biography*, 1948; *Physics Today*, vol. 28, Dec. 1975; Groueff, op. cit., especially pp. 21–22, 90–94, 182–83.

22 The sketch of Percival ("Dobie") Keith draws on Hewlett and Anderson, op. cit., especially pp. 62, 120–22; Groueff, op. cit., especially pp. 22–23, 95–100.

22 "Dobie is not the type who gets ulcers": Ibid., p. 95.

22 "You're going to have a surprise": Ibid., p. 35.

23 The sketch of Ernest Lawrence draws on all the Manhattan Project references listed above, especially Davis, *Lawrence and Oppenheimer*. In addition there is a friendly biography, Herbert Childs, *American Genius*; an obit by Glenn Seaborg, *Science*, vol. 7, Nov. 1958, and an assessment in Daniel Kevles, *The Physicists*. Interviews with a number of people who worked with Lawrence were useful, especially Sir Mark Oliphant, John Gofman, and Herbert York. In the Lawrence Papers in the Bancroft Library there is a personal memoir by Louis Alvarez, ref. UCRL 17359. Lawrence Papers, Carton 11, Folder 8. Lawrence was frugal, living in a modest home. During the war, however, he had a security chauffeur, and this led to his one personal luxury, a chauffeur-driven Cadillac convertible. Even after the success of the gaseous diffusion system had relegated his electromagnetic system to technical obsolescence, he was still pushing for more and bigger machines. As late as July 1945, he wrote to Groves about the urgency of a new electromagnetic project that he called the "Cyclone." (See Lawrence Papers, Carton 29, Folder 38.) In 1949 he used his political influence to push through yet another machine, the Materials Testing Acelerator. It was designed to produce plutonium directly from the waste left from enrichment plants. It was scrapped almost as soon as it had been completed, because it was clear other systems worked better. Lawrence, never at a loss for a reason to keep his expensive projects alive, suggested it could be used to produce deadly isotopes for use in radiological warfare. (See Davis, op. cit., p. 267, and the Lawrence Papers, Carton 32, Folders 10 and 25.) Before he had finished with his machines Lawrence had cost the American taxpayer some $2 billion.

24 "My position could well be compared to that of a caterer . . .": Groves, op. cit., p. 40.
25 "The reaction is self-sustaining": Groueff, op. cit., p. 89.
25 ". . . the Italian navigator . . .": Arthur Compton, *Atomic Quest*, p. 144.
26 ". . . we do not speak of tons of silver . . .": Groves, op. cit., p. 107.
26 "We're not looking for scientists": Groueff, op. cit., p. 62.
27 "I am no longer even enthusiastic . . .": Lawrence Papers, Carton 28, Folder 379.
29 "Just as outfielders . . .": Martin Sherwin, *A World Destroyed*, p. 59.
29 The relationship between Oppenheimer and Groves is featured in all the Manhattan Project books listed above. It is also analyzed in the books about Oppenheimer's later career: Philip Stern, *The Oppenheimer Case;* John Major, *The Oppenheimer Hearing*; Peter Michelmore, *The Swift Years*.
30 "soft-spoken major-general . . .": The Groves-approved biography is in the Lawrence Papers.
31 "if the radiance of a thousand suns . . .": Lansing Lamont, *Day of Trinity*, p. 235.
32 "Only this tightrope . . .": Brig. Nichols, official report.
32 "That corresponds to a blast . . .": Groueff, op. cit., p. 357.
32 ". . . keep this thing quiet": Lamont, op. cit., p. 240.
33 Byrnes had been briefed on the Manhattan Project as Roosevelt's "domestic czar," the director of War Mobilization. Szilard did not know of Byrnes's political fears of a postwar congressional inquiry if the project did not succeed. See Hewlett and Anderson, op. cit., pp. 339–40.
34 The minutes of the presidential committee meeting that considered whether to drop the bomb on Japan are reproduced in Sherwin, op. cit., pp. 295–304. The quotes "After much discussion . . ." and "Handling of undesirable scientists," are on p. 302. There is no doubt that Groves detested Szilard. In the middle of the project he had tried to have him transferred from Chicago to an innocuous post at Harvard, and had even offered to reimburse the university for his salary and expenses from the Manhattan Project funds. He had asked Conant, who had been president of the university before taking up his government post, to find a way of removing Szilard, but, according to Groves, "Conant's reply was that under no circumstances would he want such a character at Harvard, even if I paid him a thousand per cent profit on the deal." (See Groves Papers, personal notes, dictated Aug. 20, 1968.)
34 Groves's xenophobia was compounded in Szilard's case by a suspicion of anyone who had no deep religious convictions, especially Jews. Henry Wallace, Roosevelt's Vice-President until 1945, had no doubts about Groves; he saw him as both anti-Semitic and anti-foreigner (see John Blum, *The Price of Vision*, p. 472). Groves was quite clear about his feel-

ings about Szilard, "as far as I could determine [Szilard] did not have a shred of honour in his make-up": Groves Papers, personal notes, Aug. 20, 1968.

34 Groves memorandum on how the urgency of the project would "continue and increase" is in the Lawrence Papers, Carton 30, Folder 23.

35 Roosevelt's deferred decisions on the bomb are typical of his entire decision-making style. The classic analysis is still Richard Neustadt, *Presidential Power*. Determined always to make decisions for himself, Roosevelt was rarely explicit about his own doubts and opinions; he created a tangled web of overlapping jurisdictions and lines of communication that could be resolved by the President alone. This bureaucratic structure maximized his power and enabled him to put off final decisions until the last moment, when the pros and cons had become as clear as they were ever likely to be. He played his cards so close to his chest that he alone was able to put the final hand together.

36 There is a small library on the decision to drop the bomb. One theme builds a series of ambiguous statements into a conclusion that the real motivation was as a warning to the Soviet Union. This theme has a long history in Communist propaganda and was seized upon by revisionist historians in the orgy of American self-criticism accompanying the Vietnam War. All recent reassessments of the existing material stress the question of bureaucratic momentum and acknowledge that, insofar as personal motivations were important, the one to end the war as soon as possible was overriding. See Sherwin, op. cit., especially chap. 8; Barton Bernstein, *The Atomic Bomb*, pp. 94 ff.; Daniel Yergin, *Shattered Peace*, especially pp. 433–34. Truman himself had indicated a tough stand toward the Soviets on the question of Poland *before* he even knew of the Manhattan Project.

36 ". . . the greatest thing in history": Robert Donovan, *Conflict and Crisis*, p. 96.

36 The Roosevelt-Churchill protocol, signed at Hyde Park on September 9, 1944, said: "When a bomb is finally available it might perhaps after mature consideration be used against the Japanese." See Sherwin, op. cit., p. 111. A few days after this meeting Roosevelt indicated to Stimson and Bush that he regarded this question as open (Sherwin, op. cit., pp. 122–26 and 144–45). No one even told Truman that Roosevelt had any doubts.

37 Even before the Manhattan Project the scientific community had a huge amount of information on the delayed effects of radiation, especially from studies on X rays and radium exposure. The Frisch-Peierls memorandum of 1940—the scientific assessment of a bomb that really sparked the entire project—had said explicitly, "The radiations [from the bomb] would be fatal to living beings even a long time after the explosion": Gowing, *Brit-*

ain and Atomic Energy, p. 392. The fact that such advice might have been decisive is suggested by the action taken by Truman, on Stimson's advice and against Groves's objection, to strike the ancient Japanese capital of Kyoto from the target lists where it had appeared ahead of Hiroshima. This was done precisely because such an act would have adverse effects on postwar relations with Japan.

CHAPTER 3

The outline of American nuclear policy toward the Soviet Union is found in three main sources: Richard Hewlett and Oscar Anderson, *The New World*; Martin Sherwin, *A World Destroyed* (for the war years); and an unpublished thesis by Gregory Herken, "American Diplomacy and the Atomic Bomb" (for the postwar years). In addition the following sources provide further information: Dean Acheson, *Present at the Creation*; Bernhard Bechhoefer, *Postwar Negotiations for Arms Control*; Barton Bernstein, "The Quest for Security," p. 1003; Barton Bernstein, "American Foreign Policy and the Origins of the Cold War"; John Blum (ed.), *The Price of Vision*; Robert Donovan, *Conflict and Crisis*; David Lilienthal, *Journals: The Atomic Energy Years*; Joseph Lieberman, *The Scorpion and the Tarantula*; Joseph Nogee, *Soviet Policy Toward International Control of Atomic Energy*; Arthur Steiner, "Scientists, Statesmen and Politicians: The Competing Influences on American Atomic Energy Policy 1945–46," *Minerva*, vol. 12, Oct. 1974; Daniel Yergin, *Shattered Peace*; George Quester, *Nuclear Diplomacy*.

39 The initiative of Niels Bohr with Roosevelt and Churchill is described in R. V. Jones, *Most Secret War*, p. 475; also in Margaret Gowing, *Britain and Atomic Energy*, pp. 354–55.
40 Churchill's response to Bohr is in Jones, op. cit., p. 477.
40 "We did not speak . . .": Gowing, *Britain and Atomic Energy*, p. 355.
40 "The upper crust . . .": Groves diaries, July 2, 1945, quoted in Martin Sherwin, *A World Destroyed*, p. 222.
41 Truman's reaction to the news of Alamogordo is reproduced in ibid., especially pp. 223–24; also Yergin, *Shattered Peace*, p. 115; also Truman's personal journal, found by Robert Ferrell and Erwin Mueller, Truman Library.
42 "The bomb is too dangerous . . .": Public address, Aug. 9, 1945, quoted in Herken, "American Diplomacy and the Atomic Bomb," p. 17.
42 "in his hip pocket": Comment by Stimson in his diary, Sept. 4, 1945, quoted in ibid., p. 68.
42 Groves's attitude to the Russians is well expressed in a speech he gave to IBM executives in New York, Sept. 22, 1945. Groves said, "We are not a master race, but we have a spirit in the United States that is not du-

plicated elsewhere. It is a spirit of teamwork, a spirit of ambition to get the job done and to get it done right. When other nations have that spirit of teamwork and that ambition and willingness to work as many days in the week as they can stand up then they too can do this job in two to five years. They cannot do it without these qualities."

42 Soviet uranium supplies had been identified in a U.S. intelligence report as early as 1943. It correctly identified the location of "important sources" near Tashkent (see John Scott, *Heavy Industry in the Soviet Union*). A Canadian businessman, using Russian sources, confirmed this and told Henry Wallace, who passed it on to the War Department (John Blum [ed.], *The Price of Vision*, pp. 316, 472). The importance of the American belief in the Soviet uranium shortage is stressed in Yergin, op. cit., pp. 123–24, 136, 140, 143.

43 "When will the Russians be able to build the bomb?" Nuel Pharr Davis, *Lawrence and Oppenheimer*, p. 260.

43 ". . . a most deadly illusion . . .": Quoted in Herken, op. cit., p. 479.

44 The full title of the Smyth Report was "General Account of the Development of Methods of Using Atomic Energy for Military Purposes."

44 "Principal breach of security . . .": David Lilienthal, *Journals: The Atomic Energy Years*, p. 134.

45 Byrnes's Kremlin counterpart, Vyacheslav Molotov, minimized the bomb as a bargaining chip to such an extent during the sessions that he rendered it ineffective as a diplomatic tool. Herkin, op. cit., pp. 101–104.

45 "There is no powder in the gun . . .": In "Scientific Interchange on Atomic Energy," Bush to Truman, Sept. 25, 1945; quoted in Herken, op. cit., pp. 75 and 91.

45 "irretrievably embitter": Memo to Truman, 1946, quoted in Donovan, *Conflict and Crisis*, p. 130.

45 "The chief lesson I have learned . . .": Proposed action for the control of atomic energy, Stimson to Truman, Sept. 11, 1945, quoted in Herken, op. cit., p. 62.

45 Truman's three atomic secrets: Presidential press conference, Tiptonville, Tennessee, Oct. 8, 1945, quoted in ibid., p. 82.

46 Stimson farewell cabinet meeting is drawn from U.S. State Department, *Foreign Policy of the United States, 1945*, vol. 2, pp. 54–55; Dean Acheson, *Present at the Creation*, pp. 123–25; Barton Bernstein, "Quest for Security," pp. 1018–20; Donovan, *Conflict and Crisis*, pp. 130–33; Truman, *Memoirs: Years of Decisions*, pp. 525–28; Blum, op. cit., pp. 482–85, 487, 492; Yergin, op. cit., pp. 132–35.

46 "It seems doubtful . . .": Memorandum to Truman after cabinet meeting of Sept. 21, quoted in Herken, op. cit., p. 76.

46 "The discussion is unworthy . . .": Acheson, op. cit., p. 123.

46 "If they catch up with us . . .": Presidential press conference, Tiptonville, Tennessee, quoted in Herken, op. cit., p. 84.

47 The development of the Acheson-Lilienthal report and the 1946 American plan for international control of atomic energy—eventually misnamed the Baruch Plan to appease the vanity of Bernard Baruch—is based on the works listed in the first note to this chapter. In addition see Daniel Lang, *From Hiroshima to the Moon*, chap. 6.
49 "fall into the illusion . . .": Lilienthal, *The Atomic Energy Years*, p. 15.
49 "No fairy tale . . .": Ibid.
50 "Brilliant and profound": Ibid., p. 27.
50 ". . . too little to say . . .": Ibid., p. 29.
51 "Distinctly unhappy . . .": Richard Hewlett and Oscar Anderson, *The New World*, p. 547.
51 "not as a final plan . . .": Ibid., p. 553.
51 ". . . I was quite sick . . .": Lilienthal, op. cit., p. 30.
51 "rigged poker game": Letter, Wallace to Truman, July 23, 1945, quoted in Herken, op. cit., p. 367.
52 ". . . the worst mistake . . .": Lilienthal, *The Atomic Energy Years*, p. 59.
52 "babying the Soviets . . .": Quoted in Herken, op. cit., p. 272.
52 Churchill's Iron Curtain speech quoted in ibid., p. 279.
52 Truman was first briefed by Stimson on the existence of the Manhattan Project on April 25, 1945; see Sherwin, op. cit., p. 162. The first face-to-face Cold War argument was Truman's firm line with Soviet Foreign Minister Molotov on April 23, 1945. See ibid., pp. 158 ff.
53 "I can't hear you" and "I knew all I wanted to . . .": Baruch quoted in Herken, op. cit., p. 331.
53 "a banker's, not a believer's": Ibid., p. 331.
53 "This is a capitalistic country . . ." and "asking for adventure . . .": Ibid., p. 311.
55 "If the atomic bomb . . .": Joseph Nogee, *Soviet Policy*, p. 62. Also Herken, op. cit., p. 363.
55 The debate over the Atomic Energy Act and the civilian versus military domination of the commission is discussed in Hewlett and Anderson, op. cit., vol. 1.

CHAPTER 4

There are two full-book-length studies of the early history of atomic energy in the U.S.S.R.: Arnold Kramish, *Atomic Energy in the Soviet Union*, and George Modelski, *Atomic Energy in the Communist Bloc*. Both were published over twenty years ago, but they contain a surprising amount of detailed information. As the flow of memoirs and historical works published in the Soviet Union has increased, a number of Western scholars have brought the strands together. Of particular use were sections in Herbert York, *The Advisors*; David Holloway, *Technology, Management and the Soviet Military Establishment*, Adelphi Paper, Institute for Strategic Studies, London, 1971;

Ronald Amann, et al. (eds.), *The Technological Level of Soviet Industry*. After research for this chapter was complete, the authors received a copy of David Holloway's paper "Entering the Nuclear Arms Race: The Soviet Decision to Build the Atomic Bomb, 1939–45," Working Paper No. 9, ISSP, Woodrow Wilson Center, Washington, D.C., 1979. Drawing on the same sources the authors had already consulted, this paper revealed one quotation from Stalin in a new source. The most important Soviet sources were two biographies of Igor Kurchatov: Igor Golovin, *I. V. Kurchatov*, and P. N. Astashenkov, *Kurchatov*. The authors also used a translation of a serialization of the Golovin biography in *Sovetskaya Rossiya*, produced by the Joint Publications Research Service (ref. JPRS 37, 804). An important historical overview is provided by Kurchatov's successor, A. P. Alexandrov, "Nuclear Physics and the Development of Nuclear Technology in the USSR," *Oktyabri Nauchnyi Progress (October and Scientific Progress)*, Academy of Sciences, Moscow, 1967.

Much of the detail in this chapter comes from a series of articles written by Vassily Emelyanov, especially "Korchatov As I Knew Him," *Yonost*, nos. 4 and 5, 1968, "The Nuclear Energy Industry: Its Very Beginning," *Voprosi Iistori*, no. 5, May 1975; "About My Time and About Myself," in *Komsomolskaya Pravda*, Nov. 12, 1976, and a memoir of Vannikov in *Red Star*, Sept. 7, 1977. In an interview Emelyanov mentioned that he had completed his memoirs of the atomic energy years, but they had not been published at the time this book was completed. It has since appeared in Czechoslovakia. More detail is contained in the memoirs of three German scientists who worked on the Russian project. Heinz Barwich, *Das Rote Atome*, was published after the author defected to the West and is particularly frank. Two senior German scientists published their memoirs in East Berlin: Manfred von Ardenne, *Ein Glückliches Leben für Technik und Forschung*, and Max Steenbeck, *Impulse und Wirkungen*. Other sources that proved useful on a number of points included: Nikita Khrushchev's memoirs, *Khrushchev Remembers*; Zhores Medvedev, *Soviet Science*; A. M. Petrosyants, *From Scientific Search to Atomic Industry*. Basic biographical material is found in the two volumes prepared by the Institute for the Study of the U.S.S.R.: *Who Was Who in the Soviet Union* (Metuchen, N.J.: Scarecrow Press, 1968) and *Prominent Personalities in the USSR*, also Scarecrow Press, 1968. In addition, the third edition of *The Large Soviet Encyclopaedia* was consulted.

Unless otherwise noted, personal material on Soviet leaders of the project, particularly Vannikov, Zavenyagin, and Slavsky, comes from interviews with Soviet refugees, two of whom are willing to be acknowledged: Boris Rabbot, formerly a senior aide to a member of the Central Committee of the Communist Party, and Leonid Finkelstein, formerly a popular-science writer under the name of Vladimirov.

58 Peter Kapitsa was successively a student, protégé, and then colleague of Lord Rutherford during the most productive years of his stewardship of

the Cavendish Laboratory at Cambridge, England. Kapitsa was acutely conscious of the transition in the physics community from intellectual independence to military asset. "The year that Rutherford died [1938]," recalled Kapitsa, "there disappeared for ever the happy days of free scientific work which gave such delight in our youth. Science has lost her freedom. Science has become a productive force. She has become rich, but she has become enslaved and part of her is veiled in secrecy. I do not know whether Rutherford would continue nowadays to joke and laugh as he used to." See *Science*, vol. 153, Aug. 12, 1966, p. 726. Kapitsa had a personal foretaste of the restraints that would be placed on scientists in 1934. On one of his lecture tours to the Soviet Union he was suddenly detained on Stalin's personal orders and prevented from returning to England. He was not permitted to leave the Soviet Union for the next thirty years. Stalin built a special institute for him on a prize site in Moscow originally slated to become the United States Embassy. In late 1946, Kapitsa lost his post as director of the institute when he angered Lavrenti Beria by refusing to work on the bomb project. He remained under house arrest for seven years, returning as head of the institute in 1955. Khrushchev recalled how Kapitsa explained his refusal to work on the bomb with a "long-winded dissertation" based on "some sort of moral principle" (*Khrushchev Remembers*, pp. 64–65). For all the petitions, polemic, and heartrending that flowed from the Western nuclear scientists, Peter Kapitsa was the only physicist to suffer personally for his beliefs before 1950.

60 The sketch of Kurchatov draws on all the general Soviet references listed above, particularly the biographies by Golovin and Astashenkov and the memoir by Emelyanov. Also, thc *Large Soviet Encyclopaedia*, 3rd edition, vol. 14, pp. 48–49.

60 "you are getting too scientific": Igor Golovin, *I. V. Kurchatov*, JPRS translation, p. 19.

60 As early as 1945 the German scientists had been promised Stalin prizes for helping with the gaseous diffusion process: Manfred von Ardenne, *Ein Glückliches Leben für Technik und Forschung*, p. 89. The German teams used to check work carried out independently by Russian scientists. As in the Manhattan Project, the top priority given to the Soviet bomb project meant that all tasks were duplicated in case one of them failed. Barwich noted the duplication in *Das Rote Atome*, p. 71, and A. P. Alexandrov indicated that this was a general policy in "Nuclear Physics and the Development of Nuclear Technology in the USSR," p. 197. In 1948 when the gaseous diffusion plant ran into serious technical problems, the German scientists were brought in to help solve them and visited the top-secret plant for the first time. "What do you say now?" asked the Germans' guide of the plant. "You Germans never thought we could do it." Barwich, op. cit., p. 108.

60 Soviet sources, however, explicitly acknowledge that their spy rings in the

United States passed on the word about the successful test at Alamogordo before the bomb was dropped on Hiroshima on Aug. 6. Golovin notes, "In mid-July 1945 a report was received concerning an explosion of monstrous power at the Alamogordo atomic testing grounds in the United States." See Golovin, op. cit., JPRS translation, p. 17. The Soviet attitude toward the American bomb is, perhaps, best summed up by Vassily Emelyanov: "The only possibility of restraining these new claimants to world domination and of cooling their ardor was to create our own atomic bomb very quickly. This was well understood by the leaders of our country, and those who were drawn to work on solving the atomic problem were clearly aware that the most important factor was time." See Holloway, "Entering the Nuclear Arms Race," p. 47.

61 "A single demand . . .": Holloway, ibid., p. 41.

62 The description of the Soviet technical intelligentsia and the special managerial style of the "Red Specialists" is drawn from: Vernon V. Aspaturian, "The Soviet Military Industrial Complex," *Journal of International Affairs*, vol. 26, 1972, p. 1; Jeremy Azrael, *Managerial Power and Soviet Politics;* Kendall Eugene Bailes, "Stalin and Revolution from Above"; K. E. Bendall, "The Politics of Technology: Stalin and Technocratic Thinking Among Soviet Engineers," *The American Historical Review*, vol. 79, 1974, p. 445; Leopald Haimson, "The Solitary Hero and the Philistines," *Daedalus*, Summer 1960, p. 541; John McDonnell, "The Soviet Defence Industry as a Pressure Group," in Michael MccGwire, et al. (eds.), *Soviet Naval Policy;* John McDonnell, "The Organisation of Soviet Defence and Military Policymaking," in Michael MccGwire (ed.), *Soviet Naval Influence;* Karl Spielmann, "Defence Industrialists in the USSR," *Problems of Communism*, Sept./Oct. 1976.

62 ". . . the country's only hope . . .": *Komsomolskaya Pravda*, Nov. 12, 1976.

62 "The director is the sole sovereign . . .": Azrael, op. cit., p. 247n.

62 "They would have shot [us]": Barwich, op. cit., p. 116.

64 Vannikov's role in the bomb project is mentioned in most of the general references in the first note for this chapter. Particularly useful were Golovin's *Kurchatov*, the Emelyanov memoir in *Red Star*, and the article in *Voprosi Iistori*. Also Severyn Bialer (ed.), *Stalin and His Generals*, especially p. 153; *The Large Soviet Encyclopaedia*, 3rd edition, vol. 4, p. 291; *The Soviet Military Encyclopaedia*, vol. 2, pp. 14–15. The same sources were used for the sketch of Zavenyagin. The memoirs of the German scientists, whose work Zavenyagin supervised, were important, especially Barwich, op. cit., and von Ardenne, op. cit. Also Zavenyagin's biography in *The Large Soviet Encyclopaedia*, 3rd edition, vol. 9, p. 267. Information on the personality of each of the top Soviet administrators came from interviews with Russian refugees, especially Boris Rabbot and Leonid Finkelstein.

64 "I am ready to die for it": Emelyanov, *Komsomolskaya Pravda*, Nov. 12, 1976.
65 "The newspaper hacks. . .": Alexander Solzhenitsyn, *The Gulag Archipelago*, p. 535.
65 ". . . patriot of his own enterprise . . .": Barwich, op. cit., p. 94.
65 For Zavenyagin as protector of the intelligentsia, see anecdote in Bialer, op. cit., p. 86.
65 Zavenyagin at Magnitogorsk: John Scott, *Behind the Urals*, pp. 231–33.
66 Zavenyagin's office is described in Barwich, op. cit., p. 41.
66 "We as engineers . . .": Emelyanov, *Voprosi Iistori*, no. 5, May 1975, p. 123.
67 ". . . very valuable for the Soviet Union . . .": Interview with Heinz Barwich, *Der Spiegel*, no. 44, 1965. The "we" in this case referred to the German team, which had no access to intelligence sources. The same scientist later gave a more general estimate of the Smyth Report: it turned the Soviet planners away from "laboratory ideology" and decisively toward large-scale thinking. Publication of the report coincided precisely with the Soviet commitment to industrial development.
67 On Slavsky, see *The Large Soviet Encyclopaedia*, 3rd edition, vol. 23, p. 543. Also Arnold Kramish, *Atomic Energy in the Soviet Union*, pp. 178–79; and Golovin, op. cit., JPRS translation, p. 17, where Slavsky is described as "temperamental."
68 Solzhenitsyn on Maltsev: *Gulag*, p. 552.
68 On Soviet uranium mining in East Germany, see Kramish, op. cit., pp. 170 ff.; Anonymous: "Soviet Operation of Uranium Mines in Eastern Germany," East European Fund, Mimeographed Series, no. 11, 1952; N. Grishin, "The Saxony Mining Operation (Vismut)," in Robert Slusser (ed.), *Soviet Economic Policy in Postwar Germany*; "The Secret Mines of Russia's Germany," *Life*, vol. 29, Sept. 25, 1950; Antony C. Sutton, *Western Technology and Soviet Economic Development*, chap. 18. The reference to radiation protection practice as "criminal" is in *Life*, op. cit., p. 83, and the fact that uranium was at first air-freighted to the Soviet Union is mentioned in the anonymous article in the East European Fund Series, p. 10.
69 "discomforts of a frosty winter . . .": Golovin, op. cit., JPRS translation, p. 22.
69 Khariton was a slight and somewhat deformed man. He was nicknamed Rhakiton, a cruel pun meaning "a man affected by rickets." He spent several years at the Soviet equivalent of Los Alamos—again, its exact location is unknown. The place is referred to only by the simple name Problema, "The Problem." This fact was mentioned by Leonid Finkelstein, interview in Munich, July 1978.
70 Khrushchev's admission that the first Soviet test was not a deliverable bomb is in *Khrushchev Remembers*, p. 59.

CHAPTER 5

The basic source on the British bombing policy during World War II is Sir Charles Webster and Noble Frankland, *The Strategic Air Offensive Against Germany*; the new bombing directive of Feb. 14, 1942, is reproduced at p. 323. The analysis of strategic bombing in general has drawn on: Robert C. Batchelder, *The Irreversible Decision*; Bernard Brodie, *Strategy in the Missile Age*; E. L. M. Burns, *Megamurder*; David Divine, *The Broken Wing*; David Irving, *The Destruction of Dresden*; R. V. Jones, *Most Secret War*; C. P. Snow, *Science and Government*; A. J. P. Taylor, *English History, 1914–45*; Harold Wilensky, *Organizational Intelligence*.

73 Sources on Lord Portal include R. V. Jones, *Most Secret War*; Denis Richards, *Portal of Hungerford*; Margaret Gowing, *Independence and Deterrence*, vols. 1 and 2.
73 "consensus of informed opinion": Charles Webster and Noble Frankland, *The Strategic Air Offensive Against Germany*, p. 183.
73 "the economic basis for the hopes . . .": Gowing, *Independence and Deterrence*, vol. 1, p. 3.
74 "Is the Minister of Defence satisfied . . .": Ibid, p. 212; also House of Commons debates, *Hansard*, vol. 450, col. 2117.
75 "When we ask questions . . .": Ibid, p. 52.
75 The press had become restrained: Gowing, *Independence and Deterrence*, vol. 1. p. 52.
75 The *Economist* dismissed rumors: Ibid, p. 54.
75 "Are we so helpless": Ibid, p. 114.
75 "That stupid McMahon Act": British Broadcasting Corporation, Attlee obituary, London, lib. no. 080148.
76 "as the first memorandum . . .": Gowing, *Britain and Atomic Energy*, p. 35.
76 signing away Britain's atomic birthright: Jones, op. cit., p. 474.
77 The sketch of Clement Attlee is based on: Roy Jenkins, *Mr. Attlee*; Edward Francis-Williams, *A Prime Minister Remembers*, especially chap. 8; *Current Biography*, Clement Attlee, 1940.
78 ". . . safely-berthed schoolmaster": *Current Biography*, Clement Attlee, p. 34.
78 The sketch of Lord Bridges draws on "Baron Bridges," *Biographical Memoirs of Fellows of the Royal Society*, vol. 16, p. 37. Also Joan Bright Astley, *The Inner Circle*; Stephen Roskill, *Hankey: Man of Secrets*; Lord Moran, *Winston Churchill*.
78 ". . . frown on a stranger": Lord Moran, op. cit., pp. 221–22.
79 The bureaucratic maneuvers of Lord Bridges to ensure that no one was in a position to rock the boat on the British bomb decision are found in

the official history by Gowing, *Independence and Deterrence*, vols. 1 and 2. On p. 24 Bridges is seen promoting Anderson and resisting a special ministerial appointment; on 26–27 he proposes the new administrative arrangements; on p. 33 he ensures that Sir Henry Tizard is kept out of the decision-making process. As a former civil servant, Sir John Anderson was particularly close to Bridges.

81 The sketch of Sir John Anderson is based on: Sir John Wheeler-Bennett, *Sir John Anderson, Viscount Waverley*. Also "Sir John Anderson, A Non-Party Man's Reflections," London *Sunday Times*, July 1, 1945.

81 ". . . he has a heart after all": Beverley Baxter, "The Rt. Hon. Sir John Anderson," *Strand Magazine*, Sept. 1939, p. 485.

81 ". . . Jehovah's witnesses" Anthony Sampson, "A Career in the Establishment," *Observer*, London, Sept. 30, 1962.

82 One could have more sympathy with the British gripe that they had been let down by the Americans if it were not for two important facts. First, in 1941 the British had ignored American overtures for atomic collaboration. It was only when it became clear that Britain's technology was not as advanced and her wartime resources were too stretched that she agreed to cooperate.

Second, the British treated the Canadians in an even harsher manner than they themselves had been treated by the Americans. They ignored their wartime undertaking to Canada: that Canada's heavy-water reactor project would be the basis of a postwar joint effort in peaceful nuclear power. They removed staff from the Canadian project when they were needed. Indeed, the only British argument advanced in favor of continuing cooperation with Canada was that the Canadians might otherwise be tempted to join the U.S. project or, worse, they might even develop a combined project in the Dominions to rival the British program. The official British history refers to Anderson's "high-handed pro-consular attitude" to the Canadians. The habits of Ireland and Bengal were hard to shake. The British attitude to the Canadians is described in Gowing, *Independence and Deterrence*, vol. 1, pp. 132 ff.

82 The sketch of Lord Portal is based on: Richards, op. cit.; Gowing, *Independence and Deterrence*, vol. 1; Jones, op. cit.

83 ". . . the techniques of manipulating men": Anthony Verrier, *The Bomber Offensive*, p. 97.

83 ". . . how tenaciously Portal fought . . .": Gowing, *Independence and Deterrence*, vol. 1, p. 45. Portal was also well respected by the U.S. military commanders, and it is now accepted that, despite Groves and the McMahon Act, there was an exchange of atomic information during 1947–49 and it was due to Portal.

84 Professor Patrick Blackett: Ibid., pp. 115, 171–72, 194–206.

84 "The author . . . a layman": Ibid., p. 172.

84 Tizard sketch from Ronald Clark, *Tizard*.
85 ". . . as a Great Power . . . a great nation . . .": Margaret Gowing, *Independence and Deterrence*, vol. 1, p. 229.

CHAPTER 6

The description of nuclear decision making during the formative years of Truman's second term draws heavily on the official history of the period: Richard Hewlett and Francis Duncan, *Atomic Shield*. In addition the following sources were used: Robert Donovan, *Conflict and Crisis;* Robert Gilpin, *American Scientists and Nuclear Weapons Policy;* Samuel Huntington, *The Common Defense;* David Lilienthal, *Journals: The Atomic Energy Years;* John Major, *The Oppenheimer Hearing;* Norman Moss, *Men Who Play God;* Warner Schilling, "The H-Bomb Decision," *The Political Science Quarterly*, vol. 76, 1961, p. 24; Lewis Strauss, *Men and Decisions;* David Rosenberg, "American Atomic Strategy and the Hydrogen Bomb Decision"; Philip Stern, *The Oppenheimer Case;* Harry Truman, *Memoirs: Years of Trial and Hope;* Daniel Yergin, *Shattered Peace;* Herbert York, *The Advisors*. Two recently declassified documents were also consulted: National Security Memorandum 68 and Kenneth Conit, *The History of the Joint Chiefs of Staff: The Joint Chiefs and National Policy 1947–49*. Declassified, March 1978.

87 "probably something like that": David Lilienthal, *Journals: The Atomic Energy Years*, p. 571. Even after his retirement, Truman, on one occasion, indicated that he still did not believe that the Russians really had the bomb. See Robert Oppenheimer, *Open Mind*, pp. 70–71.
88 ". . . yearn for the relief . . .": National Security Memorandum 68, Apr. 1950 [declass. 1975], p. 4.
88 "We must keep ahead": Harry Truman, *Memoirs: Years of Trial and Hope*, p. 306.
88 "That was the era . . .": John Major, *The Oppenheimer Hearing*, p. 151.
88 "the keystone of our military policy . . .": Richard Hewlett and Francis Duncan, *Atomic Shield*, p. 183.
89 Sources for David Lilienthal's background are: His journals, *Atomic Energy Years, TVA Democracy on the March*, and *Change, Hope and the Bomb;* Groves Papers at West Point; Hewlett and Duncan, ibid.
89 "Where this will lead us . . .": Lilienthal, *Atomic Energy Years*, p. 577.
90 As early as 1953 Henry ("Scoop") Jackson stressed the special role of the Joint Committee on Atomic Energy. Turning the constitutional division between the Congress and the Executive on its head, it was, Jackson asserted, "the Committee [who] made the decisions with the advice and consent of the executive branch." Harold Orlans, *Contracting for Atoms*, p. 161.

91 "inadequate to meet the requirements . . .": David Rosenberg, "American Atomic Strategy and the Hydrogen Bomb Decision," p. 66.
91 no stockpile: Lilienthal, *Atomic Energy: A New Start*, p. 1.
91 "Whatever corrective measures . . .": Yergin, *Shattered Peace*, p. 465.
91 ". . . a grim, gray look . . .": Lilienthal, *Atomic Energy Years*, p. 165.
92 The description of American strategic bombing policy draws on the sources listed in the first note to chap. 5, especially references to Batchelder, Burns, Brodie, Verrier, and Wilensky. The guarded official history was also consulted: Wesley Craven and James Cate (eds.), *The Army Air Forces in World War II*, vols. 1, 2, and 3. In addition, George Quester, *Nuclear Diplomacy*, pp. 34 ff.; John Major, op. cit.; Perry Smith, *The Air Force Plans for Peace*; Daniel Yergin, op. cit.
92 ". . . the existence of civilization . . .": Major, op. cit., p. 151.
92 General LeMay's statement: Huntington, *The Common Defense*, p. 369.
93 The Navy did fight back, however, asserting its own right to exist in the nuclear age. With the war in the Pacific over, the Navy launched a vigorous campaign to publicize two atomic tests at Bikini Atoll in July 1946. A host of invited guests, including politicians and newspaper reporters, gathered to watch the bombs explode in the middle of a small armada of decommissioned ships. The operation was supposed to test "weapons' effects." To no one's surprise, those ships closest to the explosion were blown up and those farther away were showered with radioactive material, effectively immobilizing them. Triumphantly, the Navy planners concluded there was a role for seaborne atomic weapons; all that was required was a separate delivery system: a super-carrier capable of launching nuclear armed bombers.
93 "barest minimum for our security": Major, op. cit., p. 152.
93 "the greatest event in world history . . .": Hewlett and Anderson, *The New World*, p. 436.
93 "total power . . .": Hewlett and Duncan, *Atomic Shield*, p. 402.
93 wrong for a civilized society . . . ". . . goodwill after the war": Rosenberg, op. cit., p. 70.
94 "like mess kits and rifles": Hewlett and Duncan, *Atomic Shield*, p. 182.
94 "It is a terrible thing . . .": Lilienthal, *Atomic Energy Years*, p. 391.
95 "Don't you bring that fellow . .": Philip Stern, *The Oppenheimer Case*, p. 90.
96 The GAC report on the H-bomb was declassified in 1974. The conclusion, "We all hope . . ." is in Part III of the report. Also extracted in Herbert York, *The Advisors*, pp. 46–56.
97 "Reports from Los Angeles and Berkeley . . .": Lilienthal, *Atomic Energy Years*, p. 582.
97 ". . . blow them off the face of the earth . . .": Ibid., pp. 584–85.

97 ". . . fed up to the ears . . .": Ibid., p. 591.
98 "I want the facts . . .": Ibid., p. 594.
98 learning about the "minds in the Kremlin": Acheson, *Present at the Creation*, p. 196.
99 it was folly: Hewlett and Duncan, *Atomic Shield*, p. 400.
99 "made a lot of sense": Rosenberg, op. cit., p. 83.
101 "If we act . . .": Hewlett and Duncan, *Atomic Shield*, p. 523.
101 "peace power": Ibid., p. 548.
102 Truman considered his decision on production expansion "one of the most important matters" (see ibid., p. 577). The official historians doubt Truman's own assessment of the importance of this decision, pointing to Hiroshima and the H-bomb. In fact he only ratified these earlier decisions; the scope for presidential discretion had almost disappeared by the time he "made" them. This was less true in the case of the final huge production expansion. Despite the pressure, Truman had a real choice. In terms of importance the cumulative impact of the increases in the production capability of fissile material was to establish, definitively, an era of overkill and to create the conditions for almost total American reliance on nuclear weapons for its basic defense posture. Truman was right to characterize his decision so.

CHAPTER 7

The description of the early Eisenhower years and the personalities and processes of the Oppenheimer show trial are based on: Joseph and Stewart Alsop, "We Accuse," *Harper's*, April 1954; Robert Bacher, "Robert Oppenheimer (1904–1967)," *Proceedings of the American Philosophical Society*, vol. 116, Aug. 1972; David Caute, *The Great Fear*; Haakon Chevalier, *Oppenheimer*; Nuel Pharr Davis, *Lawrence and Oppenheimer*; Robert Donovan, *Eisenhower*; Robert Gilpin, *American Scientists and Nuclear Weapons Policy*; Harold Green, "The Oppenheimer Case: A Study in the Abuse of Law"; Leslie Groves, *Now It Can Be Told*; Samuel Huntington, *The Common Defense*; Robert Jungk, *Brighter Than a Thousand Suns*; David Lilienthal, *Journals: The Atomic Energy Years* and *Journals: Venturesome Years*; John Lear, "Ike and the Peaceful Atom"; John Major, *The Oppenheimer Hearing*; Peter Michelmore, *The Swift Years*; Norman Moss, *Men Who Play God*; Harold Nieburg, *In the Name of Science*; Harold Nieburg, "The Eisenhower AEC and the Congress," *The Midwest Journal of Political Science*, vol. 6, 1962, p. 115; Duncan Norton-Taylor, "The Controversial Mr. Strauss," *Fortune*, Jan. 1955; Robert Oppenheimer, *The Open Mind*; Herbert Parmet, *Eisenhower*; Richard Rhodes, "The Agony of J. Robert Oppenheimer," *American Heritage*, Oct. 1977; Warner Schilling, "The New Look of 1953," in Warner Schilling et al., *Strategy, Politics and Defense Budgets*; Philip Stern, *The Oppenheimer Case*; Lewis Strauss, *Men and Decisions*; Warren Unna, "Dissension in the AEC," *Atlantic Monthly*, vol. 199, 1957, p. 36; Herbert York, *The Advisors*. In addition, the report of the so-called Candor Committee, *Armaments and American Policy: A Report of a Panel of Consultants on Disarmament for the Department of State*, Jan. 1953 [Declassified Jan. 1975], Eisenhower Library.

107 "weed out the Reds": David Caute, *The Great Fear*, p. 49.
107 "bones in Hoover's throat": Harold Green, "The Oppenheimer Case," p. 16.
107 Those reluctant to go were given special treatment. In his own tortuous way, Strauss would tell a man he was looking for a new job for him because his loyalty to Strauss would soon be a burden and he would become a natural target of the next AEC chairman when Strauss resigned. If the man rejected this backhanded plan, plus a string of job offers, Strauss would finally tell him to take a job or get out. One former AEC official, recalling such an incident, said "He [the man] . . . departed with lavish tributes for his loyalty and devotion." Green, op. cit. p. 16.
109 ". . . no evidence . . . security risk": John Major, *The Oppenheimer Hearing*, p. 268.
110 "It's a nightmare": Philip Stern, *The Oppenheimer Case*, p. 195. Lovett was an early convert to strategic bombing, his views being reinforced by his wartime job as Assistant Secretary of War with responsibility for the air force. His commitment to strategic bombing was fatal. See Yergin, op. cit., p. 203.
111 "long-haired scientists": Major, op. cit., p. 164.
111 ". . . fair general observation . . .": Ibid., p. 169.
112 In the meantime, the question had to be addressed: How many bombs do you need to destroy a city—and, indeed, what exactly is meant by destruction in the atomic age? Even if the major cities of a nation were physically obliterated, that nation could still have military strength—in its bomber or missile force—a long time after the majority of its citizens were dead. In the case of an attack on the United States, some experts thought that a few hundred bombs on target would be enough to bring about a surrender; others thought the country could survive more than 2,500 bombs. Based on the then known Soviet bomb design, that number would equal 100 million tons of high explosive, or 400 times the total dropped on Germany by Allied bombers during World War II. See Candor Report, Part II, p. 3.
112 "The time when the Russians . . .": Ibid.
112 ". . . hysteria . . . complacency": Ibid., Part III, p. 7.
113 "no matter how many bombs . . .": Ibid.
113 ". . . tell your friend Oppie . . .": Stern, op. cit., p. 206.
113 "A tragic thing . . .": Duncan Norton-Taylor, "The Controversial Mr. Strauss," p. 110.

114 "I was an unctuous . . .": Richard Rhodes, "The Agony of J. Robert Oppenheimer," p. 72.
115 "well-dressed owl": John Lear, "Ike and the Peaceful Atom," p. 12.
116 "The atom is amoral": Norton-Taylor, op. cit., p. 170.
116 "The physicists have known sin": quoted in "The Scientists," *Fortune*, 38, Oct. 1948.
116 ". . . blood on our hands": Stern, op. cit., p. 90.
116 "The atomic bomb is shit": Spencer Weart and Gertrud Szilard, *Leo Szilard: His Version of the Facts*, p. 185.
117 ". . . schoolboy attitude . . .": Major, op. cit., p. 59.
118 "incredible mismanagement": Stern, op. cit., p. 128.
118 ". . . use a shovel . . .": Ibid., p. 129.
118 "Strauss' eyes narrowed . . .": Ibid., p. 129.
118 "Too well, Robert . . .": Ibid., p. 130.
118 "the terrible look": David Lilienthal, *Journals: Venturesome Years*, p. 522.
118 ". . . just too courageous . . .": Ibid., p. 522.
118 ". . . must be a traitor": Joseph and Stewart Alsop, "We Accuse," p. 354.
119 "he has acted under a Soviet directive . . .": Major, op. cit., p. 29.
119 ". . . valve-in-the-head character . . .": Memorandum to Mr. Luce about the Alsop article in *Harper's*, dated Oct. 12, 1954, C. D. Jackson Files, Eisenhower Library.
119 Oppenheimer had "cracked": Ibid., p. 6.
119 ". . . they are idiotic": Caute, op. cit., p. 477.
120 Teller's testimony in U.S. AEC: *In the Matter of J. Robert Oppenheimer* (transcripts, principal documents, and letters), p. 710.
120 "Washington's influence . . .": Robert Jungk, *Brighter Than a Thousand Suns*, p. 242.
121 "He was a changed person . . .": Stern, op. cit., p. 504; *Science*, vol. 155, 1967, p. 1084; also quoted in Major, op. cit., p. 287.
121 The basic sources on the preparation of the Atoms for Peace speech were the files in the Eisenhower Library, in Abilene, Kansas, especially the speech drafts found in the Central Files and the memoranda and letters in the C. D. Jackson Files. In addition the works by Donovan, Parmet, and Strauss in the first note to this chapter plus the John Lear article. At times, Jackson's proclivity for public relations degenerated into gimmickry. He and his men proposed building nuclear reactors in Hiroshima and Nagasaki—their citizens, according to the memos on the speech drafts, had a special right to know the good that man had found.
121 ". . . scare the country . . .": Lear, op. cit., p. 11.
121 "This leaves everybody dead . . .": Lear, op. cit., p. 12.
121 ". . . fixed at a figure . . .": Strauss to President, Sept. 17, 1953, Central Files, Eisenhower Library.

122–24 C. D. Jackson's public relations effort to reach for the speech: Memorandum to C. D. Jackson from U.S. Information Agency, Dec. 28, 1953, C. D. Jackson Files, Eisenhower Library.

CHAPTER 8

The basic source on the French bomb decision is Lawrence Scheinman, *Atomic Energy Policy in France Under the Fourth Republic.* In addition the following were used: Jacques Chaban-Delmas, *L'Ardeur;* Bertrand Goldschmidt, *The Atomic Adventure* and *Rivalités Atomiques;* Wilfred Kohl, *French Nuclear Diplomacy;* Alain Peyrefitte, *Le Mal Français;* Philippe Simonot, *Les Nucléocrates;* Spencer Weart, *Scientists in Power.*

The Commissariat à l'Energie Atomique, created by Charles de Gaulle, was the first of the special nuclear institutions. The U.S. AEC was still being debated. No French government agency, before or since, received as much administrative and financial autonomy—including the extraordinary power to act in the name of any minister in any matter related to atomic energy. It has always acted on its own initiative without public accountability, and at first, its total freedom without matching responsibility was mainly of an academic kind. This heritage continued long after the CEA grew from a laboratory-scale operation into a large bureaucracy.

127 ". . . a certain ambiguity . . .": Spencer Weart, *Scientists in Power,* p. 226.
127 "madness to try . . .": Ibid., p. 220.
127 ". . . potential spies and traitors": *Time,* Mar. 29, 1948, p. 28.
127 Joliot-Curie faced a personally agonizing choice between his science and his political affiliations. The Stalinist French Communist party pressured him to toe the Moscow line, and it was an appalling conflict for him. When his research team triumphantly extracted the first tiny amounts of plutonium, Joliot, instead of congratulating them, reacted with anger. Bertrand Goldschmidt, *Rivalités Atomiques,* p. 88.
128 "was carried away . . .": Weart, op. cit., p. 258.
128 ". . . Progressive scientists . . .": Lawrence Scheinman, *Atomic Energy Policy in France Under the Fourth Republic,* p. 41.
128 Joliot had agreed to build a bomb: Alain Peyrefitte, *Le Mal Français,* p. 83.
129 The description of l'Ecole Polytechnique and of the Corps des Mines is based on: John Ardagh, *The New France;* Dominique Dejeux, "Le Corps des Mines ou une Nouvelle Mode d'Intervention de l'Etat," Paris (mimeo), 1970; Erhard Freidberg and Dominique Dejeux, "Fonctions de l'Etat at Rôle des Grands Corps: Le Cas du Corps des Mines," in *Annuaire International de la Fonction Publique,* Paris, 1972–73; Jacques Kosciusko-Morizet, *La "Mafia" Polytechnicienne;* F. F. Ridley,

"French Technocracy and Comparative Government," *Political Studies*, 1966, p. 34; Philippe Simonot, *Nucléocrates*; Ezra N. Suleiman, "The Myth of Technical Expertise," *Comparative Politics*, vol. 10, Oct. 1977, p. 137; Jean-Claude Thoenig, *L'Ere des Technocrates*; also, interview with André Thépot, Paris, June 1978.

130 ". . . knowing everything . . .": Ardagh, op. cit., p. 99.

131 The sketch of Pierre Guillaumat draws particularly on Josetta Alia, "L'Homme Qui de Gaulle Ecoute," *Le Nouvel Observateur*, Sept. 14, 1968; *L'Express*, Nov. 1, 1965; *La Croix*, Oct. 22, 1965; *L'Expansion*, Oct. 1969; *Lè Nouvel Observateur*, Feb. 12, 1973, June 21, 1976, Mar. 12, 1977; *Le Monde*, Jan. 26, 1977, Aug. 4, 1977; Simonot, *Complot Pétrolière*; several interviews, including Pierre Guillaumat, Paris, June 1978, and Francis Perrin, Munich, Sept. 1977.

131 "A machine that never misfires": Josette Alia, op. cit., p. 17.

131 Guillaumat's extraordinary capacity for institution building was shown in his oil empire as well as in the CEA. At the time de Gaulle appointed him Directeur des Carburants (Director of Fuels), Guillaumat's authority was limited to some old legislation which empowered the state to issue licenses for imports of gasoline and the construction of refineries. By imposing conditions on these licenses, Guillaumat began the long process of creating a significant new institution that could stand beside the French Compagnie des Pétroles (distributing under the Total brand), a company that Guillaumat believed was too willing to play second string to the oil majors. He first established the infrastructure of a research and exploration group; then, on the basis of successful oil and gas discoveries, especially in France's former colonies, he pushed his institutional empire into refineries. In 1967 the petrol bowsers carried his new brand of gasoline called "Elf." From a standing start he created a significant new presence in a heavily cartelized industry. He did it despite a series of diversions to other spheres of public service—the job at the CEA between 1951 and 1958 and then four years as a Minister of State in de Gaulle's early cabinets. In the first government after the general's return to power in 1958, Guillaumat was appointed Minister for the Army, a post his father—who had once been de Gaulle's commanding officer—had held in 1926. Guillaumat was acceptable to, but not a member of, the army higher echelons, then deeply divided by what would soon become open mutiny in Algeria. With relief, Guillaumat returned to put the final touches on his oil empire in 1962. He declared, "I am first an engineer, never having been predisposed to the political life" (*L'Expansion*, Oct. 1969). His access to General de Gaulle was still immediate, and he was frequently called "the czar of energy" or "the éminence grise of de Gaulle's Elysée" on energy policy. Guillaumat was a perfect representative of Gaullist political economy. His oil group's

promotional material regularly carried a photograph of World War I Prime Minister Clemenceau with a caption "Each drop of oil is a drop of blood." This widely known but inaccurate quotation (see Philippe Simonot, *Le Complot Pétrolière*, p. 133) comes from the worst days of that war, when the French leader had ignominiously to plead with the United States to prevent the oil companies from diverting a few small tankerloads of oil from Europe to the Pacific. The incident became a powerful image in the collective folk memory of French status anxiety.

132 Perrin was a product of L'Ecole Normale Supérieure, France's great institution of liberal education which provided a third of France's physics professors. The "Normaliens" despised Polytechnicians, but as Perrin's experience in the CEA showed, they could not handle the engineers. Perrin's intellectual heritage of doubt, reasonableness, and seeing all points of view put him at a disadvantage. He always found himself arguing on someone else's terms and lost the battle to prevent France from exercising the bomb option—a battle which his grace, charm, and open-mindedness had prevented him from knowing how to fight. Even before Guillaumat had arrived at the CEA, Perrin had exposed his naïveté of the subtleties of the political power base that was wresting control from the scientists. The "godfather" of the Corps des Mines himself, Henri Lafond, had been proposed for membership on the council of the CEA. Lafond, a banker, and Minister for Energy in the early Vichy years, was struggling to restore the position of the Corps, its leadership decimated and its influence curtailed because of the collaboration of many of its members with the Nazis during the war. The liberal-minded Perrin objected strongly. Lafond, a leader of the French employers' association, Le Patronat Français, was the kind of "capitalist," said Perrin, who might introduce unacceptable economic biases into the direction of the scientific research. But Perrin had got it wrong. What really mattered was not that Lafond was a capitalist but that he was "le chef" of the Corps des Mines. When Lafond, who supported nationalist ambitions in the former colonies, was assassinated in 1963, Pierre Guillaumat took over as "le chef du Corps."

132 "undoubtedly predominant": Bertrand Goldschmidt, *The Atomic Adventure*, p. 81, and compare the French edition on p. 98. In the English translation of his book, the words "undoubtedly predominant" would be removed; such Gallic cynicism apparently inappropriate for Anglo-Saxon eyes.

132 "importance and urgency . . .": Scheinman, op. cit., p. 97.

132 ". . . a predominant role . . .": Goldschmidt, *Les Rivalités Atomiques*, p. 210.

133 The meeting with Mendès-France early in December is mentioned in Simonot, *Les Nucléocrates*, p. 228, Simonot's analyses of the available

information on just what it was that Mendès-France decided to do at the December 24 meeting, previously described by Scheinman, op. cit., pp. 112–114. Scheinman received extraordinary access to CEA records by accident. He was given cursory permission by Bertrand Goldschmidt to use the CEA library for his Ph. D. thesis. He stayed for a year, until Goldschmidt found out that he was still there poring over the old records. After Scheinman's work appeared, Goldschmidt produced his own version of the Mendès-France meeting which asserted that the Prime Minister had approved the preparation of a prototype bomb: Goldschmidt, *Rivalités*, p. 207. It is inconceivable that a decision of such importance would be made without a formal order or minute, and Scheinman asserts there was none: op. cit., p. 114. The allegation is also inconsistent with the fact that two months after the alleged decision, the CEA was still making submissions, seeking more specific terms of reference, and a later second meeting of the same ministers on the same subject closed on a totally inconclusive note: Scheinman, op. cit., p. 114. It seems that the pro-bomb lobby in the Commissariat chose to interpret an ambiguous discussion in a precise manner and wrote it up the way they wanted to; as memories faded over the years, the precision of the printed word in their internal documentation led them to believe their own distortion. Unfortunately, Mr. Goldschmidt, who had granted the authors one interview, refused to see us again when our questioning would have been better directed to this point.

134 Guillaumat's secret protocol with the chefs du cabinet of the two Gaullist ministers is mentioned in Jacques Chaban-Delmas, *L'Ardeur*, p. 205. The final agreement at the ministerial level is mentioned in Scheinman, op. cit., p. 122. It would not be surprising if Edgar Fauré did not know what was really going on. There is no record of his being involved in the decision, perhaps because Guillaumat acted on the basis that it was simply the implementation of the alleged decision to build a "prototype" made by Fauré's predecessor.

134 Guy Mollet's change of attitude after Suez is emphasized by Goldschmidt, *Rivalités*, p. 222, and the new military timetable is outlined in Scheinman, op. cit., p. 173.

135 The fact that Gaillard's order was backdated was revealed in 1975 by Chaban-Delmas, op. cit., p. 224.

CHAPTER 9

This chapter looks at the enshrining of the atom in three countries—Britain, Canada, and the U.S.S.R. It is possible to follow the development through one man in each country. In Britain, the sketch of Lord Cherwell is based on: The Earl of Birkenhead, *The Prof in Two Worlds*; Sir Roy Harrod, *The*

Prof; Vannevar Bush, "Churchill and the Scientists," *Atlantic Monthly*, vol. 125, March 1965; P. M. S. Blackett, "Tizard and the Science of War," *Nature*, vol. 185, 1960; Sir Robert Watson-Watt, "The Secret of 'the Prof,' " London *Sunday Times*, July 7, 1957. General sources include, Margaret Gowing, *Independence and Deterrence*, vols. 1 and 2. As ever, these have proved indispensable works for this period.

137 branded a warmonger: Margaret Gowing, *Independence and Deterrence*, vol. 1, p. 406.
137 "If we have to rely . . .": Ibid., p. 407.
138 "Our prosperity . . .": Ibid., p. 408.
138 "the biggest snob . . .": Letter from Lord Boothby to London *Daily Telegraph*, Jan. 24, 1967.
138 ". . . vulgariser of science . . .": Sir Robert Watson-Watt, "The Secret of 'the Prof,' " op. cit.
138 As the formal head of the civil service, Lord Bridges was custodian of its rules and its standards. Independent commissions limited the scope of this system, and he regarded each intrusion as a bad precedent. Bridges was quite accustomed to each new political whim being proclaimed as special or unique. His resistance was instinctive rather than dogmatic, just enough to test the strength of the political will behind it, but not enough to prevent him from influencing the structure and the top appointments when the will proved strong enough.
140 "wait and see . . .": Earl of Birkenhead, *The Prof in Two Worlds*, p. 307, quoted in Gowing, *Independence and Deterrence*, vol. 1, p. 432.
140 Cherwell lobbying for the first time: Ibid., p. 310.
140 "It's not a question of control . . .": *Nature*, vol. 278, Mar. 15, 1979, p. 201.
141 The sketch of C. D. Howe draws particularly on his biography, Leslie Roberts, *C. D.: The Life and Times of Clarence Decatur Howe; Current Biography*, 1945, *Fortune*, Aug. 1957. A series of specific facts were drawn from William Kilbourn, *Pipeline;* Lester Pearson, *Mike: The Memoirs of Lester B. Pearson*, vol. 3; J. W. Pickersgill and D. F. Forster, *The Mackenzie King Record*, vols. 1, 3, and 4; John Porter, *The Vertical Mosaic;* Peter C. Newman, *Renegade in Power;* G. Bruce Doern, *Science and Politics in Canada*.
142 Howe brought to politics his engineer's sense of direct personal responsibility for all his creations, but like a good engineer he would accept responsibility only if he was in complete control. To achieve this control he frequently threatened to resign or to stop overseeing a particular activity if he didn't get his way. One such incident occurred when some ministers hesitated to support his plan to send troops to break a strike at an aluminum plant; another when he could not get coal miners auto-

matic exemption from the wartime draft. The cabinet often balked at the unseemly haste with which he tried to push through legislation without debate, but if anyone attempted to slow him down, he would say, "I can no longer continue to accept responsibility." No one ever called his bluff; his personal strength was overwhelming. (See Pickersgill and Forster, op. cit., vol. 1, pp. 231, 489; vol. 4, p. 125; also Newman, op cit; p. 43.) Howe saw the cabinet as his board of directors with a right to ask questions but no right to interfere with top executives. Parliament was just a shareholders' meeting, rubber-stamping decisions already taken.

142 Howe's peaceful project: Wilfrid Eggleston, *Canada's Nuclear Story*, p. 6.

142 ". . . that's my enterprise": Newman, op. cit., p. 36.

142 "get in on the ground floor": Eggleston, op. cit., p. 4.

142 Howe suggested to the British that the best way of beginning the cooperation would be for the British government to buy up the stock of Canada's largest uranium mine at Great Bear Lake. Under wartime powers, the Canadian government could simply take over the mine, but this would attract unwanted publicity. Instead, Howe proposed to buy the shares on behalf of the Canadian government, which would then agree on a division of control of the uranium between the British and the Americans. The British agreed, the plan was endorsed by the Americans, and Howe began obtaining the stock.

In the end, however, the British never saw any of the uranium—at least, during the war. Howe apparently succumbed to greater pressure from the Americans; hard on the heels of the British delegation seeking the uranium, the Canadians had received one from the Manhattan Project. In September 1942, when General Groves took over the Manhattan Project, he stepped up U.S. orders from the mine and, during the winter of that year, effectively secured the mine's total output until nearly the end of 1945.

The British were furious. Howe had assured them that he was in charge of overseeing all uranium orders, but, it seemed, he had temporarily lost control. Churchill was outraged and complained to the Canadian Prime Minister that Howe had "sold the British Empire down the river." An indignant Howe sought to explain away his failure to stop the Americans by the enormous spread of his duties and because of his personal exhaustion. In an even more bizarre explanation, the British negotiator, Wallace Akers, sought to excuse Howe—and his own personal failure to secure the uranium—by saying that the Canadian Minister has "an unreliable memory for facts and figures" (see Gowing, *Britain and Atomic Energy*, p. 185). It seems inconceivable, however, that Howe did not know exactly what was going on. A more likely explanation for his behavior is that he was convinced of the superiority of the American effort and irritated by the obtuseness and the arrogance of the British. Even the official history (see ibid., p. 132) comments on the British "high-handed pro-consular attitude towards the Canadian industrial effort in support of Britain's defences." To a man like Howe, such arrogance would be insufferable, and he apparently found it preferable to deal with the Americans. They, in turn, did not forget him. A presidential memorandum, prepared for Eisenhower's Canadian visit in 1953, suggested he stress "particularly the sympathetic and understanding co-operation of C. D. Howe" (see Strauss letter, Nov. 10, 1953, Central Files, Eisenhower Library).

The British were indifferent to the Canadians. As Churchill's physician, Lord Moran, recalled, "The Prime Minister is not really interested in Mackenzie King. He takes him for granted": Lord Moran, *Winston Churchill*, p. 19. Even after the war the British continued to regard Canada's efforts to become independent in the nuclear matters as "disloyalty": Gowing, *Independence and Deterrence*, vol. 1, p. 132.

143 "closer at hand . . .": Eggleston, op. cit., p. 148.

144 The sketch of Vyacheslav Malyshev draws on *The Large Soviet Encyclopaedia*, 3rd edition, vol. 15, p. 295; *Munzinger Archiv*, May 4, 1957; and a memoir by Vassily Emelyanov, "Fighter for a Great Cause," *Izvestia*, Dec. 16, 1972. As in chap. 4, the personal information comes from Soviet refugees now living in the West.

144 Acknowledgments of Malyshev's role in the Soviet bomb project are found in a number of the Soviet references in the first note to chap. 4: Emelyanov, *Red Star*; Golovin, *Kurchatov*, JPRS translation, p. 29, and Mir Publishers, 1969 edition, p. 71; Alexandrov, "Nuclear Physics and the Development of Technology in the USSR," p. 200. Heinz Barwich's memoirs describe Malyshev's role as chairman of technical meetings more fully in *Das Rote Atome*, pp. 95, 131.

144 "The myriads of lights from the east . . .": A. Markin, *Power Galore*, p. 92.

145 Malyshev at Stalingrad: See G. K. Zhukov, *The Memoirs of Marshal Zhukov*, p. 376; Chuikov, *The Battle for Stalingrad*, p. 157; John Erikson, *The Road to Stalingrad*, pp. 369–70.

146 The major Soviet expansion of 1954 is indicated in the range of new appointments to the Minister for Medium Machine Building in that year. The new team included a number of senior town planners and military construction engineers. The leader of the new construction team was General Alexander Komarovsky, who was appointed deputy minister. See *The Large Soviet Encyclopaedia*, 3rd edition, vol. 12, p. 490; *Pravda*, Nov. 21, 1973; *Red Star*, May 20, 1976; *Soviet Military Encyclopaedia*, vol. 4, p. 261. In addition, to replace the checking function

of Beria's secret police, a new party apparatchik was appointed to the ministry that year to serve as the party's watchdog inside the ministry itself and also to promote the ministry within the party. See obit of Leonid Mezentsev, *Izvestia*, June 22, 1976.

146 "century of atomic energy": G. Modelski, *Atomic Energy in the Communist Bloc*, p. 115.

CHAPTER 10

147 ". . . cure the common cold": American Assembly, *Atoms for Power*, p. 15.

148 The basic source on Admiral Rickover is the official history: Richard G. Hewlett and Francis Duncan, *Nuclear Navy*. The sketch of Rickover has also drawn on William R. Anderson with Clay Blair, Jr., *Nautilus 90 North*; Clay Blair, Jr., *Admiral Rickover and the Atomic Submarine*; James F. Calvert, *Surface at the Pole*; Jimmy Carter, *Why Not the Best?*; John Kenton, "Building the Nuclear Navy"; Deborah Shapley, "Nuclear Navy"; Elmo R. Zumwalt Jr., *On Watch*; *Current Biography*, 1953; *New York Times*, Apr. 9, 1963; Washington *Post*, May 20, 1959, Feb. 2, 1963, July 28, 1975, and May 27, 1977; *Editor and Publisher*, Feb. 7, 1953.

148 Rickover frequently expressed his disagreement with the optimism about electricity costs from nuclear power. On one occasion he asked David Lilienthal, "Why do people say things that don't make sense, and mislead people?" Lilienthal, *Journals: Venturesome Years*, p. 531.

151 ". . . ladies powder room": Washington *Post*, May 27, 1977.

151 ". . . people perish" and ". . . doubts are traitors . . .": Ibid.

151 The "pinks": Hewlett and Duncan, *Nuclear Navy*, p. 127; Washington *Post*, Feb. 2, 1963, and May 27, 1977.

152 ". . . bisects six admirals": Kenton, "Building the Nuclear Navy," p. 84.

152 Rickover administered his organization the way he conducted his interviews: hectoring, cajoling, yelling, arguing. His lieutenants were expected to act the same way with contractors and employers, and they did. Within the clearly defined lines of single clear-cut projects, Rickover created tension by establishing overlapping jurisdictions and rival teams. It was a flat organization with a minimum of rank or hierarchy—technical excellence was the only criterion of authority. He ran his empire on guilt, never missing an opportunity to undermine someone's sense of achievement. Jimmy Carter recalled: "He demanded total dedication from his subordinates. We feared and respected him and strove to please him. I don't in that period remember him ever saying a complimentary word to me. The absence of a comment was his compliment." Carter, op. cit., p. 61.

152 ". . . shattering event . . .": Hewlett and Duncan, *Nuclear Navy*, p. 358.
152 ". . . no initiative . . .": Zumwalt, op. cit., p. 88.
152 "dumb or lazy . . .": Calvert, op. cit., p. 14.
152 "Are you resourceful?" William Anderson, op. cit., p. 21.
153 Special chair and venetian blinds: Ibid., p. 20.
153 "Did you do your best?" Carter, op. cit., p. 20
153 "Did you study as hard . . .": Zumwalt, op. cit., p. 90.
154 "Now you're being greasy": Ibid., p. 94.
154 earnings to father: Blair, op. cit., p. 20.
155 "cutthroat": Washington *Post*, May 27, 1977.
157 "Millionths of an inch . . .": Kenton, "Building the Nuclear Navy," p. 85.
158 "That Jew bastard . . .": Clinton P. Anderson and Milton Viorst, *Outsider in the Senate*, p. 176.
158 ". . . being on the team . . .": Washington *Post*, May 20, 1959.
159 Rickover's role in the development of commercial reactors is fully described in Hewlett and Duncan, *Nuclear Navy*. Further material on the initial programs of Westinghouse and General Electric and the commitment of electric utilities is drawn fròm Hewlett and Duncan, *Atomic Shield*; Philip Mullenbach, *Civilian Nuclear Power*; Frank G. Dawson, *Nuclear Power*; Wendy Allen, "Nuclear Reactors for Generating Electricity: U.S. Development form 1946 to 1963," Rand Corporation, R-2116-NSF, June 1977; *Business Week*, Nov. 9, 1957, Feb. 15, 1958, and March 11, 1967; John E. Kenton, "The Birth and Early History of Nuclear Power," *EPRI Journal*, July/Aug. 1978.
162 General Electric managed to outmaneuver Rickover on one occasion, and they paid the price of doing so. In 1956 they appealed over Rickover's head on a research contract for the development of a lightweight gas-cooled reactor. The domination of the scientists in GE's Atomic Products Division was reflected in this proposal. They had established an objective of a submarine reactor that would weigh less than the PWR and worked back from this target. Rickover's whole approach was to establish precise engineering goals first for readily achievable objectives. He continued to resist even when GE offered to pay a quarter of the research costs. This meant the company would "have complete freedom to select personnel to perform the study and to be solely responsible for direction of the study": Hewlett and Duncan, *Nuclear Navy*, p. 277. The study was approved despite Rickover's objections. It concluded that gas cooling did offer a prospect of lightweight reactors. Rickover ignored it and refused to pursùe the option. For the next twenty years Rickover regarded any proposal to investigate lightweight reactors as a direct personal challenge. Even in the mid-seventies when technical develop-

ments suggested that such reactors were more feasible than ever, Rickover was still terrorizing any proponent. He stopped the funding of small-scale research projects on such a reactor, tried to cancel small seminar discussion, and threatened to use his congressional contacts to abolish the Office of Naval Research if it persisted in its study of such an option.

164 These same utility companies also feared an "atomic TVA" and sought to forestall it by going nuclear themselves. American electric utilities were still bitterly divided between the investor-owned private utilities and the public ones; each had its own contingent of supporters in Congress and its own access to sectors of the Administration. Eisenhower and the chairman of the AEC, Lewis Strauss, were firm supporters of the private companies and shared their fears of a large public utility, such as the Tennessee Valley Authority, grabbing the benefits of nuclear power.

164 Paley Commission quotes are from U.S. Presidents Materials Policy Commission, *Resources for Freedom*, vol. 3, p. 220.

CHAPTER 11

The basic source on the first Geneva Atoms for Peace Conference is Laura Fermi, *Atoms for the World*. In addition the following were used: "Peaceful Uses of Atomic Energy: The Geneva Conference in Retrospect," *United Nations Review*, vol. 2, Oct. 1955; Isidor Rabi, "*My Life and Times as a Physicist*"; Lewis Strauss, *Men and Decisions*, Karl Winnacker, *Challenging Years*.

165 The bewilderment of the Swiss is mentioned in Laura Fermi, *Atoms for the World*, pp. 47, 93, 108.

166 ". . . iron and red tape . . .": Isidor Rabi, "*My Life and Times as a Physicist*," p. 21.

166 ". . . demonstrated beyond doubt": "Peaceful Uses of Atomic Energy," p. 2.

167 The Soviet Union's first reactor, at Obninsk, was moderated by graphite and cooled by water. This type was also developed under the civilian reactor program.

167 The scope of the Russian program at this stage was much greater than the Americans cared to concede (see Strauss letter to Jackson, Eisenhower Library, C. D. Jackson Files, June 11, 1956). The Russians were not only testing a wide range of reactor types but also looking at long-term plans for breeder reactors. In Apr. 1956, Russia completed a major reorganization of its nuclear administration, having created a new agency for the peaceful uses of the atom, the Central Directorate (later State Committee) for the Utilization of Atomic Energy. In keeping with the

openness and optimism resulting from the Geneva Conference, the new agency soon started its own journal, the International Atomic Energy Bulletin, and issued a stream of press releases. It also had a team of experts that went to all the international atomic conferences that followed.

169 German newspaper headlines: Martin-Endinghaus, op. cit., p. 112.

169 The sketch of Karl Winnacker is based on: Winnacker, *Challenging Years*, and Karl Winnacker and Karl Wirtz, *Nuclear Energy in Germany*; S. Martin-Endinghaus, "From the Atomic Egg to Brockdorf," *Bild der Wissenschaft*, Nov. 4, 1977; *Chemische Industrie*, vol. 15, 1963, p. 521; *Energiewirtschaftliche Tagesfragen*, Sept. 3, 1955; *Die Zeit*, no. 44, 1971; *Rheinische Merkur*, Sept. 20, 1963; *Frankfurter Allgemeine Zeitung*, Sept. 21, 1963. Also Wolfgang Cartellieri, A. Hocker, and W. Schnurr (eds.), *Taschenbuch für Atomfragen*, passim.

169 Winnacker himself wanted an entirely separate ministry that could draw up its own agreements with the United States and U. K. salesmen, under the Atoms for Peace program and without interference form the German Foreign Office. But the wily German minister for economics, Ludwig Erhard, was apprehensive about an open-ended claim on government funds. "We don't have a steam Ministry," he exclaimed. (See ibid., p. 114.)

170 Winnacker's own connection with I. G. Farben, the sprawling loosely knit confederation of German chemical companies, goes back to 1933 when he joined the Hoechst Company. The job had been found for him by his old chemistry professor at Darmstad, Ernst Berl, a Jew who was purged and fled to the United States. Winnacker had joined the Brown Shirts before his professor was forced to leave and later claimed to have been surprised by his departure. Winnacker made rapid progress into the hierarchy of I. G. Farben and, as he later admitted, was lucky that he was not charged with war crimes at Nuremberg—although Hoechst had no direct participation in the company's more unsavory work. Winnacker was the most senior executive to escape prosecution. He had taken over as technical director of Hoechst in 1943 and would normally have been appointed to the management committee, all the members of which were charged. After the war his interrogators noted the difficulty in checking his movements in the company, since he had burned all his correspondence. They concluded, "From interrogation, the impression was gained that Winnacker was engaged in some of the poison gas work at Uerdingen, but this was not confirmed by his written statement." (See Combined Intelligence Sub-Committee, Report on I. G. Farben and Hoechst, July 20, 1945.) In his memoirs many years later, Winnacker noted obliquely that the Uerdingen plant "could offer some highly interesting products which commanded considerable interest during the Hitler period, including the war" (Winnacker, op. cit., p. 103).

Unlike the scientists of the Manhattan Project who had been moved by feelings of remorse about the bomb to discuss—and some to doubt—the moral implications of their work, there is no such evidence that any of the scientists from I. G. Farben held similar discussions. As a result the name of the company has entered the annals of infamy without reservation. Except for occasional references to bureaucratic untidiness, Winnacker himself always referred to I. G. Farben with pride. He never felt public remorse or openly expressed any doubts about what the company had done. He constantly referred to the Nuremberg trials as a gross injustice and never mentioned the concept of justice when discussing the way the company's products had been used. The traditional Nuremberg defense of "obeying orders" was replaced in Winnacker's mind with "filling orders"; that seemed to him to be one step further removed from any moral responsibility. He believed in the moral neutrality of technology.

171 The sketch of Shoriki Matsutaro is based on Goto Kunihiko, "Matsutaro Shoriki: A Rising Force in Japanese Journalism," *Contemporary Japan*, vol. 9, Oct. 1940; Ralph Hewins, *The Japanese Miracle Men*; Edward Uhlan and Dana Thomas, *Shoriki: Miracle Man of Japan*; Kondo Kan-'ichi and Osani Hiroshi (eds.), *Sengo Sangyo-Shi E No Shogen*, papers by Arisawa and Ipponmatsu; Records of Allied Operational Occupation Headquarters, WW2, R. G. 331, Shoriki Matsutaro, File 181, National Archives, Washington, D.C.; *Japan Times*, Aug. 24, 1957; *Yomiuri Japan News*, Sept. 8, 1957, Jan. 3, 1960, May 4, 1962, and Feb. 11, 1968.

171 The most spectacular of Shoriki's postwar schemes was Japan's first private television company, which he started in the early fifties. Few Japanese could afford television sets, so he put them in public places; advertising rates were determined by a head count of the viewers.

172 ". . . become the greatest man . . .": Uhlan and Thomas, op. cit., p. 33.

172 Shoriki as keynote speaker: Shoriki Matsutaro, *Fifty Years of Light and Dark: The Hirohito Era*.

173 Hatoyama was expected to become Japan's first postwar prime minister, but General MacArthur placed him on the purged list just before the election; he had to wait until the end of the occupation to further his political career. Hatoyama and Shoriki had been close political allies for many years. Hatoyama had written for Shoriki's *Yomiuri Shimbun* as early as 1926. In 1933 Shoriki was the man selected by a group of Japanese businessmen to approach Hatoyama to arrange the transfer of certain shares in a deal that became one of the major economic scandals of the day. See Uhlan and Thomas, op. cit., p. 33, 100 ff., 173, 202. Also Arthur E. Teidman, "Big Business and Politics," in James N. Morely (ed.), *Dilemmas of Growth in Pre-war Japan*.

174 The historical experience with British construction in Meiji era: See *The Oriental Economist*, Apr. 1957.

174 ". . . magic key . . .": "Peaceful Uses of Atomic Energy," op. cit., p. 7.

175 The sketch of Homi Bhabha is based on Robert Anderson, *Building Scientific Institutions in India*; Jamshed Bhabha (ed.), *Homi Bhabha as Artist*; R. P. Kulkarni and V. Sarma, *Homi Bhabha*; Sir John Cockroft, "Homi Jehangir Bhabha," *Proceedings of the Royal Institution*, vol. 41, p. 142; M. G. K. Menon, "Homi Jehangir Bhabha," *Proceedings of the Royal Institution*, vol. 41, p. 430; C. Raghavan, "Nuclear Policy," *Mainstream*, New Delhi, July 23, July 30, Aug. 6, and Aug. 13, 1977; P. L. Raghu Ram, "Role of Nuclear Science in Economic Development," *Economic Times*, New Delhi, July 1977; *Illustrated Weekly of India*, Mar. 7, 1943.

176 In the early thirties, Bhabha was attracted by the exciting experiments of Lord Rutherford's Cavendish Laboratory at Cambridge, England. His father wanted the nineteen-year-old Bhabha to prepare himself for a senior post in the Tata family steel empire, but Bhabha begged to do physics, in particular the study of cosmic rays. His father agreed on condition that he gain a first-class degree in engineering before turning to physics. Bhabha complied with the family wishes. His pampered and secluded youth combined with an intensely passionate nature to produce a voraciously egocentric man. Conscious, above all, of his own feelings and desires, Bhabha seemed unable to relate to other people. He never married and had few close friends. The intensity of his feelings for art, music, and theoretical physics seemed to be a sublimation of emotions which he could not express in personal relations. His own painting, which he pursued throughout his life, displayed an inner melancholy, a sense of passionate foreboding. It was surrealist, but, he noted, "with no kidneys hanging out" (see *Illustrated Weekly of India*, op. cit.) and, overall, the bleak introspection of El Greco, whom he acknowledged as the greatest influence on his work. He displayed a preoccupation with form and structure that, despite the surrounding exuberance, revealed an inflexible mind. After all, other than music, Bhabha's greatest childhood passion was Meccano, a children's metal construction kit. (See Kulkarni and Sarma, op. cit., p. 5.) By the mid-fifties Bhabha had fulfilled his father's wishes and had become an engineer-administrator.

176 Bhabha at first shared the considerable enthusiasm in India for solar energy. In 1952 he explained the lack of attention to solar power in India: modern technology, he argued, had developed in Europe where solar energy was not an attractive option. The next burst of technological progress had come in the United States, which had abundant fossil fuel resources. Bhabha stressed that further development of solar energy was "urgent" for countries like India (see *Times of India*, Aug. 13, 1952).

India's main supporter of solar power in the early fifties was S. S. Bhatnagar, chairman of India's Council for Scientific Research. Bhatnagar died in 1955 just as the nuclear fever reached its peak. Bhabha never spoke about solar power again.

178 ". . . twice as much capital . . .": *Economic Weekly* (Bombay), Nov. 29, 1955. Bhabha's version of economic analysis prevailed in India. Because of Nehru's patronage, Bhabha was insulated from the normal checks and balances of democratic government. Other sections of the government simply did not ask questions about the Atomic Energy Commission's estimates of costs, or its projects and plans.

178 "I do not believe it": *Times of India*, Sept. 2, 1958.

CHAPTER 12

The description of the relationship between the AEC and the photographic industry draws on documents made available to the authors under the Freedom of Information Act. The cables and letters from which the quotations are drawn were brought together in a paper dated Jan. 17, 1952, "Report of the Director of Military Applications: Summary of Relations Between the AEC and the Photographic Industry."

The analysis of radiation protection standards and the research on the effects of radioactivity is drawn from: Virginia Brodine, *Radioactive Contamination*; Samuel Glasstone, *Sourcebook on Atomic Energy*; Samuel Glasstone and Philip Dolan, *The Effects of Nuclear Weapons*; Joint Committee on Atomic Energy (JCAE), *The Nature of Radioactive Fallout and Its Effects on Man*; JCAE, *Selected Materials on Radiation Protection Criteria*; JCAE, *Radiation Protection Criteria and Standards*; Karl Z. Morgan and J. E. Turner (eds.), *Principles of Radiation Protection*; National Academy of Sciences, *The Biological Effects of Atomic Radiation*; Jack Schubert and Ralph Lapp, *Radiation*; Lauriston S. Taylor, "History of the ICRP," *Health Physics*, vol. 1, 1958; Lauriston S. Taylor, *The Origins and Significance of Radiation Dose Limits for the Population*.

180 Truman's decision to use Nevada test site: Harry S. Truman, *Memoirs: Years of Trial and Hope*, p. 312. The President had asked the chairman of the AEC, Gordon Dean, about Nevada: "Can this be done in such a way that nobody will get hurt?" Dean, evading the question asked, replied, "Every precaution would be taken" (ibid., p. 312). The early apprehension about the political consequences of establishing a test site within the United States proved groundless. In Mar. 1949, one AEC commissioner stated flatly that a test site within the United States was politically unacceptable except in the case of a national emergency. The Korean War provided the emergency. Meeting on Nov. 14, 1950, a few

weeks after Chinese troops crossed the Yalu, the National Security Council formally recommended a continental location. Speaking for the historical record, the NSC said that it was necessary in case of an emergency, because an enemy might invade the Pacific test sites; it was also time-consuming and expensive to continue testing in the Pacific. Nevada was cheaper.

181 "Good bangs . . ." and "Bigger bombs . . .": Daniel Lang, *From Hiroshima to the Moon*, p. 281.

181 ". . . keep them confused . . .": *New York Times*, Apr. 20, 1979, p. 1.

181 Radiation Effects Study, U.S. AEC. Letter from Dr. Shields Warren, director, Division of Biology and Medicine, to Carroll Wilson, Nov. 23, 1949. Doc. No. AEC 278 12/12/79. Part of Project Gabriel on bomb debris hazards. Declassified June 19, 1979.

185 "After the action of the rays . . .": Jack Schubert and Ralph Lapp, *Radiation*, pp. 19–20.

185 The United States was not alone in its fascination with radium as an all-purpose cure. Other countries also had their share of radioactive medicines and beauty creams. For French advertisements from the 1930s see Yves Lenoir, *Technocratie Française*, pp. 69 ff.

186 The description of the radium dial painters case draws heavily on Lang, op. cit., chap. 24.

186 ". . . pictures of moonlight . . .": Ibid., p. 385.

187 ". . . doughboys . . .": Ibid., p. 401.

187 *Time* quote: Ibid., p. 393.

188 The roentgen unit is the standard measurement that applies to X or gamma rays. Different measurements are used for the effects of radiation. A "rad" (radiation absorbed dose) measures the actual radiation absorbed by the object being penetrated. Yet another unit, the "rem," is used to measure the effect of the absorption. The "rem" stands for "roentgen equivalent man." It is the precise unit for biological damage caused by radiation. One thousandth of a "rem" is a "millirem." In order not to confuse the reader the simple unit "r" will be used throughout this book, although it may refer to roentgens, rads, or rems. The exception will be when reference is made to the smaller unit, the millirem. This will be used in full.

188 ". . . those dial painters . . .": Lang, op. cit., p. 388.

189 The description of the test at Alamogordo draws on Lansing Lamont, *Day of Trinity*, pp. 236–42.

189 "a Hearst propagandist?": Ibid., p. 127.

189 Monitor not allowed to leave camp: Ibid., p. 252.

190 ". . . must be respected": Samuel Glasstone, *Sourcebook on Atomic Energy*, p. 735.

191 The central role of the geneticists in the decision-making process that

led to new standards of 1956 has been emphasized by health physicists who participated in the international deliberations. See G. K. Falia testimony before the JCAE, *Radiation Protection Criteria and Standards*, pp. 210–11, and Lauriston S. Taylor, "History of the ICRP," pp. 64, 66. The geneticists were the first to propose that any amount of radiation, no matter how small, was dangerous. Their theories told them that a mutated gene would show up sooner or later as a physical defect. During the Manhattan Project the radiation limits were known as "tolerance doses," but the views of the geneticists made the radiation protection community look for a new concept which did not imply a threshold, which they called "permissible doses." See Karl Z. Morgan in Karl Z. Morgan and J. E. Turner (eds.), *Principles of Radiation Protection*, p. 51. The geneticists' alarm grew in 1951 and 1952 as the first results came from the mice experiments at Oak Ridge. See Richard Hewlett and Francis Duncan, *Atomic Shield*, pp. 506–09; Schubert and Lapp, op. cit., pp. 188–91. The geneticists activated the committee of the ICRP—which had only just set new occupational standards in 1950. When the ICRP met in 1952 in Stockholm the deliberations were dominated by geneticists who pressed their colleagues to adopt a standard for public and not just occupational exposures. At Stockholm American experts were proposing a limit of 20r average per capita dose, whereas British experts were saying it should be as low as 3r. No agreement proved possible at this meeting, though a compromise figure of 10 rads emerged in informal talks.

Of the experts, the geneticists had played an increasing role from the early fifties in pressing for lower standards. They had fought for two things: first, a total permissible cumulative radiation dose during the peak child-bearing years from twenty to thirty; and, second, a standard that applied to the population as a whole, not just to radiation workers. They achieved little progress until the 1955 Atoms for Peace conference in Geneva, which emphasized the huge expansion in the use of manmade radioactive products. The geneticists warned of the possibility of an increase in radiation-induced mutations. Under the .3r per year standard of the day, a radiation worker could be exposed to a total dose of 600r during a forty-year working life. If administered at one time, this would be fatal, but over forty years it was estimated that there would be no ill effects. The geneticists considered, however, that the standard still allowed too much exposure during the child-bearing years, and they managed to force a new standard of 50r accumulated dose up to the age of thirty. Thereafter, the dose could rise to 100r at age forty and 200r at age sixty. In effect, they cut the old standard by one-third, and the ICRP agreed on an occupational exposure level of 5r per year. The geneticists had won a battle, but lost a war. In addition to the new standard, they had pressed for an entirely new approach to setting radiation limits; they wanted an average permissible amount for the whole population. Thus, if the number of workers in the nuclear power industry grew, as was expected, the exposure standards would have to be lowered to keep the average the same. The geneticists saw this as the best way of limiting the total number of mutated genes in the broad stream of human heredity. But, in an act that can only be regarded as professional cowardice, the ICRP refused to place limits on the amount of radiation produced by the medical profession. Doctors, it was argued, were responsible enough to set their own standards. Radiologists had always resisted any regulation of their activities on the basis that they knew best, and they had continuously rejected suggestions that individuals should be encouraged to keep a record of their own X rays—despite acknowledged widespread abuse of X-ray equipment. The ICRP found the idea of placing restrictions on the medical profession unacceptable: it was a "matter of practical necessity" that medical radiation exposure should be considered separately. For many years X-ray exposures—not just by radiologists, but also by dentists, beauticians, and from X-ray machines in shoe stores—remained the most significant source of radiation to the general public. It would not only be unregulated; it would not even be counted.

By rejecting the geneticists' case for an average population dose, the health physicists were left without a scientific basis for determining levels of permissible public exposure. Instead, they adopted a rule of thumb. As early as 1949 the American National Committee on Radiation Protection had suggested that, because of the known greater sensitivity of children to radiation, the exposure of the general public should be lower than the standard regarded as acceptable for radiation workers. The figure chosen was one-tenth and this soon hardened into a principle of calculating radiation protection. Thus the new level of occupational exposure—of 5r per year—was reduced to .5r for maximum exposures for the public.

The ICRP then superimposed yet another rule of thumb on the .5r. As it was unlikely, they argued, that any citizen would be exposed to more than three times the standard by lowering the standard to one-third—making it .17r instead of .5r per year—the acceptable maximum of .5r would not be exceeded in any case. The .17r standard for public exposure, excluding background and medical radiation, remains in force. The consensus among health physicists was that the health effects of such levels of exposure were low enough to be acceptable. That was not a scientific judgment; it was a political one.

192 Los Alamos Seminar, Sept. 1950, LAMS-1173. Declassified Jan. 29, 1979.

192 "extreme uncertainty": Los Alamos Seminar, p. 23.
192 ". . . the probability that people will receive . . .": Ibid.
194 The material on the effects of the Nevada tests, and specifically the trouble with the Upshot-Knothole series, draws heavily on a series of original documents made available during the spring of 1979 by the Department of Health, Education and Welfare. The authors have made extensive use of this documentation but especially of two reports: "Report of the Public Health Service in the Offsite Monitoring Program," Nevada Proving Grounds, Spring 1953, U.S. Department of Health Education and Welfare; Memorandum from Nathan H. Woodruff to A. R. Luedecke, "Radiation Protection Guides for Nuclear Weapons Testing," U.S. AEC, Jan. 11, 1963. The release of these documents and the accompanying controversy was covered in the press, notably the Washington *Post* and the *Deseret News* of Salt Lake City, Utah. See Washington *Post*, Dec. 8, 1978, and Jan. 8, Feb. 16, Mar. 2 and 3, Apr. 14, 19, 20, 24, and July 2, 1979. Also *New York Times*, May 13, 1979.
195 "Effects on the Population," in Woodruff-Luedecke memorandum subsection, Lt. Col. R. P. Campbell, Jr., to Brig. Gen. K. E. Fields.
196 Harry test report: In statistical consideration in field studies on thyroid disease in schoolchildren in Utah-Arizona, Dec. 3, 1965, pp. 3–4.

CHAPTER 13

198 The description of Dr. Ronald Richter's fraud is based on an article in *Pagina Un*, Aug. 1, 1967, "Buenos Aires, A History of Perónism"; *New York Times*, Mar. 27, 28, Apr. 1, May 24, 1951; Dec. 5, 1952; and Sept. 18, 1954; *Time*, Dec. 5, 1952. Interview, Jorge A. Sabato, Argentine AEC, June 1979. The Soviet reaction is described in Manfred von Ardenne, *Ein Glückliches Leben für Technik und Forschung*.
198 ". . . drown him . . .": *Pagina Un*, op. cit.
199 ". . . ask your forgiveness . . .": Ibid.
200 "Fantasy and scientific reality . . .": Manfred von Ardenne, op. cit., p. 215.
200 ". . . bull of the Pampas": *New York Times*, Mar. 27, 1951. Article by William Lawrence.
200 "slightest chance": "Lilienthal Scouts Claim," *New York Times*, Mar. 28, 1951.
200 "The Argentine claim . . .": William Lawrence, op. cit.
200 "That's the secret . . .": *Pagina Un*, op. cit.
200–202 The formative years of American nonproliferation policy in the mid-fifties have received little attention from scholars. Few have sought to penetrate the public rhetoric to assess the covert objectives that shaped

the international nuclear regime. This analysis has drawn on a number of documents released after Freedom of Information Act requests by the authors as well as interviews, especially with Phillip Farley, Myron Kratzer, Alan Labowitz, and Harold Bengelsdorf. In addition, Bernhard Bechhoefer, "Negotiating the Statute of the IAEA," *International Organization*, vol. 13, 1959; Bernhard Bechhoefer and Eric Stein, "Atoms for Peace," *Michigan Law Review*, vol. 55, 1957; Bernhard Bechhoefer, "Historical Evolution of International Safeguards," in Mason Willrich (ed.), *International Safeguards and Nuclear Industry*; Warren Donnelly, et al., *Western European Nuclear Energy Development*; Bertrand Goldschmidt, "The Origins of the International Atomic Energy Agency," IAEA Bulletin, vol. 19, Aug. 1977; Arnold Kramish, *The Peaceful Atom in Foreign Policy*; Harold L. Nieburg, *Nuclear Secrecy and Foreign Policy*.

A number of specific facts are guardedly revealed by Myron Kratzer, America's longest-serving representative in the international negotiations, in a section he wrote for the Atlantic Council report, *Nuclear Power and Nuclear Weapons Proliferation*, part 3.

202 No one had seriously grappled with the problems of nuclear proliferation before Eisenhower delivered his Atoms for Peace speech in Dec. 1953. The speech, after all, was primarily a disarmament proposal directed at the Soviet Union: a plan for an international pool of fissionable material to be supplied by the U.S. and Russia, which would deplete their stocks available for bombs and, instead, offer it to the rest of the world for the development of peaceful nuclear reactors. When the Russians rejected this idea, the U.S. was left to carry out the plan on its own. The first step was the creation of an International Atomic Energy Agency—which Eisenhower had proposed would distribute and regulate the fissionable material for peaceful uses. The second step was for the U.S. to arrange bilateral agreements of cooperation with the countries that wanted to take up atoms for peace. The 1954 amendments to the McMahon Act approved both kinds of collaboration—multilateral and bilateral—on one condition: a declaration would be for peaceful purposes only.

Anything more than this paper declaration would require some sensitive negotiations. The U.S. would, in effect, have to persuade other nations to accept restrictions on the military uses of the atom which it could no longer conceive for itself—the breakdown of the Geneva disarmament talks had made that abundantly clear.

202 U.S. National Security Council memorandum NSC-68, pp. 40–43.

202 "Atomic colonialism" was an accusation made by Homi Bhabha, the head of the Indian delegation in the negotiations for the statute of the IAEA. His most strenuous objections were against the provision in the United States draft of the statute that the international agency should be

empowered to monitor the growth of the stockpile of fissionable material, like plutonium, and limit the amount that any nation could accumulate. The United States proposed that the agency rather than the nation should decide how much it was legitimate for any country to accumulate for its stated peaceful program of plutonium use, like breeder reactors. If the Americans had stuck to their proposal the entire international system of safeguards would have been much more creditable, because it ensured that no nation would be able to accumulate a stockpile of weapons-grade material that could quickly be transformed into bombs.

Led by the Indians, the fight over the safeguards clause threatened to deadlock the IAEA Statute Conference in which the United States had invested a lot of diplomatic prestige. According to Bertrand Goldschmidt, the head of the French delegation at the conference, the American delegation remained firm until the last weekend of the conference when, after top-level consultations with Strauss and Dulles, they caved in. The explicit reference to the Agency's deciding how much plutonium a country could keep was deleted. Everyone understood that in practice this meant that the countries would decide that for themselves. The final "form of words" did not say so; there was simply no reference to who would decide. In theory, therefore, it was an objective question, something that future American negotiators could force an account of. The secrecy of the American delegation about its real intentions permitted the system to be created without reference to the need to limit the stockpiles of weapons-grade material. They reconciled themselves with the prospect of raising the matter in the future—admittedly by tortuous interpretation. Accordingly, they did not regard it as a "defeat" for the United States, but it was. The American negotiators then forgot about it, and because no one had publicly stated the objective, they were never called to account.

203 The sketch of Gerard Smith is based on the above interviews and *Current Biography*, 1970; Nomination of Gerard C. Smith to be director of the Arms Control and Disarmament Agency, 91st Congress, U.S. GPO, Washington, D.C., 1969.

203 The existence of a Western Suppliers Group did not become public knowledge until 1975. It existed in the fifties, however. The authors discovered it after interviews with representatives from five different states who had attended the meetings. It was confirmed by detailed Freedom of Information requests to the United States Department of Energy, responsible for the old AEC files. The primary job of the group was to oversee the development of the international safeguards system, and every new development in this field was first referred to it in order to ensure a united front at the international negotiations that followed. After a series of technical meetings during 1957 and 1958, held mainly in Ottawa,

a full safeguards system for the international agency was finally negotiated in London at the end of Feb. 1959. This meeting consisted only of the Anglo-Saxons: the U.S., the U.K., Canada, South Africa, and Australia. The "London Proposals" drafted at this meeting became the basis of the IAEA's first safeguards system. Later in 1959, France, Belgium, and Portugal were invited to join the club, which they did. In the years ahead the secret group was activated by the Americans whenever they believed a new step was needed with regard to safeguards. The technical basis for the group's work drew heavily on the list of prohibited exports to Communist countries known as the COCOM list. Under American leadership, all the NATO countries and Japan formed a group known as the Coordinating Committee for Export Controls to Communist Countries, which harmonized the export restriction to the Communist bloc of the Western allies.

204 The Russians were never reconciled to the American assertion that nonproliferation could be ensured by mere inspections. Their position had been made clear as early as Sept. 1955 when, under the cover of the first Geneva Atoms for Peace Conference, a top-level meeting was held. An American scientific delegation, with Gerard Smith as chief political adviser, tried unsuccessfully to convince a Russian delegation over four days of talks that the new concept of "safeguards" through inspection was enough.

204 The description of the Chinese nuclear program is based on: Lewis A. Frank, "Nuclear Weapons Development in China," *Bulletin of the Atomic Scientists*, Jan. 1966; Harry Gelber, "Nuclear Weapons and Chinese Policy," *Adelphi Papers*, no. 99, London, 1973; Morton Halperin, *China and the Bomb*; "Ch'ien San-ch'iang," in Donald W. Klein and Anne B. Clark, *Biographical Dictionary of Chinese Communism 1921–65*, pp. 188–90; Michael S. Minor, "China's Nuclear Development Program," *Asian Survey*, 1977, p. 571; William L. Ryan, *The China Cloud*; Richard P. Suttmeier, *Research and Revolution*; Donald P. Whitaker et al., *Area Handbook for the People's Republic of China*; Richard Wilson, *Anatomy of China*.

205 ". . . A-bomb prototype . . .": *Khrushchev Remembers*, p. 269.

205 "We had to draw the line . . .": Ibid., pp. 268–69.

205 "The Soviet government unilaterally tore up . . .": *China Quarterly*, April–June 1964, p. 88.

206 "cosmopolite . . .": Ryan, op. cit., p. 164.

206 The sketch of Nieh Jung-chen is based on: Harrison Forman, *Report from Red China*; Haldore Hanson, *Humane Endeavour*; "Nieh Jung-chen," in Klein and Clark, op. cit.; Edgar Snow, *Red Star Over China*; William H. Whitson, *The Chinese High Command*, 1973.

207 ". . . chopped off their tails": Forman, op. cit., p. 136.

207 ". . . best dressed man . . .": Hanson, op. cit., p. 249.

209 "Learning from the Soviet Union . . .": Wilson, op. cit., p. 155.

209 "We should and absolutely can master . . .": *China Quarterly*, April–June 1964, p. 163.

211 The sketches of Strauss and Nakasone are based on: Thomas Dalberg, *Franz Josef Strauss*; Erich Kuby et al., *Franz Josef Strauss*; S. Martin-Endinghaus, "From the Atomic Egg to Brockdorf," *Bild der Wissenschaft*, Nov. 4, 1977; Anthony Sampson, *The New Europeans*; John Emmerson, *Arms, Yen and Power*; Takimoto Michio et al., "Nakasone Yasuhiro No Nongen Kenkyu" (Personal Research about Nakasone Yasuhiro), *Sunday Mainichi*, Sept. 2, 1976; *Japan Times*, July 25, 1959, Sept. 11, 1970, Nov. 29, 1973, May 13, 1976, Aug. 19, 1976, Apr. 7, 1977, and May 4, 1978; *Yomiuri*, Oct. 12, 1963; May 18, 1970; *Asahi Evening News*, Sept. 18, 1974.

213 Nakasone at war veterans meeting: *Nippon Times*, Dec. 9, 1954.

213 ". . . atom-conscious . . .": Martin-Endinghaus, op. cit., p. 114.

214 "linguistics misunderstanding": *New Statesman*, Apr. 12, 1958, p. 460.

214 Strauss has issued a variety of conflicting denials and explanations of his discussions with Chaban-Delmas in 1957 and 1958. There seems no doubt, however, that nuclear weapons were discussed in the context of overall Franco-German collaboration in the development of military technology. The final decision to actually proceed with the French bomb, of which Chaban-Delmas was a strong advocate, had not then been taken. The available information on these talks is gathered in Catherine Kelleher, *Germany and the Politics of Nuclear Weapons*, pp. 149–51, 344; Wilfred L. Kohl, *French Nuclear Diplomacy*, pp. 54–61.

214 Nakasone never conformed to the stereotype of a "right wing" politician: he supported a positive policy toward China, he criticized the continued presence of some U.S. bases, he urged the return of sovereignty over Okinawa, and he advocated a more normal treaty of friendship to replace the special U.S.-Japan Security Treaty. The common element was his clear perception of his patriotic duty to enhance Japan's independence and self-reliance. Later, as he manipulated his faction through the convoluted politics of Japan's ruling party, he was called a "political weathercock." His changes of allegiance and direction, designed to maximize his immediate political advantage, were said to show the way the political wind was blowing. He had matured considerably from his stormy days as a Young Turk, but he was not a simple opportunist. The strength of his patriotism was a constant factor, and in that regard it stood alone. Everything else was negotiable. Everything else could be weighed coolly and rationally to assess the balance of advantage. Nakasone was from the same mold of Japanese statesmanship as the men who, in 1941, carefully weighed all the pros and cons and decided that on balance Japan had to go to war.

Nakasone learned nothing from Japan's defeat in World War II or from its postwar interdependence with the world economy. His nationalism was a throwback to an earlier era. This was most clearly revealed in the way, as a candidate for the Diet, he outlined his background and policies in the formal electoral literature. One of the spaces is reserved for a list of people whom the candidate admires most. Nakasone gave two names. The first was Tokugawa Ieyasu, the great Shogun who completed the unification of Japan, began the process of driving out all foreign influences, and established a dynasty that would rule in isolated grandeur for two and a half centuries. The second was Saigo Takamori, known as "Saigo the Great." He was one of the founders of the Meiji Restoration but split with the new government over its modernization program and its refusal to invade Korea. He led a rebellion in the name of the traditional values and prerogatives of the samurai class. The spiritual dimension of his treachery would be invoked by the militant factions of the 1930s (Takimoto et al., op. cit.).

214 Nakasone and AEC appointments. See Sakaguchi Akira, *Ishikawa Ichiro*, p. 201.

215 "Since the scientists were too immobile . . .": Hideo Sato, "The Politics of Technology Importation in Japan: The Case of Atomic Power Reactors," unpublished paper (mimeo), Yale University, New Haven, 1978.

215 Nakasone and submarines: *Yomiuri*, May 18, 1970.

CHAPTER 14

217 Screwworm fly: Samuel Glasstone, *Sourcebook on Atomic Energy*, p. 726. See also Glenn T. Seaborg, *Science, Men and Change*, p. 90.

219 The description of the nuclear airplane draws on: *Review of Manned Aircraft Nuclear Propulsion Project*, Comptroller General of the United States, Washington, D.C., Feb. 1963; W. Henry Lambright, *Shooting Down the Nuclear Plane*; H. Peter Metzger, *The Atomic Establishment*; Walter Patterson, *Nuclear Power*; 1976; Roger Williams, *The Nuclear Power Decisions*; Herbert York, *Race to Oblivion*; interview with Lord Hinton, London, May 2, 1979.

219 Use of older pilots in nuclear airplanes: See Metzger, op. cit., p. 204.

219 "test fly": Ibid., p. 206.

220 ". . . worldwide shipping": Seaborg, op. cit., p. 83.

222 "the final reward . . . immeasurable": British government publication, 1955 (Cmd 93989), a program for nuclear power.

223 "tremendous confidence": Ibid.

223 ". . . evoke fantasies": London *Times*, Feb. 6, 1957. See also Williams, op. cit., p. 82.

223 ". . . national situation depends on it": Williams, op. cit., p. 82.

223 ". . . danger in its train . . .": "Warning on Development of Nuclear Energy," London *Times*, Sept. 20, 1955.

224 ". . . enlarged program . . .": Hinton interview, London, 1979.

224 "pig-headed and bloody-minded": Ibid.

224 The sketch of Lord Hinton is based on Margaret Gowing, *Independence and Deterrence*, vol. 2, p. 7; authors' interview with Lord Hinton, London, 1979; Profile, London *Observer*, Feb. 14, 1965; Anthony Sampson, *Anatomy of Britain*, p. 547; Leonard Berten, *Atom Harvest*, p. 116.

225 ". . . you buggers . . .": Profile, London *Observer*, op. cit.

225 The description of the Windscale fire is based on Walter Patterson, op. cit., p. 162.

226 Unlike their English and Soviet colleagues, the U.S. policymakers had no centralized government electricity system that could pursue formal targets. However, the informal projections that came from the staff of the Atomic Energy Commission reflected the spirit of optimism that had become universal.

226 The projections by the U.S. Atomic Energy Commission are found in Mullenbach, op. cit., pp. 59, 102. The turning point in confidence about nuclear power in 1957 is shown in the titles of the reports. The optimistic cost figures presented by United States delegates at Geneva in 1955 were largely based on a 1953 report called Project Dynamo. A new independent appraisal of costs in 1957 showed the unrealistic nature of the Dynamo figures. The report was called "Project Size-up" (Mullenbach, op. cit., pp. 59–60). An AEC advisory committee noted in 1959 that "a cost estimator's confidence in his figures usually varied inversely with the degree of actual technological experience" (ibid., p. 59).

227 The conflict between the AEC and the Joint Committee on Atomic Energy is described in the following: Clinton P. Anderson and Milton Viorst, *Outside in the Senate*; Frank G. Dawson, *Nuclear Power*; Mullenbach, op. cit.; Lewis Strauss, *Men and Decisions*; Harold L. Nieburg, "The Eisenhower AEC and the Congress," *Midwest Journal of Political Science*, vol. 6, May 1962; Aaron Wildavsky, *Dixon-Yates*.

The conflict between Congress and the AEC led to the first full-scale challenge of a nuclear plant proposal on safety grounds. As part of the promotion of private power, Strauss had enthusiastically backed the breeder project near Detroit pushed by Walker Cisler, head of Detroit Edison. However, the Advisory Committee on Reactor Safeguards had criticized the safety of the plant. The report was leaked to Clinton Anderson by a dissident AEC commissioner, and Strauss reacted by trying to keep the report secret. Anderson attacked the AEC licensing process as "star chamber proceedings." When the controversy became public a Detroit group, financed by the United Automobile Workers Union, began the first full-scale legal attack to stop a reactor. The legal challenge failed, but Anderson laid the basis for many future challenges when he wrote into the 1957 legislation limiting insurance liability a new set of licensing requirements which ensured that all future hearings and reports had to be public.

228 ". . . international considerations": Mullenbach, op. cit., p. 306.

228 The description of the foundations of Euratom and the report of "The Three Wise Men" is based on: Irvin C. Bupp and J. C. Derian, *Light Water*; Warren Donnelly, *Commercial Nuclear Power in Europe*; Bertrand Goldschmidt, *Rivalités Atomiques*; Henry R. Nau, *National Politics and International Technology*; Lawrence Scheinman, *Atomic Energy Policy in France Under the Fourth Republic*.

229 "These three plants . . .": Bupp and Derian, op. cit., p. 40.

230 Germany's Eltville Plan is discussed in: Wolfgang Cartellieri et al., *Taschenbuch für Atomfragen*; Paul L. Joskow, "Research and Development Strategies for Nuclear Power in the United Kingdom, France, and Germany," unpublished paper (mimeo), Boston, 1976; S. Martin-Endinghaus, "From the Atomic Egg to Brockdorf," *Bild der Wissenschaft*, Nov. 4, 1977; Nau, op. cit.; K. Pruss, *Kernforschungspolitik in der Bundesrepublik Deutschland*; J. Pretsch, "Ten Years of Nuclear Energy Policy in the Federal Republic of Germany," in J. Sabotta (ed.), *State, Science and Economy as Partners*; Jurgen Peter Pesch, "Staatliche Forschungs und Entwicklungspolitik"; Karl Winnacker and Karl Wirtz, *Nuclear Energy in Germany*.

230 The peaking of the period of optimism in Japan is based on: Rodney L. Huff, "Political Decision Making in the Japanese Civilian Atomic Energy Program," unpublished doctoral dissertation; Japan Atomic Industrial Forum, *Nihon No Genshiryoku*; Hideo Sato, "Politics of Technology Importation in Japan," unpublished paper (mimeo), Yale University, New Haven, 1978; Kondo Kan'ichi and Osani Hiroshi (eds.), *Sengo Sangyo Shi*. Also numerous interviews in Japan, especially those conducted with Arisawa Hiromi, Murata Hiroshi, Ipponmatsu Tamaki, Nasu Hayao, Mukaibo Takashi, Kawakami Koichi, and Tsutsumi Yoshitatsu. For the historical experience with British construction in the Meiji era, see *The Oriental Economist*, April 1957.

230 The Soviet Plans for nuclear power are described in detail by: Philip Sporn, *Vistas in Electric Power*, vol. 3. The 1959 speech by the Minister for Power Stations is noted by Sporn on pp. 982, 991. The increase in scale of proposed plants is noted in George Modelski, *Atomic Energy in the Soviet Bloc*, especially p. 119. The criticism by Frol Koslov of the scientists' misleading cost estimates is in Arnold Kramish, *Atomic Energy in the Soviet Union*, p. 145.

231 ". . . radiation cataracts . . .": *Trud* (Moscow), Dec. 19, 1968.

231 The description of the Kyshtym disaster draws particularly on the work of Zhores Medvedev, "Two Decades of Dissidence," *New Scientist*, vol. 72, no. 1025, 1976, p. 264; "Facts Behind the Soviet Nuclear Disaster,"

New Scientist, vol. 74, no. 1058, 1977; Zhores Medvedev, *Nuclear Disaster in the Urals*. Further material was drawn from: Andrew Cockburn, "The Nuclear Disaster They Didn't Want to Tell You About," *Esquire*, Apr. 25, 1978; British Granada TV, World in Action, "The Accident" Transcript, Manchester, Nov. 7, 1977; London *Times*, Feb. 8, 1977, Dec. 8, 1977, and Apr. 2, 1977; *New York Times*, Dec. 9, 1976, and Nov. 26, 1977; Washington *Post*, Nov. 17, 1976, Nov. 26, 1977, and May 25, 1979.

232 "diplomatic sources" . . . "no intelligence": *New York Times*, April 14, 1958, and April 16, 1958.

232 Radio Moscow report January 9, 1958. Institute for the Study of the USSR, Soviet Affairs Analysis Service No. 53, May 13, 1958.

233 "rubbish": Medvedev, *Nuclear Disaster*, p. 5.

234 ". . . particularly dangerous contingent": Alexander Solzhenitsyn, *The Gulag Archipelago*, p. 407.

235 ". . . The land was dead . . .": *New York Times*, Dec. 9, 1976.

235 "graveyards of the earth" . . . "giant mushrooms": Cockburn, op. cit., p. 43.

236 The importance of safety considerations in addition to costs is stressed by later reports of Soviet nuclear advocates. Kurchatov's successor A. P. Alexandrov admitted in 1962 that during the first plan "practically all stations turned out to be more expensive than planned . . . due mainly to the inexperience of industry and changes in plans, but not due to technical reasons." The "changes in plans" probably referred to the technological gamble of substantially increasing the scale of the plants. Alexandrov asserted, however, that "the main problem is the development of safe methods for utilizing nuclear energy. Its cost is the second most important problem" (*Ekonomiskaya Gazetta*, no. 12, May 19, 1962). In 1965 another key Soviet nuclear bureaucrat, Georgi Yermakov, displayed all the exasperation of a prophet in the wilderness that characterized nuclear advocates throughout the world. After referring to the doubts about reactor safety, he said: "Now no one has any doubts about the reality and complete safety of receiving electric energy by splitting atomic nuclei. Now the 'opponents' of the construction of atomic electric power stations (and there are those among the electric power specialists) raise new objections; atomic energy is too expensive and, therefore, such power stations cannot play an essential role in the development of our power system" (*Ekonomiskaya Gazetta*, no. 29, July 21, 1965).

236 Khrushchev's handling of the Soviet technocrats is outlined in Jeremy Azrael, *Managerial Power and Soviet Politics*; Wolfgang Leonard, *The Kremlin Since Stalin*; Roy A. Medvedev and Zhores Medvedev, *Khrushchev*.

CHAPTER 15

239 The sketch of Edward Teller is based on Stanley Blumberg and Gwinn Owens, *Energy and Conflict* (and reviews of this biography in *Science*, vol. 194, Oct. 29, 1976; *Commentary*, Apr. 1977, *Bulletin of the Atomic Scientists*, May 1978); Nuel P. Davis, *Lawrence and Oppenheimer*; Freeman Dyson, *Disturbing the Universe*; Robert Jungk, *Brighter Than a Thousand Suns*; John Major, *The Oppenheimer Hearing*; Norman Moss, *Men Who Play God*; Philip M. Stern, *The Oppenheimer Case*; Edward Teller and Allen Brown, *The Legacy of Hiroshima*; Edward Teller, "The Work of Many People," *Science*, vol. 121, Feb. 25, 1955; Stanislaw Ulam, *Adventures of a Mathematician*; Herbert York, *The Advisors*; Herbert York, "The Origins of the Lawrence Livermore Laboratory," *Bulletin of the Atomic Scientists*, Sept. 1975.

239 ". . . the Hungarians . . .": Ulam, op. cit., p. 164.

240 ". . . did not want to cooperate . . .": Jungk, op. cit., pp. 269–70.

240 ". . . monomaniac . . .": Ulam, op. cit., p. 147.

240 "a visual and also an almost tactile . . .": Ibid.

241 ". . . the sole promoter . . .": Ibid., p. 210.

241 "Ulam triggered nothing" and ". . . under the patent . . .": Blumberg and Owens, op. cit., p. 280.

241 The sketch of Andrei Sakharov is based on: Andrei Sakharov, *Sakharov Speaks* (foreword by Harrison Salisbury); Andrei Sakharov, *Alarm and Hope*; Nikita Khrushchev, *Khrushchev Remembers*; Zhores Medvedev, *Soviet Science*; John Barry, *Sunday Times* (London); *Time*, Aug. 2, 1968; *Newsweek*, Aug. 5, 1968; *New York Times*, Apr. 14, 1971.

241 ". . . atmosphere of decency . . .": Sakharov, *Sakharov Speaks*, p. 29.

241 "For the next eighteen years . . .": Ibid., p. 30.

242 "I had no doubts . . .": Ibid.

243 The description of the debate over nuclear tests draws on: Robert A. Divine, *Blowing on the Wind*; Robert Gilpin, *American Scientists and Nuclear Weapons Policy*; Morton Grodzins and Eugene Rabinowitch (eds.), *The Atomic Age*; Harold K. Jacobson and Eric Stein, *Diplomats, Scientists and Politicians*; George B. Kistiakowsy, *A Scientist at the White House*; Daniel Lang, *From Hiroshima to the Moon*; Norman Moss, *Men Who Play God*; Harold L. Nieburg, *In the Name of Science*; George Quester, *Nuclear Diplomacy*; Herbert York, *The Advisors*; Herbert York and G. Allen Greb, "The Comprehensive Nuclear Test Ban," unpublished paper, San Diego, 1978.

244 The description of the Bravo test and the *Lucky Dragon* incident draws on: Robert Divine, op. cit.; Lang, op. cit.; Moss, op. cit.; York, *The Advisors*; Ralph Lapp, *The Voyage of the Lucky Dragon*; Roger Rapoport,

The Great American Bomb Machine; Samuel Glasstone and Philip Dolan, *The Effects of Nuclear Weapons.*

245 ". . . too important to be left to the generals . . .": Divine, op. cit., p. 18.

246 ". . . take out a city . . .": Ibid., p. 13.

246 ". . . strategically important city . . .": *Newsweek*, Apr. 5, 1954, p. 28.

246 ". . . Shall we put an end to the human race . . .": See Joseph Rotblatt, *Scientists in the Quest for Peace*, appendix 1.

247 ". . . a frightful row . . .": Moss, op. cit., p. 92.

248 ". . . the pure stream of human heredity": Jack Schubert and Ralph Lapp, *Radiation*, p. 199. The career of the American-born geneticist was marked by the politically controversial nature of his discipline. In 1932 he had gone to pursue his research at the Kaiser Wilhelm Institute in Berlin, only to be expelled by the Nazis the following year. A victim of censorship from the right, with its strongly held views on the importance of genetics, he was soon to be censored again from the left. From Berlin, Muller went to the then prestigious Institute of Genetics in Moscow. But he had to leave there too, in 1937, because of the emergence of left-wing biology under Trofim Lysenko, which denied the existence of genes. In 1948, with the charlatan Lysenko reasserting his control over Soviet science, Muller was expelled from membership of the Soviet Academy of Sciences for "activity aimed at harming the Soviet Union." Even in the United States, Lewis Strauss, chairman of the AEC, vetoed a paper Muller had prepared to give at the first Atoms for Peace Conference in Geneva. Nothing, Strauss thought, must be allowed to detract from the harmony of that event. At this stage Muller, perhaps conscious of his left-wing past at the height of the McCarthy purges, was being very cautious. His paper, independently published, emphasized the genetic dangers from X rays and noted that fallout from weapons tests could be justified in the interests of national security. Nevertheless his definite views of the nature of human beings had given Muller, like many geneticists, a distinctive social philosophy of humanitarianism and a sense of responsibility for the destiny of the human race. His early speeches ridiculed "the dangerous fallacy that what cannot be felt or seen need not be bothered with" and emphasized man's genetic inheritance as his "most irretrievable possession." Within a year Muller sharpened his attack and expressed increasingly pointed concern about the long-term hazards of radioactive fallout. By 1957, in congressional testimony, he called continued testing a "monstrous mistake of policy on both sides." Robert Divine, op. cit., pp. 52, 136.

248 ". . . a terrific explosion . . .": York, *The Advisors*, p. 92.

249 Pauling's estimates: Robert Divine, op. cit., p. 123.

250 ". . . no immediate hazard . . .": Ibid., p. 33.

250 ". . . ounce overweight . . .": Ibid., p. 187.

250 "The world is radioactive . . .": Lang, op. cit., p. 379. Libby supervised an AEC monitoring and assessment system for Strontium-90. In one of the most tasteless euphemisms of nuclear advocacy he called it "Project Sunshine." He adopted a new public relations–oriented unit for measuring quantities of Strontium-90, called "sunshine units." This was quickly denounced as propaganda by the Joint Committee on Atomic Energy during the hearings that destroyed many of Libby's assumptions about the lack of danger from the substance. Undaunted, Libby attempted to promote an antiradiation pill which included a heavy dose of bone-seeking calcium that might lessen human uptake of Strontium-90. This technological fix was never heard of again, and public concern grew as state government monitoring continued to show increases in Strontium-90 fallout.

251 "bone splinters . . .": Walter Patterson, *Nuclear Power*, p. 133.

251 ". . . temporary suspension . . .": Robert Divine, op. cit., p. 146.

252 ". . . fallout-free nuclear bomb . . .": York and Greb, op. cit., p. 18, Teller was shocked by one statement Eisenhower made during this meeting. The President asked whether it would be possible to share the secret of the "clean bomb" with the Russians. The implication was clear—that it would be better for the Russians to attack the United States with such a weapon—but Teller was appalled at the idea of giving any secrets to the Russians, especially secrets which might enable them to continue to perfect through testing while the United States stopped its own test programs. See Divine, op. cit., pp. 148–50.

252 The Carbon-14 problem had been raised during the original H-bomb debate in 1950. Robert Bacher, the former Los Alamos physicist, had raised the prospect of worldwide contamination as a result of merely testing the proposed H-bomb. Hans Bethe showed that testing alone would not cause the degree of Carbon-14 contamination that Bacher alleged. Pauling overlooked this earlier debate in making his calculations on the Carbon-14 problem. Whatever the harmful effects of tests, widespread use of the "clean bomb" during a war would cause a significant increase in Carbon-14 levels. The "clean bomb" debate is outlined in Robert Divine, op. cit.; York and Greb, op. cit.; York, *Race to Oblivion;* Freeman Dyson, "The Neutron Bomb," *Bulletin of the Atomic Scientists*, Sept. 1961; Nigel Calder, "Notes on the Neutron Bomb," *New Scientists*, no. 244, July 20, 1961; Freeman Dyson, "The Future Development of Nuclear Weapons," *Foreign Affairs*, vol. 38, Apr. 1960; Freeman Dyson, *Disturbing the Universe;* George Kistiakowsky, *A Scientist at the White House;* George Kistiakowsky, "Enhanced Radiation Weapons, Alias the Neutron Bomb," *Technology Review*, May 1978.

252 ". . . the so-called clean bomb . . .": Sakharov, "Radioactive Carbon in Nuclear Explosions and Nonthreshold Biological Effects," in A. V. Lebedinsky (ed.), *Soviet Scientists on the Danger of Nuclear Tests.*

252 "Beginning in 1957 . . .": *Sakharov Speaks*, p. 40.
252 "The total number of victims . . .": in Lebedinsky, op. cit., p. 48.
253 "underground propaganda": Medvedev, *Soviet Science*, p. 92.
253 The way nuclear physicists established havens for biology in their institutes is described in Mark B. Adam, "Biology in the Soviet Academy of Science," in John R. Thomas and Ursula M. Kruse-Vaucienne, *Soviet Science and Technology*.
253 ". . . stand-still agreement . . .": William Epstein, *The Last Chance*, p. 49.
253 The first suggestion of a test ban came from Vannevar Bush even before the Bravo test. Bush proposed to Dean Acheson that a deferral of the first H-bomb test could lay the basis for a test ban agreement. See Major, op. cit., p. 143. The reason behind this initiative remained obscure until 1976, when Herbert York, former director of the Livermore Laboratory, revealed one crucial fact. York noted that careful reading of the fallout from an H-bomb test could give knowledgeable scientists a clue about the special Reller-Ulam design trick which was the key to a thermonuclear weapon. Even the Mike test could have given the secret away. York, *The Advisors*, p. 100.
254 "I would be a slob . . .": *Sakharov Speaks*, p. 32. Khrushchev's memoirs substantially confirm Sakharov's account, although the former Premier telescopes three widely separated meetings with the scientist into one time frame and misleads in the process—much as he did with his suggestion that the Americans had broken the test moratorium in 1961, when the events he related occurred in 1958.
254 ". . . bureaucratic interests . . .": Ibid., p. 33.
254 "We're not holding you . . .": Ibid.
255 "Ethically speaking . . .": In Lebedinsky (ed.), op. cit., p. 49.
256 "God . . . sense of humor": Interview with Gertrud Weiss Szilard, San Diego, Apr. 1978.
256 ". . . plowshares": Ibid., p. 82.
257 ". . . impractical and immoral": Gilpin, op. cit., p. 280. Teller's political power base, especially with the U.S. Air Force, had been based on his discovery of big bombs. It was precisely Oppenheimer's advocacy of tactical weapons that brought down the wrath of the Air Force on him. Teller's act of tribal disloyalty during the Oppenheimer hearings had isolated him from his scientific colleagues. Now he was advocating the positions that Oppenheimer had taken a decade before.
258 ". . . Like a bunch of lunatics . . .": Gilpin, *American Scientists*, p. 269.
258 ". . . the doubtful honor . . .": Ibid, p. 291.

CHAPTER 16

263 "A rare find . . .": Peter Faulkner (ed.), *The Silent Bomb*, p. 224.

264 The emphasis on the fact that GE's Oyster Creek bid required no government subsidy was very misleading. General Electric had scorned any direct official subsidies for most of its bids, including Dresden 1 near Chicago, Humboldt Bay, and Bodega Bay in California. There was a self-fulfilling element in the propaganda, however. The public relations success reinforced GE's determination to attempt the achievement for which it had received credit after the Oyster Creek announcement.

265 The analysis of the corporate strategy of the two giant electric manufacturers in the sixties draws on the following: (1) General Electric: Ronald G. Greenwood, *Managerial Decentralization*; Alvin A. Butkus, "The GE Puzzle," *Dun's Review*, July 1970; Allan T. Demaree, "GE's Costly Ventures into the Future," *Fortune*, Oct. 1970; also two sketches of Fred Borch in *Current Biography*, 1971, p. 47, and *Forbes*, Feb. 15, 1969. (2) Westinghouse: John MacDonald, "Westinghouse Invents a New Westinghouse," *Fortune*, Oct. 1967; "A Dynamo in the Generator Market," *New York Times*, Sept. 28, 1969; "The Lustre Dims at Westinghouse," *Business Week*, July 20, 1974; Edmund Faltmeyer, "Westinghouse Comes Home to Electricity," *Fortune*, Aug. 1976; Donald Burnham in *Current Biography*, 1968, p. 71. (3) Miscellaneous: "Nuclear Power Goes Critical," *Fortune*, Mar. 1967; "Comparing Two Electric Giants," *The Magazine of Wall Street*, Nov. 9, 1957; "Nuclear Power Turns up the Juice," *Business Week*, Mar. 11, 1967; "Utilities Turn to the Atom," Washington *Post*, Aug. 24, 1970; "Neck and Neck—and Breathing Hard," *Forbes*, Feb. 1, 1973; "The Opposites: GE Grows While Westinghouse Shrinks," *Business Week*, Jan. 31, 1977; Philip Mullenbach, *Civilian Nuclear Power*; Frank Dawson, *Nuclear Power*; Irvin C. Bupp and Jean-Claude Derian, *Light Water*; Robert Perry et al., "Development and Commercialization of the Light Water Reactor, 1946–1976," RAND Corporation, R-2180-NSF, June 1977, Santa Monica, Calif.

266 ". . . Christian soldiers" and ". . . tears in your eyes . . .": *Fortune*, Oct. 1970, p. 91.

266 Borch, "to let the outfit know . . .": Ibid., p. 92. In 1969 Borch recalled how the strategy turned sour. "Our plan," he said, "was to first have nuclear power losses peak, then computers, then jet engines. Instead computers peaked in 1966 and then nuclear in 1968, and the losses were way above estimates." *Forbes*, Feb. 15, 1969, p. 53.

267 ". . . first in performance": *Fortune*, Oct. 1967.

267 ". . . haven't got a wire . . .": Ibid.

268 ". . . lump of butter . . .": *Fortune*, Oct. 1970, p. 93.

268 "to ram this thing . . .": Ibid. Just as GE launched its nuclear drive, the Westinghouse Division responsible for all sales to utilities—both conventional and nuclear—came under the control of the Westinghouse nuclear team. John Simpson, project manager of the *Nautilus* and, at first, of Shippingport, took over the division with his team of Rickover graduates. Naturally they were determined to capitalize on the head start their company had received in the nuclear field. They were especially proud of their nuclear expertise and poured scorn on the GE tendency to switch its executives around. "Nobody at GE headquarters had any nuclear experience," Simpson pointed out. "Even after the turnkeys, GE still had a transformer man responsible to be a switchgear expert." *Forbes*, Feb. 1, 1973.

269 The TVA deal draws on Tom O'Hanlon, "Atomic Bomb in the Land of Coal," *Fortune*, Sept. 1966; "Nuclear Power Goes Critical," *Fortune*, Mar. 1967; Philip Sporn, *Vistas in Electric Power*; Joint Committee on Atomic Energy: *Nuclear Power Economics 1962–1967*, Washington, D.C., U.S. GPO, 1968 (henceforth JCAE 1962–67); Bupp and Derian, op. cit.

269 The sketch of Sporn draws on JCAE 1962–67; Bupp and Derian, op. cit.; Sporn, op. cit.; *Time*, Mar. 12, 1956; *Business Week*, Oct. 30, 1965; *Current Biography*, 1966, p. 388.

269 "These manufacturers . . .": JCAE 1962–67, p. 8.

270 "come-on bid": Ibid., p. 7.

270 ". . . most charitable interpretation . . .": Ibid., p. 9.

270 "the Henry Ford of power": *Business Week*, Oct. 30, 1965.

271 "unduly conservative" and ". . . more optimistic": Bupp and Derian, op. cit., p. 47.

271 The sketch of Donald Cook draws on: *Current Biography*, 1952; *Business Week*, Oct. 30, 1965; *Dun's Review*, Dec. 1965; *Fortune*, Nov. 1969. Arturao Gandara, "Electric Utility Decision Making and the Nuclear Option," R-2148NSF, RAND, Santa Monica, Calif., June 1977.

271 "From the dawn . . .": Gandara, op. cit., p. 68.

271 "Revolution . . .": *Dun's Review*, op. cit.

271 ". . . a power supermarket": Ibid.

271 "restless, demanding . . ." and ". . . lot of challenge": *Business Week*, op. cit.

272 ". . . could not count on the nuclear plants . . .": Washington *Post*, Aug. 24, 1970.

272 ". . . delighted to have two . . .": *Business Week*, Nov. 17, 1975.

273 "Howe's boys": Canadian nuclear investments were determined by a very narrow group of individuals from the extraordinary personal empire of C. D. Howe. From November 1953 the chief executive of Atomic Energy of Canada, Ltd. was the most personal protégé the minister ever

had, William Bennet. He had worked as Howe's private secretary and when, in 1946, the thirty-five-year-old Bennet felt the need to strike out on his own, Howe gave him an Order of the British Empire and appointed him head of the nationalized uranium monopoly, Eldorado Mining and Refining Ltd. The other half of the Canadian nuclear equation was provided by Richard Hearn, who had been one of "Howe's boys" during the war. He was general manager of the Hydro-Electric Power Commission of Ontario, the nation's largest utility. Ontario Hydro had a strong sense of regional chauvinism and it could not draw on the huge undeveloped hydroelectric resources of neighboring Quebec. It was regarded, not simply as a foreign country but as a result of a political controversy during the great Depression about purchases from Quebec, it was regarded almost as enemy territory. Ontario Hydro turned to nuclear power because it offered provincial independence in the long term. Howe's empire survived his own electoral defeat in 1957—indeed it took an organic life of its own. In Feb. 1959 the new Canadian government abruptly canceled one of Howe's pet projects, an advanced jet fighter on which the high-technology aircraft industry was depending. The political controversy about the loss of jobs and the waste of hard-won know-how meant the government needed an immediate display of commitment to high technology. The government therefore precipitously approved the first large 200 Mw reactor at Douglas Point even though basic design work had only just begun and the proposed test reactor, the NPD, was still three years from completion. They might have lost the election, but "Howe's boys" were still running the show. On "Howe's boys" see: "W. J. Bennet," *Current Biography*, 1954, p. 78; Leslie Roberts, *C. D.: The Life and Times of Clarence Decatur Howe*, especially pp. 145–46; Wilfrid Eggleston, *Canada's Nuclear Story*, especially pp. 309, 311; John Porter, *The Vertical Mosaic*, especially pp. 430–32, 551; Merrill Denison, *The People's Power*, pp. 236, 250.

274 The description of the reactor choice decision in the U.K. during the mid-sixties is based on: Roger Williams, *The Nuclear Power Decisions*; Duncan Burn, *Nuclear Power and the Energy Crisis*; Walter Patterson, *Nuclear Power*; Walter Patterson, *The Fissile Society*; interview, Lord Hinton, London, May 1979; "Britain's GE," *Barron's*, Jan. 8, 1957.

275 "Ultimately everyone . . .": Hinton, *Three Banks Review*, Dec. 1961, pp. 3–18.

275 "ultimate arbiter of commercial policy": Williams, op. cit., p. 95.

275 The CEGB put it this way to a government committee inquiring into the electricity supply industry: ". . . the Atomic Energy Authority is primarily a research and development organization and is staffed by people who have been trained for these activities; the picture that they paint [of the current reactor debate] is naturally the view as seen by an enthusi-

astic research staff. The picture painted in the CEGB memoranda and evidence shows that same view as it appears to an organization carrying heavy financial and commercial responsibilities." When Hinton was asked if the AEA's R and D program was guided by the CEGB's requirements he replied, "I think that their activities are guided by what they think our requirements ought to be." See Patterson, *Fissile Society*, p. 25. The row between the two was deep and bitter.

276 ". . . the greatest breakthrough . . .": House of Commons debates, *Hansard*, May 25, 1965.

276 ". . . extrapolate data . . .": *Sunday Times* (London), June 27, 1978.

276 There was a need for greater explanation, however. How come, for example, the price of the LWR's in Britain was so much greater than it was in America? How come the fuel fabrication costs as quoted by the AEA were lower than U.S. prices? The details were not there—at least not immediately. The technical summaries of the seven tenders for the contract—including the LWR's—were, as Roger Williams has observed, "so brief, their account of the methodology of comparison so sketchy, as to be tantamount to asking the technical world to take the result virtually on trust. It was at best naïve of the Board to suppose that the specialist reader would be satisfied with the outline of the assessment process given in the summaries: Yet for what other readership were these summaries produced?" Williams, op. cit., p. 126.

277 The conflict over reactor choice in France between the atomic bureaucracy and the nationalized utility is well documented: Bupp and Derian, op. cit.; F. de Gravelaine and S. O'Dy, *L'Etat EDF*; Philippe Simonot, *Les Nucléocrates*; Dominique Saumon and Louis Puiseux, "Actors and Decisions in French Energy Policy," in Leon Lindberg (ed.), *The Energy Syndrome*; Michel Herblay, *Les Hommes du Fleuve et de l'Atome*; Bertrand Goldschmidt, "Les Principales Options Techniques du Program Français de Production d'Energie Nucléaire," *Revue Française de l'Energie*, Oct. 1969; Louis Puiseux, "EDF et la Politique Energétique," *Les Entreprises Publiques*, Mar.–Apr. 1978.

The fact that the disagreement invoked the century of rivalry between two corps—Mines vs. Ponts et Chaussées—is highlighted in some of these works, notably Simonot, *Les Nucléocrates*. The historical background to this dispute has been provided in part by André Thépot (interview, Paris, June, 1978). In 1963 a group of Young Turks had taken over the Corps des Ponts et Chaussées precisely to make it more territorially aggressive. See Jean-Claude Thoenig, *L'Ere des Technocrates*, especially pp. 67 ff. On Pierre Massé, in addition to the above, see René Gaudy, *Et la Lumière Fut Nationalisée*; J. H. Dreze, "Some Postwar Contributions of French Economists to Theory and Policy," *American Economic Review*, supplement, 1964.

The artificial division between prototype and commercial plant sowed the seeds of the future conflict. When the time for major investments came, the CEA had created a large division to build the prototype plants—a division that would have to be dismantled if it didn't have a role. The CEA was making a bid to become the architect engineer for all the nuclear plants. At first the CEA cooperated with a particular unit of EDF known as the Direction des Etudes et Recherches (Studies and Research) Minières—which had a mandate as vague as its title suggests. Under the foundation head of this section, Pierre Ailleret—brother of the foremost military proponent of the French bomb, Charles Ailleret—the CEA had dominated EDF's program. Etudes et Recherches had problems with its institutional identity and the long-term future of the "nuclear age" gave them something to believe in. Another section of EDF was responsible for actually building all power plants—this was the empire within an empire called the Direction de l'Equipement. This group had an extremely strong sense of esprit de corps and total faith in its own technical excellence. The men from Equipement dismissed Etudes et Recherches as "dreamers"—the ultimate term of abuse from a practical engineer. (Gravelaine and O'Dy, op. cit.; p. 82.) The foundation head of this powerful Direction—when EDF was formed after the war—was Pierre Massé himself. It was he who created its special stature. When the CEA tried to become the actual builder of nuclear plants it was issuing a challenge to the traditional bureaucratic turf of this Direction—*droit acquis*, as the French call it. The conflict was bound to be sharp.

278 "intellectual terrorism" and "foreign experimental reactor": Simonot, *Les Nucléocrates*, p. 57. Later when Horowitz found he had to retreat on the gas-graphite system, he developed a passing enthusiasm for the Canadian heavy-water system—anything but an American model. In accordance with the spirit of de Gaulle's "Vive le Québec Libre," Horowitz invariably refered to the CANDU as "the Quebecois reactor system" (ibid., p. 59).

278 ". . . CEA said white . . .": Ibid., p. 126.

278 "no taboos . . .": Ibid., p. 127.

278 Pompidou suspected engineers: Ibid., p. 76.

279 "too late . . .": Ibid., p. 80.

CHAPTER 17

280 The failure of the Fermi breeder has passed into antinuclear folklore with the detailed critique by John Fuller, *We Almost Lost Detroit*, though a few factual inaccuracies in that study have been corrected; see also "We Did Not Almost Lose Detroit," Earl Page, Detroit Edison, May 1976. On Walker Cisler see also *Fortune*, Mar. 1952; *Current Biography*, 1955.

281 The term "advanced reactors" relates to the process by which a reactor changes or converts the isotopes of U-238 in uranium fuel rods into plutonium while it is actually "burning" the isotope U-235. A light-water reactor converts about half as much U-238 as it uses in the form of U-235. Breeders convert more than they use; in the nuclear industry jargon, they have a "conversion ratio" of more than one. "Advanced reactors" have a ratio closer to one than the LWR's. A number of different types of "advanced reactors" were built in the sixties amid debates in each advanced industrial nation about whether there would be a "power reactor gap" between LWR's and the final success of overcoming the problems with the breeder. At the time everyone thought they knew the parameters and, in the face of almost complete technical ignorance—and complete ignorance about costs—a series of detailed computer studies were done to "prove" the role of the "advanced reactors." These were studies that had precise estimates for equipment costs, precise construction schedules, and precise figures for the "value" of plutonium. The companies and the reactor teams involved with advanced convertors set about "proving" the existence of a "reactor gap" which required economizing on uranium. In 1965 the General Atomic Company received the go-ahead from the U.S. AEC to develop its high-temperature gas-cooled reactor; in 1971 the Gulf Oil Company, which took over the project, decided to push it through. Having formed a major international alliance with Shell Oil to develop the reactor, Gulf announced an offer of fixed-price deals on highly competitive terms at the fourth Geneva Atoms for Peace Conference.

The Gulf drive improved the image of advanced convertors throughout the status-conscious nuclear community—until four years later (1975), when Gulf and Shell decided to cut their losses, buy their way out of the contracts they had signed, and get out of the nuclear reactor business. In the same period, the British formally adopted and then canceled their own advanced convertor, a heavy-water moderated system using slightly enriched uranium. In Germany, a reactor system similar to Gulf's enjoyed a brief Indian summer of enthusiasm. The project had been kept alive only because it became the reactor idea of the Julich Nuclear Research Center, the rival of Karlsruhe, and it was politically impossible for the German government to support the Karlsruhe breeder and not support the Julich high-temperature reactor. The creator of the concept, physicist Rudolph Schulten, had justified it in as many different ways as there had been intellectual fads in the nuclear community and he was never short of "precise" cost estimates—to the second decimal place of pfennigs per kilowatt hour for his 1,000-Mw model. And this, years before his strife-torn 300-Mw model was built. By the time it was obvious that the idea was leading nowhere, the government had spent too much to refuse to finish the prototype. It was the same story in Japan, where good money was thrown after bad to avoid a public admission of waste. The heavy-water Japanese reactor, named Fugen, was built because the Hitachi Company insisted on its own project to balance the fact that its rival electric manufacturer, Toshiba, was leading the breeder consortium. Like Westinghouse in the United States, Hitachi was known as the "engineers' company" and its political clout was enough to ensure that the Power Reactor Development Committee of 1964–66 (the group that recommended the breeder) also decided to build Fugen. Hitachi sent its key nuclear executive, Kiyonari Susumu, into the new corporation and he soon became president. Fugen was safe at the top. Just as well, because before the construction had started, its entire raison d'être disappeared. The original reactor design had a "burn-out" problem, which meant that the conversion ratio had to be lowered—after which there was nothing "advanced" about it. Nevertheless, the prototype was built, with no prospect of serial production. It joined the others on the scrap heap: Gulf's high-temperature reactor in Colorado, the Germans' "pebble-bed" system at Julich, and the British model of a so-named steam-generating heavy-water reactor. They all became monuments to past uncertainty.

284 The description of Milton Shaw draws on: "The Next Step Is the Breeder Reactor," *Fortune*, Mar. 1967; Robert Gillette, "Nuclear Safety," *Science*, vol. 176, 1972, p. 496, and vol. 177, 1972, pp. 771, 867, 970, 1030; Peter Metzger, *The Atomic Establishment*; Joel Primack and Frank von Hippel, *Advice and Dissent*; Irvin C. Bupp and Jean-Claude Derian, *Light Water*.

285 ". . . infant technology": Washington *Post*, Aug. 24, 1970.

285 "no assurance . . .": Ralph Nader and John Abbotts, *The Menace of Atomic Energy*, p. 105.

285 Ford cross-examination of Shaw, ECCS Rule-Making Hearings, transcript, pp. 7382–83, quoted in Primack and von Hippel, op. cit., p. 224.

286 "trying to avoid the problem . . .": Ibid., p. 223.

286 Academy of Sciences on waste disposal, from Metzger, op. cit., p. 153.

286 Ducks at Hanford: See UPI dispatch, "Dangerous Radiation in Washington Ducks," Oakland *Tribune*, Mar. 14, 1970, quoted in Metzger, op. cit., p. 154.

287 Shaw's confident assertion refers to Shaw's testimony before a congressional committee asking for a complete appropriation for the waste disposal project so that he would not have to come back and justify the next year's expenditure. "Another year's work of research and development in this area on top of fifteen years work would not be particularly productive. We need the project and we are ready to proceed with it," Shaw declared (see Metzger, op. cit., p. 156). Within a year Shaw, the man

who made a fetish of technical precision in his breeder program, would have to admit that the AEC had selected the only wet salt mine in the world.

289 The description of the Japanese advanced reactors is based on interviews in Tokyo in Sept. 1977 and Aug. 1978, especially with Arisawa Hiromi, Murata Hiroshi, Ipponmatsu Tamaki, and Imai Ryukichi. In addition the article by Arisawa in Kondo Kan'ichi and Osani Hiroshi (eds.), *Sengo Sangyo-shi E No Shogen*; the semiofficial history of the Japan Atomic Industrial Forum, *Nihon No Genshiryoku*, vol. 1, pp. 170 f. and vol. 12, pp. 26 ff.; Hideo Sato, "The Politics of Technology Importation in Japan," unpublished paper (mimeo), Yale University, 1978; Rodney Louis Huff, "Political Decisionmaking in the Japanese Civilian Atomic Energy Program"; Chalmers Johnson, *Japan's Public Policy Companies*.

289 In addition to authors' interviews, the material on Mitsubishi and Niwa Kaneo draws on a series in the *Mainichi Daily News*, Oct. 13, 17, and 27, 1970; *Newsweek*, Apr. 23, 1973; Kamatani Chikayoshi, "The Role Played by the Industrial World in the Progress of Japanese Science and Technology," *Journal of World History*, vol. 9, no. 2, 1965. Niwa's background in the Nagasaki shipyards gave him his formula for success in the breeder debate. As early as 1904, Mitsubishi had established a separate research organization at the Nagasaki shipyards; it was one of the first in any segment of Japanese industry. It became a center for basic research and engineering development with its sophisticated tanks for model ships and its material testing laboratories.

During the war Niwa was the chief engineer on the construction of one of the largest battleships ever built, the *Musashi*. It was an engineering achievement that he was never to match, but it gave him his modus operandi for the breeder: a clear government priority commitment, a domestic research and development program from basic design through to materials testing, and close integration between research and practical engineering teams. During the meetings of the Power Reactor Development Committee, Niwa would point to the *Musashi* as an example of how Japan could leap to the forefront of international technology. The omens were not that good, however. With its armored decks and subdivided hull, the *Musashi* was supposed to be unsinkable, but in Oct. 1944 it took American planes only a few hours to send it to the bottom of the Sibuyan Sea in the center of the Philippine archipelago. Nevertheless, Niwa wanted to build "a new Musashi."

291 Walker Cisler's influence on Japan's breeder program was not quite what the Japanese had expected. Japanese utility executives had assumed that Cisler, as president of the Atomic Industrial Forum, would have the same influence on his government's policy as his Japanese equivalent. But Cisler never went out of his way to explain his different status, nor

to let them know that his influence was rapidly declining precisely because of his single-minded pursuit of the breeder. The Japanese thought it was a great honor when Cisler let them fund part of Detroit Edison's Fermi I project, and in exchange they received technical information, personnel training, and, after the accident, some parts of the dismantled reactor.

291 The Hafele breeder team at Karlsruhe inhabited an institute that was torn by personal status anxieties and jealousies. Karl Wirtz, a cold, aloof Prussian out of the mold of the German god-professor, was determined, from the day the institute opened in 1955, to be the dominant personality. Others with a piece of bureaucratic turf to protect found themselves with a common interest in resisting Wirtz's claims, and Wirtz himself did not have the political skill to withstand the alliance of mediocraties against him; each decision ended in a stream of petty squabbles. The breeder project was the biggest prize in the tense atmosphere at the institute, and, eventually, Wirtz lost it. On the German breeder project see Henry Nau, *National Politics and International Technology;* Karl Winnacker and Karl Wirtz, *Nuclear Energy in Germany*; interviews with Dr. Klaus Traube and Kurt Rudzinski, Frankfurt, Aug. 1978.

292 The basic analysis of the 1964 cost estimates is drawn from Otto Keck, "Fast Breeder Reactor Development in West Germany," and Otto Keck, "The West German Fast Breeder Reactor Program, Energy Policy," 1979.

CHAPTER 18

293 The description of the Israeli bomb program draws in part on: John J. Fialka, "How Israel Got the Bomb," *Washington Monthly*, Jan. 1979; S. Flapan, "Israel's Attitude Toward the NPT," in Stockholm International Peace Research Institute, *Nuclear Proliferation Problems;* Robert E. Harkavy, *Spectre of a Middle East Holocaust;* Avigdor Haselkorn, "Israel: From an Option to a Bomb in the Basement," in Robert M. Lawrence and Joel Larus (eds.), *Nuclear Proliferation Phase;* Aubrey Hodes, *Dialogue with Ishmael;* Fuad Jabber, *Israel and Nuclear Weapons;* George Quester, *The Politics of Non-Proliferation.* A variety of both intentional and accidental leaks from the CIA on the Israeli bomb have been reported; see Boston *Globe*, July 31, 1975; *New York Times*, July 18, 1970, Jan. 27, 1978, and Mar. 2, 1978. The mid-1970 leak conveniently coincided with Nasser's visit to Moscow and was a clear warning to the Soviets.

294 Western Suppliers Group: The fact that the Americans were convinced they could rely on the secret indirect monitoring system of uranium sup-

plies is confirmed in the memoirs of the French Foreign Minister of that time, Maurice Couve de Murville. He reported two meetings he had in the White House in May and Oct. 1963, during each of which, he said, President Kennedy "expressed his serious anxieties" about Israel's nuclear program. Couve added, "They [the United States] were intending to organize a strict surveillance by all the means at their disposal—and they used to have no shortage of them. They were particularly endeavoring to reveal Israeli purchases of natural uranium from other countries." Maurice Couve de Murville, *Une Politique Etrangère*, p. 99. The policy worked for a while. Just three months before Couve's first meeting with Kennedy, the Western Suppliers Group had met in London on Feb. 15, 1963. Those attending were the U.S., Britain, France, Canada, Belgium, South Africa, and Australia. During the meeting South Africa reported in detail on its recent uranium sale to Israel. Documents relating to this meeting were made available to the authors after Freedom of Information Act requests.

294 "Don't tell anyone else": Not "even Dean Rusk and Robert McNamara," the President added. Inquiry into the Testimony of the Executive Director for Operations, U.S. Nuclear Regulatory Commission, vol. 3, Feb. 1978, p. 178. This is page three of four pages of testimony by Dr. Carl Duckett of the CIA. All other pages of his testimony and Duckett's name in the table of contents had been deleted from the report as released to the Natural Resources Defense Council. The release of this page seems to have been a clerical error (interview with Thomas Cochrane of the NRDC).

295 "We believe that Israel . . .": Memorandum "Prospects for Further Proliferation of Nuclear Weapons," CIA, Sept. 1974, paragraph 3. This document was released under a Freedom of Information Act request by Jacob Scherr, an attorney for the NRDC. Later the CIA said it was released "by mistake."

296 The admission by Francis Perrin of French assistance for the Israeli reprocessing facility came in an interview with the authors in Paris, June 1978. Perrin added that the collaborations in this field continued until 1961—three years after de Gaulle's return to power. French Foreign Minister Couve de Murville was generous to de Gaulle when, in his memoirs, he noted in regard to the 1957 atomic agreement with Israel, "Very quickly after the return of de Gaulle we had taken steps to prevent this collaboration in extending itself into other fields, that is to say the extraction of plutonium for military ends." Couve de Murville, op. cit., p. 98. Perrin's recollections are supported by the hints contained in three interviews published in 1976. Three French leaders made indirect admissions behind the cover of anonymity for Philippe Simonot's book *Les Nucléocrates*. French observers have identified one as Robert Galley. In

the late fifties, he directed the French enrichment plant at Pierrelatte and became a Gaullist minister. He admitted that the Mollet government was the only one ever consciously to assist proliferation. When asked directly whether France had assisted Israel in the nuclear weapons field he replied, "Yes. Beyond what the world thinks" (Simonot, *Les Nucléocrates*, p. 81). The second man—identified as George Besse, also in the Commissariat's enrichment program in the fifties and later the head of the French fuel cycle group, simply said "No comment" when asked whether France had given Israel the bomb. He did, however, offer the following cryptic comment: "I built the plant at X. It wasn't for making preserves. I obeyed the order of Mollet and then of de Gaulle. I did it without hesitation and I regret nothing" (ibid., p. 125). The third man was head of one of the French companies that was asked to supply material to Israel. He recalled how he sent the Commissariat's chief, Pierre Guillaumat, a copy of the then recently published book *On the Beach*. It described the destruction of the world after a nuclear conflict in the Middle East. Guillaumat could only reply, "They are the Government's orders" (ibid., p. 75).

296 Several Israelis died: Interview with Israeli nuclear weapons expert who wanted to remain anonymous, Oct. 1980.

297 Although Levi Eshkol only froze—rather than destroyed—Israel's bomb option, he made his hostility quite clear. In the spring of 1966, Eshkol sacked Dr. Ernst Bergmann from his post at the Atomic Energy Commission and simultaneously from the position he held in parallel as Scientific Director in the Israeli Ministry of Defense. Responsibility for the AEC was shifted from Defense to the Prime Minister's office, and a new Commission was appointed with an explicitly industrial and academic orientation. It was under the chairmanship of one of the scientists who had resisted in 1957. The Committee for Denuclearization of the Israeli-Arab conflict stated that although Eshkol would not openly support their nuclear free zone idea, he was acting consistently with that objective. The incident with the *Scheersburg* A is based on: Elaine Davenport et al., *The Plumbat Affair*; *New York Times*, Apr. 30, 1978; Enrico Jacchia, *L'Affaire "Plumbat."* Technically, the missing uranium was subject to "safeguards" by Euratom, but natural uranium was not very "sensitive" and shipments had to be logged but not controlled. Only a handful of people knew that it was actually important to account for all the uranium that Israel acquired. American intelligence did not find out about the loss for a year, and by then the last and most unsatisfactory "inspection" of Dimona had been done. No one had any doubt the uranium went to Israel, but the disappearance of the cargo remained a secret until unofficially leaked in the spring of 1977.

298 1978 CIA Briefing on NUMEC: See Duckett testimony in "Inquiry into

the Testimony of the Executive Director for Operations, U.S. Nuclear Regulatory Commission," p. 3.

299 The Gilpatric Report quotes are found in "Committee on Nuclear Proliferation Report to the President," Jan. 1965, pp. 2–3: Gilpatric Papers, John F. Kennedy Library, Boston.

299 ". . . the central priority": quoted in Hodes, op. cit., p. 232. Besides its overall review the Gilpatric committee studied each potential proliferator—India, Argentina, and Israel. The Israeli sections have not been released.

302 "A second Yalta": Karl Winnacker and Karl Wirtz, *Nuclear Energy in Germany*, p. 201.

302 "A Versailles . . .": *Capital*, Oct. 1968.

302 "Against Germany's Breeder": *Volkswirt*, Aug. 30, 1968. Karl Winnacker, who had ruled off the processing area as a job for the chemical industry and specifically for his own company Hoechst, saw at first hand the intrusiveness of safeguards in such a plant. In Oct. 1967 he visited the reprocessing plant in Wets Valley, New York, which was being safeguarded by the IAEA as an experiment. He didn't like what he saw. See Winnacker and Wirtz, op. cit., p. 207.

302 impossible to develop the breeder: Japan Atomic Industrial Forum, *Nihon No Genshiryoku*, vol. 3, p. 193.

303 ". . . devil is hidden . . .": *Volkswirt*, Aug. 30, 1968.

303 ". . . approximate problem": IAEA Proceedings, fourth Atoms for Peace Conference, Geneva, 1971, vol. 9, p. 303.

304 The analysis of the development of the IAEA's safeguards system has three components: the two different starting points for a system represented by the "strategic points" approach on the one hand and the "critical time" approach on the other and finally the triumph of the former in the 1970 meetings of the safeguards committee. There are few secondary sources that assess this period, and the authors have had to interpret primary documents and conduct extensive interviews. The basic collections of documents are found in two symposia organized by the IAEA in Karlsruhe and Vienna in 1970 and 1975 respectively. The proceedings have been published by the agency as *Safeguards Technique*, Vienna, 1970, and *Safeguarding Nuclear Materials*, Vienna, 1976. Further papers have been delivered in the Atoms for Peace series of conferences: See *Peaceful Uses of Atomic Energy*, IAEA, Vienna, 1971, vol. 9, and *Nuclear Power and Its Fuel Cycle*, IAEA, Vienna, 1977, vol. 7. Our understanding of the political implications of this technical literature was developed in interviews with Myron Kratzer, Allan Labowitz, and Harold Bengelsdorf of the United States, Jan Prawitz from Sweden, Claude Zanger from Switzerland, Imai Ryukichi from Japan, and, from the IAEA, David Fischer, Slobadon Nakicenovic, and Gerry Hough. In

addition Allan McKnight, formerly at the AEA, and John Kennekjens of Canada provided some unpublished papers. Secondary sources include *The Atlantic Council Nuclear Power*—Part 3 of which was drafted by Myron Kratzer, who headed U.S. safeguards negotiating teams at the IAEA during the sixties and early seventies. Kratzer provided some further historical insights in Bertrand Goldschmidt and Myron Kratzer, *Peaceful Nuclear Relations: A Study in the Creation and Erosion of Confidence*, International Consultative Group on Nuclear Energy, Rockefeller Foundation, New York, and Royal Institute of International Affairs, London, 1979. Other secondary sources on safeguards: Imai Ryukichi, "Nuclear Safeguards," *Adelphi Papers*, no. 86, 1972; Wolf Hafele, "NPT Safeguards," in *Nuclear Proliferation Problems*, Stockholm International Peace Research Institute; Allan McKnight, *Atomic Safeguards*; Robert Pendley and Lawrence Scheinman, "International Safeguarding as International Organization," vol. 29, 1975; "Safeguards: Five Views," *IAEA Bulletin*, vol. 13, no. 3, 1971; Lawrence Scheinman, "Nuclear Safeguards, the Peaceful Atom and the IAEA," *International Conciliation*, vol. 572, Mar. 1969; Stockholm International Peace Research Institute, *Safeguards*; Paul Szasz, *The Law and Practices of the IAEA*; Mason Willrich (ed.), *International Safeguards*; Winnacker and Wirtz, op. cit.

The analysis of the concept of "critical time" is based on access to internal documents of the IAEA—notably the consultant report of 1967 prepared by Frank Morgan of the U.K. AEA for the Safeguards Secretariat of the IAEA, and the series of papers and panel meetings held in Tokyo and Vienna in 1969 indicate clearly how the experts were heading in the direction of making "critical time" the centerpiece of the safeguards system. This was reflected in the first documents prepared by the safeguards secretariat for the 1970 safeguards committee—see, in the IAEA's internal coding system, the document identified as GOV/COM.22/62/REV 1. A perspective on history written by the victors, with reference to the IAEA panel meetings, is found in W. Hafele, "Systems Analysis in Safeguards of Nuclear Material," *Peaceful Uses of Atomic Energy*, vol. 9, IAEA, Vienna, 1971. This Hafele paper, delivered at the fourth Atoms for Peace Conference, reflected the triumph of the strategic points approach. At the same conference the Safeguards Secretariat published, for the first time, their "pragmatic" compromise formula for at least four wash-outs per year. They were, however, resigned to the defeat of the critical time approach. "For plutonium of highly enriched uranium," the paper by the IAEA safeguards secretariat said, "a critical time of ten days has been proposed: this would require a physical inventory taking every ten days for material balance areas processing these materials and such a verification requirement clearly hampers the operation of the plant. For low-enriched uranium or highly irradiated material a critical time of one month is proposed, requiring a frequency of physical inventory taking of one per month; this may also be too burdensome. The critical time concept is thus not directly useful for defining a maximum time interval for closing material balances for international safeguards" (ibid., p. 503). Not unnecessary, just "too burdensome." In the years after the 1970 Safeguards Committee, nuclear advocates would point to the safeguards system as solving the problem of proliferation. At the same time they would point to the economic advantages of nuclear power and the future of breeders. But safeguards had been seriously compromised for reasons of economics.

306 ". . . ten tedious months . . .": IAEA Bulletin, vol. 13, no. 3, 1971, pp. 5, 7.

CHAPTER 19

307 The return and the second evacuation of the Bikini Atoll is based on Giff Johnson, "Micronesia: America's 'Strategic Trust,'" *Bulletin of the Atomic Scientists*, Feb. 1979; Washington *Post*, Mar. 19 and 27, Apr. 3, 1978, and May 22, 1979.

309 ". . . best available source of data . . .": Washington *Post*, Apr. 3, 1978.

309 "Listen, world . . .": James Cameron, *Point of Departure*, p. 63.

309 The Bikini Islanders were not the only group whose geographical isolation made them politically manageable in terms of radioactive exposure. When France lost her Saharan testing range with the independence of Algeria, the French government switched to French Polynesia in the South Pacific. They began testing in 1966 and continued until 1974 when international pressure forced them underground. Determined not to lose its testing range, the French government actively suppressed political moves for self-government and independence in French Polynesia. The Tahitian political leader, a World War I hero called Poovanaa a Ooopa, was jailed on trumped-up charges and secretly removed to a decade of exile in France. French colonial officials ruthlessly manipulated electoral laws and rigged local elections to suppress democratic movements and especially the growing criticism of the nuclear tests. Information was kept to a minimum—especially about patterns of radioactive fallout. Some rudimentary material was available to the United Nations in New York, but these reports hid more than they revealed; the real effect of the tests on the Polynesians could not be assessed. From 1963, the French stopped separate publication of public health statistics on leukemias and cancers in French Polynesia.

310 There is a considerable literature on the atomic disasters of the sixties: the effects of uranium tailings, cancers among uranium miners, waste-

disposal failures, plutonium contamination at Rocky Flats. See generally H. Peter Metzger, *The Atomic Establishment*; Roger Rapoport, *Great American Bomb Machine*; Walter Patterson, *Nuclear Power*; Joel Primack and Frank von Hippel, *Advice and Dissent*, chap. 12; Arell S. Schurgin and Thomas C. Hollocher, "Radiation Induced Lung Cancers Among Uranium Miners," in Union of Concerned Scientists, *The Nuclear Fuel Cycle*. The early failure to give appropriate emphasis to the dangers of Iodine-131 contamination led to congressional hearings in 1963; see Joint Committee on Atomic Energy, *Fallout, Radiation Standards and Countermeasures*, especially the testimony of Harold Knapp. He conducted the internal reassessment of radiation standards in the AEC, ibid., vol. 2, p. 1078. The attempt to suppress these results is outlined in Metzger, op. cit., pp. 103 ff., 276 ff. In the spring of 1979 the Department of Health, Education and Welfare released thousands of pages of documents about this incident and especially about the research program to investigate thyroid cancers in the area close to the Nevada test site. These documents were reviewed extensively in the press; see Washington *Post*, Apr. 14, 19, 20, 1979; also *New York Times*, May 13, 1979. In addition see the testimony of Harold Knapp at the hearings (unpublished at the time of writing) of the Senate Health and Scientific Research Subcommittee of the Senate Judiciary Committee, held at Salt Lake City, Utah, Apr. 19, 1979.

311 Windscale accident: See Patterson, *Nuclear Power*, pp. 162–66.

311 "All things considered . . .": Letter from Harold Knapp to John Conway, executive director, JCAE, Sept. 9, 1963, p. 4.

312 "most interesting": From minutes of a meeting held on Apr. 6, 1965, and headed "Second Meeting of Review Group; Utah-Nevada Population Study," p. 3. The report, as published, was Edward S. Weiss, "Surgically Treated Thyroid Disease Among Young People in Utah, 1948–62," *American Journal of Public Health*, vol. 57, Oct. 1967, p. 1812.

313 Radon gas levels in U.S. mines examined in Ralph Nader and John Abbotts, *The Menace of Atomic Energy*, pp. 178–79.

314 "It got so that I did not need to wait . . .": *New York Times*, May 20, 1979.

314 The development of the knowledge about the effects of radioactivity during the 1960s is described in a number of works: Virginia Brodine, *Radioactive Contamination*; Samuel Glasstone and Philip J. Dolan, *The Effects of Nuclear Weapons*; John Gofman and Arthur Tamplin, "Epidemiological Studies of Carcinogenesis by Ionizing Radiation," in *Proceedings of the Sixth Berkeley Symposium on Mathematical Statistics and Probability* (San Francisco: University of California Press, 1971); Karl Z. Morgan and J. E. Turner (eds.), *Principles of Radiation Protec-*

tion; Iwao Moriyama et al., *Radiation Effects on Atomic Bomb Survivors,* Technical Report Series, n. 6–73, Hiroshima and Nagasaki, Japan Atomic Bomb Casualty Commission, 1973; Symposium, "Review of Thirty Years Study of Hiroshima and Nagasaki Atomic Bomb Survivors," *Journal of Radiation Research,* supplement, 1975; Lauriston Taylor, *The Origins and Significance of Radiation Dose Limits.*

314 "threshold theory": Metzger, op. cit., p. 135.

315 X rays and leukemias: Jack Schubert and Ralph Lapp, *Radiation: What It Is and How It Affects You,* pp. 180 and 208.

316 The case of Dr. John Gofman and Arthur Tamplin is drawn from Richard S. Lewis, *Nuclear Power Rebellion,* chap. 4; Walter Patterson, *Nuclear Power,* pp. 149 ff; John Gofman and Arthur Tamplin, *Poisoned Power.* Gofman and Tamplin were nicknamed by a British AEA safety official "gin and tonic"; see Duncan Burn, *Nuclear Power and the Energy Crisis,* p. 62. Gofman is quoted in Richard Lewis, op. cit., as saying that the AEC's attempts to censor him and Tamplin made the Lawrence Livermore Laboratory "look like a scientific whorehouse. . . ."

317 ". . . why do you criticize . . .": Gofman and Tamplin, *Poisoned Power,* p. 255.

317 In launching their attack on the 0.17r standard Gofman and Tamplin were well aware that radiation emissions from power plants were lower than that, but they believed that if it was technically possible to keep radiation emissions lower than the present standard, then the present standard should be lowered, too. They had their eye on the future when the nuclear industry would be much bigger and the total releases of radiation into the atmosphere would also be larger. In the wake of the controversy the two scientists caused the AEC to take steps to lower the standards. For the first time, the AEC decided that whatever the legal limits of exposure were the operators of nuclear plants should keep radiation releases "as low as practicable." Once again, this was a political decision, not a scientific one. It assumed that there was a level of risk to the public that was acceptable; if the public could accept the risks of driving cars or flying in airplanes, then they could accept the risks of nuclear power. The growth of the antinuclear groups in the early seventies, however, was evidence that a sizable section of the public was not prepared to accept such risks with nuclear power; they felt deceived by the nuclear community.

318 "Dr. Michael May . . .": John Gofman, *Irrevy.*

CHAPTER 20

321 Fuel oil was sold at artificially low costs because the oil companies made their big profits not in refining and distribution, but by getting the oil

out of the ground as quickly as possible. The faster demand grew the more money they made. With gasoline demand for automobiles representing a secure market, the oil companies' drive was directed at substituting oil for other fuels, especially coal. They priced fuel oil—the heavy end of the refining process—as if it were a mere by-product of gasoline production. Despite a series of allegations and public inquiries in many countries about the "dumping" of fuel oil, coal markets were substantially cut back on the assumption that oil prices would stay low forever. This illusion was fostered by the need of the oil companies to take advantage of their temporary total control of the oil-producing fields—so coal mines were closed everywhere, and once a working mine is closed it becomes much more expensive to reopen it because of the dangers of partially used underground shafts. In both Japan and France—which have substantial high-cost coal reserves—the official statistics stopped counting many previously mined reserves because they were assumed to be too uneconomic to bother with.

323 The description of Japan draws on: Rodney L. Huff, "Political Decision-making in the Japanese Civilian Atomic Energy Program"; Imai Ryukichi, "The Political Outlook for Nuclear Power in Japan," *The Atlantic Community Quarterly*, vol. 13, summer 1975; Japan Atomic Industrial Forum, *Nihon No Genshiryoku*; Chalmers Johnson, *Japan's Public Policy Companies*; Tano Kamakura, *Denryoku Sangoku-shi*; Kondo Kan'ichi and Osani Hiroshi (eds.), *Sengo Sangyo-shi E No Shogen*; Ohtani Ken, *Denryoku*; Hideo Sato, "The Politics of Technology Imposition in Japan," unpublished paper (mimeo), New Haven, Yale University, 1978; Chitoshi Yanaga, *Big Business*, and interviews with Arisawa Hiromi, Murata Hiroshi, Ipponmatsu Tamakai.

323 The sketches of Kikawada and Ashihara draw on interviews and Kamakura, op. cit.; Ohtani, op. cit.; Shin Takamiya, *Kikawada; Yomiuri*, Sept. 2, 16, 20, 25, 1974; *Japan Economic Journal*, Nov. 19, 1974; *Japan Times*, June 25, 1967; *Yomiuri*, Mar. 14, 1977.

324 The political spearhead for Kikawada's group of Young Turks in the immediate postwar debate was Matsunaga Yasuzaemon. He had a definite political philosophy that prevented him from cooperating with the militarist state of the thirties. As late as 1938 he could declare: "Development of industry depends on private enterprise. We cannot trust bureaucrats. They are the scum of the earth" (Ohtani, op. cit., p. 13). After the war these views found a willing audience in GHQ Tokyo where, apparently, there were no TVA men. The fiat that created the nine-company structure was issued in 1952. It was a controversial use of the plenary powers, coming, as it did, so late in the occupation. But after that, not even Matsunaga himself could change the system—as he tried to do in 1957. See *Oriental Economist*, Oct. 1957.

325 Ashihara Yoshishige, president of Kansai Electric since 1968, had rea-

son to be disappointed with the JAPCO choice of a PWR. The president of JAPCO was an old Kansai man, Ipponmatsu Tamaki. Ashihara had joined the Kansai Distribution Company in 1942, when his former company was merged with it in state-ordered reorganization of that year. He found himself junior to Ipponmatsu, who had been one year behind him at Kyoto Imperial University's School of Electrical Engineering. This was the training ground of the Osaka business elite, just as Tokyo Imperial University was the center for Tokyo's top men. In the manner of the Japanese corporate hierarchy, Ashihara and Ipponmatsu were selected at an early stage as men who could rise to the top and were carefully groomed for the job. They engaged in a quiet but bloodless power struggle for the succession, but in the late fifties Ashihara received the nod from the retiring president. Ipponmatsu, an early convert to nuclear power, had built up a strong nuclear technical group in the company. He left it behind him when he departed to become vice-president and, soon, president of JAPCO. The gentlemanly struggle for succession at Kansai was in sharp contrast with the strife-torn executive battleground at Tokyo Electric. Ashihara announced Kansai's first PWR order to considerable local acclaim in Osaka. The target date for completion was set as 1970, and nuclear power, he declared, would fuel the major Expo of 1970. This was a high-prestige factor for the Osaka business community, for the Expo would give their firms international exposure of a kind reserved for Tokyo-based groups. Ashihara was vice-president of the Osaka Expo Organizing Committees.

326 The description of the EDF conversion to nuclear power draws on: Irvin C. Bupp and Jean-Claude Derian, *Light Water;* Frédérique Gravelaine and Sylvie O'Dy, *L'Etat E.D.F.;* Michel Herblay, *Les Hommes du Fleuve et de l'Atome;* Ministère de l'Industrie et de la Recherche, *Les Dossiers de l'Energie*, 14 vols., Paris, 1977; Louis Puiseux, *Le Babel Nucléaire;* Louis Puiseux, "EDF et la Politique Energétique," *Les Entreprises Publiques*, Mar.–Apr. 1978; Philippe Simonot, *Les Nucléocrates;* D. Saumon and Louis Puiseux, "Actors and Decisions in French Energy Policy," in Leon N. Lindberg (ed.), *The Energy Syndrome.* Also interview in Paris, June 1978, with Jean-Claude Derian and Louis Puiseux.

326 "the electric revolution . . .": Gravelaine and O'Dy, op. cit., p. 152.

329 "the shock troops": Ibid., p. 78.

329 "heroic age": Herblay, op. cit., pp. 59, 162.

329 The sketch of Michel Hug draws on interviews and Gravelaine and O'Dy, op. cit.; Herblay, op. cit.; *La Vie Electrique*, Mar.–Apr. 1975.

330 The original nuclear regions pushed their self-interests in promoting the reactor types they allocated and then in resisting the forced-pace scale up to the 1,300-Mw reactor size. See Herblay, op. cit., pp. 194, 206–08.

331 It was in France alone that the normally cost-conscious electricity utility leaders embraced the breeder so completely. Hug's advocacy did not wane as the large-scale breeder—the 1,200-Mw Superphénix—lost sight of its original claim to the "first" commercial breeder because it would not be cost competitive and became another "prototype." Hug believed that the traditional utility faith in economies of scale from building bigger and bigger units applied to the most technically complex tasks of nuclear construction. Even as the CEA planners curbed their ambitions to building only a 1,400-Mw successor to Superphénix, Hug still pushed them to build two 1,800-Mw monsters. On his own responsibility he even ordered the construction of a special railway line to the proposed site on the Saône River. As one colleague put it, "Saône 1 and Saône 2, that's not an EDF project, but Hug 1 and Hug 2." See Gravelaine and O'Dy, op. cit., p. 154.

CHAPTER 21

The analysis of the fuel cycle developments in the seventies draws generally on: Duncan Burn, *Nuclear Power and the Energy Crisis*; Congressional Research Service, *Commercial Nuclear Power in Europe* and *Nuclear Proliferation Factbook*; Spurgeon Keeny et al., *Nuclear Power Issues and Choices*; Henry R. Nau, *National Politics and International Technology*; Office of Technology Assessment, *Nuclear Proliferation and Safeguards*; Donald R. Olander, "The Gas Centrifuge," *Scientific American*, Aug. 1978; Karl Winnacker and Karl Wirtz, *Nuclear Energy in Germany*; Edward F. Wonder, *Nuclear Fuel and American Policy*; *Nuclear Power and Its Fuel Cycle*, Proceedings of an International Conference, vols. 2 and 3, IAEA, Vienna, 1977.

333 All of the Soviet Union's European satellites expressed an interest in natural uranium reactors. Some, like Czechoslovakia and Germany, had their own supplies of uranium and sought to avoid dependence on enrichment services from the Soviet Union. They expressed a preference for heavy-water reactors, which the Soviets had built for plutonium production but not developed for electricity generation. However, the Soviets also had a fully developed graphite reactor which used natural uranium, in addition to their pressurized-water reactors which required enrichment. In the mid-sixties each Warsaw Pact country extracted pledges of Soviet nuclear support, but the pledges were followed by repeated delays, Soviet indifference, and broken promises, resulting in a number of bitter arguments. They continued until the early seventies until the satellites tied themselves to the Soviet Union for the supply of enriched uranium by adopting light-water reactors. See Jozef Wilczynski, *Atomic Energy for Peaceful Purposes in the Warsaw Pact Countries*.

333 The sketch of André Giraud is based on: *Le Monde*, Sept. 23, 1970, Dec. 6, 7, 1970, Feb. 16, 1971, and Apr. 7, 1978; *L'Express*, Aug. 17, 1970, Mar. 22, 1971; *L'Aurore*, Sept. 22, 1970, Oct. 22, 1973; *La Croix*, Feb. 17, 1971, Oct. 29, 1975; *Vie Française*, Oct. 9, 1970; *L'Expansion*, Jan. 1976; *Figaro*, Apr. 6, 1978; *Le Point*, Apr. 10, 1978.

In 1964, when Guillaumat was leaving politics to complete the task of building his oil company, he chose Giraud to take his old place as Director of Fuels in the Ministry of Industry. His oil plans needed the full backing of the French state, and the Director of Fuels was the central position to ensure this. The post was one of the most central and traditional preserves of the Corps des Mines of which Guillaumat was then "le chef"; it had real power. "When you are Director of Fuels," Guillaumat recalled, "you know everyone in the state" (interview, June 1978). Guillaumat's concept of the French state was tangible and bounded. It consisted of a series of key positions with established relations, spheres of competence, and traditions of behavior among a few hundred persons. The CEA top job was another post with real power. In 1969 Giraud left the ministry because of the French political turmoil. Just as Guillaumat had accepted the job of Minister for the Army when the French state was faced with pending mutiny in Algeria, so Giraud was called to higher national service with the civil disturbance of May 1968. The conflict had centered on the educational system, and the new Minister for Education was Olivier Guichard, who, with Michel Debré and Georges Pompidou, was one of the "barons" of de Gaulle's years of exile. In 1955, with de Gaulle's return a pipe dream, Guillaumat had given Guichard a job as public relations director of the CEA, but he remained one of de Gaulle's closest personal confidants.

In mid-1969 Giraud was appointed as Guichard's "Chef du Cabinet," the minister's chief of staff position, so powerful in the French system of government. For a year educational reform had been the center of French public debate, focusing on participation and the curriculum. Giraud dismissed much of the educational debate as "metaphysics" and asserted that he enjoyed his blooding in public politics as the Minister of Education. Indeed, unlike Guillaumat, who refused overtures to return to politics, Giraud came back as Minister for Industry under Giscard d'Estaing in 1978.

333 ". . . Shell of the Atom": *L'Express*, Mar. 22, 1971.

334 "The Guillaumat of the Atom": *L'Expansion*, Jan. 1976.

334 ". . . my administrative knowledge . . .": *L'Aurore*, Oct. 22, 1971.

338 The analysis of the uranium cartel is based on the original documents from the RTZ subsidiary in Australia and made available to the authors by the Australian branch of Friends of the Earth. Some of these documents, in addition to many more from the files of Gulf Oil, are re-

produced in the congressional hearings on the cartel. See *International Uranium Supply and Demand*, Hearings before the Subcommittee on Oversight and Investigations of the Committee on Interstate and Foreign Commerce, 2 vols., House of Representatives, U.S. GPO, 1977 (henceforth referred to as Cartel Hearings).

340 The club had almost failed at the Apr. 1972 meeting in Paris when the brash, aggressive Australian producers had demanded double the quota they had tentatively been given in March. The problem was overcome by agreeing to extend the club from 1977 to 1980 and offering the Australians a substantially increased quota in the second period. The Australian case had been resisted by Pierre Taranger's threat to undersell them from existing French stockpiles and the French ability to quickly increase production in its former colonies in Africa. At the April meeting in Paris, which he chaired, Taranger pointed out to the gathering—all from the white British Commonwealth—how he had kept the governments of Gabon and Niger out of their meetings and how he had resisted demands to increase production. He said the French had "grudgingly agreed" to a small increase in Niger but, he added, "would try to delay bringing the production in by one year on technical grounds" (Cartel Hearings, vol. 1, p. 496).

340 ". . . 'new boy on the block' . . ." and ". . . at least as important . . .": Cartel Hearings, vol. 1. pp. 466, 542.

341 RTZ is one of the few English commercial success stories since the war. When Sir Val Duncan took over the Rio Tinto company in 1948 it had a small copper mine in southwest Spain. Two decades later its elegant headquarters in St. James's Square near Buckingham Palace was the financial heart of a mining empire with almost $3 billion in sales. A graduate of Harrow and Oxford, Duncan, a suave British businessman, was driven to develop the world in the model of Cecil Rhodes. Duncan had a simple view of the world—dividing nations into stable and unstable, he concentrated his resources in areas controlled by the white members of the British Commonwealth. His first major coup, which started Rio Tinto on the road to becoming one of the world's biggest mining houses, was the takeover of a number of Canadian uranium mines in the fifties. He also acquired uranium mines in Australia during that period. In the uranium slump, Rio Tinto had continued to expand in copper, iron ore, and bauxite. It had also acquired what was potentially the biggest uranium mine in the world, at Rossing in Namibia. The problem with Rossing was that in 1971 the International Court of Justice declared South Africa's occupation of Namibia illegal and the United Nations had carried resolutions denouncing any economic development directed by the occupation authorities. Nevertheless, RTZ was developing the mine, hoping it could make a deal with whatever independent govern-

ment emerged in Namibia. It had to—a new nationalistic Australian government had just blocked its moves to take over some of the deposits in that country.

341 On the attempt to cut out middlemen, all the producers thought that it was logical for them to deal directly with utility companies—although the records indicate that Australian government officials objected to the rule, but the real drive to cut out Westinghouse came from the CEA and Gulf. They were the only potential nuclear majors in the cartel. As early as the Johannesburg meeting in May 1972, the French and Gulf talked about the Westinghouse problem. Gulf agreed to check where they were getting the uranium they promised. In September the Gulf man telexed his CEA contact, saying, "We have not yet been able to pinpoint the source but will continue to work on the problem." That week a meeting of Canadian producers discussed the Westinghouse bids. "The consensus finally reached," the official minutes record, "was that if the club was to survive as a viable entity, it would be necessary to delineate where the competition was and the nature of its strength as a prelude to eliminating it once and for all" (Cartel Hearings, vol. 1, p. 594). The next month, at a club meeting, the French delegate emphasized that "something had to be done" about Westinghouse (ibid., p. 281).

The General Atomic Division of Gulf Oil, formally distinct from its Canadian uranium mining subsidiary, was at this stage aggressively competing against Westinghouse in the reactor market. They were angry that Westinghouse was still offering low-priced uranium with its reactor bids. The next January an executive in Gulf's reactor division noted that Westinghouse was seeking a tie-up with one of the huge new Australian deposits. If consummated, the Gulf man said, it "would provide Westinghouse with a potential source for US reactor sales. It would also provide a source for their substantial foreign shortage. We will work with Gulf Minerals of Canada Ltd. to put pressure on the Australians to block the proposed arrangement" (ibid., p. 609).

342 ". . . member of the oligopoly": *Barron's*, Oct. 19, 1977.

343 The classification of the centrifuge work in Germany came as a shock to the German company, Degussa, which was developing the idea. It had had a special allergy to classified work ever since it partnered I. G. Farben in a company that produced Zyklon-B gas for the death camps during the war. In 1970, Degussa led a German industrial consortium in a joint centrifuge venture with Britain and the Netherlands.

344 The challenge for Giraud was a technical one; because the centrifuge technology had not yet been definitely proven, Urenco was not able to satisfy the growing fears of an "enrichment gap." This was an opportunity, perhaps the last the French diffusion technology would ever have. At this stage predictions of installed nuclear capacity throughout the

world were at their highest levels. There was a real fear that the Americans would not be able to satisfy the demand. For some years the planned expansions to American enrichment capacity had been delayed by the insistence of the Nixon administration that the next plant should be built by private enterprise. The private consortia were still continuing their investigations of the commercial practicality of a private plant, and, until they concluded their work, American plans were effectively paralyzed. Meanwhile fears had grown that an enrichment gap would exist, perhaps as early as 1983, but definitely by 1985.

André Giraud had shown his determination to become less reliant on the United States enrichment monopoly when, in 1971, he had signed the first international contract for the supply of low-enriched uranium with the Soviet Union. Like the United States, the Russians found that their enrichment plans were now far in excess of their military needs. Giraud's example was quickly followed by others who signed up for Russian enrichment services. No one, however, wanted to become significantly dependent on the Soviet Union. The case for a European enrichment plant in the medium term became more definite. In 1971 the Americans tried to head off the growing European interest in their own enrichment plant by offering to build a multinational plant based on American technology. It was a hamfisted effort to perpetuate the old monopoly and was scorned by the newly confident Europeans.

CHAPTER 22

347 For an excellent description of the British advanced reactor decision, see Roger Williams, *The Nuclear Power Decisions*, pp. 226–42. In the build-up to the decision made in July 1974, the Americans pressed on the British the advantages of their PWRs and the Canadians, the CANDU. Light-water reactors were certainly the cheapest in the world, but the British had doubts about their safety. They also wanted to develop their own technology. The choice eventually before the Energy Minister, Eric Varley, was between a British-made reactor using heavy water, called the Steam Generating Heavy Water Reactor, and SGHWR, a third generation reactor similar to the CANDU. One of the major problems, however, was that the heavy water had to be imported, adding considerably to the cost. Those in favor of the SGHWR pointed to the success of the British prototype and the prospect of cooperation with Canada. They also doubted that the PWR could be developed quickly in Britain. Those in favor of the PWR cited cost advantage and operating experience—plus the possibility of an export market. The government decided in favor of the SGHWR. Eric Varley said, after the

decision, that it had been made on the grounds of reliability and acceptability to the public—with the added consideration of "giving a boost to British technology"; see Roger Williams, op. cit., p. 234.

348 The description of the Siemens company is drawn from Siegfried von Weiher and Herbet Goetzler, *The Siemens Company;* Robert Heller and Norris Willatt, *The European Revenge; Frankfurter Allgemeine Zeitung,* July 14, 1971; *Financial Times* (London), Sept. 16, 1975; *International Management*, Feb. 1971; *Capital*, Sept. 1969; *Der Spiegel*, no. 42, Oct. 1977; *Fortune*, Apr. 10, 1978.

348 "The House of Siemens": The chief executive who watched the company grow into a multinational with $4 billion in sales could say without a trace of irony: "What else should we call it? 'Corporation' sounds a little too capitalistic to all of us" (*Capital*, Sept. 1969). This paternalistic tradition was reflected in the rigidity of the company's internal organization. It took six years to negotiate a reorganization into a divisional structure with clearly defined areas of responsibility and rational central service departments. When it was completed in 1969, company spokesmen acknowledged that the reorganization was needed to get rid of "rigidly isolated kingdoms within its general framework" (ibid.). By this stage, however, the nuclear kingdom was mature.

348 The sketch of Wolfgang Finkelnburg is based on *Atomwirtschaf*, July–Aug. 1965; *Handelsblatt*, June 4, 1965, Nov. 9, 1967; Nov. 22, 1968; *Siemens Zeitschrift*, no. 42, 1968; *Physikalische Blatter*, vol. 21, 1965, vol. 23, 1967; Alan D. Beyerchen, *Scientists Under Hitler*; Further material on Erlangen and Heinz Goeschel comes from *Das Neue Erlangen*, no. 14, Mar. 1969; *Kaltetechni-Klimatesrung*, vol. 18, no. 2, 1966; *Frankfurter Allgemeine Zeitung*, May 25, 1965.

349 When Germany's largest private utility, the Rheinisch-Westfälisches Elektrizitätswerke (RWE) decided to opt for light-water reactors, Siemens knew they had to drop their heavy-water projects and make their own light-water types. The first contact went to their rivals, AEG, for a 250-Mw boiling-water reactor. It was ordered by the utility RWE and built under license from General Electric.

350 The Siemens relationship with Westinghouse is based on interviews with Wolfgang Braun and Hans-Joachim Preuss at Erlangen in July 1978. Also *Industriekurier*, Oct. 31, 1967; June 13, 1970; *Suddeutscher Zeitung*, June 12, 1970.

351 The sketch of Bernhard Plettner is based on Manzinger Archives; *Frankfurter Allgemeine Zeitung*, July 17, 1971; *Der Spiegel*, Nov. 8, 1971; *Handelsblatt*, Nov. 1974; *Die Welt*, Nov. 1974; *Capital*, May 1975; *Die Welt*, Nov. 19, 1976. Plettner's veto of the plan to drive AEG out of the market by grabbing two reactor contracts led to the sacking of the head of Siemens's power plant division. They would have done AEG a big favor, since one BWR proved an expensive lemon.

352 ". . . shed a tear": *Der Spiegel*, Nov. 25, 1974.

352 The Biblis order which stretched the Siemens technology to the 1,200-Mw mark was the key to its technical success. Siemens had been driven to this task by the new senior executive of the purchasing utility, Heinrich Mandel. Long the resident nuclear advocate inside RWE, Germany's largest utility, Mandel was about to succeed Karl Winnacker as Germany's "nuclear pope." RWE's early investments in nuclear had been determined by a political fear of public power, but in the process a strong pro-nuclear team had developed inside the Technical Division. Its head was Mandel, whose academic bent had led him to take doctorates in both engineering and physics. His intellectual fascination with nuclear energy remained constant from the time that RWE sent him to the first Atoms for Peace Conference in Geneva in 1955. After his appointment as technical director on the RWE Management Committee, he reorganized the separate institutional fiefdoms in his section which built and operated the RWE power plants and effectively centralized decision making. The Biblis order of 1969 coincided with the worldwide upsurge in utility interest in nuclear power, and in Germany, where RWE was a technical herd leader, new orders came in a steady stream. In 1975 six orders were placed by German utilities, in order to take advantage of special capital investment depreciation tax breaks allowed for that year. Also in 1975 came the Brazil order for two reactors and six options and in 1976 the Iranian order for two plants, quickly followed by a letter of intent for another four. On Mandel and RWE see *Die Atomwirtschaft*, Oct. 1956 and Aug. 1964; *Der Volkswirt*, Nov. 20, 1964; *Der Spiegel*, Dec. 15, 1968; *R. W. E. Verbund*, no. 86, May 1974; *Munzinger Archives*, Otto Keck, "Government Policy and Technical Choice in the West German Reactor Program," 1979.

352 "We wish that competition . . .": *Der Spiegel*, no. 47, Nov. 15, 1977.

353 The sketch of Baron Empain is based on *L'Express*, Feb. 22, 1976; *Le Monde*, July 8, 1969, Feb. 5, 1974, Mar. 28, 1978; *Figaro*, March 27, 1978; *Paris Match*, Feb. 1978; P. Allard et al., *Dictionnaire des Groupes Industriels et Financiers en France*. Analysis of Framatome domestic political triumph draws on Philippe Simonot, *Les Nucléocrates*; Frédérique de Gravelaine and Sylvie O'Dy, *L'Etat E.D.F.*; Michel Herblay, *Les Hommes du Fleuve et de l'Atome*; Louis Puiseux, *La Babel Nucléaire*; Irvin C. Bupp and Jean-Claude Derian, *Light Water*.

355 "'. . . agreement of the President . . .'": Simonot, *Les Nucléocrates*, p. 161.

355 The Westinghouse interest in Framatome was acquired in 1972. Its 45 percent holding was significant because under French law, any shareholder who owns one-third of the shares can veto any company decision. Framatome needed Westinghouse very badly at that stage. Unlike Siemens, it had no independent source of technology; indeed, when Fra-

matome received its first order from EDF in 1970 it was a company with only two hundred employees. As one French commentator noted, "EDF has placed its nuclear plant orders with an industry that did not exist" (Herblay, op. cit., p. 225). In the reorganization of 1975, Framatome showed its political strength by restricting the Commissariat under the aggressive Giraud to less than the one-third "blocking" share interest. Giraud had a major political fight on his hands to retain for the CEA a blocking majority in the new Framatome-controlled consortium that would build breeder reactors—a technology wholly developed by the CEA itself.

CHAPTER 23

Basic sources used in tracking the growth of the antinuclear movement were: Anna Gyorgy, *No Nukes*; Peter Faulkner (ed.), *The Silent Bomb*; Mans Lonnroth and William Walker, *The Viability of the Civil Nuclear Industry*, International Consultation Group on Nuclear Energy, distributed by Royal Institute of International Affairs, London, July 1979; Ralph Nader and John Abbotts, *The Menace of Atomic Energy*; "The Nuclear Debate: Opponents and Proponents," a compilation of articles in a *Nucleonics Week, Special Report*, McGraw-Hill, 1976; George Vickers, "Showdown over Nuclear Power," *Seven Days*, vol. 11, no. 4, Mar. 24, 1978; Union of Concerned Scientists, *The Nugget File*; Christoph Hohenemser et al., "The Distrust of Nuclear Power"; International Atomic Energy Agency, *A Study of the Early Nuclear Controversy in the United States*; Arthur Murphy (ed.), *The Nuclear Power Controversy*.

357 The analysis of the three GE engineers and the NRC engineer who resigned is based on *Nucleonics Week*, Jan. 29, 1976, and Feb. 5, 12, 19, 26, 1976. Also Peter Faulkner (ed.), *The Silent Bomb*, especially chaps. 14 and 15; Duncan Burn, *Nuclear Power and the Energy Crisis*, pp. 87 ff.; Ralph Nader and John Abbotts, *The Menace of Atomic Energy*; interview with Dale Bridenbaugh, San Jose, Apr. 1978.

358 "sheer good luck": Letter to William Anders (NRC chairman), 2nd enclosure, Report on NRC Safety Review Process, Feb. 6, 1976, quoted in Nader and Abbotts, op. cit., p. 112.

358 "increasingly alarmed . . .": *Nucleonics Week, Special Report*, op. cit., p. 10.

358 "covered up or brushed aside . . .": Nader and Abbotts, op. cit., p. 56.

358 "Four Resignations. . . .": *Nucleonics Week, Special Report*, op. cit. p. 12.

359 "The cumulative effect . . .": JCAE hearings, Feb. 1976.

360 Nuclear power accidents: see compilation by Union of Concerned Scientists, *The Nugget File*.

361 The pattern of public reaction was universally similar in part because the pattern of advocacy had been so similar. The nuclear communities in each industrial nation had regarded themselves as a special technical elite and had shared a grandiose vision of their technology and its contribution to human welfare. The arrogance spawned by this belief—nourished as it had been in the flow of international meetings—gave rise to a fundamental distrust of the nuclear community in a significant minority. The belief that the public had been denied information about nuclear power was quickly transformed into an assumption that the nuclear community had something to hide.

362 There were very few nuclear power sites, especially in the nations of high population density, that were not close to significant concentrations of populations. The Japanese, and later the French, took the problem to its logical conclusion and instituted a nationwide system of compensation payments for those who lived near reactors. In Japan, a special levy for the funds was made on all electric utilities, the money being used to finance new public works projects for the affected communities and, in many cases, for direct subsidies for local industries—particularly fishing. Fishermen feared the heated water discharges from the plants on coastal sites would affect their catches. In France, those who lived near nuclear plants were given electricity at cheaper rates.

364 The description of the *Thresher* inquiry is drawn from: Norman Polmar, *Death of the Thresher*; Norman Polmar and D. A. Paolucci, "What Killed the Scorpion," *Sea Power*, May 1978; Rowland F. Pocock, *Nuclear Ship Propulsion*; Loss of the U.S.S. *Thresher*, Hearings before the Joint Committee on Atomic Energy, 88th Cong., 1st and 2nd Sessions, 1965; *New York Times*, Apr. 12, 1963, Jan. 10, 1965. Rickover's approach to the *Thresher* is reminiscent of the politically motivated technical assertions of another U.S. naval disaster, the sinking of the U.S.S. *Maine* in 1898. Ironically, that was subject of a devastating critique by Rickover himself. See H. G. Rickover, *How the Battleship Maine Was Destroyed*.

367 The Union of Concerned Scientists, a Boston-based group, was created in 1968 as a result of student protests against academic research for the military. The group opened its campaign by issuing an alarming statement indicating severe doubts about the safety of nuclear reactors. The reactor safety team was headed by Henry Kendall, a professor of physics at MIT. "A major nuclear accident," said the team, "has the potential to generate a catastrophe of very great proportions, surely greater than any peacetime disaster this nation has ever known" (Boston *Globe*, Aug. 22, 1976).

367 "Test 485 . . .": Quoted in Nader and Abbotts, op. cit., p. 101. By 1979, when the tests were successfully conducted, the public damage had been done.

368 ". . . met in hotel rooms . . .": Ford and Kendall evidence.
368 "unverified": George Brokett et al., "Loss of Coolant/Emergency Core Cooling Augmented Program Plan, August 1971," quoted in Nader and Abbotts, op. cit., p. 105.
369 The Rasmussen study: Official title is Reactor Safety Study (An Assessment of Accident Risks in U.S. Commercial Nuclear Power Plants), WASH-1400. For critics of, see especially Anna Gyorgy, *No Nukes*, pp. 114–17; Nader and Abbotts, op. cit., pp. 119–25.
369 Rasmussen purported to analyze the full range of accident possibilities, but the fire at the Brown's Ferry plant, triggered by a worker's use of a naked candle to inspect electrical wiring, was exactly the kind of unpredictable error that was not considered. Many other incidents that actually occurred were also ignored.
372 The effectiveness of the protest groups also depended on the kind of authority vested in central and local government to approve nuclear plant sites. It varies considerably. In Germany, Canada, Japan, Sweden, and the U.S. these decisions are subject to local or provincial government approval; in Britain and France central government is in control. Moreover, regulatory and licensing decisions can be challenged in the courts in Germany and the U.S., but not in Canada, France, Japan, Sweden, and Britain. In both Sweden and Germany—and in an increasing number of states in the U.S.—operating licenses are issued only to those utility companies which have been seen to be making satisfactory progress on the problems of handling spent fuel rods and the disposal of toxic wastes.
372 A general description of the Whyl demonstrations appears in Gyorgy, op. cit., pp. 324–25, 347 ff. Also in John Berger, *Nuclear Power*, pp. 341 ff.
372 Seabrook demonstration is written up in Gyorgy, op. cit., especially p. 395.

CHAPTER 24

There is a small library on the development of nonproliferation policy during the seventies. The authors have relied extensively on press reports and the periodical literature, both of which are too voluminous to list. Two congressional compilations of this material are *Peaceful Nuclear Exports and Weapons Proliferation: A Compendium Committee on Government Operations*, U.S. Senate, 94th Congress, 1st Session, U.S. GPO, Washington, D.C., 1975; *Reader on Nuclear Nonproliferation*, prepared for the Subcommittee on Energy, Nuclear Proliferation and Federal Services of the Committee on Governmental Affairs, U.S. Senate, 95th Congress, 2nd Session, U.S. GPO, Washington, D.C., 1978. Three congressional documents were also valuable: in Mar. 1976 the Congressional Research Service compiled a report, *Nuclear Weapons Proliferation and the International Atomic Energy Agency*, and in Sept. 1977 the Service produced *Nuclear Proliferation Factbook*. The Office of Technology Assessment prepared a study for Congress entitled *Nuclear Proliferation and Safeguards* that was republished by Praeger in 1977. Other sources include Atlantic Council, *Nuclear Power and Nuclear Weapons Proliferation*; William Epstein, *The Last Chance*; Brian Johnson, *Whose Power to Choose?*; Ashok Kapur, *International Nuclear Proliferation*; Robert M. Lawrence and Joel Larus (eds.), *Nuclear Proliferation*; Alva Myrdal, *The Game of Disarmament*; George Quester, *The Politics of Non-Proliferation*; Stockholm International Peace Research Institute, *Safeguards Against Nuclear Proliferation* and *Nuclear Proliferation Problems*; Mason Willrich (ed.), *International Safeguards and Nuclear Industry*; Albert Wohlstetter et al., *Moving Towards Life in a Nuclear Armed Crowd* and *Swords From Plowshares*. In addition the authors have profited from interviews with: John Palfrey, Myron Kratzer, Allan Labowitz and Harold Bengelsdorf, from the U.S.; Ivan Head of Canada; Bertrand Goldschmidt of France, M. A. Velodi and P. N. Haksar of India; David Foscher of IAEA, Alan Wilson and Neil MacDonald of Australia, and Imai Ryukichi of Japan.

374 "The Buddha is smiling": *Motherland*, June 1, 1974.

375 Two years after the Indian test, when the public and congressional concern about proliferation was more acute, Henry Kissinger could tell a congressional committee about the U.S. reaction to the test. "We deplored it strongly," he said. What he had in fact said at the time was: "I do not believe that the Indian nuclear explosion changes the balance of power, though if India had asked our advice we would probably have not recommended it" (Secretary of State press conference, June 7, 1976, and Committee of Government Operations, U.S. Senate Export Reorganization Act, 1976 Hearings, U.S. GPO, 1976, p. 793).

377 The analysis of the Indian nuclear explosion is based on interviews in New Delhi in July 1978—especially with M. A. Velodi of the Ministry of Foreign Affairs and P. N. Haksar, former principal private secretary to Indira Gandhi. In addition J. P. Jain, *Nuclear India*; Ashok Kapur, *India's Nuclear Option*; A. G. Noorani, *Aspects of India's Foreign Policy*; T. T. Poulose (ed.), *Perspectives of India's Nuclear Policy*; K. Subrahmanyam, "India: Keeping the Option Open," in Robert Lawrence and Joel Larus (eds.), *Nuclear Proliferation*.

377 ". . . reactor was naked": Walter Stewart, "How We Learned to Stop Worrying and Sell the Bomb," *Macleans*, Nov. 1974.

378 "The only defense . . .": *Nuclear India*, Nov. 1964.

378 "artillery testing range": *Indian Express*, June 2, 1974; *Motherland*, June 2, 1974.

380 The London Suppliers Group of the seventies was a lineal descendant

of the old Western Suppliers Group of the fifties and sixties. The new emphasis was on technology rather than uranium supplies. The transition had been telegraphed as long ago as 1962 when a committee chaired by the former AEC commissioner, Henry De Wolfe Smythe, noted, without saying a suppliers group existed, how safeguards policy was complicated by the fact that it was no longer true that "the supply of uranium . . . [was] . . . extremely limited and easily controlled." See *Peaceful Nuclear Exports and Weapons Proliferation: A Compendium*, p. 236. In Feb. 1967 the U.S. convened a suppliers meeting in their London embassy: it was the last attended by the uranium suppliers (like Australia and South Africa) and the first attended by technology exporters (like Germany, Sweden, and Switzerland). The frequency of meetings of the group in the sixties remains secret.

381 The contract to supply Iraq with light-water reactors using enriched uranium proved a little more difficult. As the second-largest oil supplier to France, Iraq had to be treated more circumspectly. In May 1979, three weeks before the reactors were due to be shipped to Iraq, they were conveniently blown up outside Marseilles by five well-placed charges of plastic explosive. Everyone naturally blamed the Israelis, but the more observant sections of the French press, stressing the convenience of the event, indicated that French security might have done the job themselves. See *Time*, May 7, 1979; *Le Nouvel Observateur*, Apr. 14, 1979.

383 ". . . it is too late": Washington *Post*, Apr. 2, 1977.

384 In Japan the refusal to accept the new American policy which asserted that waste disposal did not require reprocessing was in part a product of linguistic confusion. The Japanese used the same basic word to describe *both* reprocessing and waste disposal. The Japanese word *shori* is equivalent to "treat" or "deal with," with strong overtones of final treatment or disposal. The word for reprocessing was *sai-shori* and the word for waste disposal was *haikitsu-shori* (the term *sai* is equivalent to the English prefix "ré-" and *haikibutsu* is the same as "waste"). One can imagine the confusion of Japanese political leaders who were being told that *sai-shori* was totally unrelated to *haikitsu-shori*.

385 The Non-Proliferation Act of 1978 required that the principle of "timely warning" would receive "foremost consideration" in making a determination of "no significant increase of risk of proliferation." A tougher, more binding clause was deleted in Congress under strong administration pressure. Nevertheless U.S. experts continued to assert that in the case of highly sensitive facilities there was very little real discretion in the legislation. See, e.g., Victor Gilinsky, "Plutonium, Proliferation and the Price of Reprocessing," *Foreign Affairs*, winter 1978/79, p. 381.

385 "It is doubtful . . .": Imai Ryukichi, "Nuclear Safeguards," *Adelphi Papers*, no. 86, 1972, p. 4.

385 The final assessment of the safeguards system draws on the references listed in chap. 18 note for p. 304. In addition the further development of the system in the seventies and the response to the new U.S. policy is traced in the IAEA's Standing Advisory Committee on Safeguards Implementation, the documents are coded as AG/43, to which the authors have had access.

385 ". . . inherent safeguardability . . .": IAEA 43/11, p. 7.

387 Desai's efforts to restrain the nuclear hawks included the action he took against Dr. R. Ramana in 1978. Ramana and Homi Sethna, the chairman of the AEC, had been Bhabha's key lieutenants, and the two had been in charge of the test in 1974. Sethna was a realistic engineer preoccupied with the practical problems of the nuclear power problem—which international hostility to the explosion threatened. Ramana, head of the research institute, had no responsibility to deliver on a program, and he could indulge his proclivity for flights of rhetoric about the future of peaceful nuclear explosions. At the Pokaran test in 1974 he had been stirred: "The explosions produced an artificial hill," he exuded, "a most beautiful sight which came on the skyline from nowhere" (*National Herald*, May 20, 1974). Sethna and Ramana had fallen out personally years before, and every meeting of the AEC had been dominated by their bickering. In 1977, Ramana tried to prevent any compromise with the U.S. and a series of leaks began appearing in the press about the possibility of substituting the American fuel in the Tarapur reactor with a "wholly Indian" mixture of plutonium and uranium. Soon the stories identified Ramana as the only scientist fighting for this policy, but they went too far in Mar. 1978 when one leak referred to a meeting with Prime Minister Desai in which he was depicted as resisting the patriotic line and "pleading for forbearance." The next day Desai summoned an ad hoc cabinet meeting and Ramana was sacked. *Financial Times*, Jan. 10, 1978; *The Hindu*, Mar. 16, 1978; *Indian Express*, Mar. 15, 1978; *Times of India*, Mar. 20, 1978.

388 On Pakistan see D. K. Palit and P. K. S. Namboodiri, *Pakistan's Islamic Bomb*; *Time*, July 9, 1979; *New York Times*, Apr. 29, 1979, July 1, 1979; *Nucleonics Week*, June 28, 1979. Egypt once sought a bomb from India (see Palit and Namboodiri, p. 62).

CHAPTER 25

389 The IAEA gave assistance for nuclear programs only so the amounts allocated were not available for other energy sources. Third World economic planners—who were not present in Vienna—were not able to put up alternative uses for the aid. All of the delegations in Vienna—or in any nuclear conference—were heavily influenced and many were com-

pletely dominated by representatives of the national nuclear bureaucracies. The Third World delegations did not differ from the norm in terms of the way they used their international access to improve their domestic bureaucratic positions. Nuclear aid was on a guaranteed escalator during the annual budget fight at each IAEA general conference. The Third World delegations banded together to force an increase in the aid budget by threatening to use their voting power to cut the budget allocated for the safeguards function—the IAEA activity in which the Western nations were most interested. As concern with proliferation grew in the seventies, the Third World blackmail worked and the amount promised for nuclear aid increased year by year. In the treasury departments of each aid-giving nation, these amounts came out of the agreed total aid budget, but the economic planners in the Third World countries had no influence on how it was spent. Because of the special status of the IAEA, the money went to nuclear projects alone.

391 ". . . social benefits etc. . . .": A. M. B. Nural Islam et al., "Problems faced by Bangladesh in Introducing a Nuclear Power Programme," in *Nuclear Power and Its Fuel Cycle*, Proceedings of an International Conference, vol. 6, IAEA, Vienna, 1977, p. 220.

392 The outline of the AECL bribery incidents is drawn from: *New York Times*, Nov. 26, 1976, Dec. 3, 1976, Dec. 18, 1976, July 8, 1977, and Aug. 1, 1977; *New Statesman*, Aug. 12, 1977. The sensitivity of any "questionable payments" in the nuclear field was well known. In 1978, when Westinghouse pleaded guilty to a bribery charge and received judicial blessing to suppress the circumstances, the company was careful to emphasize that the case was not a nuclear sale.

394 "We leave it to Hermie . . .": *Time*, Jan. 23, 1978.

394 ". . . standard advertising brochures . . .": *New York Times*, Jan. 14, 1978.

394 ". . . political decision": Washington *Post*, Jan. 18, 1978.

394 Disini had no intention of remaining a mere "agent." He took control of the traditional Westinghouse distribution company in the Philippines and organized a construction consortium for the nuclear plant. His own insurance company, which had reported an income of some $20,000 the previous year, handled the massive $668 million insurance contract on the plant. Furthermore, Disini's intervention secured the government financing package for New York's Citibank against the rival group of the American Express bank—for fees that have never been disclosed. Put together, the various commissions were quite enough to meet anyone's golf club dues—or to pass on a share to a collaborator, if that was necessary. *See Wall St. Journal*, Jan. 12, 1978, p. 1. Also *Time*, Jan. 23, 1978, pp. 35–36.

395 One advantage that the less capital intensive electricity plants, such as hydroelectric, did not have was the offer of large subsidized interest loans

from the American Export-Import Bank, which put together a $650 million financing package for the Westinghouse export. The deal did, however, prevent the Philippines from raising capital on the international money market for other projects.

395 "referenced" and "sub-minimal": *IAEA Bulletin*, Apr. 1977.

395 The description of the Philippines reactor decision is based on: *Wall Street Journal*, Jan. 12, 1978; *New York Times*, Jan. 14, 1978; Washington *Post*, Jan. 18, 1978, Feb. 8, 1978; *Time*, Jan. 23, 1978; *Australian Financial Review*, Apr. 28, 1978; *Philippine Daily Express*, Jan. 19, 1978; W. Hersch et al., *Reactor Safety Mission to the Philippines Report*, IAEA, Vienna, 1978; *Science*, vol. 205, July 6, 1979. Also the file of cables from the State Department and trip reports from the U.S. AEC made available to Jacob Scherr of the Natural Resources Defense Council under the Freedom of Information Act.

395 "credible event": Hersch et al., op. cit., p. 12.

396 The analysis of the Indian program draws on a number of interviews in New Delhi, especially with Lavaj Kumar and T. L. Shankar of the Planning Commission, also with K. C. Khanna of the *Times of India* in Bombay. The following sources were also consulted: Fuel Policy Committee Report, New Delhi, 1972, especially para 6.13; Draft Five Year Plan Planning Commission, New Delhi 1978; P. L. Ragu Ram, "Role of Nuclear Science in Economic Development," *Economic Times* (Bombay), July 26, 27, 28, 1977; H. N. Sethna and M. R. Srinivasan, "India's Nuclear Power Programme and Constraints Encountered in Its Implementation," in *Nuclear Power and Its Fuel Cycle*, op. cit., p. 3; *Times of India*, July 21, 1970, June 23, 1972, Aug. 8, 1974, Nov. 11, 1974, Feb. 24, 1976, Mar. 24, 1976, Apr. 2, 1977, and June 19, 1978; *The Hindu*, Mar. 23, 1978; *Evening News of India*, Jan. 22, 1972; *National Herald*, May 26, 1974, Feb. 2, 1976; *Hindustani Times*, May 27, 1974, Feb. 28, 1978.

397 On Sarabhai see C. Raghavan, "Nuclear Policy," *Mainstream* (New Delhi), July 23, 30, Aug. 6, 13, 1977; Erik Erikson, *Gandhi's Truth*; *The Hindu*, May 30, 1966; *Times of India*, Nov. 25, 1962, June 3, 1966.

398 "Tarapur is so heavily contaminated . . .": Quoted in *Nature*, June 7, 1979, p. 469. The Tarapur reactor had been "referenced back" to the canceled GE project at Bodega Bay in California, just as the Philippines plant was "referenced back" to the Puerto Rico plant.

CHAPTER 26

The description of the nuclear recession of the mid-seventies and the analysis of the costs of nuclear power represents the authors' general conclusions from a wide range of material. The following sources were consulted: Irvin C.

Bupp, "Energy Policy in the United States: Ideological BTU's," in Leon Lindberg (ed.), *The Energy Syndrome;* Irvin C. Bupp and Jean-Claude Derian, *Light Water;* Irvin C. Bupp, "The Nuclear Stalemate," in Robert Stobaugh and Daniel Yergin (eds.), *Energy Future;* Duncan Burn, *Nuclear Power and the Energy Crisis;* Herman E. Daly, "Energy Demand Forecasting: Prediction or Planning," *Journal of the American Institute of Planners,* vol. 42, Jan. 1976; Edmund Faltermeyer, "Nuclear Power After Three Mile Island," *Fortune,* May 7, 1979; IAEA, *Nuclear Power and Its Fuel Cycle,* Proceedings of an International Conference, vol. 1, Vienna, 1977; Paul L. Joskow and Martin L. Buaghman, "The Future of the U.S. Nuclear Energy Industry," *Bell Journal of Economics,* vol. 7, spring 1976; Charles Komanoff, *Power Plant Performance;* Spurgeon Keeny et al., *Nuclear Power Issues and Choices;* Warren H. Donnelly, "Nuclear Power Through 1990," in Congressional Research Service, *Project Interdependence;* Amory Lovins, *Soft Energy Paths* and *Is Nuclear Power Necessary?;* W. E. Mooz, "Cost Analysis of Light Water Reactor Power Plants," R-2304-DOE, RAND Corp., Santa Monica, Calif., 1978; Arthur W. Murphy (ed.), *The Nuclear Power Controversy,* chaps. 2 and 3; Congressional Research Service, *Nuclear Proliferation Factbook;* Walter Patterson, *Nuclear Power* and *The Fissile Society;* Robert Perry et al., *Development and Commercialization of the Light Water Reactor, 1946–1976,* R-2180 NSF, RAND Corp., Santa Monica, Calif., 1977; Gus Speth, "The Nuclear Recession," *Bulletin of the Atomic Scientists,* Apr. 1978; Vince Taylor, *The Myth of Uranium Scarcity,* PanHeuristics, a division of Science Applications Inc., Los Angeles, 1977; Vince Taylor, *The Easy Path,* PanHeuristics, Los Angeles, 1979.

400 Rare as a comet: quoted in Gus Speth, "The Nuclear Recession," *Bulletin of the Atomic Scientists,* Apr. 1978, p. 24.

401 Coal/nuclear capital costs: Ibid.

403 "The Uranium Thing" and ". . . most stupid performance . . .": *Newsweek,* Feb. 16, 1977, p. 35.

403 ". . . one of the world's leading toy puzzle solvers . . .": See *Business Week,* Jan. 31, 1977, p. 62.

403 "A Westinghouse nuclear power plant . . .": quoted in "Uranium Short Sale," by Richard Karp, *Barron's,* Oct. 17, 1977.

404 uranium cartel: Based on "Uranium Short Sale," ibid; International Uranium Cartel, Hearings Before Subcommittee on Oversight and Investigation of the Office on Interstate and Foreign Commerce, House of Representatives, 95th Congress, 1st Session, May–Aug. 1977; *Business Week,* Jan. 31, 1977, "The Opposites: GE Grows While Westinghouse Shrinks"; David Burnham, *New York Times,* Dec. 4, 1979; "The Great Uranium Flap," Anthony Parisi, *New York Times,* section 3, July 9, 1978.

406 Disposal of radioactive wastes draws from: "The Disposal of Radioactive Wastes for Fission Reactors," *Scientific American*, vol. 236, no. 6, June 1977; Alan Jakiaio and Irvin C. Bupp, "Nuclear Waste Disposal: Not in My Backyard," Harvard Graduate School of Business Administration, *Technical Review*, Mar./Apr. 1978; "Radioactive Wastes, Some Urgent Unfinished Business," *Science*, vol. 195, Feb. 18, 1977, p. 661.

409 The sketch of Marcel Boiteux draws on: Frédérique de Gravelaine and Sylvie O'Dy, *L'Etat E.D.F.*; Philippe Simonot, *Les Nucléocrates; Entreprise*, Sept. 22, 1967; *L'Expansion*, Oct. 1967; *L'Express*, Sept. 1, 1969, Dec. 23, 1974; *La Croix*, Sept. 27, 1969, Mar. 23, 1977; *Jours Français*, Feb. 5, 1971; *Le Management*, May 1972; *L'Aurore*, Oct. 7, 1974; *Figaro*, Aug. 9, 1977; *Libération*, June 8, 1978.

410 The revival of Soviet interest in nuclear power is based on: Philip R. Pryde, "Nuclear Energy Development in the Soviet Union," *Soviet Geography*, vol. 19, Feb. 1978; Jozef Wilczynski, *Atomic Energy for Peaceful Purposes in the Warsaw Pact Countries*; A. M. Petrosyants, "Nuclear Power and Its Significance in the USSR as a Source of Energy"; A. P. Alexandrov et al., "Structure of the Nuclear Power Industry with Allowance for Energy Production Other Than Electricity"; and A. N. Gregoryants et al., "Development of the Nuclear Power Industry in the USSR": all three papers are found in *Nuclear Power and Its Fuel Cycle*, Proceedings of an International Conference, vol. 1, IAEA, Vienna, 1977; *New York Times*, Aug. 22, 1965, Dec. 27, 1972, Oct. 21, 1973, May 30, 1974, and Jan. 14, 1977; *Christian Science Monitor*, Aug. 23, 1966, Sept. 28, 1977; *The Financial Times* (London), May 11, 1977; *Business Week*, Aug. 2, 1976.

410 A Western report of Russia's atomic energy program noted, "The present commentators stress that the lessons of that error have been well understood and no repetition of the cycle of public boasting and then quiet withdrawal will take place" (*New York Times*, Aug. 22, 1965).

411 ". . . 'the opponents' . . .": *Ekonomyskaya Gazetta*, no. 29, July 21, 1965. Yermakov had been a foundation member of the original special atomic division of the Electricity Ministry which had been absorbed. The new Soviet commitment was made clear in 1968 when the separate Atomic Power Division was revived at a status that assured its chief of a position on the collegial board of management for the whole ministry. That job was given to A. N. Gregoryants, a former head of the power stations section of the nuclear bureaucracy, the State Committee on the Uses of Atomic Energy.

411 For all their determination, the Soviet planners faced the same kinds of problems as their Western counterparts. Nuclear power offered cost advantages only in certain places; there was no absolute shortage of energy—the Soviet Union had huge coal and hydroelectric potential—but

the coal and hydro sites were mainly in Siberia, far from the heavily populated and industrialized part of the Soviet Union west of the Urals. It was expensive to carry coal or transmit electricity over such distances. For the same reasons, different regions of Europe or the United States went nuclear to varying degrees. Soviet plans to shift population and industry to the East had not worked, and so the commitment to nuclear power intensified.

CHAPTER 27

412 Sketch of Paracelsus: See Isaac Asimov, *Biographical Encyclopedia of Science and Technology*, pp. 115–17. Asimov's description of Paracelsus is the most pithy: "Despite a mystical obscurity of statement, he marks the beginning of the transition from alchemy to chemistry. . . . The purpose of alchemy, he decided, was not to discover methods for manufacturing gold but to prepare medicines with which to treat disease. . . .

"Nor did Paracelsus in any way give up the mysticism of alchemy and astrology. . . . He sought increasingly for the *philosophers stone*, which he believed to be the elixir of life, and even claimed to have found it, insisting that he would live forever. . . . For the rest of his life he kept furiously inveighing against his enemies and predecessors. He seized medicine by the scruff of its neck, so to speak, and if it wasn't entirely sense that he shook out of it, the shaking was beneficial just the same."

414 "The tractors of the farmers . . .": Helmut Wustenhagen, German delegate, at press conference at Salzburg alternative conference, April 1977.

414 "Renounce nuclear energy?": IAEA, *Nuclear Power and Its Fuel Cycle*, 1977, vol. 2, p. 232.

415 Sketch of Weinberg: *Science*, Jan. 14, 1977.

415 From the time that nuclear energy stepped out of the scientific laboratories into the realm of the industry, Oak Ridge was the international nuclear capital of the world. As the headquarters of the Manhattan Project, the specially selected site on one of the five pine-covered ridges around the Clinch River in Tennessee became the first—and is probably still the largest—atomic city. The first pilot plutonium-producing reactor was built here, as were all the systems for enriching uranium. The fissile material for the Hiroshima bomb was manufactured at Oak Ridge, and it remains a major center for producing low-enriched uranium for light-water reactors throughout the world. It was also the site of the American prototype for a fast breeder reactor. The project was threatened by President Carter's policy change. When work at Oak Ridge began in late 1942, security guards were bemused by the pleas of a bearded religious fanatic who was convinced that a new Vatican was being built

there and wanted to stake his claim as the first American Pope. "Beat it," the security man said. "We don't know what's getting built here." With all the acuteness of innocence the man replied: "Well, if you don't know what's being built, how do you know it isn't a Vatican?" (Daniel Lang, *From Hiroshima to the Moon*, p. 21). In a sense, it was Alvin Weinberg who became "Pope." As soon as he became director of the Oak Ridge National Laboratory in 1955, Weinberg began his campaign to extend the laboratory's charter beyond the nuclear sphere. He was convinced that the concentration of multidisciplinary talent in the AEC-founded institutions could be put to broader use. As early as 1955 he focused on the eventual consumption of the world's supplies of fossil fuels like oil and, eventually, even coal. He believed that nuclear power would play an important role in the future but emphasized that a focus on energy in the broadest sense, not just nuclear, was required to make the choices that the world would eventually face. He also wanted to launch major studies of food and water supplies and research on health and environmental effects of industrial growth. In May 1955 he noted, "Eventually the Atomic Energy Commission might be renamed simply the Energy Commission, with all that such a renaming would imply. I suppose one could go farther and think eventually of a 'natural resources' or 'geological engineering' commission to take responsibility for these long term environmental problems, which we hope are tractable to large scale project research" (Harold Orlans, *Contracting for Atoms*, p. 230). Under his leadership the Oak Ridge teams extended their nuclear horizons into a variety of major research projects. The optimism about nuclear power led them to study the prospects for large-scale purification of salt water. This resulted in a short-lived enthusiasm in the 1966 program launched by President Johnson under the label "Water for Peace," in conscious imitation of the American public relations triumph with "Atoms for Peace." Oak Ridge also expanded its work on the biological effects of radiation—the largest program of genetic research had been done there. It studied other causes of cancer—and possible cures. Weinberg's early environmental consciousness also resulted in a range of studies—using the expertise that had focused on nuclear power—on a variety of dangers that Weinberg called "physical insults to the biosphere."

415 Shortly before the Salzburg conference, Weinberg and his colleagues at Oak Ridge had produced a study on the future of energy sources in the twenty-first century. It was the first major report from inside the nuclear community suggesting that the opportunity existed for a moratorium on the building of nuclear plants; specifically, it suggested such a halt for thirty years, from 1980 to 2010. The report assumed that the breeder would be developed; but to the dismay of the more militant nuclear ad-

vocates, it also basically agreed with the energy demand estimates of the Ford Foundation Energy Project study, *A Time to Choose*, which had concluded that nuclear power was not crucial to world energy needs until well into the next century.

415 ". . . fragile monuments . . .": Daniel Kevles, *The Physicists*, pp. 417–18.

415 "We nuclear people . . .": *Science*, vol. 172, July 7, 1972.

416 "We are unaccustomed . . .": IAEA, *Nuclear Power and Its Fuel Cycle*, vol. 1, p. 760.

416 "Our enterprise . . .": Ibid., p. 762.

416 "Our technology . . .": Ibid., p. 771. Weinberg's candor continued. "If one is worried about proliferation, how can we seriously contemplate a world in which 100 tons of plutonium may be reprocessed every day?" The fact that he believed it could be managed was little consolation to his restive audience. Even worse was his insistence on emphasizing the special nature of nuclear power, when the community's whole public debating posture was one of stressing the normalcy and benign nature of their craft.

Weinberg had been enthusiastic about the post–Oyster Creek reactor orders in the mid-sixties. At the time he had declared that all the problems of nuclear power would be solved by the prospects of a cure for leukemia and other cancers. By the late seventies this seemed to be the only thing that could save the nuclear industry in the long term. Public reaction, based on fears of damage to health, was frustrating nuclear plans everywhere. In a sense, a large and growing number of people were willing to pay higher energy costs to avoid nuclear power. Weinberg himself admitted that he would prefer solar if it cost only twice as much as nuclear—he didn't believe it would.

416 ". . . audience so disconcerted . . .": Jungk interview, Aug. 1977.

417 "If the world forswears . . .": IAEA, *Nuclear Power and Its Fuel Cycle*, vol. 1, p. 770.

418 Lovins's energy strategy: "The Road Not Taken?" *Foreign Affairs*, Oct. 1976.

418 The article in *Foreign Affairs* was extensively quoted in the international press, entered in the *Congressional Record*, discussed in *Business Week*, and subject to the most requests for reprint ever received by *Foreign Affairs*.

420 *Nuclear Power: Issues and Choices*, 1977, Nuclear Energy Policy Study Group.

421 Weinberg, "I put the previous scenarios forward . . .": IAEA, op. cit., p. 770.

CHAPTER 28

Sources used include: Daniel Martin, *Three Mile Island;* Richard Curtis and Elizabeth Hogan, *Nuclear Lessons;* Robert Jungk, *The New Tyranny;* David Lilienthal, *Atomic Energy;* Anthony Nero, A *Guidebook to Nuclear Reactors;* Union of Concerned Scientists, *The Nugget File;* Report of the President's Commission, The Accident at TMI, the so-called Kemeny Commission, Oct. 1979; "Crisis: Three Mile Island," *Washington Post Special Report,* 1979; interviews presidents TMI and doctors; transcript of emergency meeting of NRC, Mar. 30, Apr. 1, 1979, in Conference Room, Washington, D.C.; an NRC Incident Response Center, Bethesda, Maryland, Rockefeller Foundation Report.

423 stay off the streets: see *Columbia Journalism Review,* July–Aug. 1979, p. 58 ref. Atlanta *Constitution* reporter, Barry King.
423 Middletown residents' reactions in interviews with authors at time of accident and one year later. Also see *Columbia Journalism Review,* op. cit., pp. 57–58.
424 "Race with Nuclear Disaster" and "No Injuries Reported in Nuke Mishap": Ibid., p. 58.
424 "evil" steam: Ibid.
424 Drop in reactor output: *New York Times,* Mar. 16, 1980.
425 Description of accident: see Kemeny Commission Report, pp. 90–141.
427 Walter Creitz on TV: *Washington Post Special Report,* op. cit., chap. 4. See also for quotes of John Herbein, Mayor Robert Reid.
427 ". . . is your plant a lemon?": Ibid.
427 NRE calculation on dose: NRC transcript, Mar. 30, 1979, p. 5.
428 Edited transcript of Mar. 30–Apr. 1, 1979.
433 For account of demonstrations, see *Newsweek,* May 7, 1979.
433 "The unfortunate events . . .": Washington *Post,* May 20, 1979.
434 ". . . the average consumer . . .": *Newsweek,* Apr. 7, 1980.
435 "We are tempted to say . . .": Kemeny Commission Report, p. 10.
436 Rockefeller Foundation Report: op. cit.
436 "If we don't substitute . . .": *New York Times,* Mar. 16, 1980, p. F9.
436 Best survey on European reaction by Walter Patterson. *The Accident at Three Mile Island*—a paper prepared for the Congressional Research Service, U.S. Library of Congress, Sept. 1979.
438 "I can assure . . .": Apr. 3, 1979, House of Parliament.
438 ". . . maimed the brains . . .": *New York Times,* Feb. 11, 1979.
438 "nuclear neurosis": Joseph Leaser, interview with authors, Apr. 1980.
439 50 percent increase in infant deaths: Dr. Gordon Macleod, former Secretary of Health for Pennsylvania. Rebuffed by Dr. George Tokuhata,

director of epidemiology, Pennsylvania Health Dept. Debate reported in *New York Times*, Apr. 3, 1980.

439 ". . . no one really knows . . .": Leaser interview, op. cit.

440 "The emotional trauma . . .": Ibid.

442 "It now appears . . .": *British Journal of Radiology*, vol. 49, no. 583, July 1976.

442 "Radiation of Hanford Workers Dying from Cancer and Other Causes": *Health Physics*, vol. 33, 1977, p. 369.

Acknowledgments

In writing this account, we owe a great deal to many people in many parts of the world who have been generous with their time and their assistance. We owe special gratitude to the late Alan Cooper, in London, for research and translation of Russian material. Significant sections of the book could not have been written without Jacob Scherr, of the Natural Resource Defense Council in Washington; Walter Patterson in London; Armin Selheim of *Der Spiegel* in Hamburg; Yoshitatsu Tsutsumi in Japan; and, especially, our chief researcher, Therese Stanton, in New York.

We thank the staffs at the following libraries and archives for their help: *The Sunday Times*, London, *The New York Times*, the Washington *Post*, *Newsweek; Le Monde* and *L'Express; Der Spiegel, Die Suddeutscher Zeitung*, Die Deutsche Museum, Die Institut der Deutschen Wirtschaft; the Foreign Correspondents Club in Tokyo; *The Times of India*, the Press Trust of India, and the Institute for Defence Studies and Analysis in New Delhi; Radio Free Europe and Radio Liberty; The Library of Congress, the British Museum, the Bancroft Library, University of California at Berkeley, the Eisenhower Library at Abilene, Kansas, the West Point Military Academy Library, the New South Wales Public Library, Sydney; and, in particular, Roger Anders at the U.S. Department of Energy and Ed Reese at the U.S. National Archives.

We are grateful to many historians, scientists, journalists, politicians, and government officials who have given us their time and helpful suggestions. A few have wanted to remain anonymous. The others are:

United States: Bob Alvarez, Dale Bridenbaugh, Irvin Bupp, R. E. Cunningham, Warren Donnelly, John Gofman, Paul Joskow, John Kenton, Lew Kowarski, Solomon Levy, Frank Marriot, John Palfrey, Robert Pollard, James Ridgeway, David Rosenberg, Gertrud Weiss Szilard, Alice Kimball Smith, Bruce Tucker, Edward Wonder, and James Vincent.

USSR: Reuben Ainsztein, Leslie Dienes, Vassily Emelyanov, Leonid Finkelstein, John Hardt, David Holloway, Arnold Kramish, Mark Kuchment, Zhores Medvedev, Boris Rabbot, and T. Harry Rigby.

United Kingdom: Lorna Arnold, Margaret Gowing, Lord Hinton, Mark Imber, Tony Magrew, Bruce Page, John Simpson, Lord Sherfield, Sir Kelvin Spencer, William Walker, and Roger Williams.

France: Daniel Charlades, Jean-Claude Derian, Bertrand Goldschmidt, Pierre Guillaumat, Francis Perrin, Louis Puiseux, Ezra Suleiman, Anthony Terry, and André Thépot.

Germany: Wolfgang Braun, Robert Gerwin, Jutta Kamke, Otto Keck, Gerhard Matz, Hans-Joachim Preuss, Kurt Rudzinski, Heinz Schimmelbusch, Klaus Traube, and Siegfried von Weiher.

Japan: Arisawa Hiromi, Hori Ichiro, Imai Ryukichi, Ipponmatsu Tamaki, Kaneko Koji, Kawakami Koichi, Mukaibo Takashi, Murata Hiroshi, Nasu Hayao, Ohmae Kenichi, Ohtani Ken, Jenny Phillips, Hideo Sato, Murray Sayle.

India: P. R. Chari, P. N. Haksar, K. C. Khanna, Lavaj Kumar, T. L. Shankar, M. A. Velodi.

Canada: Patrick MacFadden, Ian Ball.

Latin America: José Fonseca Filho, Jorge Sabato.

On international proliferation and safeguards, our thanks are due to Phillip Farley, David Fisher, Ivan Head, C. G. Hough, Myron Kratzer, Al Labowitz, Johan Lind, Alan McKnight, Slobodan Nakicenovic, Alan Wilson.

On general overviews of the development of atomic energy, we thank Robert Jungk, Amory Lovins, Marvin Miller, and Sir Mark Oliphant.

We are indebted to Wendy Cooper, George Drost, Ulf Arnold, Keiko Shirono, and Yoriko Woodward for their translations. Jane Cousins, Peter Curtis, Phillip Jacobson, Kim Jones, Miles Kupa, Michal Levin, Neil McDonald, Alan Renouf, William Shawcross, and Peter Watson helped to arrange interviews and gave us hospitality. People who refused to see us: Admiral Hyman Rickover.

Our thanks are due to Harold Evans who, as editor of the London *Sunday Times* during the four years it took us to complete this book, generously allowed us access to the facilities of that newspaper.

In the New York office of Times Newspapers, we owe much to Betsy Blatz, Mike Lavolpe, Ed Miller, George Moliterno, Nora Niego, Tamara Pittman, Maureen O'Toole, Rosemary Vlasto and Eileen Wojno, plus Deborah Kasouf in the *Times* Washington office.

At our publisher, Holt, Rinehart and Winston, we wish to acknowledge a deep debt to Jill Gutelle Weinstein for her careful attention to detail and her concern for every stage of the complex process of turning a manuscript into a book.

In various stages the manuscript was read by John Barry, Marvin Miller, Tom Oliphant, and Eleanor Randolph. We are most grateful to them for their helpful suggestions.

We owe our editor, Marian Wood, an immense debt for her care, judgment, and tolerance; and, finally, our special thanks to Robert Ducas, our agent, for his unbounded enthusiasm and support.

Index